LEONARDO DA VINCI'S
ELEMENTS
OF THE
SCIENCE OF MAN

LEONARDO DA VINCI'S
ELEMENTS
OF THE
SCIENCE OF MAN

Kenneth D. Keele

1983

ACADEMIC PRESS

A Subsidiary of Harcourt Brace Jovanovich, Publishers

New York   London

San Diego   San Francisco   São Paulo   Sydney   Tokyo   Toronto

ACADEMIC PRESS, INC.
111 Fifth Avenue, New York, New York 10003

United Kingdom Edition published by
ACADEMIC PRESS, INC. (LONDON) LTD.
24/28 Oval Road, London NW1 7DX

Library of Congress Catalog Card Number: 83–71978

ISBN: 0-12-403980-4

PRINTED IN THE UNITED STATES OF AMERICA

To my wife

Contents

Acknowledgements

This book has had a long gestation. First conceived some thirty years ago when writing the work *Leonardo da Vinci on the Movement of the Heart and Blood,* it lay in abeyance until 1970, when illness necessitated an early retirement from my professional clinical responsibilities. In that year the Wellcome Trust enabled me to continue my work on Leonardo as a Wellcome Research Fellow. The extent of my indebtedness to the Wellcome Trustees is symbolised by their provision of a workroom in the Wellcome Library containing their collection of literature on Leonardo da Vinci. Within the Wellcome Library I have continually been stimulated by the interest and encouragement of the Staff. In particular, I owe a great deal to the late Director, Dr. Noël Poynter, whose guidance is sorely missed, and to Mr. Eric Freeman, Director. In 1974 the publication of the Madrid Codices provided a further source for study and insight into Leonardo's scientific methods.

Five years were spent in harvesting and classifying the riches contained in the six to seven thousand pages of all Leonardo's extant notes. In this labour I received invaluable help from Sir Robin Mackworth-Young, the Royal Librarian at Windsor Castle, who made available to me Leonardo's manuscripts and the extensive literature possessed by Her Majesty the Queen in that Library. The drawings in the Royal Library, Windsor Castle, are reproduced by gracious permission of Her Majesty Queen Elizabeth II.

In 1975 I was approached by Sir Robin regarding the production of a first complete edition of Leonardo's notes on anatomy. This *Corpus of the Anatomical Studies of Leonardo da Vinci,* made in collaboration with Professor Carlo Pedretti, was completed and published by Harcourt Brace Jovanovich in New York in 1980. I had perforce to make another very detailed study of Leonardo's anatomical notes, coordinating them with the principles of his science. This is the subject of the present work. It is my hope that both works will be seen as part of an integrated whole.

From no point of view are Leonardo's notes simple. The interpretation of drawings and the deciphering of his script nearly always present difficult and frustrating problems – problems clearly reflected in the differing translations and commentaries of such scholars as Jean Paul Richter, Irma Richter, Edward MacCurdy, Donald O'Malley and others. Regarding Leonardo's scientific outlook I received much generous help in interpretation from the late Ladislao Reti, and in the field of anatomy I am indebted to Professor Ruth Bowden of the Royal Free Hospital Medical School, London, and Professor Robert McMinn at the Royal College of Surgeons of England. I would like to acknowledge, also, the continued interest in my work shown by Professor Sir Ernst Gombrich, particularly when he was Director of the Warburg Institute. His broad outlook on the significance of Leonardo's work to both artist and scholar is exemplified by the extensive Leonardiana in the Library of the Warburg Institute, which was made available to me.

Mr. William Jovanovich's valuable help with publication has been much appreciated. I would like to express my warm appreciation of the pleasant and fruitful association with Mr. Charles Hutt. Mr. Hutt has not only given me invaluable help regarding the design, format and text detail of this volume, but has skilfully undertaken the task of photographing the many illustrations. The importance of the illustrations in this book can hardly be overestimated since Leonardo often expresses himself so much more clearly in drawings than in words.

At all stages of this work I have received the greatest help from my wife, Mary Keele. Every page of the manuscript has been typed and retyped with creative comment and exemplary patience.

KENNETH D. KEELE

List of Abbreviations

Abbreviated titles used in text and quotation references

	In text	*In quotation*
Windsor Drawings in the Royal Library, Windsor Castle	Windsor Drawings	W with inventory number
The Anatomical Studies by Leonardo da Vinci. The Queen's Collection at Windsor Castle. 3 vols. Keele and Pedretti*	Anatomical Studies	K/P with folio number
Codex Arundel 263 in the British Museum	Codex Arundel	BM
Codex Atlanticus	Codex Atlanticus	CA
Codex On the Flight of Birds	Codex On the Flight of Birds	Sul Vol
Forster Manuscripts. 3 vols.	Forster I, II, III	Forster I, II, III
Codex Leicester (Hammer)	Codex Leicester	Leic
The Madrid Codices I, II	Madrid Codices [collectively]; Madrid Codex I, II	Madrid I, II
Manuscripts A, B, C, D, E, F, G, H, I, K, L, M	MSS A–M	A–M
MS Bibliothèque Nationale (BN 2038) [This MS has been included as part of MS A by the Reale Commissione Vinciana. It runs from A 81 r to A 114 v, corresponding to BN 1 r to BN 34 v. Previous references, including Richter (1970), use 'BN'.]	BN 2038	BN 2038
Codex Trivulziano	Codex Trivulziano	Triv
Treatise on Painting, translated by Philip McMahon	Treatise on Painting	TP
Vol I: translation with *paragraph* number	Treatise on Painting	TP I with paragraph number
Vol II: facsimile of Codex Urbinas	Codex Urbinas	CU

---

*Volume 3 of this work contains a complete concordance including the Anatomical Folios A and B and the Quaderni d'Anatomia Vols. I–VI.

Leonardo's Alphabet and Numbers

∧ ∂ A	=	a	ı ı	=	1
ꓯ ϑ ꓭ b	=	b	ᴤ 2	=	2
ɔ	=	c	₹ ₹ ₹ ₹	=	3
ꝑ ꝑ	=	d	ꝑ 4	=	4
ꞁ ꞁ ꝺ ꝺ	=	e	ᴣ ꞓ	=	5
ꞕ	=	f	ꝺ 6	=	6
8	=	g	‹ ›	=	7
ꝺ	=	h	8	=	8
ı ꞁ	=	i	℮ 9	=	9
ꓘ	=	k	o	=	0
ꝺ	=	l			
ꞛ ꞛ Ꞌꞝ	=	m			
ɴ	=	n			
ο	=	o			
ꝯ	=	p			
ꝼ	=	q			
ᴠ ꓤ ᴠ	=	r			
s ꞁ	=	s			
ꞁ ꞏ	=	t			
ꞣ	=	u			
ꝭ	=	v			
✕	=	x			
ꝩ	=	y			
ξ ꞔ	=	z			

Introduction

This book, *Leonardo da Vinci's Elements of the Science of Man,* has taken its shape from an injunction of Leonardo himself on a page of his anatomical notes: 'Arrange it so that the book On the Elements of Mechanics with its practice shall precede the demonstration of the movement and force of man and of other animals' (W 19009 r; K/P 143 r).

Such an arrangement is in fact essential to the appreciation of Leonardo's anatomical drawings, for he designed these to reveal the mechanism of 'Man, the Machine'. He himself endorses this fact when he adds the explanation that by means of studying his Elements of Mechanics 'you will be able to prove all your [anatomical] propositions' (W 19009 r; K/P 143 r).

Some thirty years ago when writing my book, *Leonardo da Vinci on the Movement of the Heart and Blood,*[1] the full truth of Leonardo's injunction became evident to me. I have therefore accepted the challenge of presenting Leonardo's principles of physical science which preceded his physiological investigations as represented by his Elements of Mechanics. The enterprise has been even more difficult than I had anticipated, for I have found that Leonardo's Elements of Mechanics existed only in scattered fragments, many of which appeared to be contradictory; in their great confusion no general principle was detectable. In such circumstances I at first felt that the difficulties of relating Leonardo's mechanics to his anatomy were insuperable. However, many of these difficulties were resolved by a systematic survey of the whole of Leonardo's extant notes on the physical sciences and the science of man arranged by subject and in approximate chronological order. This revealed the context within which he performed his anatomical studies at different periods of his life and his motives for doing so.

In 1966 I was able to contribute the first gleanings from such a study to the International Symposium on Leonardo da Vinci held at Los Angeles.[2] Here I briefly uncovered the dynamics underlying Leonardo's physiology of vision, his earliest anatomical exploration. The present work comprises an integration of Leonardo's Elements of Mechanics with his investigations on the movements of man and animals as well as his anatomical drawings. The enterprise has been unexpectedly assisted by the fortuitous appearance of the long-lost *Madrid Codices. Madrid Codex* I contains the nearest approach to a systematic explanation of Leonardo's principles of mechanics which we possess. Systematic study of the *Madrid Codices* since their publication in 1974 has not only confirmed my previous conclusions but has also enabled me to show how Leonardo integrated his innumerable mechanical observations and experiments in mechanics into underlying principles. It has also confirmed how consistently he applied these principles to human anatomy and physiology as revealed in the *Corpus of the Anatomical Studies.*[3]

This book presents an outline of the results of such studies. Although it may not appear brief, it is in fact a much abbreviated sketch of the thousands of pages of Leonardo's notes on mechanics, anatomy, etc. I have strictly confined my efforts to clarifying the principles underlying Leonardo's researches in each major field and have illustrated these with relevant examples and figures. Much secondary material that is not strictly relevant to my theme has been omitted. I would like these chapters therefore to be looked upon as a set of guide-posts to Leonardo's Elements of the Science of Man rather than as a complete exposition of the subject.

It is often not realised how detailed and thorough were Leonardo's investigations. The innumerable observations and experiments, for example, on the phenomena of 'movement' in relation to each of the 'four elements', particularly water, constitute a vast accumulation of data, all skilfully integrated into his general principles. This can only be touched upon in the space available in such a book as this; yet the whole body of this work on water lies behind Leonardo's investigations of the movements of the blood in the human heart and aorta.

There is great interest in relating Leonardo's work to that of his medieval predecessors. However, as I delved deeper into Leonardo's own notes I found their relation to both his predecessors and successors ever more elusive and difficult to evaluate. Leonardo's thought-patterns emerge increasingly more clearly as he weaves his own personal web of ideas. He drew widely on past sources as his own library lists and the fine works of Solmi,[4] Duhem[5] and Zubov[6] prove. But what he took he digested, absorbed and transformed into his own personal creative form of science, often rendering such gleanings almost unrecognisable.

With regard to his successors the position is similar. His work contains many remarkable 'anticipations', in the realms of both physics and physiology, but they are embedded within his own personal intricate concept of science. It has been only too easy to extract them piecemeal and give them false positive or negative values.

Thus, finding Leonardo's science subtle and difficult for a twentieth-century mind to grasp, I have given its intelligibility first priority. In order to enter the world of Leonardo's physics we have to strip ourselves of such firmly preconceived ideas as those based on the Newtonian concept of gravity. Also, when it comes to Leonardo's anatomical and physiological studies, Galenic concepts combined with geometrical schemata demand equally radical differences of approach. And the gap between Leonardo's view of the movement of the heart and blood and that of his successors, such as William Harvey, expresses a fundamentally different method of investigation.

Such considerations bring one face to face with the significance of Leonardo's isolation in the history of science. The fact that his work falls outside the main current of the flow of recognised progress adds, in my opinion, freshness to his claim for contemporary interest. For here was a thinker who can without exaggeration be classed as brilliant, who integrated observation and experiment into a fusion of subjective and objective expression in the form of what we now call 'art'. Such an integration has not been achieved since his day. So closely does Leonardo fuse visual expression and verbal idea that many of his statements appear only in the form of illustration without any verbal explanatory text. Sometimes one single figure represents a whole series of logical concepts. In one such figure on perspective I have found at least four stages of developing thought. It is not surprising to find that some of Leonardo's figures are so full of thought-content as to defy interpretation.

Similar difficulties encompass his use of words, so much so that a glossary of his terms is desirable – but even this is complicated by his use of the same term with different meanings at different periods. For example, '*luce*' means light, eye, pupil or cornea, in different contexts. Many traps beset the translation of such terms, let alone difficulties arising from his left-handed script, his code of abbreviations and his failure to punctuate or divide his lettering into clear-cut words.

It rapidly becomes very evident to the reader of Leonardo's notes that his favourite language is in terms of geometrical forms. Here again, though Leonardo gave intensive study to Euclid's *Elements,* his own geometry taking off from this base concentrates on the 'transformations' or 'mutations' of the geometry of movement. This consists largely of geometrical figures which 'equal' or are related to one another by geometrical duplication, rotation and proportional perspective alterations. It is of interest therefore to read that in schools today Euclid's geometry is being

'replaced by the study of symmetry and the effects of transformations such as reflections, rotations and enlargements'.[7] Leonardo's geometry certainly could not be more accurately and concisely described.

I have felt this introduction necessary to define the nature of the task undertaken in subsequent chapters. The book is designed as an attempt to present Leonardo's physical science in comprehensible outline in order to show how he welded it into his science of man. I hope that by rendering this science intelligible, glimpses of its beauty will be obtained – enough to show that this rises into the same category as his paintings. Perhaps, too, the Herculean strivings of this 'loner' 500 years ago will be found relevant to our twentieth-century strivings to bridge the chasm we have created between 'art' and 'science'.

LEONARDO'S NOTEBOOKS AND THEIR INTERPRETATION

Since this study is based upon the examination of all the existing notebooks of Leonardo, to which repeated reference will be made, it is desirable to identify these and briefly summarise their chequered history.

When Leonardo died at Amboise on 2 May 1519 he left to 'Messer Francesco Melzi, gentleman of Milan, in remuneration of his services gratuitously performed by him in the past, whatever books the testator may at present possess and all other instruments and manuscripts touching and relating to his art and profession of painting'.[8] Though Melzi took these precious works back to Milan and treasured them until the end of his life, about 1570, he made them readily available to visitors, some of whom it appears, with or without Melzi's permission, carried manuscripts away. So reports Vasari, he himself having visited Melzi at Vaprio near Milan to obtain details for his biography of Leonardo.[9]

Melzi himself made a systematic study of Leonardo's notes from which he compiled the *Treatise on Painting*. This was published in an abbreviated form in 1651. It was not published in full until 1817. The first facsimile edition with English translation was produced by McMahon in 1956.[10] This, Leonardo's one publication, contains a great deal of scientific observation on such subjects as light, vision and botany. A whole section of the treatise is devoted to the movements of man.

After Melzi's death his heir, Orazio Melzi, permitted the dispersal of the manuscripts. A tutor, Lelio Gavardi, removed thirteen manuscripts from the attic of Orazio Melzi's house. In spite of protests from a friend, Mazenta, these volumes found their way into the hands of Mazenta's brothers. Here they were located by Pompeo Leoni, sculptor to King Philip II of Spain. Pompeo Leoni obtained possession of ten of the manuscripts and followed up his success by approaching Orazio Melzi, who let it be widely known that his attic contained many more drawings, clay models, anatomical studies and other relics of Leonardo. From this time, about 1590, the fate of Leonardo's legacy to posterity is far from clear. Most of it appears to have been completely lost. Of what remained, Pompeo Leoni apparently obtained the largest share. He took many manuscripts and drawings to Spain with the avowed intention of either giving or selling them to King Philip. Over a thousand loose pages of manuscript, probably found in Melzi's attic, Pompeo Leoni pasted into a huge volume of 400 folios, so large and heavy as to earn for itself the name *Codex Atlanticus.*[11] This work was published in full in 1894–1904. A third edition was published in 1974.

In 1609 when Pompeo Leoni died he was still in possession of his collection of Leonardo's manuscripts. In 1613 Giovan Battista Leoni, Pompeo's legitimised son, tried to sell at least part of his father's collection to the Grand Duke Cosimo II de' Medici in Florence. Among the items offered was 'a book of about 400 pages . . . on each page are pasted various drawings of devices of a secret art and other things by the above-mentioned Leonardo'. Another item was 'fifteen small books of observations and studies of various subjects by the same [Leonardo] and particularly of anatomy'.

Whilst the former can be identified as the huge *Codex Atlanticus,* the fifteen books of anatomical drawings have entirely disappeared. The whole of this offer was refused by Duke Cosimo after being advised by his experts that the drawings and paintings were 'not worthy of such a Prince'. The *Codex Atlanticus* was described as 'a very insignificant thing'.[12]

Thus Pompeo Leoni's collection of Leonardo's manuscripts descended to his daughter's husband, Polidoro Calchi, from whom the Count Galeazzo Arconati bought the *Codex Atlanticus* and ten notebooks, presenting them to the Ambrosian Library in Milan in 1636. The huge *Codex Atlanticus* bears the title *Designs of Machines and Secret Arts and Other Things of Leonardo da Vinci Collected by Pompeo Leoni.* The other volumes are now known as MSS A, B, E, F, G, H, I, L and M and the *Codex Trivulziano.* Cardinal Borreomeo had given the Ambrosian Library another manuscript, now called MS C, in 1609. The *Codex Trivulziano* was later exchanged for MS D, and another, MS K, was presented to the library in 1674 by Count Archinti.[13]

About 1634 Luigi Maria Arconati, Count Galeazzo Arconati's natural son, compiled from Leonardo's notes a work, *On the Movement and Measurement of Water.* This, like the *Treatise on Painting,* is valuable for containing passages from Leonardo's notes otherwise lost. It was published in 1923 by Carusi and Favaro.[14]

Another competitor for Pompeo Leoni's Leonardian booty was Thomas Howard, Earl of Arundel. Though details are lacking, it is almost certain that the Earl of Arundel obtained possession of a volume entitled *Designs of Leonardo da Vinci Restored by Pompeo Leoni.* This is now in the Royal Library at Windsor Castle and has been catalogued.[15] It includes nearly all Leonardo's anatomical drawings. Though possibly seen in 1690, these would appear to have been unappreciated until found in a chest by the Royal Librarian to King George III. In 1773 the anatomist William

Hunter wrote to Albrecht von Haller, the great physiologist, about them.[16] These were published in fragments as *Anatomical MS A*[17] (1898), *Anatomical MS B*[18] (1901) and the *Quaderni d'Anatomia*[19] (1911–1916). Of these fragments only the *Quaderni d'Anatomia* has an English translation. None gives any attention to chronological arrangement, anatomical commentary, or correlation of Leonardo's own text with his drawings.

These defects have recently been remedied by the production in 1978–1980 of all Leonardo's anatomical drawings as *Leonardo da Vinci; Corpus of the Anatomical Studies in the Collection of Her Majesty the Queen at Windsor Castle,*[20] by Kenneth D. Keele and Carlo Pedretti.

The Earl of Arundel also obtained a large collection of miscellaneous scientific notes by Leonardo in the manuscript called *Codex Arundel.*[21] In 1666 this was presented to the Royal Society by his heirs; in 1831 it found its present home in the British Museum. It was published in 1926–1930. The Earl of Arundel's acquisitive eye also caught sight of the great *Codex Atlanticus* in the Ambrosian Library. When John Evelyn visited the Ambrosian Library in 1646 he relates in his diary how 'my Lord Marshall [the Earl of Arundel] who had seen them, told me all but one book are small, that a huge folio contained 400 leaves full of scratches of Indians, etc. But whereas the inscription pretends that our King Charles had offered 1,000l for them – the truth is, and my Lord himself told me, that it was he who treated with Galeazzo for himself, in the name and by permission of the King : . . but my Lord having seen them since, did not think them of so much worth'.[22]

The recent discovery of the *Madrid Codices* has once more brought the Earl of Arundel into the story of Leonardo's manuscripts. In 1623, the Prince of Wales, later to become King Charles I, whilst staying in Spain saw two books by Leonardo da Vinci which Don Juan de Espina had obtained from Pompeo Leoni. The prince's offer to buy them was refused. In 1629 and 1631 the Earl of Arundel received two reports from his agents, each telling of failure to obtain the books of drawings by Leonardo da Vinci. In 1637 the earl wrote from Hampton Court somewhat impatiently to the British Ambassador in Spain, 'I beseech You be mindful of D. Jhon de Spina's booke, if his foolish humor change'.[23] Don Juan de Espina's foolish humour did not change, and in 1642 after his death the two books of Leonardo's drawings were placed in the Royal Palace Library whence, it is thought, about 1830, they were removed to their present home in the Biblioteca Nacional in Madrid.

Until 1796 the great collection of Leonardo's manuscripts in the Ambrosian Library rested there only too undisturbed. In that year Napoleon had them transferred to Paris. This turned out to be a blessing in disguise, for in Paris these manuscripts for the first time came under the appreciative eye of a scientist, G. B. Venturi, the Italian inventor of the pressure flowmeter. Venturi was thrilled to the core by what he had found. It was he who lettered the individual manuscripts alphabetically with the letters which they bear today. He also proposed to publish three treatises on Leonardo's work on mechanics, optics and hydrostatics. Venturi was the first to have the insight that Leonardo was 'guided by the geometrical spirit in all his studies, whether he wished to analyse an observation, to link together his reasoning, or generalise his own ideas'.[24] In short, Venturi recognised a fellow scientist in Leonardo.

After 1815 the *Codex Atlanticus* was returned to the Ambrosian Library in Milan, but the lettered manuscripts somehow remained in Paris.

During the nineteenth century the reputation of Leonardo as a scientist grew so rapidly that it assumed mythological proportions. By 1900 the *Codex Atlanticus* and all the manuscripts lettered by Venturi had been published.[25] Leonardo manuscripts now appeared from all sides. The *Codex Leicester* (Hammer),* devoted to the study of water, found in Rome at the end of the eighteenth century, was published in 1909 by Calvi.[26] *Forster* I, II and III, almost entirely devoted to mechanics, appeared mysteriously in Vienna and found their way to the Victoria and Albert Museum in London in 1876. They were published in 1936.[27]

This was thought to be the last existing scientific fragment of Leonardo's work until 1967, when the two outstanding *Madrid Codices* appeared and were published in 1974.

THE INTERPRETATION OF LEONARDO'S SCIENTIFIC NOTES

Why have Leonardo's obviously important contributions to the science of man been so long overlooked? This question has been ever more frequently asked during the past century since his notebooks have been increasingly coming to light.

Whilst many reasons can be cited for this tragic neglect of his works, three causes stand out from the rest. Most important of all is the loss of his manuscripts. Even now, although we possess some 7,000 sheets of his writings and drawings, it is calculated with good reason that only about one-third of his notes have survived. Our assessment of his science has, therefore, to be based upon gross abbreviation of his works – and Leonardo himself emphatically condemns all abbreviations – 'Abbreviators of works do harm to knowledge . . . for certainty springs from a complete knowledge of all the parts which united compose the whole' (W 19084 r; K/P 173 r).

This salient cause leads almost inevitably to the second; confusion of the notes we do possess. In these Leonardo makes frequent references to completed 'books' on many subjects, often citing chapter and section. The promised orderly presentation of his works which this suggests has never materialised. Almost without exception such 'books' have not survived. Presumably they have been lost, and with them the logical sequence into which he shaped his work. Thus we are left with a vast collection of what may

be called his laboratory notes – and anybody who has made rough notes in the course of laboratory work will appreciate how chaotic those can be.

To loss and confusion must be added the factor of mutilation. Many of Leonardo's notes contained in Melzi's attic seem to have been tied up in bundles of loose sheets. Some we know to have been torn out of notebooks. Some notes were actually cut out of one page and inserted by Pompeo Leoni into another. Such enormous collections as the *Codex Atlanticus* are composed of sheets piled together with little consideration of date or content; thus a note written by Leonardo on one subject in 1489 may well be found next to one on another subject written in 1516.

A further obstacle lay inherent in the form of Leonardo's genius itself. His chosen mode of expression of his thoughts was visual, particularly through geometrical figures. On occasions he expresses a whole train of thought in a sequence of geometrical figures without any verbal explanation at all. As a left-handed writer his script runs from right to left on the page; so too do his drawings. This presents relatively little difficulty in itself. Far more difficulty in interpretation arises from his complete lack of punctuation and malformation of words.

One further source of confusion arises from Leonardo's habit of using his notes as a filing system. He repeatedly surveys his whole work at different periods, even adding later views sometimes on the same page years later; as it were, a spiral form of progress. This inevitably produces apparently contradictory remarks on one page, for as his outlook changes so his interpretations of natural phenomena change. This applies to his views of the macrocosm of nature as well as to the microcosm of man. Such spiral changes persist throughout his notes. In short, Leonardo's scientific work, like so much of his artistic work, was never 'finished'. With every 'solution' to a problem he consistently found more questions to be solved.

If this formidable list of obstacles accounts for the lack of appreciation of Leonardo's scientific work shown by his contemporaries and successors, it also opens up possible ways of reaching an understanding of his work. Amongst these the need for its arrangement in chronological order is paramount. Only by such arrangement can the changing sequences of his scientific thought be revealed.

In order to carry out such a project satisfactorily two requirements are obviously necessary. First, the whole of Leonardo's extant manuscripts must be explored. Secondly, every note must be arranged in chronological order as far as possible. This I have attempted to do and in doing so have encountered unexpected as well as expected difficulties upon which it would be unprofitable to elaborate here. Suffice it to say that working through Leonardo's enormous corpus of notes produces the feeling of someone confronting a vast jigsaw puzzle consisting of more than 7,000 pieces, knowing in addition that several more thousand missing pieces are necessary to complete the picture.

In deciphering, transliterating, translating and interpreting Leonardo's script I have naturally made full use of those English translations available to me, those of Richter,[28]

---

*In 1981 the *Codex Leicester* was purchased by Dr. Armand Hammer and the name of the *Codex* changed to *Codex Hammer*. For ease of reference I have kept to *Codex Leicester* and the original folio numbers throughout this book.

MacCurdy,[29] and O'Malley and Saunders[30] in particular. It has to be borne in mind that these very valuable English translations all have serious limitations from the point of view of my present objective, the chief of which is their incompleteness.

Since Richter and MacCurdy were literary men, they were not concerned with Leonardo's creation of scientific terms or his efforts to construct his own consistent rational science. For example, neither appreciated the significance of his use of the term *percussion* as the 'greatest' of his 'four powers'. Since Richter reproduces a transcription of Leonardo's text with his English translation I have found his work invaluable; however, apart from the sections on perspective and vision, the scientific notes of Leonardo are woefully lacking from his selection. MacCurdy translated these more extensively but without the accompanying Italian text and without illustrations, omissions which often render interpretation of Leonardo's science impossible, since Leonardo's text and drawings most often form an indissoluble entity.

O'Malley and Saunders have confined themselves to the greater part of Leonardo's anatomical notes. In this field their English translation is very valuable, but the anatomical part of Leonardo's science is amputated from the rest of his work. No attempt is made to trace the chronological development of anatomy or the physical or physiological basis of his concepts of the movements of the human body. In this last subject the work of J. Playfair McMurrich in his *Leonardo da Vinci the Anatomist* continues to convey valuable insight.[31]

Efforts to arrange Leonardo's notes in chronological order date back to Richter's great pioneering volumes, *The Literary Works of Leonardo da Vinci,* of 1883. Much of his dating of the individual lettered manuscripts in the second and third editions of 1939 and 1970 remains valid. Calvi's meticulous studies of the changes in Leonardo's orthography and script at various periods of his life not only rendered dating of some manuscripts more firm but also opened the way for Carlo Pedretti's dating of sheets in those vast heterogeneous volumes, the *Codex Atlanticus* and the *Codex Arundel.*[32] Calvi's dating was accepted in 1935 by Kenneth Clark in the first edition of his catalogue *The Drawings of Leonardo da Vinci in the Collection of His Majesty the King at Windsor Castle.*

Since this time the extensive work of Carlo Pedretti on the dating of the individual sheets of the *Codex Atlanticus,*[33] the *Codex Arundel*[34] in the British Museum, the *Treatise on Painting,*[35] the second edition of Kenneth Clark's *The Drawings of Leonardo at Windsor* in 1968 and in 1977 his *The Literary Works of Leonardo da Vinci: A Commentary to Jean Paul Richter's Edition*[36] have been of the greatest value to me in arranging Leonardo's scientific notes chronologically. It is realised that a number of sheets in all these works still can be dated only approximately, but Pedretti has steered through many traps and snares by virtue of his unique detailed knowledge of the whole corpus of Leonardo's manuscripts.

Allowing for some inevitable inaccuracies, I have for the purpose of this book arranged Leonardo's scientific notes into approximately quinquennial steps from 1482 onwards to 1519. This approach gives a reasonably firm set of chronological sequences. I have found that it throws a most revealing light on the spiral movement of Leonardo's studies of the macrocosm in general, as well as the science of man in particular. By this method of examining his works one can see Leonardo emerging from the artist-engineer through the phase of art and invention into the creative scientific artist of maturity so characteristically embodied in his science of man.

REFERENCES

1. Keele, Kenneth D. *Leonardo da Vinci on the Movement of the Heart and Blood.* London, Harvey and Blythe Ltd., 1952.
2. Keele, Kenneth D. 'Leonardo da Vinci's Physiology of the Senses', in *Leonardo's Legacy,* ed. by C. D. O'Malley. Los Angeles, University of California Press, 1969, pp. 35–36.
3. Keele, Kenneth D., and Pedretti, Carlo. *Leonardo da Vinci; Corpus of the Anatomical Studies in the Collection of Her Majesty the Queen at Windsor Castle.* New York, Harcourt Brace Jovanovich, 1978–1980. Vols. 1–3.
4. Solmi, Edmondo. 'Le Fonti dei Manoscritti di Leonardo da Vinci', *Giornale Storico della Letteratura Italiana,* Torino, 1908. Supplemento no. 10–11.
5. Duhem, Pierre, *Etudes sur Leonard de Vinci.* Paris, A. Hermann, 1906–1913. Vols. 1–3.
6. Zubov, V. P. *Leonardo da Vinci,* trans. by David H. Kraus. Cambridge, Mass., Harvard University Press, 1968.
7. *Everyman's Encyclopaedia,* ed. by R. L. Carter. Bedford, Orbis Publishing Ltd., 1976. Vol. 5, p. 2369.
8. Brown, John W. *The Life of Leonardo da Vinci.* London, William Pickering, 1828. Appendix, IV, p. 218.
9. Vasari, Giorgio. *Lives of the Painters, Sculptors and Architects,* trans. by A. B. Hinds. London, Dent & Sons, 1927. Vol. 2, p. 163.
10. Leonardo da Vinci. *Treatise on Painting (Codex Urbinas Latinus 1270),* trans. and ed. by A. Philip McMahon. Princeton, N.J., Princeton University Press, 1956.
11. Leonardo da Vinci. *Codex Atlanticus.* Il Codice Atlantico di Leonardo da Vinci della Biblioteca Ambrosiana da Milano, riprodotto e pubblicata della R. Academia dei Lincei. Milan, 1894–1904.
12. Leonardo da Vinci. *The Madrid Codices,* with commentary by Ladislao Reti. New York, McGraw-Hill, 1974. Vol. 3, pp. 15, 16.
13. *Ibid.,* p. 17.
14. Carusi, E., and Favaro, A. *Il Trattato del moto e misura dell'acqua.* Bologna, Istituto di Studi Vinciani, 1923.
15. Clark, Kenneth, and Pedretti, Carlo. *The Drawings of Leonardo da Vinci in the Collection of Her Majesty the Queen at Windsor Castle.* 2d edition, London, Phaidon, 1968.
16. Kemp, M. 'Dr. William Hunter on the Windsor Leonardos and His Volume of Drawings Attributed to

Pietro da Cortona', *The Burlington Magazine,* 1976. Vol. 118, p. 144.

17. Leonardo da Vinci. *Anatomical MS. A.* I manoscritti di Leonardo da Vinci della Reale Biblioteca di Windsor. Dell Anatomia, Foglio A, pubblicata da Teodoro Sebachnikoff, trascritti e annotati da Giovanni Piumati. Paris, 1898.

18. Leonardo da Vinci. *Anatomical MS. B.* I manoscritti di Leonardo da Vinci della Reale Biblioteca di Windsor. Dell Anatomia, Foglio B, pubblicati da Teodoro Sebachnikoff, trascritti e annotati da Giovanni Piumati. Turin, 1901.

19. Leonardo da Vinci. *Quaderni d'Anatomia,* pubblicati da Ove C. L. Vangensten, A. Fonahn, H. Hopstock. Christiania, J. Dybwad, 1911–1916. Vols. 1–6.

20. Keele and Pedretti. *Leonardo da Vinci; Corpus of the Anatomical Studies, op. cit.*

21. Leonardo da Vinci. *Il Codice Arundel 263* (British Museum). Rome, Reale Commissione Vinciana, 1926–1930.

22. Evelyn, John. *Diary and Correspondence of John Evelyn,* ed. by William Bray. London, George Routledge & Sons, n.d., p. 152.

23. Da Vinci, *The Madrid Codices, op. cit.,* vol. 3, p. 19.

24. Venturi, J. B. *Essai sur les ouvrages Physico-Mathematiques de Leonardo de Vinci.* Paris, Duprat, 1797, p. 172.

25. Leonardo da Vinci. *Les Manuscrits de Leonard de Vinci.* Manuscrit A . . . M de la Bibliotèque de l'Institut publiés par M. Charles Ravisson-Mollien. Paris, 1881–1891.

26. Calvi, G. *Il Codice di Leonardo da Vinci della Biblioteca di Lord Leicester in Holkham Hall.* Milan, Cogliati, 1909.

27. Leonardo da Vinci. *Il Codice Forster.* Rome, Reale Commissione Vinciana, 1930–1936.

28. Richter, J. P., and Richter, Irma A. *The Literary Works of Leonardo da Vinci.* 3d edition, London, Phaidon, 1970.

29. MacCurdy, E. *The Notebooks of Leonardo da Vinci.* London, Jonathan Cape, 1938.

30. O'Malley, C. D., and Saunders, J. B. de C. M. *Leonardo da Vinci on the Human Body.* New York, Henry Schuman, 1952.

31. McMurrich, J. Playfair. *Leonardo da Vinci the Anatomist.* Baltimore, Williams & Wilkins, 1930.

32. Calvi, G. *I Manoscritti di Leonardo da Vinci.* Bologna, Pubblicazioni dell'Istituto Vinciana in Roma, 1925.

33. Pedretti, Carlo. *Studi Vinciani: saggio di una cronologia dei fogli del Codice Atlantico.* Geneva, Bibliotèque d'Humanisme et Renaissance, 1957.

34. Pedretti, Carlo. *Saggio di una cronologia dei fogli del Codice Arundel di Leonardo da Vinci.* Geneva, Bibliotèque d'Humanisme et Renaissance, 1960.

35. Pedretti, Carlo. *Leonardo da Vinci on Painting: A Lost Book,* Libro A. London, Peter Owen, 1965. Part 3, concordance.

36. Pedretti, Carlo. *The Literary Works of Leonardo da Vinci: A Commentary to Jean Paul Richter's Edition.* Oxford, Phaidon, 1977.

Chapter 1

A Scientific Biography of Leonardo da Vinci

It has been asserted that judging from his notes Leonardo spent far more time on science than on art and that therefore he was in fact more a scientist than an artist. This assertion may be true in terms of our present-day definitions of science and art; it is quite untrue in terms of Leonardo's view of the 'science of painting'. He himself saw what we call 'science' as the first stage of a unified creative activity. 'The scientific and true principles of painting', he writes, 'are principles which are comprehended by the mind alone . . . This is the pure science of painting which is in the mind of those who reflect upon it. From it is born creative action, which is of much more value than the reflection or science just mentioned' (TP I 19).

An account of Leonardo's life of science must accept his own evaluation. In this chapter a brief description of his life will be given, laying emphasis on the development of his predominant scientific interests and views at different periods. Almost without exception previous biographical accounts of Leonardo's life have revolved around his artistic productions. Not until the firmer dating of his scientific manuscripts was recently achieved has it become possible to describe the progress of his scientific efforts.

THE YOUNG ARTIST–INVENTOR IN FLORENCE (1452–1483)

Leonardo da Vinci was born on 15 April 1452 at Vinci, a small town on the wooded slopes of Monte Albano some twenty miles west of Florence. His mother was a peasant girl named Caterina; his father was Ser Piero da Vinci, a notary. Leonardo was illegitimate. His mother was married off to a local peasant, Acattabrigha, whilst the father's family brought up the little boy. Ser Piero's brother Francesco became particularly fond of his small nephew and at his death bequeathed Leonardo some land which became the cause of a long legal battle between Leonardo and his many legitimate half-brothers. Local tradition has it that the actual house where Leonardo was born is at Anchiano, a short distance from Vinci and a little higher up the slopes of Monte Albano. Be this as it may, the little grey stone cottage with its window looking out over the olive groves is an appropriately beautiful place for Leonardo's début. An-

other local tradition attributes Leonardo's lifelong fascination with mountains and their streams to the nature of the countryside of his birth. There is considerable internal evidence in his works to support this. Not only does Leonardo's earliest known drawing consist of the neighbouring landscape but also several of his remarkable maps of the Arno basin include Vinci; on some he writes the name of the little town. Both the landscape and the maps were the first works of their kind. A relief map of the Arno illustrates well the hills and valleys of the region between Vinci and Pisa where Leonardo spent many years of his life.

Brought up at Vinci for the first years of infancy and youth, Leonardo at about the age of fifteen had shown so much artistic promise that his father brought him to Verrochio's studio in Florence. Maybe Ser Piero's childlessness by his first wife, Albiera Amadori, helped in concentrating his affection on his young illegitimate offspring. However, Ser Piero does not appear to have taken much trouble over the early education of his only child. Apart from a rudimentary knowledge of arithmetic, reading and writing he learnt little in those early years in Vinci. All through his life indeed, Leonardo continually taught himself. It was a pattern of learning of which he was proud, yet at the same time he never ceased to regret his lack of that kind of learning in those subjects thought necessary to a 'man of letters'. In particular he was relatively ignorant of those portals to ancient learning, Latin and Greek, so he could not easily approach directly the great natural philosophers of antiquity who founded the sciences. Thus it comes about that he defies his critics, declaring, 'If indeed I have no power to quote from authors as they have, it is a far bigger and more worthy thing to read by the light of experience which is the instructress of their masters. They strut about puffed up and pompous, decked out and adorned not with their own labours but with those of others, and they will not even allow me my own' (CA 117 rb).

Vasari tells a tale about Leonardo's early life which casts too vivid a light on Leonardo's creative use of natural phenomena to be ignored. On being asked by a peasant for a painting on a round piece of wood Ser Piero allotted the task to Leonardo. Leonardo decided to paint a monstrous, terrifying head. 'To a room, to which he alone had access', writes Vasari, 'Lionardo took lizards, newts, maggots, snakes, butterflies, locusts, bats, and other animals of the kind, out of which he composed a horrible and terrible monster, of poisonous breath, issuing from a dark and broken rock, belching poison from its open throat, fire from its eyes, and smoke from its nostrils, of truly terrible and horrible aspect. He was so engrossed with the work that he did not notice the terrible stench of the dead animals, being so absorbed in his love for art'. When Ser Piero came to see the finished picture Leonardo dimmed the light and arranged the picture on the easel. Ser Piero, 'taken unaware, started back, not thinking of the round piece of wood, or that the face which he saw was painted, and was beating a retreat when Lionardo detained him and said, "This work has served its purpose; take it away, as it has produced the effect intended'".[1] Thus early in his life Leonardo reveals himself as an artist–inventor, using observations of nature to create illusions of monstrous reality. Throughout his life he used this method of composing 'monsters' such as dragons, masks for his allegorical drawings, or 'wild men' for festive occasions. The so-called caricatures were composed of deliberately disproportioned parts of faces, heads and bodies. When these artistic inventions were composed into wholes according to his geometrical laws of proportion, Leonardo created symbols of 'beautiful and true' reality. With the same guidelines in the field of statics he designed buildings and cathedrals, and in the field of dynamics, working mechanical inventions. He applied the same methods to composing the bodies of animals and men in his early 'anatomies'. Later he always drew these living 'machines' or 'instruments' as anatomy and physiology working according to his mechanical principles.

Leonardo entered Andrea Verrochio's workshop about the age of sixteen. There he found scope for the development of his creative, artistic and scientific powers. He became a skilled craftsman in painting, sculpture and in a wide field of the mechanical arts of invention. This was the period when Leonardo's twin gifts of art and science expressed themselves in intuitive inventions. It was whilst he was Verrochio's pupil that Leonardo painted the angel and the landscape background on his master's picture of the *Baptism of Christ*. A comparison of Leonardo's angel with that of Verrochio in the same picture reveals an uncanny difference in grace and beauty which Verrochio himself noticed so that, according to Vasari, he 'resolved never to touch the brush again, because Lionardo, though so young, had so far surpassed him'.[2] Further examination of the background of this picture shows Leonardo's hand in the revealing lines of the flow of water from the far distance down little waterfalls up to the waves in the foreground rippling around the legs of Christ.

Both Verrochio and Leonardo after him showed particular interest and skill in making plaster casts. Andrea Verrochio used such casts 'to form natural objects, . . . such as hands, feet, legs, arms and busts'.[3] Leonardo, too, made casts of 'heads of women smiling . . . and children's heads also executed like a master'.[4] Later in life Leonardo was to extend this technique into his anatomical investigations, making casts of the cerebral ventricles and the heart.

Another interest common to Verrochio and Leonardo was the problem of perspective. This, of course, was not new. Broached in practice by Brunelleschi, it had been developed in theory by Leon Battista Alberti and Piero della Francesca. Vasari relates how Verrochio (whose interest in horses was also shared by Leonardo) made drawings of horses 'with the method of enlarging things in proportion without errors',[5] a method which developed from his studies of 'science and especially geometry' in his youth.[6] Though it would appear that the geometrical problems involved in perspective were too complex for Verrochio, they provided Leonardo with the starting point for a lifelong development of perspectival geometry which was to permeate all his subsequent art and science. It may well have been this interest in perspective that induced Leonardo to apply his inventive genius so early to the complex problems of grinding concave mirrors.

Two events in his early years in Florence introduced Leonardo to the Hospital of Santa Maria Nuova. One was Verrochio's high opinion of the hospital whilst a patient there. Visited by his friends while recovering from an illness, he told them, 'I want a little fever to enable me to remain comfortably here in hospital'.[7] Another event which brought Leonardo into contact with the hospital occurred in 1472 when at the age of twenty he was admitted to the Company of Painters. This Company was included in the Guild of Physicians and Apothecaries which was based on the Hospital of Santa Maria Nuova. For many years Leonardo used it as a bank for his savings. On his later return to Florence it was on another patient who had died in the hospital that Leonardo performed some of his most successful anatomical dissections.

Close to Verrochio's workshop was that of the brothers Antonio and Piero Pollaiuolo. These two exerted a marked effect on Leonardo's development through their studies of human action as revealed by their dissections of muscles. Whereas Verrochio seems to have confined his anatomical studies to flayed corpses such as that produced in his statue of the flayed Greek god, Marsyas, Antonio Pollaiuolo's studies were designed to resolve the muscular anatomy underlying human actions. The results are seen vividly in his altarpiece of the martyrdom of St. Sebastian. Here, as Vasari relates, he has 'represented an archer drawing the bowstring to his breast and bending down to charge it, putting all the force of his body into the action, for we may see the swelling of his veins and muscles and the manner in which he is holding his breath'.[8] This vigorous picture (now in the National Gallery, London) was painted whilst Leonardo was in Verrochio's neighbouring studio. It provided him with an example of human action, a subject he was to study intensively for many years. Equally influential for Leonardo was Pollaiuolo's picture *Hercules Crushing Antaeus,* the figure of Antaeus being, as Vasari relates, 'remarkable as all the life is being crushed out of him by the grasp of Hercules, and he expires with open mouth'.[9] This picture clearly impressed Leonardo. A little sketch of it is to be found in his *Treatise on Painting* in *Codex Urbinas* 128 v. Here Leonardo uses it for a study of the production of what he calls the 'compound' centre of gravity of a man carrying a heavy weight (see Figure 6.10, p. 169).

One cannot ignore the emotional crises of the life of a man so versatile in the expression of his emotions as Leonardo. Perhaps the unhappiest of these occurred in 1476 when an anonymous charge of homosexuality was brought against him. Although the charge was dismissed by the Court, there is evidence in Leonardo's notes of intense distress at that period. He felt friendless and disillusioned. Scrawled over one page (CA 4 vb) are the words, 'You know, I have told you this before, I am without any of my friends . . . If there is no such thing as love, what is there?' (CA 4 rb).

It is difficult to avoid associating this intense distress with an unfinished picture which Leonardo painted about this time, that of *St. Jerome* (Figure 1.1), for here pain, physical and mental, is blended into one despairing human figure. Even in his own misery Leonardo did not forget the expressive muscular anatomy of face and neck which he learnt from his early dissections. The wide-flung arms with the right hand holding a jagged stone, even the painfully bruised chest, all are drawn into a human figure enclosed by hard, unsympathetic rocks. And in the foreground lies one of those lions in which Leonardo showed so much interest, its open jaws expressing the only possible sign of living sympathy.

In the year 1476 Ser Piero, now married to a third young wife, produced his first legitimate son and heir. He also set up his own independent business in a house near the Badia in Florence. This house was within earshot of the lions kept behind the Palazzo Vecchio. Leonardo closely observed their behaviour and on occasion dissected them after they died. In this connection the contemporary sketch in Figure 6.1 (p. 160) is revealing. Here Leonardo compares the roaring heads of man and animals; one at least is derived from that of a lion. Leonardo's persistent tendency to compose these monsters recurs in the form of dragons with twisted necks (W 12370 r).

Few notes of a scientific character appear to have been written during Leonardo's first period in Florence. The dominant theme throughout is that of 'inventions', some examples of which will be described here. Many of these 'inventions', however, were derived from the work of predecessors. It is not easy to know in which class to put them because when Leonardo took a mechanical device from the manuscript of some earlier inventor he almost always modified and improved it beyond recognition.

Since both Leonardo and Verrochio were painters intensely concerned with perspective it is not surprising to find that Leonardo's inventive genius concentrated early on the problems involving the control of light. 'How to make a great light', he writes, following the statement with a sketchy drawing of light penetrating a convex lens (CA 26 ra). The claim is more firmly established in Figure 1.2, where he describes 'A lamp that makes a beautiful and great light', accompanied by a sketch of a candle in a box fronted by a convex lens. This interest in lenses appears on a number of other pages on which he depicts an apparatus for grinding lenses or mirrors. Concave mirrors, in fact, attracted most of his attention. In Figure 1.3a he reveals why this is so. Here he gives a detailed drawing of an apparatus for grinding concave mirrors in order 'to make rays converge and make things burn'. The drawing is accompanied by a parabolic section through a cone, suggesting that he has a parabolic curvature for his mirror in mind in order to obtain that pyramidal (or conical) concentration of rays to the point which we call the focus (Figure 1.3b). Leonardo does not use that term.

Fire, so closely related to light, he creates, not only by the use of burning mirrors and lenses but also by friction of wheels, even under water (CA 10 ra). Noting that fire used up air in its burning, he invented a method of creating a vacuum to raise water by means of a fire burning in a closed bucket at the top (Figure 6.27, p. 183). A better-known invention was Leonardo's self-regulating turnspit (CA 5 va). In this the rate of turning the spit is cybernetically controlled by the convection current turning a set of vanes in the chimney above. Therefore as the fire dies down the rate of rotation of the spit slows, and vice versa. It is of interest that Robert Southey in 1807 attributed this 'recent' invention to Count Rumford.[10] Until then in England turnspits were apparently most often turned by dogs.

Water fascinated Leonardo from the first as a readily available, visible source of power. In these early years he spent a great deal of energy attempting to invent a form of Archimedes' Screw which would work by perpetual motion, as seen in Figure 4.34 (p. 119). He produced many bizarre-looking modifications before deciding that perpetual motion was impossible (Figure 4.34 and CA 386 rb). It was during these early years that Leonardo produced his first versions of a diving apparatus (see Figures 1.20 and 6.27, pp. 29, 183).

Leonardo's interest in the countryside round his home in Vinci may be reflected in his invention of an olive press with much more efficient leverage than that in contemporary use (CA 14 va).

Leonardo's inventions during this period were, however, by no means confined to peaceful purposes. Following the Pazzi conspiracy in 1478, when a plot to assassinate the brothers Lorenzo and Giuliano de' Medici in the cathedral was partially successful, Florence found herself in desperate conflict with the armies of Pope Sixtus IV. This may well have turned Leonardo's thought to inventions of war. These included mechanisms for making and overturning scaling ladders (see Figure 1.5, p. 12). He also planned the construction of gun barrels and multiple-barrelled guns (see Figure 1.10, p. 16). This last invention Leonardo would appear to have elaborated upon the crude concept of the German engineer, Kyeser. Leonardo himself added vertical movement to the series of gun barrels and a gun carriage for mobility (Figure 1.10a, p. 16). Assault bridges for attacking

Figure 1.1.   *St. Jerome,* a picture of pain (Vatican Gallery, Rome).

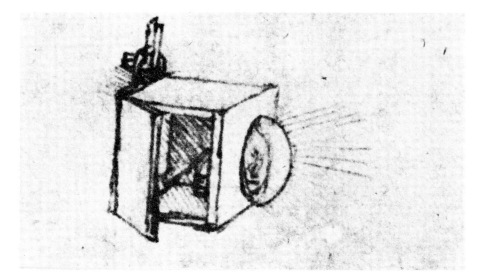

Figure 1.2. Leonardo's invention of a projector by placing a convex lens in front of a candle contained in a camera obscura (dark box). He describes it as 'a lamp that produces a beautiful and great light' (CA 9 vb).

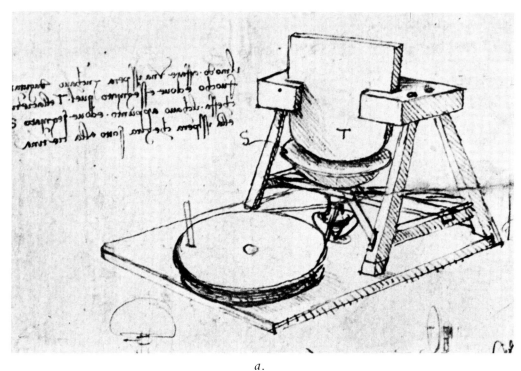

a.

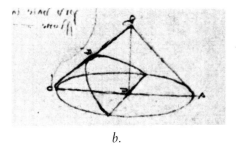

b.

a. Apparatus for grinding mirrors and lenses. Leonardo calls it 'A method of grinding a concave sphere in order to cast fire'. The part *T* makes the concavity in *S* beneath it, which rotates (CA 4 va).

b. The parabolic section of a cone upon which the shape of the grinder *T* is based (CA 32 ra).

Figure 1.3.

fortified castles were also derived from previous inventors (CA 392 va). Many weapons of war were copied by Leonardo from a work of Roberto Valturio, *De Re Militari*.[11] This appeared in a Latin edition in 1472 and so was available to Leonardo at this time, though his quotations from it appear only after the Italian edition of 1483.

In the five years after Leonardo left Verrochio's workshop and set up on his own, he found increasing frustration in Florence. His lack of 'letters' together with his own 'mechanical' outlook barred his entry into intellectual circles of influence. The arrogance of the 'intellectuals' infuriated him. He expresses himself in bitter, well-chosen words: 'Those who are inventors and interpreters between Nature and Man, as compared with the reciters and trumpeters of the works of others, are to be judged and not otherwise esteemed, as an object in front of the mirror compared with its image seen in the mirror. For the first is something in itself, the other nothing' (CA 117 rb).

This interpreter between nature and man sought a new field in which to do his work. Milan was rich, less 'cultured'; its ruler, Ludovico Sforza, was threatened with war by Venice. Leonardo saw a possible way of using his inventions profitably to himself and wrote a letter to Ludovico accordingly.

'Most Illustrious Lord, having now sufficiently seen and considered the experiments of all those who are reputed masters and contrivers of instruments of war, and

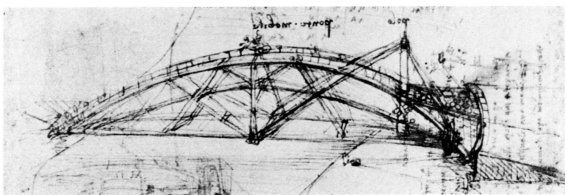

Figure 1.4. 'Mobile bridge'.
Leonardo writes this title above the curve of the drawing of a bridge which rotates around the axis, labelled 'polo' (CA 312 ra).

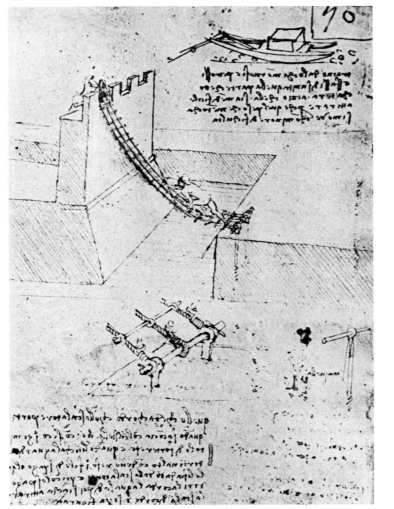

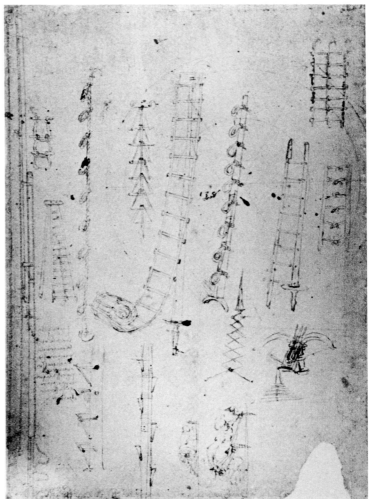

Figure 1.5. 'The most varied types . . . of scaling ladders' (B 50 r and CA 316 va).

that the invention and operation of the said instruments are nothing different from those in common use, I shall endeavour without prejudice to anyone else to explain myself to your Excellency, showing your Lordship my secrets, and then offering at your pleasure to undertake to carry into execution in due course all the things which I am about to enumerate briefly and summarily to you.

1) I have methods of constructing very light and strong bridges which can be transported with the greatest ease, offering the means of pursuing an enemy, and if necessary of fleeing from him; and others safe and indestructible by fire or battle, easy and convenient to lift and place [Figure 1.4]. Also methods of burning and destroying those of the enemy.

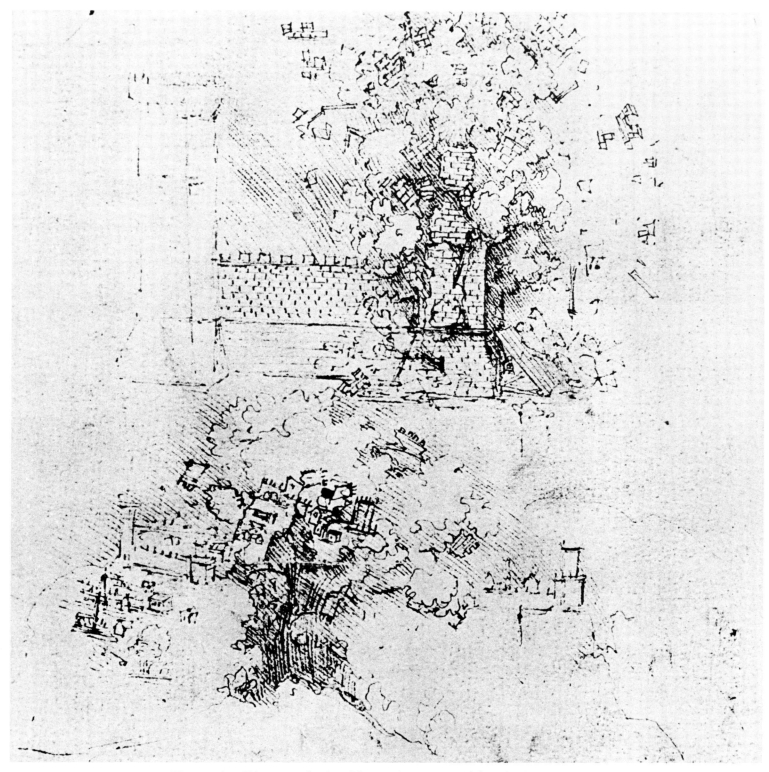

Figure 1.6. 'I have methods of destroying any citadel or fort' (W 12652 v).

2) When a place is besieged I know how to take the water out of moats, and how to construct the most varied types of bridges, covered ways and scaling ladders, and other machines required for such expeditions [Figure 1.5].

3) Item. If owing to the height of the ramparts or the strength of the position it is impossible for the besiegers to make use of bombards, I have methods of destroying any citadel or fort which is not built of rock [Figure 1.6]'.

As so often Leonardo illustrates his case more vividly than he writes it down. A contemporary drawing in Figure 1.6 shows the devastating effect of 'destroying any citadel' which he had in mind.

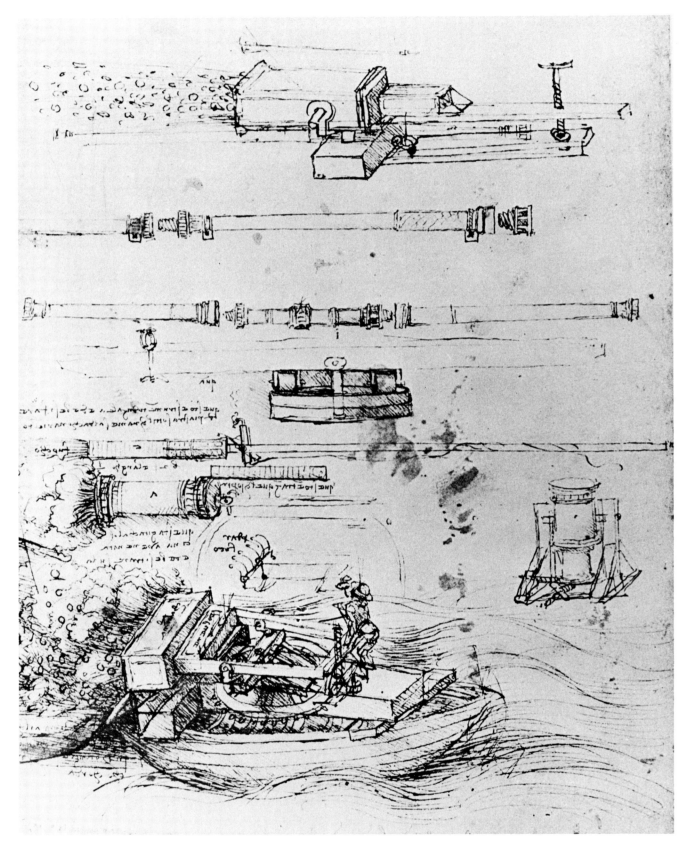

Figure 1.7. Mortars which 'fling small stones' and smoke, on land and sea (W 12652 r).
Leonardo's conical screw method for breech-loading guns is also illustrated on this page.

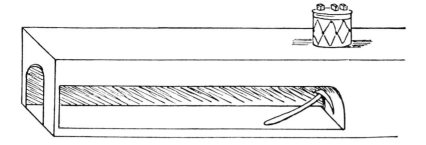

Figure 1.8. Excavation of an underground passage 'without a sound'.

Sound is detected by placing a drum on the ground above the secret passage. If there is noise, then 'the dice will jump a little on the drum' (BN 2037 1 r; B 91 r).

Figure 1.9. 'I will make covered wagons, safe and unattackable' (B 59 v).

The first tanks to be drawn are shown propelled by crankshafts from within.

'4) Again I have kinds of mortars, most convenient and easy to carry, and with these I can fling small stones almost resembling a storm, and with the smoke of these cause great terror to the enemy with great damage and confusion'.

This is also described in the item enumerated 9. Such a mortar is dramatically illustrated in Figure 1.7. Here one such mortar is mounted on a boat for action at sea. This Leonardo describes thus:

'And if the fight should be at sea I have many kinds of machines most efficient in offence and defence and vessels which will resist the attack of the largest guns and powder and fumes.
5) Item. I have methods of reaching an indicated point by secret winding passages excavated without a sound, even if it were necessary to pass under a trench or a river [Figure 1.8].
6) Item. I will make covered wagons, safe and unattackable, which entering among the enemy with their artillery, there is no body of men so great that they would not break. And behind these infantry could follow, quite unhurt and without any hindrance [Figure 1.9]'.

This particular invention of the tank is realistically drawn on a manuscript in the British Museum.

'7) Item. In case of need I will make big guns, mortars and light ordnance of fine and useful forms, differing from those in common use [Figures 1.10 and 1.11].

8) Where the bombards would fail in their effect I would contrive catapults, ballistas, mangonels and mortars and other instruments of marvellous efficacy, not in common use [Figure 1.12]. And in short, according to the variety of circumstances I can contrive endless means of offence and defence'.

Clearly by the time he reached items 7 and 8 Leonardo was getting a little tired of enumerating the details of all his devices. He may even have suspected that his reader, the Duke Ludovico Sforza, was getting tired of reading them too. However, he still has a rich store of military devices 'not in common use' up his sleeve. Two deserve mention: the giant ballista in Figure 1.12 which has a cord some eighty-six feet long and is fired by a hammer catch used by a man whose relative size gives some idea of the whole 'instrument' and a big gun. The 'big gun not in common use' was that invention of Leonardo which he called the 'architronito'. It was a steam cannon, described in Figure 1.13, wherein water is injected into a heated chamber and the force of its expansion into steam expels the cannon ball. That this invention came into Leonardo's mind intuitively rather than as the result of scientific study is reflected by the fact that he carried his investigation of the phenomenon no further for many years. That the invention was not impracticable was demonstrated by its principle being put into use in the nineteenth century during the American Civil War.

Having thus finished his account of military inventions, Leonardo devoted the last two paragraphs of his letter to peacetime accomplishments.

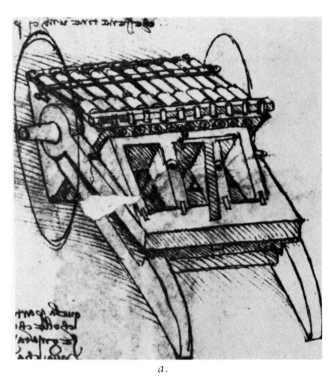

a.

a. 'Light ordnance of fine and useful forms' (CA 56 va).

This multiple-barrelled gun consists of three rows with eleven guns in each. Each row constitutes the side of a triangle. The whole prism of guns is mounted on a gun-carriage, thus markedly 'differing from those in common use'.

b.

b. Model No. 76, Museo Nazionale della Scienza & della Tecnica, Milano.

Figure 1.10.

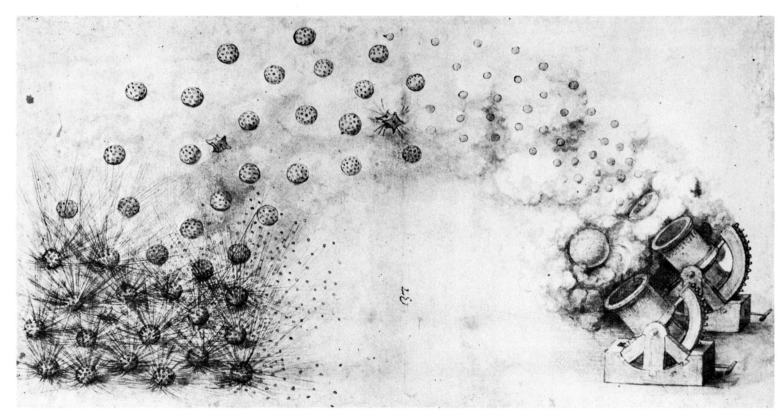

Figure 1.11.   Mortars with explosive shells (CA 9 va).

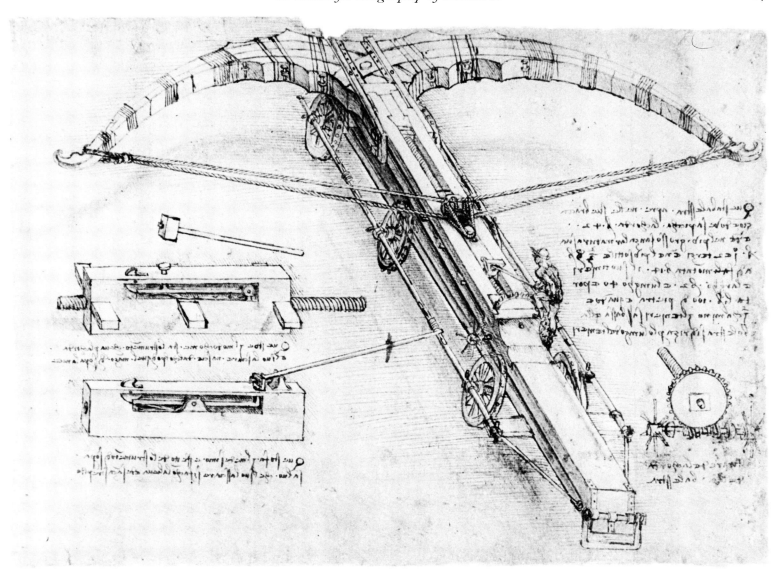

Figure 1.12. The gigantic size of this ballista is shown by the drawing of the man mounting it (CA 53 vb).

'10) In time of peace I believe I can give perfect satisfaction bearing comparison with anyone else, in architecture, in the designing of buildings, public and private, and in conducting water from one place to another [Figure 1.14].
Item. I can carry out sculpture in marble, bronze or clay and I can also do in painting whatever can be done to stand in comparison with anyone else, be he who he may.

It will also be possible to put in hand the bronze horse which is to be the immortal glory and eternal honour of the prince your father, and of the illustrious house of Sforza.

And if any of the above-mentioned things seem to anyone to be impossible or unfeasible I am most ready to make the experiment of them in your park or in whatever place may please your Excellency, to whom I most humbly commend myself' (CA 391 ra).

After reading this letter, one finds Vasari's explanation of Leonardo's departure from Florence to Milan odd. 'Lionardo', he writes, 'was invited to Milan with great ceremony by the duke to play his lyre, in which the prince greatly delighted. Lionardo took his own instrument, made by himself in silver, and shaped like a horse's head, a curious and novel idea to render the harmonies more loud and sonorous, so that he surpassed all the musicians assembled there' [Figure 1.15].[12]

There is still doubt as to the exact year in which Leonardo despatched his letter, as well as that of his departure from

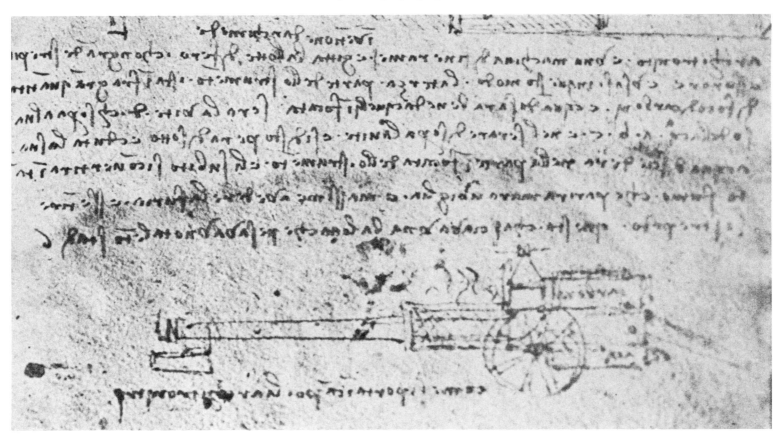

Figure 1.13.   The steam cannon, which Leonardo called 'architronito' (B 33 r).

The supply of coal is labelled 'carbone'; below this is the tank of water, labelled 'acqua'. This is injected into the hot chamber in front, evaporates into steam under pressure, and discharges the cannon ball.

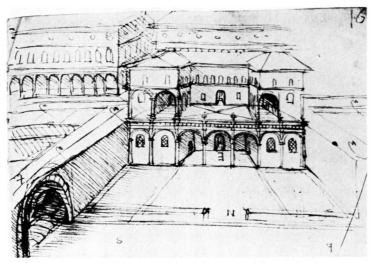

Figure 1.14.   'The design of buildings, public and private' (B 16 r).

Here Leonardo depicts his basic scheme of city planning. Houses are to be built on two stories; the upper stories are connected to high-level roads used by pedestrians only. These roads have central drains. They arch over lower-level means of access to the houses consisting of canals used for the delivery of goods, as at the gate marked *N*. The arches contain spiral staircases provided with urinals.

Florence to Milan, a doubt which is reflected also in the exact dating of the drawings of this period, but they demonstrate the truth of every claim made in the letter.

Leonardo, however, was now about thirty years old, and a brief description of his appearance and personality will render more human this sensitive man whose obsession for science exposes him to the risk of making him appear to be a coldly mathematical calculating machine.

LEONARDO'S PERSONALITY

In physical appearance Leonardo was strong, well built and handsome. According to Vasari his 'personal beauty could not be exaggerated, his every movement was grace itself . . . He possessed great personal strength, combined with dexterity, and a spirit and courage invariably royal and magnanimous . . . The grace of God so possessed his mind, his memory and intellect formed such a mighty union, and he could so clearly express his ideas in discourse, that he was able to confound the boldest opponents . . . His charming conversation won all hearts, and although he possessed nothing and worked little, he kept servants and horses of which he was very fond, and indeed he loved all animals, and trained them with much kindness and patience. Often, when passing places where birds were

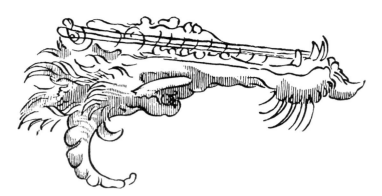

Figure 1.15. A musical instrument like a lyre, made in the shape of the head of a horned animal.

Strings are drawn across the upper jaw from the upper incisor teeth to the palate. Presumably the hollow skull would act as a resonating sound box. This curiously monstrous head, not unlike a horned horse, may well be the instrument to which Vasari refers. This is described by J. P. Richter in *The Literary Works of Leonardo da Vinci,* 3d edition (New York, Phaidon, 1970), vol. 1, pp. 69–70.

sold, he would let them out of their cages and pay the vendor the price asked. Nature had favoured him so greatly that in whatever his brain or mind took up he displayed unrivalled divinity, vigour, vivacity, excellence, beauty and grace. His knowledge of art, indeed, prevented him from finishing many things which he had begun, for he felt that his hand would be unable to realise the perfect creations of his imagination, as his mind formed such difficult, subtle and marvellous conceptions that his hands, skilful as they were, could never have expressed them. His interests were so numerous that his inquiries into natural phenomena led him to study the properties of herbs and to observe the movements of the heavens, the moon's orbit and the progress of the sun . . . By the splendour of his magnificent mien he comforted every sad soul, and his eloquence could turn men to either side of a question. His personal strength was prodigious, and with his right hand he could bend the clapper of a knocker or a horseshoe as if they had been of lead. His liberality warmed the hearts of all his friends both rich and poor, if they possessed talent and ability. His presence honoured the most wretched and bare apartment'.

Leonardo, says Vasari, 'spoke freely of his art, and explained how men of genius really are doing most when they work least, as they are thinking out ideas and perfecting the conceptions, which they subsequently carry out with their hands'.

Speaking of Leonardo's anatomical drawings, Vasari comments, 'Whoever succeeds in reading these notes of Lionardo will be amazed to find how well that divine spirit has reasoned of the arts, the muscles, nerves and veins, with the greatest diligence in all things . . . To Lionardo we owe a greater perfection in the anatomy of horses and men. Thus

by his many surpassing gifts, even though he talked much more about his works than he actually achieved, his name and fame will never be extinguished'.[13]

These extracts from Vasari's life of Leonardo may sound to some more like a eulogistic obituary notice than the true life portrait of a person. However, it should be remembered that they were written by a man who admired Michelangelo much more than Leonardo. Another important point to be noticed is the relevance to Leonardo's science of many of Vasari's descriptions in his life of Leonardo. Leonardo, we know, was a master of the art of painting and sculpture; he was a master of many other arts as well. He elevated the work of mechanics to an art of engineering, as expressed by his many inventions, but from the point of view of his science, devoted to the interpretation of nature and man, it is most important to recognise that Leonardo was a master of a still but little appreciated art – the art of experiment. When, for example, Leonardo expressed the idea that men of genius think out ideas and perfect their conceptions which they subsequently carry them out with their hands, he might as well be referring to his genius for the art of designing and carrying out experiments as to his art of painting. With Leonardo, indeed, even his paintings often consisted of experiments. How often has he been reproached for the fate of *The Last Supper* because it was in some ways an experimental failure? As with so many truly original experiments, some were successful, some were not. In this experimental sense Leonardo included his painting among the 'creative sciences'. In this field we applaud his brilliant successes and reproach him for his unhappy failures.

Paolo Giovio gives some further insights into Leonardo's courageous explorations when he tells us that 'Leonardo da Vinci . . . added great lustre to the art of painting, denying that this could be properly carried out by those who had not attained the noble sciences and liberal arts necessary to the disciples of painting . . . The science of optics was to him of paramount importance and on it he founded the causes of beams of light and shade most diligently down to the minutest details. He dissected the bodies of criminals in the medical schools, a dehumanised and filthy work, in order that the variations of the joints of the limbs flexed by the actions of the nerves of the vertebrae should be painted according to the laws of nature . . . He drew on panels the shape of all these parts down to the smallest veins and the marrow of the bones with wonderful skill, in order that out of this work of so many years innumerable examples should be published from copper engravings for the use of art . . . But while he was thus spending his time in the close research into subordinate branches of his art he carried only very few works to completion'.[14] Giovio also, like Vasari, says that Leonardo's 'charm, his brilliance and generosity were not less than the beauty of his appearance'. It is to be noted that Giovio's short account was written in 1527, only eight years after Leonardo's death, whereas Vasari's first appeared in 1551. It is abundantly clear that Giovio, who probably obtained his information from Leonardo himself, appreciated to some extent his unique view of the relation between art and science. Training in the sciences was almost unknown in those days in the sense that we understand the phrase today. In fact, perhaps the best example of it consisted in Leonardo's own creation of methods of research. Study of the past literature, observation, experiment and mathematical generalisation were techniques which he himself evolved into scientific method.

In spite of his physical attractiveness and conversational charm Leonardo was obsessed with a scientific curiosity which was not easily communicable to his fellow men. Thus he was a lonely man with but few close friends. His thousands of pages of notes contain very few affectionate references to anyone, be it father, mother, or any other man or woman. His capacity for objectivity was such that he was able to use his own body for experiments. His agonised cry, 'If there is no such thing as love what is there?' (CA 4 v) is one of the very few signs Leonardo has left of his deep need for affection.

Francesco Melzi, Leonardo's pupil and disciple in the last decade of his life, also reveals signs of deep affection in Leonardo. Writing to Leonardo's half-brothers after his death, Melzi declares, 'I believe you have been informed of the death of Master Leonardo your brother, who was also at the same time the best of fathers to me. It would be impossible for me to express the pain which has gripped me because of his death; until my own limbs have been laid to rest I will have endless sorrow to bear; and this is only right for he daily showed toward me the most sincere and ardent affection. Everyone is stricken with sorrow at the loss of such a man for it is not within the power of nature to bring forth such another man'.[15]

There can be little doubt that Leonardo's expression of affection was largely suppressed by his extraordinary powers of objective observation. There is one important exception to this. Throughout his life there are stories of his affection for animals. Not only did he treat horses 'fondly' and release birds from their cages but also, most significantly of all, he never (with one exception) performed experimental vivisection, thereby depriving himself of an important source of physiological knowledge.

This capacity for objectivity or detachment goes some way perhaps to explain another remarkable feature of Leonardo's character, the absence of arrogance. That he himself was aware of his uniqueness is indisputable, but nowhere in his notes does he boast of his exceptional capacities. Only once does he write of the rarity of such a man as himself, and that quite objectively. One cannot help wondering whether Francesco Melzi in his grief-stricken letter was indeed referring to that passage in the *Madrid Codex* I 6 r in which Leonardo declares, 'Read me O reader if you find delight in me because I am very rarely reborn in the world. Because the patience of such a profession is found in few who may again want to reconstruct [*riconpore*] such things anew. And come O men to see the wonders which may be discovered in Nature by such studies'.

LEONARDO'S FIRST PERIOD IN MILAN (1483–1499)

THE SEARCH FOR WORDS

The personality of the Leonardo just described was still immature when he journeyed from Florence to Milan with his lyre in his hand about 1482–1483. At this time he was still predominantly a skilled craftsman and artist.

His first years in Milan are shrouded in mists of doubt; so much so that it has been suggested that about this time he made a visit to the Near East, to Armenia and Syria. This suspicion arose from some 'letters' describing 'mount Taurus and the river Euphrates'. One begins, 'To the Devatdar of Syria, lieutenant of the Sacred Sultan of Babylon' (CA 145 va). As it turns out, the 'letter' is an example of Leonardo's ability to form realistic fictional images by convincingly composing details in words much in the same way as he composed in painting the detailed structures of animals into imaginary 'monsters'. Parts of his description of Asia Minor and the light of the sun on the peaks of the Taurus mountains are derived from Ptolemy and Aristotle.

Words seem to have preoccupied his attention in these early years in Milan. Acutely aware as he was of his limited vocabulary, the greater part of his earliest notebook, *Codex Trivulziano,* is filled with lists of Italian words. In his list of books in his library, Leonardo refers to this as 'My book of words'. These words are of particular interest in that they include a number of terms used by Leonardo in his science; some (only too few) are accompanied by his own definitions of their meaning, thereby providing a useful glossary of his terms. Some definitions bear the stamp of his own personal outlook. One feels that his definition of '*barbarismo*' was only too sensitively rendered as 'incorrect words badly pronounced'. A 'syllogism' (*siligismo*) Leonardo defines as 'ambiguous speech'. Some words which recur frequently in his later scientific studies are '*infuso*', defined as 'scattered within and mixed'; '*condenso*' he defines as 'compact and imporous'; and '*detrimento*' is defined as 'damage and pain' (Triv 12 r to 13 v). In view of Leonardo's own character it is interesting to see him defining the word '*curioso*' as 'great thoughtfulness and diligence'. Perhaps the most significant definition from our present point of view is his description of science itself. 'Science', he writes, 'is knowledge of things possible, present and past' (Triv 17 v). He studies Latin grammar at length in MS I.

Other subjects which show the direction of his attention during those years when he was writing the *Codex Trivulziano* are notes on light, vision and perspective. Studies on perspective constituted in fact Leonardo's portal into the 'knowledge of things possible and present'. There is no doubt that he commenced this line of enquiry during his early years in Florence. The well-known study for the *Adoration of the Magi* in the Uffizi in Florence demonstrates this, as does the unfinished picture, also in the Uffizi Gallery, showing the crowd of gesticulating spectators arranged in a concave curve behind the Virgin and Child. This curve follows the line of equal visual power, or '*virtu visiva*', as

Leonardo calls it. He frequently uses it in his studies of perspective.

Leonardo came to Milan as a 'foreigner' and was treated as such. The lack of definite information about him during these years may well be due to the obscurity of his fate. Ludovico Sforza's plans for war against Venice, upon which Leonardo's application had presumed, faded away, leaving Leonardo with lean and hungry years. He found refuge with Ambrogio de Predis, whose half-brother, Cristoforo, was a deaf-mute. Leonardo was fascinated by his replacement of speech by eloquent gestures and refers to deaf-mutism several times in his *Treatise on Painting*.

THE *MADONNA OF THE ROCKS*

In 1483 Leonardo with Ambrogio de Predis signed a contract with the fraternity of the Immaculate Conception for an altarpiece, every detail of which was designed by the prior of the community to fit into a carved wooden frame already prepared. It was to represent seraphim, angels and prophets according to the theological beliefs of the fraternity. Moreover, it was to be finished within a year. Leonardo produced the *Madonna of the Rocks,* a picture far removed from the prior's specifications and, of course, still unfinished by the date of the contract. Some twenty years of subsequent litigation finally ended with this original picture finding its home in France and a second, modified version being delivered to the fraternity. This eventually found its way to London. A mere glance at either version confirms Leonardo's intense concern with light and shade in perspectival distances, using the geological structure of the rock formations to accentuate these effects. The remarkable foreshortening of the hand of the Virgin held over the head of the Christ child reminds one of his emphasis on the importance of perspective in representing parts of the body. Gone were the seraphim and prophets. Instead Leonardo depicted how 'the Earth has a spirit of growth . . . its bonds are the successive strata of the rocks which form the mountains . . . and the dwelling place of its creative spirit' (Leic 34 r). No wonder the fraternity sued him.

LEONARDO'S VISUAL SCIENCE

Leonardo remained an inventor throughout his life, but he was not only an Edison of his time, for he became dissatisfied with empirical inventions and during these early years in Milan began to grope around for methods of finding out what lay behind them and his observations of nature. What lay behind the experience of seeing things? What is vision, and how do objects act on the sense of sight? Which senses of the body convey most information about the world around us, and how do they do so? In the *Codex Trivulziano* he enters on a survey of the human senses: 'Senses are of the earth, whilst reason stands apart when it contemplates' (Triv 33 r). 'All our knowledge originates in our sensibilities' (Triv 25 v). Then, striving to comprehend the relationship between the objective and subjective aspects of experience, he writes, 'The lover is moved by the thing loved as the sense is by that which is sensed, and it unites with it, and they become one and the same thing. The work is the first thing born of the union. If the thing loved is base the lover becomes base. When the object taken into union is in harmony with that with which it unites there follows rejoicing, pleasure and satisfaction. When the lover is united to the loved one it finds rest there; when the weight has fallen [to earth] it is at rest' (Triv 6 r). The last sentence is quite typical of Leonardo's close weaving of physical with psychological phenomena. In this he was following the example of Empedocles, who equated 'love' with 'attraction' in all its forms, and 'hate' with 'repulsion'.

Leonardo's survey of the five senses is thorough and 'diligent'. We shall return to it later. At this stage it will suffice to point out the route by which he comes to the conclusion that of all the senses that of sight is pre-eminent in acquiring information about the world around us. Predisposed as he is by possessing the visual powers of the painter, he nevertheless considers in turn the other senses as sources of information. But as a painter he has become acquainted with the experiences presented to the eye; he has learnt to see the world in perspective, and perspective is geometry, and geometry means not only proportion and measurement but also analysis of the nature of surface, line and point. So it comes about that the overwhelming majority of Leonardo's first speculations about science, and in particular the science of man, concentrate on light and shade, vision, perspective, proportion and measurement. Through experiences so derived he rapidly approaches nature as a set of forms or geometrical figures. This geometry he applies not only to the surfaces of things which the painter sees and paints but also to the forces, or 'Powers of Nature', which produce all action and movement.

The close involvement between Leonardo's physics and the science of man is particularly emphasised when his 'thought and diligence' are applied to the subject of vision. Here he follows inexorably the path that leads from the object seen outside the body through the eye to the 'seat of the soul', where the experience of seeing takes place. In this quest he embarks upon the first phase of his anatomy which is particularly notable for his drawing of skulls. His choice of heads for dissection is shown in these drawings to depend upon the fact that they reveal the path of vision. Nearly all these beautifully drawn skulls depict the optic nerve and locate the '*senso comune*', i.e., the sensory centre where vision enters consciousness. His dissections of other parts of the body at this time also concentrate on the nervous system. Thus by 1489 (a date which Leonardo writes on one of the anatomical sheets) Leonardo's focus of attention has divided into two separate streams. One flows along the course of a deepening comprehension of his many technical inventions for transmitting movement and power; the other follows up the problems of the nature of observation, of vision, and its relation to both nature and his inventions. In this latter field he already strongly suspects that perspective holds the key.

In the summer of 1490 there occurred an event which triggered off Leonardo's glowing curiosity into explosive activity. Ludovico Sforza despatched Leonardo and Francesco di Giorgio Martini to Pavia in order to obtain their opinions on the progress of the work on the cathedral there. This journey had a threefold impact on Leonardo's development. First, it constituted public recognition of his capacities as an architect. Previously he had submitted a model for the cupola of Milan cathedral, but after much unsatisfactory quibbling on the part of the cathedral authorities he had withdrawn it. Nevertheless Leonardo's architectural interests, blending so fundamentally with his geometrical awareness, produced many drawings of plans and elevations for cathedrals, palaces and even a brothel. Most notable of all was his town plan based on principles of hygiene (see Figure 1.14, p. 18).

Secondly, this journey gave Leonardo one of the few lasting friendships of his life, that of Francesco di Giorgio Martini. Some thirteen years older than Leonardo, Francesco, who had been born in Siena, developed there as a painter, sculptor and architect. At the age of thirty, in 1469, he was responsible for the maintenance of the water supply of Siena and for the erection of the fountain (Fonte Gaia) so prominent a feature of the famous Piazza del Campo there. Later he went to Urbino as architect and military engineer where he availed himself of the rich library collected by Federigo Montefeltre, the Duke of Urbino. Following this distinction he was called into consultation by various potentates all over Italy. Thus he came to Milan to contribute to the irksomely undecided problem of the cupola for Milan cathedral and went with Leonardo on a similar mission to Pavia. Francesco's one great work, a *Treatise on Civil and Military Architecture,* greatly influenced Leonardo. Not only does one manuscript of this work contain marginal notes in Leonardo's handwriting but also some twelve folios of his *Madrid Codex* II written in 1504 consist of an abridged transcription of part of Francesco's treatise.

The third and most important result of this journey to Pavia was Leonardo's discovery of the magnificent library in the castle. This great hall, its walls lined with shelves full of manuscripts, was one of the most notable features of Italy. One visitor had declared that he had greater happiness on seeing it than on visiting the Holy Places of Jerusalem.[16] Such a comment, a little surprising coming from a procurator of the Carmelites in Rome, assuredly reflected Leonardo's opinion, for whilst Francesco di Giorgio Martini went back to Milan after two weeks in Pavia, Leonardo stayed on for six months, exploring the library. A good sample of his activities there is supplied by his note, 'In Vitelone there are 825 conclusions on perspective' (B 58 r); later he reminds himself, 'Look up Vitelone in the library at Pavia' (CA 225 rb). Leonardo is here referring to the thirteenth-century Silesian investigator, Witelo, whose work on optics was largely derived from the Arab, Alhazen. Leonardo's interest in perspective and optics was further stimulated by meeting Fazio Cardano, a professor of mathematics at the University of Pavia who had translated John Pecham's *Perspectiva Communis*. From it Leonardo made a note of the passage, declaring, 'Among the various studies of natural processes, that of light gives most pleasure to those who contemplate it . . . Perspective therefore is to be preferred to all the discourses and systems of the schoolmen . . . In it you will find the glory not only of mathematics but of physics, adorned as it is with the flowers of both' (CA 203 ra).

These months in Pavia seem to have orientated Leonardo in his quest for knowledge. Immediately after his return to Milan he began two new notebooks, MS A and MS C. The former inaugurated a systematic study of the phenomena of perspective and the powers of weight, movement, force and percussion. The latter was almost entirely devoted to the study of light and shade. Both are characterised by his introduction of 'pyramidal' laws of perspective. These, as we shall see, he carried over from the perspective of light into the fields of movement, force, weight and percussion.

LEONARDO AS FESTIVAL PRODUCER AND ENTERTAINER AT THE COURT OF MILAN

These abstract studies, however, did not bring in any money, and the next few years of Leonardo's life in Milan were marred by poverty. Since he had to live, he devoted his practical genius to constructing the machinery for such festive occasions as the marriage of the young rightful Duke of Milan, Gian Galeazzo Sforza, to Isabella of Aragon. This occasion was to become famous throughout Italy for the ingenious machinery by which the seven planets circled through the heavens, the creaking of their wooden gear wheels being cleverly concealed by music, some of which may well have emerged from instruments invented by Leonardo himself. It was at this time that he wrote a brief note under the date 1490 saying that he had 'recommenced the horse' – a reference to the statue of Francesco Sforza mentioned in his letter of application to Ludovico Sforza.

Another kind of activity which won Leonardo favour at the Court of Milan was his capacity for telling tales, fables and 'prophesies'. Many pages of his notebooks at this time are filled with such tales, most of them, like Aesop's fables, amusingly serious with a moral. It is interesting to see how many of them reflect his views on the natural sciences. For example, he tells a tale of water disdaining its natural element, the ocean, and wanting to climb up high into the sky. With the aid of fire it turns into fine vapour until it seems as fine as air, but in the cold its fine particles unite and become heavy, its pride deserts it, it falls from the sky to the dry earth where, imprisoned for a long time, it does penitence for its sin (Forster III 2 r).

How many in Leonardo's audience would recognise in this little moral tale the whole core of his theory of the circulation of the four elements? This theory being derived from Aristotle in the first place, Leonardo at that time was also applying it to the movement of sap in plants and the movement of blood in man.

Many of these tales shed similar gleams of light on Leonardo's integration of the psychology of man with the physical world around him. One such tale should be cited here since with it Leonardo complements the psychology of the previous tale. It may be summarised thus: A patch of snow on the topmost peak of a high mountain thought to itself, 'Shall not I be thought vain and proud for having placed myself – such a small patch of snow – in so high a place whilst so large a quantity of snow as I see around me is placed lower than me? Surely my small quantity does not merit this elevation, and surely I shall be melted by the sun, as happened to so much of the snow around me . . . I will flee from the anger of the sun and lower myself to find a place more suitable to my small quantity'. Then throwing itself down, it began to descend, rolling on the snow as it went, but the more it sought a low place the more it grew in size, so that when it came to rest it found itself as large as the hill on which it stopped. 'This is told for those who in humbling themselves become exalted' (CA 67 va).

Both these tales illustrate Leonardo's interest at this time in the 'powers' of heat and cold in relation to the movements of the four 'elements', earth, air, fire and water.

LEONARDO'S 'ELEMENTS OF MACHINES'

Over this same period his almost obsessional studies of what he called 'the four powers' reached their peak. In *Forster* II page after page is devoted to detailed studies of the balance. Numerous experiments are performed in order to find out the relation between the length of the arms and the weights on a balance. With obsessional zeal he extends his studies to balances with bent arms, curved arms, 'circular balances', as he calls pulleys, combined into fantastic complexity. Screws, stress on beams and arches – all receive equally thorough experimental attention in which every variation which his 'thought and diligence' have suggested to him is literally weighed up. Its 160 folios (320 pages) presage the *Madrid Codex* I which covers a similar field over the same period of years, about 1492–1497. The difference between the two, however, is striking. *Forster* II is a laboratory notebook; the *Madrid Codex* I presents these notes more tidily arranged; here the figures are beautifully and carefully drawn. The whole lay-out is clearly designed to illustrate the principles behind the findings in *Forster* II.

It is known that Leonardo contemplated the production of a treatise on the 'Elements of Machines' constructed on the logical pattern of the *Elements of Geometry* by Euclid. Leonardo intended to demonstrate how machines are built up from basic principles by the transmission of movement and power just as Euclid's *Geometry* was built up from a set of axioms. Leonardo had analysed machines into some twenty basic elements, as shown by Reti.[17] At first sight this *Madrid Codex* I would appear to constitute such a treatise, but as it progresses Leonardo, as was his wont, continues to raise unanswered questions and forgets his original intention. Thus the work gradually loses its formal shape as a treatise as Leonardo concentrates on these new problems. In this way *Madrid Codex* I provides a good example of the fate of so many of Leonardo's treatises by showing how they, like his pictures, are almost inevitably 'unfinished'.

THE GREAT HORSE

At the same time as he was compiling these notebooks Leonardo was creating his great horse. On *Madrid Codex* II 157r he writes, 'In the evening, May 17th 1491. Here a record will be kept of everything related to the bronze horse on which I am at present working'. The notes continue rather disjointedly until he writes, 'This day 20th December 1493 I have decided to cast the horse without tail and lying on its side' (Madrid II 151 v). Numerous details are given regarding the armature, the mould and the furnaces for casting. Each stage of the procedure is freely sprinkled with suggested 'tests' and 'experiments'. Meanwhile Leonardo had made an 'earthen' model of the great horse. Just as the 'paradise festival' had been produced by Leonardo as part of the celebrations for the young Duke Gian Galeazzo's marriage to Isabella of Aragon in 1490, so this enormous model of the horse was produced for the festivities at the marriage of Ludovico's niece, Bianca Maria Sforza, to the Emperor Maximilian I in 1493. The story of this model is briefly told by Vasari. 'In truth, those who have seen Lionardo's large clay model aver that they never beheld anything finer or more superb. It was preserved until the French came to Milan with King Louis of France, and broke it all to pieces. A small wax model, considered perfect, was lost, as well as a book of the anatomy of horses, done by him'.[18] This horse, about twenty-four feet high, was never to be cast in bronze. The huge quantity necessary, some 200,000 pounds, was instead turned into guns. Leonardo in a letter to the Duke Ludovico comments, 'Of the horse I say nothing, for I know what the times are like' (CA 335 va). In another letter he tells how 'for thirty-six months I have fed six mouths, and all I have had is fifty ducats' (CA 315 va).

Such poverty turned his thoughts to a fresh outburst of technical invention. Milan was a city famous for its metal and textile industries; Leonardo focussed his attention on the latter. Beautiful drawings of spinning machines and looms appear, but his greatest hopes were based on a needle-sharpening machine. He dreamed of obtaining fantastic profits from it – 'A hundred times an hour with 400 needles each time makes 40,000 an hour, and in twelve hours 480,000 . . . that is 1,000 lire every working day, and with twenty days worked in the month it is 240,000 lire or 60,000 ducats a year' (CA 318 va). This for a man who had received 50 ducats in the past three years was clearly a plan to get rich quick. Stung by his poverty, Leonardo seems to have forgotten his own adage, 'He who wants to get rich in a day will be hanged in a year' (W 12351 r). There is no evidence that he attained his objective. What did however happen was that Ludovico Sforza gave him a commission. This offered him the chance of demonstrating the fruits of his tree of knowledge of painting, of perspective and of the movements of man in the expressions of different emotions. He was asked to paint *The Last Supper*.

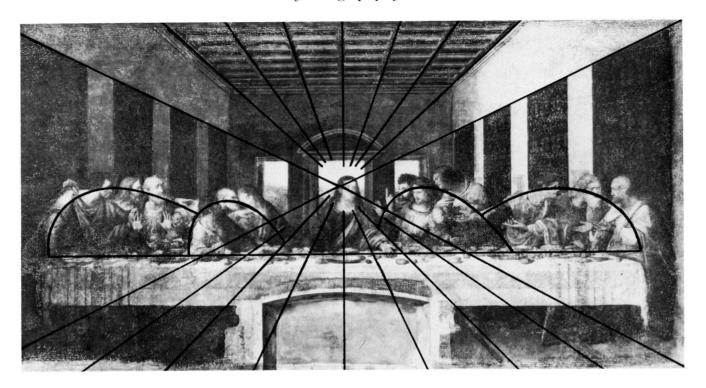

Figure 1.16.   The Last Supper (Santa Maria delle Grazie, Milan).

The picture has been marked to show the lines of perspective of which it is composed. These pass through the head of Christ at the vanishing point. The emotional waves of sound emanating from Jesus Christ's announcement of his betrayal are portrayed with almost geometrical precision as they spread from their centre to the periphery at the ends of the table. Figure adapted from Paul Hamlyn, *The Life and Times of Leonardo* (London, 1968), p. 72.

THE LAST SUPPER

One of the main reasons for Leonardo's low output of completed works of art and science was the meticulous diligence in detailed composition which he applied to all his works. The commission for the large fresco of the Last Supper came, however, at a time when he had spent nearly twenty years constructing a 'science' of perspective and of the movements of man. These constructions we shall return to later. Here it is relevant only to recall that Leonardo painted this fresco to provide a perspectival illusion of the room in which the Last Supper was held. It appeared as an actual extension of the refectory in which the monks ate their meals. He strengthened the perspectival illusion by lighting his painted 'room' from the light of a window which actually existed (and still does) on the left-hand wall of the refectory. Here indeed he successfully fused his perspective of light and shade with the perspective of distance and colour.

With regard to the movements of man, Leonardo had by this time made innumerable studies of the three main types of human movement. Movement in space or 'local motion' constituted movement from place to place as a result of the shift of the centre of gravity of the body beyond the feet. Movements of the body without change of place Leonardo called '*moto actionale*', i.e., actions or gestures of the limbs. Both forms of movement might be involved in what

Leonardo called '*moti mentali*', or the emotions of the soul, in which he insists that the movements of the hands and arms must in all their actions display the intention of the mind or soul. Unless a painting does this, says Leonardo, it is 'twice dead'. These are the movements displayed by Leonardo in *The Last Supper;* the picture bursts with emotional life.

Lastly, true to his scheme of interweaving physical and psychological events, Leonardo has expressed the psychological shock waves emanating from the central sound of the words of Christ, 'One of you shall betray me', in the form of physical waves passing through the grouped heads of the twelve disciples towards the periphery, like waves in water (Figure 1.16).

Leonardo's period of preparation for this great work was relatively short, and he completed it within two or three years.

THE INFLUENCE OF THE MATHEMATICIAN LUCA PACIOLI

About this time Leonardo came into contact with the man who was to expand the vast range of his efforts into their final mathematical form. Luca Pacioli was a man of great ambition, doubtful integrity and a remarkable ability for utilising the works of his mathematical predecessors. Born in Borgo San Sepolcro about 1445, Luca learned much of his mathematics from Piero della Francesca. Vasari in his biography of Piero della Francesca makes withering comments about Fra Luca, to whom he refers thus: 'The man who should have done his utmost to increase Piero's glory and reputation, who had learned from him everything which he knew, impiously and malignantly sought to annul his teacher's fame, and usurp the honour due to him, publishing under his own name of Fra Luca dal Borgo all the results of the labours of that good old man'.[19] Luca had also as a young man lived in Rome with Leon Battista Alberti, there no doubt acquiring some knowledge of Alberti's mathematical works also. Later he became a Franciscan monk and was attracted to Milan by the wealth and power of Ludovico Sforza. Here he met Leonardo.

Whilst at Perugia Luca had produced a vast compilation called the *Summa de aritmetica geometria proportioni et proportionalita*. It was published in Venice in 1494 in Italian, not Latin. It was an irresistible magnet for Leonardo, and its attraction for him was all the stronger because the book was almost entirely unoriginal. Luca had gleaned it from the works of such men as Plato, Aristotle, Euclid, Archimedes, Leonardo of Pisa, Blaise of Parma, Albert of Saxony, Jordanus Nemorarius, as well as his teacher, Piero della Francesca. Thus Leonardo was introduced to an Italian version of all those great mathematicians of the past as well as to the works of some of his contemporaries. Moreover, he had Luca Pacioli himself there to explain them to him. About 1496 Leonardo makes a meticulous note recording the purchase of the *Summa aritmetica* for 119 soldi (CA 104 ra).

Leonardo makes few references to Euclid's optics and geometry before 1496; the vast majority of his notes on Euclid's works occur after he had met Luca Pacioli and obtained his *Summa aritmetica*. There can be little reasonable doubt that Leonardo went systematically through Euclid's thirteen books of geometry with Luca Pacioli. With Leonardo's aid, Luca Pacioli produced the greater part of his *De divina proportione*. For this work Leonardo drew the geometrical figures including those of the five regular solid bodies (see Figure 5.16, p. 144). The whole work is said by Koyré to be 'based almost word for word on an (unpublished) book of Piero della Francesca'.[20] Koyré thus confirms Vasari's comment.

In the portrait of Luca Pacioli at Naples by Jacopo de' Barbari, Luca is shown pointing with his left hand to Euclid's Book XIII, open at Proposition 9, which deals with the sides of a hexagon and a decagon inscribed in the same circle and a straight line cut in extreme and mean ratio, i.e., 'the divine proportion'. The young man standing by his side, however, does not look at all like Leonardo da Vinci.

Thus Luca Pacioli not only brought to Leonardo the geometry of Euclid, much arithmetic and 'the multiplication of roots', as Leonardo himself notes (CA 120 rd), but also enriched Leonardo's acquaintance with the works of Leon Battista Alberti and Piero della Francesca.

If Luca was indeed deficient in acknowledgement to his master, Piero della Francesca, he did not repeat this sin against Leonardo. For he acknowledges Leonardo's drawings of the figures of the solids and refers to Leonardo's

'inestimable work on local motion, percussion, weight and all kinds of forces, that is accidental weight'.[21] In this sentence he reveals a greater understanding of Leonardo's 'natural philosophy' or science than any other contemporary – and few commentators since – for not only does he enumerate Leonardo's 'four powers' including percussion, which Leonardo by this time had come to consider as the most powerful of natural phenomena, but he also recognises Leonardo's conclusion that 'force' and 'accidental weight' are equivalent terms (see Chapter 4).

Luca at the same time refers to Leonardo 'having already with all diligence finished his praiseworthy book on painting and human movement'. It would appear that Luca when he wrote this sentence had seen at least some of the large number of notebooks which Leonardo had already written on both these subjects and, reasonably enough, considered them finished. He evidently did not appreciate the extension to infinity of Leonardo's quest for knowledge or that he, Luca himself, had just propelled Leonardo's investigations still further into the sciences of physics and man.

Between 1496 and 1499 (the period when Leonardo and Luca Pacioli were both in Milan) Leonardo completed four copious notebooks, MS I, MS M, *Forster* II and *Madrid Codex* I. Their concentration on movement, weight and force has been mentioned already. As might be expected, mathematical notes, mostly based on Euclid's geometry, become common in MS M and MS I. They are immediately applied by Leonardo to his observations on the 'four powers'. It is at this time that Leonardo begins to make a systematic application of the cone which he calls a 'pyramidal' or perspective figure to the increase or decrease of all the 'continuous quantities' of the 'powers' of nature. He now, presumably with Luca Pacioli's help, realises that the 'uniform difformity' of his pyramidal figure can be equivalent to 'the rule of three' or arithmetical proportion for 'discontinuous' quantities. All this is emphasised in MS M.

His increasing appreciation of the curvilinear aspects of Euclid's geometry, and possibly that of Archimedes, carries Leonardo from rectilinear pyramids to 'pyramids' with curved sides, which he calls 'falcates'. This appreciation coincides with his increasing interest in the visible movements of water and the formation of eddies, which thus become more amenable to his geometrical approach (see Chapters 4 and 5).

By 1499, however, Ludovico Sforza was in deep trouble. What he had dreaded for so long came to pass – Louis XII, King of France, asserted his hereditary right as descendant of the Visconti (from whom the Sforzas had usurped power) to the dukedom of Milan. In July 1499 the French invaded Lombardy. Ludovico fled to Emperor Maximilian for help. Milan was occupied by the French, led by Ludovico's bitter enemy, Trivulzio. In a note dated 1 August 1499 Leonardo, apparently oblivious to the political chaos around him, wrote on 'Movement and weight'. This note expresses subtle thoughts on the movements produced by forces acting on a weight hanging from a beam. By November of that year the whole of Milan, including the

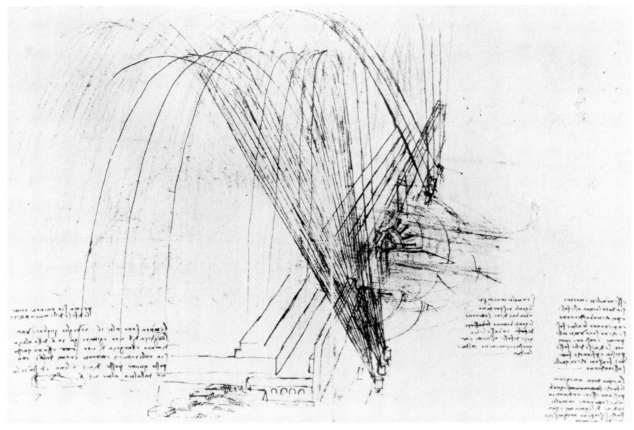

a.

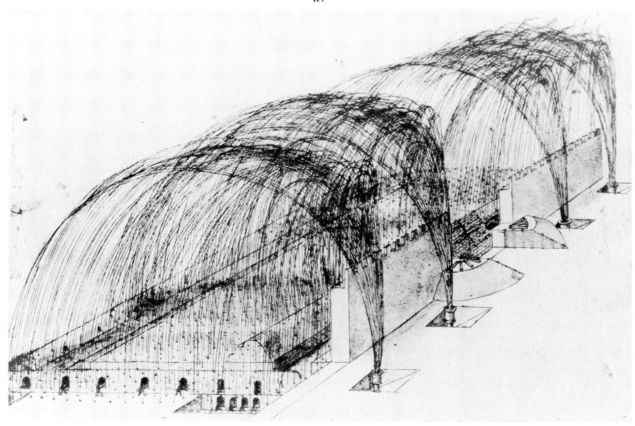

b.

Figure 1.17.

c.

Figure 1.17 (Continued)

a. Bombardment of a city (W 12337 v).

'Let all the defences of the walls and the towers be destroyed first', writes Leonardo on this page. Notice that the trajectories of the projectiles conform to Leonardo's law of curved pyramids.

b. The arrangement of mortars for bombarding a city (W 12275).

Here Leonardo displays the use of the mortars in the cannon foundry drawn on Figure 6.5 (p. 164). The trajectories of the mortar shells are systematically displayed to saturate the area of the bombarded city with an efficiency comparable with the beauty of their representation.

c. The trajectory of a cannon ball (A 43 v).

Here Leonardo illustrates how, 'the central part of the path of a cannon ball is the most powerful', and has the greatest power of percussion. He shows, too, how the waves of percussion spread out as blast and sound in the air. Finally he illustrates the power of gravity on the trajectory of the cannonball, curving it downwards to the surface of the ground.

castle, was in French hands. In the courtyard of the castle still stood Leonardo's great clay model of the horse commemorating the Sforzas. Leonardo may well have watched it being used for target practice by the Gascon archers. Before the year was out he and Luca Pacioli had left Milan.

INTERLUDE IN MANTUA

Leonardo, with other 'important' persons in Milan, took refuge from the French at Mantua. Here Isabella d'Este invited all suitable exiles, a category that included those of political and artistic distinction. Mantua attracted all the latest political news, and it was soon known that Ludovico was preparing to take back Milan. Leonardo awaited the outcome of this attempt, for if it were successful, he would be able to return to Milan. Ludovico did return, retook Milan and in February 1500 advanced on Novara. Here he received effective help in the form of guns and mortars from the Emperor Maximilian. He deployed these so intelligently in the bombardment of Novara that Leonardo's plan for systematic bombardment of the city (as drawn in

Figure 1.17, W 12275 and CA 24 ra) is thought to have been used. Was Leonardo there? It is not impossible that he joined Duke Ludovico on his briefly victorious campaign; however, there is no direct evidence of this.

After the fall of Novara, Ludovico's Swiss troops deserted and Ludovico himself was eventually captured by the French on 10 April. He was destined to end his days in prison at Loches, eight years later.

By the end of February it was obvious to all that Ludovico's bid for power had failed. Leonardo at Mantua acceded to Isabella's request for a portrait to the extent of making a sketch of her.

In the banqueting hall of her castle he would have seen Andrea Mantegna's *Triumph of Caesar* in all its pristine freshness – a freshness which has now largely been restored (some 475 years later) in the orangery at Hampton Court. In her famous 'studiolo' also hung Mantegna's *Triumph of Parnassus* with Isabella herself well to the fore as elegantly beautiful as Mantegna could make her. Leonardo, however, saw her differently. She was pregnant. A pregnancy of six or seven months would be obvious to any observant eye, let alone Leonardo's, and Isabella was to give birth to her son and heir, Federico, on 17 May 1500. So Leonardo portrayed a plump full-breasted Isabella, rather fleshy-faced, even when seen in profile (Figure 1.18). The sharp nose, firm chin and wide-open eyes convey the signs of her lively political and cultural acumen. The drawing of the hands and arms suggests indecision on the part of the artist. They are, none the less, of interest, for the right hand resting on the left wrist assumes the same elegant posture as the hands of the *Mona Lisa*. Was Leonardo deciding to use this position of the hands in displaying the beauty of pregnancy? Perhaps this sketch of Isabella displayed too obviously her pregnant state, for her husband, Francesco Gonzaga, disliked it intensely and quickly gave away the copy of the drawing which Leonardo left in Mantua. However, in Venice where he arrived in March 1500 Leonardo showed his sketch of Isabella to Lorenzo Gusnaco, who wrote to Isabella, 'Leonardo da Vinci is in Venice and has shown me a portrait of Your Highness which is exactly like you, and is so well done that it is not possible for it to be better'.[22]

IN VENICE AS HYDRAULIC ENGINEER

Venice provided Leonardo with quite a different set of problems; the Venetians had called on Leonardo, the technician in hydraulics. From his earliest days Leonardo had found the 'element' water a fascinating medium for invention. His long sojourn in Milan had been noteworthy for his work on canals, locks and irrigation. His work on lock gates in that city, particularly at San Marco and on the canal system within the city, still provides material evidence of his efforts. His new design of turning a pointed arch on its side to obtain a mitred lock gate which closes more firmly as the pressure of water on it increases played its part in the evolution of canal construction. This mitrilisation Leonardo also applied to closure of the mitral valves of the heart (see Chapter 15).

The Venetians had just suffered a defeat at sea at the hands of the Turks. On land the Turks were threatening invasion of her northeastern borders near Friuli and along the river Isonzo. With the defensive aid of such a river as the Isonzo, Leonardo points out, 'a few men are the equivalent of a larger number', and he considers the problem of increasing its size by building sluice gates near Friuli. But the main concern of the Venetians at that time was their recent naval defeat at the hands of the Turks. They feared naval attack in the Adriatic. Leonardo responded to this with designs for diving apparatus invisible from the surface, one-man diving boats navigable under water, which look like mini-submarines, and a method of holing vessels by frogmen equipped with flippers (Figures 1.19 and 1.20). These plans, however, seem to have been past inventions drawn up years before (Figure 1.20 and CA 346 va).

It seems curious that Venice with its great Arsenal as well as the newly founded Aldine Press exerted no attraction for Leonardo. He stayed there only a few weeks. All hope of returning to Milan was now gone. Leonardo notes, 'The

Figure 1.18.   Portrait of Isabella d'Este, when she was six to seven months pregnant (Louvre, Paris).

A number of features, including the posture of the hands, are comparable with the portrait of Mona Lisa.

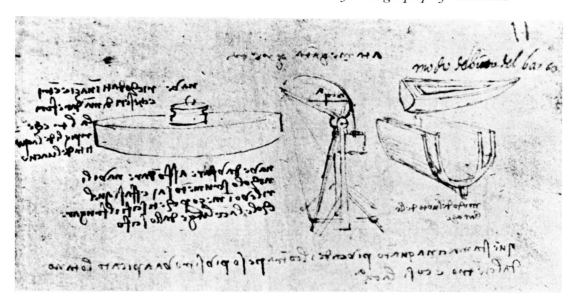

Figure 1.19. 'A ship to be used to sink a fleet of ships' (B 11 r).

These words are written beneath a small submarine with a conning tower; a boat with double hull structure is drawn on the right.

Figure 1.20. A diver equipped with protruding air bag and goggles (CA 333 v).

He bores a hole in the planks of a ship using a special fixing frame.

duke has lost his state, his property and his liberty; and none of his enterprises has been completed' (L 0). So by the end of April 1500 he had returned to his native Florence.

LEONARDO'S SECOND SOJOURN IN FLORENCE (1500–1506)

Leonardo's personality was now, at the age of forty-eight, mature. Its three strong characteristic strands of creative painting, engineering and mathematical science had been finally interwoven. The rest of his life was to be de-voted to the task of integrating and expressing these strands in designs at once beautiful and true. Perhaps the truest symbol of Leonardo's efforts is provided by the way intricately interwoven figures are scattered through his drawings. Many of these demonstrate natural objects and processes such as the movements of water, flowers, leaves and branches of trees; finally they become more abstractly mathematical. Rightly have many authors of books on Leonardo, for example, Richter, MacCurdy and Popham, chosen this well-known interlacing pattern of the engraving in the British Museum, Figure 1.21, as symbolic of

Figure 1.21. An intricately interwoven drawing in which is inscribed 'Academia Leonardi Vinci' (British Museum, London).

The delicate complexity well symbolises Leonardo's mind. Durer made similar figures in his 'six knot' series, on a visit to Italy.

Leonardo's creative striving to unite the strands of his art and science.

Two strong motives lay behind Leonardo's decision to return to his native Florence: the mundane need to earn a living was undoubtedly one, for it must be realised that he was now homeless and jobless. The other was the opportunity of once more renewing his fruitful friendship with Luca Pacioli. Luca had not accompanied Leonardo to Venice, but he had obtained the chair of mathematics at the University of Florence.

Leonardo's first objective was obtained, probably with the aid of his father, Ser Piero, when he expressed his wish to undertake painting an altarpiece in the monastery of the Servite monks. Here he was given board and lodging, but, as Vasari relates, 'for a long time he never began anything. At length he drew a cartoon of the Virgin and St. Anne with a Christ, which not only filled every artist with wonder, but, when it was finished and set up in the room, men and women, young and old, flocked to see it for two days, as if it had been a festival'.[23] At this time Leonardo's fascination

with the pyramidal law was being repeatedly expressed in his MS M, MS L and *Codex Arundel*. Can it be considered, therefore, that it is only by chance that his studies of this subject, such as the drawing in the National Gallery, London, and the picture in the Louvre, are all based on a pyramidal design? Moreover, its 'complex centre of gravity' is also illustrated in the little sketch of Hercules crushing Antaeus in his *Treatise on Painting*.[24] Within the frame of this 'dry' geometry Leonardo infuses the figures of St. Anne and the Virgin with most moving expressions of the love of two mothers.

Nevertheless, it appears that once the cartoon was finished so was Leonardo's interest in it. Isabella d'Este's emissary pestering Leonardo for a picture had to report, 'Leonardo has lost all patience with the brush and is now working entirely at geometry'.[25] Here the influence of Luca Pacioli can be sensed.

LEONARDO WITH CESARE BORGIA AND MACHIAVELLI IN THE ROMAGNA AND PISA

It was not long, however, before Leonardo, the inventive, artist–engineer, was called in again, this time by Cesare Borgia. Since 1500 this son of Pope Alexander VI had been successfully carving out possessions for himself in the Romagna. These extended from Piombino on the west coast to Rimini on the east. Such conquests were vulnerable to attack from hostile neighbours. Cesare, well aware of the need for their systematic fortification, called in the man with the greatest reputation in this field, Leonardo da Vinci. Almost all we know of the part played by Leonardo in this scheme is contained in MS L, a series of maps of the Romagna, and studies of fortifications in the *Codex Atlanticus*. Leonardo's drawings show a gradual transition from the high-walled, angulated and turretted castles illustrated in MS B (1489), through the more squat, bastioned fortresses derived from Francesco di Giorgio Martini, to the low, convex casemates in a series of retreating curves around a central core, reminiscent of the Maginot Line of 1939 (Figures 1.22 and 1.23). The raison d'être of these changing schemes was Leonardo's appreciation of the increasing velocity of cannon balls and the destructive power of their 'percussion', or impact. In his view the velocity of percussion and the resistance of the percussed object were directly proportional to the damage done. He therefore shaped the walls of the fortifications in low curves to diminish their resistance to the impact of the cannon balls. 'Percussion is less strong the more oblique it is', explains Leonardo (CA 48 rb).

The maps Leonardo made during these journeys were compiled from hodometer readings. Perhaps the most famous is the plan of the town of Imola in which Leonardo was trapped with both Cesare Borgia and Niccolo Machiavelli during the winter of 1502–1503. On the same sheet as this colourful map Leonardo records the distances between Imola and neighbouring towns. And on the map of the Val di Chiana (Figure 1.24) which includes Arezzo one can still see faintly traced the lines and the mileage between various points from which he made his observations of distances and elevations. The heights of the hills he shaded in like our present-day Ordnance Survey maps.

Leonardo's maps were not made for merely military purposes. Two other motives were at work. Always aware of the possibility of turning the Arno into a navigable river, he made detailed maps of the whole Arno basin. The Val di Chiana containing Lake Trasimene he saw as a possible source of water for the Arno when it dried up in the summer. He even considered the possibility of taking water from the Tiber by a tunnel into this lake. This tunnelled diversion can be clearly seen in the right upper corner of Figure 1.24.

His other motive for map-making was more theoretical; it concerned his geological view that the earth had grown out of the sea, and as it grew it left inland seas or lakes pocketed between its mountains and hills.

In Piombino on the western coast Leonardo devoted his attention to improving the fortifications of the castle and to draining the marshy land in that region. His plan for drainage of marshes was to bring 'turbid' water containing earth and gravel into the area. After it had dropped its turbid deposit this water, now clear, was to be taken off by multiple shallow surface channels, so gradually raising the land level. In this design Leonardo was repeating the process by which he considered that the land had emerged from the sea in past geological ages. A little drawing on *Codex Atlanticus* 139 rc shows his plan. In the valley of the Arno, near Florence; in the country round Milan where the rivers Ticino and Adda join the Po; in the Pontine Marshes south of Rome (see Figure 1.28, p. 39); and finally in the valley of the Loire, Leonardo was continually concerned in controlling waterways to make them navigable. In addition by drainage and irrigation as at Piombino he also planned to make these regions more fertile and prosperous. This was one of his lifelong preoccupations for the betterment of the lot of his fellowmen.

However, having returned to Florence from his excursions through the Romagna, Leonardo was immediately approached once more as a military engineer, this time by Machiavelli. Since 1494 Pisa had broken away from Florence and claimed independence. Florence had been waging

Figure 1.22. A fortress with high walls and towers (B 69 r).

To the left a double spiral staircase is drawn, designed to be included in the tower. It resembles the double spiral staircase at the château of Blois.

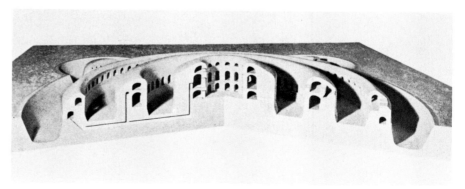

Figure 1.23. Fortress consisting of concentric low casemates; a model of a rough drawing (CA 48 ra). Figure from Ludwig H. Heydenreich, *Leonardo da Vinci* (London, George Allen & Unwin, 1954), plate 141.

Figure 1.24. Map of the Val di Chiana showing Chortona, Perugia and Lake Trasimene, on the right-hand side of which is a suggested tunnel running beneath the hills to the Tiber (Tevero) (W 12278).

a desultory and unsuccessful campaign ever since to regain control of this important port in which she had placed her University. It appears that Machiavelli had the bright idea of diverting the river Arno away from Pisa to the sea during the summer of 1503 and involved Leonardo in the project. In *Madrid Codex* II 53 r Leonardo drew a beautifully coloured map almost certainly devoted to the details of such a project, though no trace of the proposed course of the diversionary canals is to be found on it. Indeed, this map might have been equally useful to Leonardo in furtherance of his long-contemplated scheme for constructing a canal from Florence to Pisa, a project much closer to his heart than the very opposite manoeuvre planned by Machiavelli. This latter scheme is drawn out in colour equally beautifully in *Madrid Codex* II 22 v and 33 r, and on a number of other occasions.

Meanwhile amidst all this practical activity as military engineer the strand of theoretical science in Leonardo's character had by no means been dormant. Archimedes, as glimpsed in the works of Luca Pacioli, had given an inspir-

ing example of the union of physical events with geometry. Leonardo searched feverishly for manuscripts of his works. As Cesare Borgia's conquest of towns in Romagna increased Leonardo took particular note of their libraries with their contents. For example he writes, 'Borges will procure for you the Archimedes which was in the possession of the bishop of Padua, and Vitellozzo the manuscript which was in Borgo San Sepolcro' (L 2 r). And in a long note about the life of Archimedes, probably derived from Plutarch, he describes the part played by Archimedes in the siege of Syracuse by the Romans (BM 279 v). This tale he may well have obtained from the libraries in Urbino or Pesaro which he visited.

LEONARDO'S VISIT TO PIOMBINO

Back in Florence after the abortive attempt to divert the Arno, Leonardo, once more at the behest of Machiavelli, was sent on another mission, this time to Piombino in November 1504. This journey has been revealed in the recently recovered *Madrid Codex* II. Cesare Borgia fell from power at the time of his father's, Pope Alexander's, death in 1503. Piombino was then restored to its previous ruler, Jacopo Appiani. Lying as it did between the Papal States to the south and Pisa to the north, Machiavelli saw it as vital to Florence that Piombino should be friendly. Leonardo became a pawn in the game, and in November 1504 he was sent to Piombino to give advice about the fortifications of the town. Ironically enough, Leonardo had performed this same service for Cesare Borgia who had ousted Jacopo Appiani only some three years before. So Leonardo used his same earlier plans and was immediately in a position to write, 'All Saints Day 1504. I made in Piombino for the lord this demonstration' (Madrid II 25 r). Amongst the many pages of this *Madrid Codex* containing drawings of his plans

for improving the fortifications Leonardo includes some twelve folios taken out of the work of his friend, Francesco di Giorgio Martini. But it reflects Leonardo's outlook at this time to note that in this same notebook there are more pages devoted to geometry than any other subject, and this geometry is predominantly concerned with pyramids and their relationship with other geometrical figures such as cubes and spirals.

ON THE FLIGHT OF BIRDS

Studies of the flight of birds, so numerous throughout Leonardo's notes (see, e.g., Figure 1.25), are also present in *Madrid Codex* II. They come to a climax in the year 1505 when he summarised his understanding of the principles of

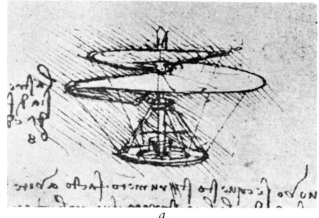

a.

Figure 1.25.

a. Helical screw helicopter (B 83 v). Just as a screw can be turned so as to enter wood, so the same principle could in Leonardo's view be applied to a screw entering the air. The turning was to be performed by a steel spring.

b. This scrappy sketch illustrates the nearest Leonardo approached to flight (CA 309 va). It will be seen that movement is confined to the outer half of the wing, and the man hangs down through the central hole. In these features this wing design anticipated that of Otto Lilienthal in 1895. Compare Figure 6.36.

b.

flight in a little codex of eighteen folios, *On the Flight of Birds*. Here he records his observation of the flight of a hawk from a hill called Monte Cecero (the hill of the swan) as he walked up from Florence to Fiesole. On the cover of this notebook, too, he writes, 'The first flight of the great bird [Leonardo's term for his flying machine] from the summit of Monte Cecero will fill the universe with wonder; all writing will be full of its fame, bringing eternal glory to the place of its origin'. Leonardo very rarely allowed himself outbursts of enthusiasm; unhappily there is no evidence that this one was justified.

Leonardo's well-known interest in flight began as early as 1485 when he saw it simply as a problem of mechanical invention. One can follow his progress through his attempts to use the powers of man for flight to his analysis and comparisons of the bird and man with regard to flight

(about 1500–1505). It is perfectly logical but somewhat ironical that his latest notes on the force of the impetus of a bird's wing against a 'moving body' sustaining it, i.e., the wind, forms part of a long investigation of the causes of the movement of air as wind (W 12672). Thus his long researches into human flight were left unfinished after some thirty years of intensive investigation. Like so many of Leonardo's scientific studies his progress twisted round like the curve of a spiral with its conclusive apex unreached. Flight was left in the realm of what Leonardo saw ever more clearly as 'a science of things possible'.

LEONARDO'S VERDICT ON WAR

Leonardo's experiences in the Romagna with Cesare Borgia had brought him, perhaps for the first time, into full realisation of what he called 'the most bestial madness' of

Figure 1.26. *The Battle of Anghiari* (copy by Rubens) (Louvre, Paris).

In this enactment of 'bestial madness' the two powerful opposing forces of horses and men have clashed, or 'percussed'. Leonardo chooses the moment of stillness before the rebound of such percussion to depict this violence.

man as enacted in war. In Florence on the other hand it was seen differently. There was great relief at Cesare Borgia's disappearance from the scene. Piero de' Medici, who had constantly threatened the Florentine Republic with a return of the Medici to power, had also disappeared. The Florentines felt like celebrating victories and set Leonardo in competition with Michelangelo to glorify with battle scenes the walls of the council chamber in the Palazzo Vecchio.

Leonardo was deputed to represent a Florentine victory at Anghiari in 1440. He was given a long official description of the battle to instruct him. He himself however had described 'How one ought to represent a battle' in his *Treatise on Painting*. Amidst an atmosphere of smoke and settling dust he describes the bodies of dead men covered with dust, their blood turned to red mud. 'Others, dying, teeth grinding, their eyes starting forth, their fists clutching their bodies, and their legs distorted'. 'And', he adds, 'do not leave any open place without footprints full of blood' (TP I 283). From the copies that remain of the picture, e.g., Figure 1.26, one can see that Leonardo depicted battles as maddened horses and men meeting in violent 'percussion', both psychological and physical. Percussion increases in violence with the velocity and impetus of percussing bodies. Under these conditions its impact is most damaging. Here he has chosen that moment of motionless ferocity when both man and horse come to the 'end of the movement' in percussion. To this he adds violent leverage as exerted by the different forces of men and horses on the pole of the standard, vividly expressed by the grotesquely distorted shape of the standard-bearer's body. In this way Leonardo built his science of mechanics into his picture of man's bestial madness. However, ruined by his experimental methods, all we have left is a copy made by Rubens.

LEONARDO'S STUDIES OF THE MACROCOSM AND MICROCOSM

During these same years Leonardo intensified his studies of the movements of water and the formation of the earth out of the waters of the seas. Not only did he produce a new version of the causes of 'circulation' of the 'watery humours' of the macrocosm but he also examined closely the process of growth of both the earth and man. His examination of the earth revealed to him its strata of rock formation. Through the fossil content of the strata he glimpsed the picture of life in far distant ages – he glimpsed, too, the possibility of geological periods rather than biblical time. Inland lakes he saw as portions of primeval seas cut off and locked between mountain ranges as these grew upwards. Later, with the breakdown of these barriers, water flowing through narrow gorges found its way back to its primal sea. This process he illustrated in his maps of the Arno watershed where the lakes formed at different levels are clearly shown breaking through narrow gorges as their water descends. The whole process is also clearly and beautifully illustrated in the background of his most famous of paint-

Figure 1.27. The *Mona Lisa* (Louvre, Paris).

In this picture Leonardo portrays genesis; the birth of the Earth out of its waters and within the body of the woman the generation of an infant is suggested. This also grows out of the waters in the womb. Here Leonardo would be depicting the genesis of both the macrocosm and microcosm.

ings, the *Mona Lisa* (Figure 1.27). Why, one may ask, does Leonardo circumscribe this woman within such geological formations? His answer in my opinion is clear.[26] Within her body is a new living world in the form of a babe growing out of the amniotic waters just as the great world grows out of the waters of the sea. This is a picture of the genesis of both man and the world in the womb of time.

There is a similarity between this portrait and that of the pregnant Isabella d'Este. One observes the rounded face and figure and the absence of jewelry and rings on the hands. Especially significant is the similarity of posture of the smooth graceful hands in these two portraits. It is noteworthy that about the same time as Leonardo was painting the *Mona Lisa* he was making his dissections of the pregnant uterus and his famous beautiful drawings of the babe in the womb; studies which we shall return to later (see Chapter 17). The anatomy of the smile for which this painting has been so famous was carefully analysed by Leonardo in a contemporary note (W 19046 r; K P 51 r). Here he identifies the lips of the mouth with the actual muscles. These close and open themselves and move into a smile or laughter. And he ends his description of the movements of a smile with the typical remark, 'These I intend to describe and illustrate in full, proving these movements by means of my mathematical principles'. My interpretation of the smile of Mona Lisa is that it subtly expresses the secret, which she has successfully kept for so long, that she is pregnant (see Figure 9.8, p. 222).

It may be found surprising, even slightly repellant, that Leonardo should speak of mathematical analysis of so hauntingly beautiful a creation as his Mona Lisa's smile. This was far from being so to him. His geometry revealed forms and movements however transient and subtle; it revealed to him the secrets which he shared with Luca Pacioli in his book, that of Divine Proportion. For Leonardo, art was built on the scaffolding of science. It leapt over the bounds of quantitative analysis and measurement into the realms of harmony and beauty. His mathematics were to him like the keyboard of a piano. Though this is purely mechanical, a composer can create 'divine music' out of its simple mechanism.

FAMILY AFFAIRS

'On the 9th day of July 1504, Wednesday at seven o'clock died Ser Piero da Vinci notary at the Palazzo del Podesta, my father, at seven o'clock. He was eighty years old and left ten sons and two daughters' (BM 272 r). Ser Piero left none of his property to his one illegitimate son, Leonardo. Three years after, however, Ser Piero's brother, Leonardo's uncle, also died. This uncle, Francesco, had always favoured Leonardo and left everything to him, ignoring Ser Piero's other children who in their anger united to contest Francesco's will. Leonardo, incensed by this, went to law, not only to justify his uncle Francesco's bequest but also to claim a share of his father's estate. The case, of course, dragged on for years.

LEONARDO'S SECOND SOJOURN IN MILAN (1506–1513)

After the ruination of the painting of the Battle of Anghiari Leonardo abandoned it, but he had already been paid considerable sums of money and the Florentine Signoria wanted to keep him to his contract. It was therefore with great relief that Leonardo received an invitation amounting to a command from the French king Louis XII to go to Milan. In May 1506 he went, and in spite of repeated protests from Florence, he stayed there. Here in Milan the French viceroy, Charles d'Amboise, greeted Leonardo with a warmth he had never before experienced, a warmth that was all the more welcome to him in contrast with the cold disapproval he had left in Florence. In claiming Leonardo's presence in Milan and forbidding his return to Florence, Charles d'Amboise wrote to the Signoria a letter saying, 'The magnificent works with which your fellow-citizen Maestro Leonardo da Vinci has endowed the cities of Italy and especially Milan, have aroused a particular love for him in all those who have seen them, even in those who have never met him. And we too confess that we were among those who loved him before ever we knew him personally'.[27] And Charles d'Amboise seems to have seen that Leonardo was something more than a technically skilled painter, for he added, 'After our conversations with him and our experience of his many excellences we see in truth that the fame he has won through his painting is as obscurity to the praise he deserves to receive in other respects in which he is of the highest excellence . . . And if it is necessary to commend a man of such qualities to his own people we commend him to you with all our heart'.[28] Leonardo responded with plans for a house and garden for his protective patron.

Here in Milan Leonardo took under his wing a young pupil, Francesco Melzi, who afforded him a devotion and companionship like that of a loving son to an ageing, lonely father for the rest of Leonardo's life. Leonardo's Milanese sojourn was, however, interrupted by the litigation against his brothers in Florence. He made no bones about writing to influential friends to try to accelerate the procedure. Whilst in Florence on this irksome matter Leonardo records how he stayed in the house of Piero di Braccio Martelli. Here he used his time in trying to gather together the vast disordered mass of his notes of the past twenty years. Leonardo writes of them thus, 'Begun in Florence in the house of Piero di Braccio Martelli on the 22nd day of March 1508. And this is to be a collection without order taken from many papers which I have copied here, hoping afterwards to arrange them in order in their proper places according to the subjects of which they treat'. These are the opening words of the manuscript obtained by the Earl of Arundel, now in the British Museum.

GEOMETRICAL TRANSFORMATIONS DOMINATE
LEONARDO'S NOTES

About 1505 Leonardo opened a notebook, *Forster* I, with the words, 'A book entitled, On the transformation of one body into another without diminution or increase of substance'. For forty folios he describes transformations of cubes into pyramids, spheres into cubes, et cetera. This procedure of geometrical 'transformation' is carried further in *Codex Madrid* II where he applies it to transforming rectilinear figures into curvilinear figures.

From this time analysis of the shapes of waves and eddies in flowing water increasingly occupies his attention. These are particularly to be found in the *Codex Leicester* and MSS E, F and G and some of the drawings in the Royal Library, Windsor, where they are elaborately illustrated. Many are scattered in the *Codex Atlanticus* and *Codex Arundel*.

What Leonardo calls 'transformation' or 'transmutation' consists of replacing what has been taken away, by its equivalent elsewhere. A subtle and vivid example of this is provided by the alterations in the shape of the human body brought about by its movements, for whatever movements a man makes, the surface area of his skin and the weight of his body, though 'transformed' geometrically, remain constant. We shall see that Leonardo was acutely aware of this in his extensive studies of attitudes and gestures, that is to say, the external movements of the whole human body. He applied it again later to the internal movements of different anatomical organs of the body, particularly to the contractions of abdominal muscles and the transformations of the moving shapes of blood in the heart and blood vessels.

Leonardo was greatly impressed with the similar transmutations of the movements of water and air. It is therefore not surprising to find him writing, 'In order to give the true science of the movement of birds in the air it is necessary first to give the science of the winds, and this we shall prove by means of the movements of water. This science in itself is capable of being received by the senses; it will serve as a ladder to arrive at the perception of flying things in the [invisible] air and wind' (E 54 r).

THE RETURN TO HUMAN ANATOMY
AND PHYSIOLOGY

It was whilst trying to arrange his notes in Martelli's house in Florence that Leonardo felt impelled to apply his accumulated knowledge of the powers of nature to the internal mechanisms of the human body. One day he was talking to an old man in the hospital of Santa Maria Nuova. The old man suddenly died. Leonardo describes the incident thus, 'The old man, a few hours before his death told me that he had lived a hundred years and that he felt nothing wrong with his body other than weakness. And thus while sitting upon a bed in the hospital of Santa Maria Nuova in Florence, without any movement or other sign of any mishap he passed out of this life. And I made an anatomy of him in order to see the cause of so sweet a death. This I found to be a fainting away through lack of blood to the artery which nourishes the heart, and other parts below it, which I found very dry, thin and withered. This anatomy I described very diligently and with great ease owing to the absence of fat and humours which greatly hinder the recognition of the parts' (W 19027 v; K/P 69 v).

Thus after some twenty years Leonardo returned to his researches on the structure and function of the human body. This time not concentrating his investigation on the physiology of vision and the senses but making a full-scale study of the whole body as a mechanical instrument, intensely fascinating to this pioneer bio-engineer.

Though he now felt himself well equipped by all his past artistic and scientific experience to undertake this exploration of this vast new world of the human body, he did not achieve it without a hard struggle to overcome his own sensitive repugnance to the task. He describes both aspects of his problem vividly, 'And though you have a love for such things you will perhaps be impeded by your stomach, and if this does not impede you, you will perhaps be impeded by the fear of living through the night hours in the company of quartered and flayed corpses fearful to behold. And if this does not impede you, perhaps you will lack the good draughtmanship which appertains to such representation; and even if you have the skill in drawing it may not be accompanied by a knowledge of perspective; and if it were so accompanied, you may lack the methods of geometrical demonstration and methods of calculating the forces and strength of muscles; or perhaps you will lack patience so that you will not be diligent. Whether all these things were found in me or not the hundred and 20 books composed by me will give the verdict, yes or no. In these I have been impeded neither by avarice or negligence, but only by time. Farewell' (W 19070 v; K/P 113 r).

For some years Leonardo struggled with this task on his own. Some of it was done in Florence, some in Milan. But wherever he prepared these 120 'books' of drawings, most of them have been lost. From them he compiled the more finished anatomical drawings, which finally found their home in Windsor Castle.

His plan for attacking this analysis of the structure and function of the human body was not merely descriptive, it was a dynamic extension into anatomy and physiology of his science of the 'four powers' of nature working on the four 'elements'. Leonardo describes this mechanical approach more than once. For example, in a passage entitled 'On Machines' he writes, 'Why nature cannot give movement to animals without mechanical instruments is demonstrated by me in this book on the active movements made by nature in animals. And for this reason I have drawn up the rules of the 4 powers of nature without which nothing through her can give local motion to these animals' (W 19060 r; K/P 153 r). This passage, written on a page which also contains his drawings of the human uterus and foetal circulation, shows how intensely and consistently Leonardo felt that he was focussing his science of the inorganic world onto the organic world of the human body. It is impossible therefore to appreciate his anatomy and physiology without understanding his investigations of the 'four powers'.

About 1510 Leonardo had the good fortune to meet a young, brilliant anatomist, Marcantonio della Torre. According to Vasari, Marcantonio was 'one of the first who began to illustrate the science of medicine, by the learning of Galen, and to throw true light upon anatomy . . . In this he was marvellously served by the genius, work and hands of Lionardo who made a book about it with red crayon drawings outlined with the pen, in which he foreshortened and portrayed with the utmost diligence . . . Whoever succeeds in reading these notes of Lionardo will be amazed to find how well that divine spirit has reasoned of the arts, muscles, the nerves and veins with the greatest diligence in all things'.[29] Unhappily Marcantonio della Torre had but little time to work with Leonardo, for he died of plague in 1511.

During this second period in Milan the sympathetic understanding of Charles d'Amboise saw to it that Leonardo was given freedom to work as he wished. His newly aroused geometrical approach to the movements of water found application once more to the waterways of Lombardy, in particular, to a project for controlling the waters of the turbulent river Adda. He made drawings of the most difficult region where he proposed to construct a dam, a construction which like so many of Leonardo's ideas was put into practice only after his death.

WAR ONCE MORE AND MIGRATION TO ROME

By 1509 war once more interfered with Leonardo's life-pattern. In this year Louis XII joined the League of Cambrai against Venice and conducted a brilliantly successful campaign. It was too successful, for it earned the envy and suspicion of the warrior-pope, Julius II, who transferred his allegiance to Venice and mounted a campaign to expel the French invaders from Italy. This he achieved by 1512 when Milan once more changed hands, this time coming under the puppet rule of Ludovico Sforza's son, Maximilian. Once more Leonardo found himself unwelcome in Milan; once more he left the city, this time for Rome where the new pope, Leo X, was a Medici and his brother Giuliano friendly to Leonardo. He saw Rome now as a promised land, the cultural centre of all Italy. Leonardo writes, 'I left Milan for Rome on the 24th day of September 1513' (E 1 r). With him he took a small band of pupils, among them Francesco Melzi.

ROME – CITY OF FRUSTRATION
(1513–1516)

In Rome he was lodged in the Belvedere, a luxurious summer palace on the top of the Vatican Hill. Here Giuliano saw to it that he had roomy quarters, including a workroom. At last Leonardo was being treated like a research worker with an adequate grant and full laboratory facilities. But somehow Leonardo's period in Rome turned out to be unhappy and relatively unproductive. As a painter he was ignored because the pope had found Raphael entirely to his taste as both a person and an artist. He did once approach Leonardo for a picture, but, as Vasari reports, 'he straightway began to distil oil and herbs to make the varnish, which induced Pope Leo to say: "This man will never do anything, for he begins to think of the end before the beginning!"'[30] Pope Leo clearly had no comprehension of Leonardo or his methods.

As an inventor his activities continued, for this strand permeated his whole life – even to unfortunate experiments with varnish. Here in Rome Leonardo performed chemical experiments largely because his patron Giuliano de' Medici was a devotee of alchemy. This subject Leonardo had earlier rejected. It is not unexpected to find him inventing improved water-cooled retorts, nor perhaps to find him producing coloured salts such as ferric oxide and copper acetate since these might, like the varnish, be utilised in his technique of painting. But he also devised a rolling-mill for producing metal strips from which coins could be stamped in the mint at Rome (G 70 v and 43 r). This appears to have been his last invention.

In the circumstances Leonardo turned back to anatomy. But there again he met with frustration. 'The Pope has found out that I have skinned three corpses', he writes anxiously and then refers to Giovanni the mirror maker who 'has hindered me in anatomy, blaming me before the Pope and likewise at the hospital' (CA 182 vc). Once more in the person of Pope Leo, son of Lorenzo de' Medici, Leonardo was confronted with that learned neoplatonic culture which he had found sterile as a young man in Florence. Once more he was made aware of being an 'unlettered' man, ignorant of Latin, whose interest in mechanics was considered laughable if not despicable. He was now ignored and neglected. He tried to live up to his own aphorism, 'Patience serves as a protection against wrongs as clothes do against cold. If you put on more clothes as the cold increases it will have no power to hurt you. In the same way patience must grow when you meet with great wrongs and they will be powerless to vex your mind' (CA 117 vb). Perhaps Leonardo had heard of Andrea Mantegna's recent experience in Rome. Having painted Faith, Hope and Charity on the walls of the papal chapel in the Belvedere, Mantegna awaited payment. All that happened was that he was told to add the figure of Patience.

So Leonardo went to the abattoir to obtain bullocks' hearts and lungs. On these he performed brilliant dissections and drew fine illustrations of those parts. Into these hearts he inserted glass models of the aorta in order to see the eddies of blood flowing through the aortic valve.

Putting on more 'clothes of patience', Leonardo retired to his workshop to produce concave mirrors of extremely long focal length. Such mirrors would be enormous and unwieldy. One suspects he intended to use them for observing the moon and stars. His plans were, however, continually frustrated by his German assistant Giorgio, with his crony, Giovanni delgi Specchi (John the mirror maker). 'He [Giorgio] never did any work without discussing it every day with Giovanni who then spread news of it everywhere', complains Leonardo, adding, 'I cannot make anything secretly because of him; and if I should set him to make my model he would publish it' (CA 182 vc). Far from there being any patent rights in those days, industrial espionage was the order of the day. Leonardo could only complain bitterly to his patron, Giuliano de'

Figure 1.28. Plan for drainage of the Pontine Marshes (W 12684).

Leonardo's plan was to take the water from the valleys into two main canals, one marked along the Via Appia (Nympha) emptying at Badino, the other at right angles, the Rio Martino. This main design, much elaborated, has now been put into effect.

Medici, who was at that time severely ill with pulmonary tuberculosis. Leonardo's letter begins, 'So greatly did I rejoice, most illustrious Lord, at the desired restoration of your health that my own malady almost left me' (CA 182 vc). This and the address of a physician in Rome on another contemporary sheet tell us that Leonardo was ill as well as frustrated.

His cloak of patience now wearing very thin, Leonardo turned to the beautiful gardens of the Belvedere which had been recently laid out by the previous pope, Julius II. Beyond the pine garden containing the recently discovered statues of Apollo and the Laocoon was a section containing rare plants sent from all over the world. Here Leonardo found consolation in botanical discovery. With his mind still saturated with the geometrical figures of curved pyra-

mids and spirals, he detected the spiral arrangements of plant leaves, revealing the pattern of phyllotaxis in exposing leaves to sun and rain. His observation of rings spreading in water from percussion, or impact, of a stone made him ready to see the rings of annual growth in trees, so anticipating the whole subject of dendrochronology. These findings he had no difficulty in keeping secret, for no one was interested in them.

Frustrated, ill and neglected as he was, Leonardo was not reluctant in 1515 to accept a mission which took him out of Rome to the Pontine Marshes in order to plan the drainage of this unhealthy area. The only direct evidence we have of Leonardo making this journey is the spectacular map he drew of the region with his plan for drainage clearly displayed (Figure 1.28). The principle underlying it was the same as that which he had devised for draining the marsh-land around Piombino, with certain modifications. What is perhaps most remarkable is the fact that the area is now healthy as a result of a drainage plan carried out as recently as 1936–1939 which elaborated Leonardo's principles and design.

At the end of the year 1515 Leonardo travelled with Pope Leo to Bologna. Italy had once more been invaded by the French, this time under Francis I who had rapidly evicted the young Maximilian Sforza from Milan and threatened to advance further into Italy, even to Rome. Here in Bologna Francis I parleyed with Pope Leo and met Leonardo. Once more Leonardo decided to enter into the service of the French king, and so it came about in 1516 that he found himself ensconced in the pleasantly unobtrusive château of Cloux at Amboise on the river Loire.

THE LAST YEARS AT AMBOISE
(1516–1519)

Leonardo's last sad years at Cloux present the picture of a man of steadily failing powers struggling with his fate. Life at Amboise in the court of the brilliant young King Francis I was gay. Leonardo felt anything but gay but he struggled successfully to bring to bear his required arts of festival making. On one occasion he produced a replica of his famous Paradise festival; on another he constructed a robot lion which walked and opened its breast to eject a bunch of French lilies in front of the king.

Either here, or more probably in Rome, Leonardo had produced his last painting, *St. John the Baptist*. Probably Leonardo painted this last symbolic figure when he was ill with cerebrovascular disease. Quite apart from its artistic merit, it is striking for the fact that it betrays no sign of Leonardo's recently acquired anatomical knowledge. Though St. John is represented as a youth, the outlines of muscles are conspicuous by their absence, a feature which combined with the smooth-skinned features of the face, creates its indubitably intended androgynous character. Such a beautiful youth in classic times represented Death. And St. John pointing with an exquisitely graceful fore-finger to the cross above seems to accentuate Leonardo's identification of himself here, both as perceiving the approach of his own death and as forerunner of a greater perfection to come. No one was more acutely aware than he of the incompleteness of his achievements, and no one was more convinced than he that these achievements had opened up a path into the future.

In October 1517 Leonardo was visited by an Italian cardinal, Luigi d'Aragona. With him was his secretary, who wrote the following account of the visit: 'In one of the outlying parts [of Amboise] Monsignor and the rest of us went to see Messer Leonardo da Vinci, the Florentine, who is more than seventy years old, and is one of the most excellent painters of the day. He showed His Eminence

three pictures, one of a certain Florentine lady done from life for the late Magnificent Giuliano de' Medici, another a young John the Baptist, and a third the Madonna and Child seated on the lap of St. Anne, all perfect. Nothing more that is fine can be expected of him, however, owing to the paralysis, which has attacked his right hand. A Milanese, who was educated by him and paints excellently, lives with him. Although Leonardo can no longer paint with his former sweetness he can still draw and teach others. This gentleman has written of anatomy with such detail, showing with illustrations the limbs, veins, tendons, intestines and whatever else there is to discuss in the bodies of men and women in a way that has never yet been done. All this we have seen with our eyes; and he said that he had dissected more than thirty bodies of men and women of all ages. He has also written on the nature of water, on various machinery and on other matters, which he has set down in an infinite number of volumes all in the vulgar tongue; which if they were published would be useful and very delightful'.[31]

From this informative record one can diagnose that Leonardo had had a stroke affecting his right hand and that he looked older than his years, for he was but 65 and looked 'more than seventy'. One learns also that his right-handed paralysis affected his colouring but not his drawing. We know he drew and wrote with his left hand all his life. This he continued to do in notes known to have been made after this observation in 1517. Thus Leonardo's right-handed paralysis alone is sufficient to account for the absence of any paintings from 1517, and probably from a year or so before, that is, during the whole of his residence in Amboise. Moreover, the nature of the illness to which he refers in his letter to Giuliano from Rome in 1515 is probably justifiably diagnosed as cerebrovascular disease.

In spite of this disability Leonardo's last years in Amboise were not unfruitful. He was still able to talk, and feeling 'time's winged chariot drawing near', Leonardo talked a great deal to King Francis, apparently to some effect. For, if Benvenuto Cellini is to be believed, Francis told him, 'Inasmuch as Leonardo possessed so vast a genius in such abundance . . . King Francis was strongly enamoured of his great virtues and took such pleasure in hearing his discourse that there were few days on which he parted from him'.[32] The king's 'exact words', as reported by Cellini, were that 'he did not believe that there had ever been another man born into the world who had known so much as Leonardo, and this not only in matters concerning Sculpture, Painting and Architecture, but because he was a great Philosopher'.[33]

Architectural plans would have seemed to have been one of the main subjects of their discourse, for Leonardo not only produced plans for rebuilding the great château of Amboise but also projected a palace at Romorantin and a system of canals in the Loire watershed (BM 270 v). One hears the accustomed voice of Leonardo when he declares, 'If the tributary *mn* of the river Loire were turned with its turbid waters into the river at Romorantin this would fatten the land which it would water and would render the country fertile to supply food for the inhabitants and would make navigable canals for merchant vessels' (BM 270 v). He accompanies the proposed plan with a sketch. On the same page he applies his principle of transmutation or 'transformation' to the building of houses. Under the heading, 'Mutation of houses' [*Mutatione di case*], he writes, 'Let the houses be transmuted and put in order [*le case sieno trasmutate e messe per ordine*] and this will be done with facility because such houses are first made in pieces on the open places and then they can be fitted together with their timbers in the site where they are to be permanent'. And then he adds, 'The river at Villefranche may be conducted to Romorantin . . . and the timber of which their houses are built may be carried in boats to Romorantin'.

Nor is it wrong to see Leonardo's lifelong concern with hygiene being put into his plans for rebuilding the château of Amboise. Here he proposed a set of water closets connected by flushing channels within the walls and ventilating shafts reaching to the roof. The doors of the closets were automatically closed by counterweights hanging on hinged pulleys.

The great halls where balls and festivities were to be held were to be on the ground floor, because 'I have seen many rooms collapse and bury numbers of dead'. And all timberwork was to be bricked in to minimise the risk of fire. Such considerations had been introduced into his architectural planning ever since 1484 when he had designed cities in an attempt to prevent the devastations of the plague.

It is characteristic of Leonardo that within a few years of his death he wrote down in his notebooks plans for at least half a dozen new treatises or discourses. Among these were a 'Treatise on Painting', of which Melzi's compilation was a mere assembly of bits and pieces. A 'Treatise on Light and Shade' is promised on several occasions, as also a 'Book on Perspective'. In the field of mathematics Leonardo promised a 'Treatise on Continuous Quantity', and a book entitled 'De Ludo Geometrico', no doubt revealing the still undeciphered significance of his studies in lunules (see Figure 5.34, p. 157). In anatomy his last plans were to propose a 'Discourse on the nerves, muscles, tendons, membranes and ligaments', and in addition a 'Special book on the muscles and movements of the limbs'.

These titles reveal the progressive consistency and integration of Leonardo's thought throughout his life, his perpetual quest for interweaving the complexity of minutely detailed observations into an integrated whole. This is symbolised by the intricate pattern of the interlacing band around the circle within which is inscribed the words, 'Academy of Leonardo da Vinci', sometimes appropriately referred to as Leonardo's mandala (see Figure 1.21, p. 30).

Thus at the end of his physical life Leonardo felt himself still at the beginning of his insight into his spiritual, intellectual life. On 24 June 1518 he wrote his last words, 'I shall go on'.

On 23 April 1519 Leonardo made his will bequeathing 400 ducats of his savings to those 'brothers in the flesh' (*fratelli carnali*) with whom he had gone to law. All his scientific and artistic works he left to Francesco Melzi.

On 2 May 1519 Leonardo da Vinci died.

REFERENCES

1. Vasari, Giorgio. *Lives of the Painters, Sculptors and Architects,* trans. by A. B. Hinds. London, Dent & Sons, 1927. Vol. 2, p. 158.
2. *Ibid.,* p. 98.
3. *Ibid.,* p. 100.
4. *Ibid.,* p. 156.
5. *Ibid.,* p. 97.
6. *Ibid.,* p. 95.
7. *Ibid.,* p. 100.
8. *Ibid.,* p. 81.
9. *Ibid.,* p. 82.
10. Southey, R. *Letters from England,* ed. by J. Simmons. London, The Cresset Press, 1951, p. 473.
11. Valturio, Roberto. *De Re Militari.* Leonardo almost certainly used the Italian translation, Di Roberto Valturio di Arimono, *Opera de l'arte militare . . . ,* traslata per el spectabile doctor misier Paulo Ramusio di Arimino, Verona, Bonon di Bononis, 1483, as stated by Augusto Marinoni in his *Gli Appunti grammaticali e lessicali de Leonardo da Vinci,* Milan, 1944–1952.
12. Vasari, *op. cit.,* vol. 2, p. 160.
13. *Ibid.,* p. 167.
14. Giovio, Paolo. *Leonardo Vincii Vita,* in J. P. Richter and Irma A. Richter, *The Literary Works of Leonardo da Vinci.* 3d edition, London, Phaidon, 1970. Vol. 1, p. 2. The translation from the Latin has been modified.
15. Letter from Messer Francesco Melzi to Leonardo's brothers, informing them of his death, quoted in Italian in John W. Brown. *The Life of Leonardo da Vinci.* London, William Pickering, 1828, p. 229. Translated by author.
16. Vallentin, Antonina. *Leonardo da Vinci: The Tragic Pursuit of Perfection.* London, W. H. Allen, 1952, p. 149.
17. Reti, Ladislao. 'Elements of Machines', in *The Unknown Leonardo,* ed. by Ladislao Reti. London, Hutchinson, 1974, pp. 264–287.
18. Vasari, *op. cit.,* vol. 2, p. 162.
19. *Ibid.,* vol. 1, p. 331.
20. Koyré, Alexandre. 'Mathematics', in *The Beginnings of Modern Science,* ed. by René Taton, trans. by A. J. Pomerans. London, Thames and Hudson, 1964, p. 23.
21. Pacioli, Luca. *De divina proportione.* Venice, Paginus Paganinus, 1509, p. 33.
22. Gusnesco, Lorenzo, quoted by Julia Cartwright. *Isabella d'Este.* London, John Murray, 1915. Vol. 1, p. 172.
23. Vasari, *op. cit.,* vol. 2, p. 164.

24. Leonardo da Vinci. *Treatise on Painting (Codex Urbinas Latinas 1270)*, trans. and ed. by A. Philip McMahon. Princeton, N.J., Princeton University Press, 1956. Vol. 2, p. 128 v.
25. Vallentin, *op. cit.*, p. 284.
26. Keele, Kenneth D. 'The Genesis of Mona Lisa', *Journal of the History of Medicine and Allied Sciences* (Yale University), 1959. Vol. 14, p. 135.
27. Vallentin, *op. cit.*, p. 373.
28. *Ibid.*, p. 376.
29. Vasari, *op. cit.*, vol. 2, p. 163.
30. *Ibid.*, p. 166.
31. Richter, Irma. *Selections from the Notebooks of Leonardo da Vinci*. London, Oxford University Press, 1952, p. 384.
32. Vallentin, *op. cit.*, p. 517.
33. Richter, Irma, *op. cit.*, p. 383.

Chapter 2

Leonardo's Gateway to Science:
To Know How to See

Part I. Physics

THE NATURE OF LIGHT AND PERSPECTIVE

Wide differences of opinion have been expressed with regard to Leonardo's position as a 'scientist'. For some 300 years his efforts to create a scientifically organised body of knowledge were either completely ignored or misunderstood. In 1796 G. B. Venturi caught vivid glimpses of Leonardo's scientific methods. Venturi's pioneer studies, necessarily incomplete, opened an era of rapidly increasing appreciation of Leonardo's achievements. Unhappily these led to imaginative exaggerations of their significance. These in their turn created an aura of mystery which resulted in attributing to Leonardo some almost divine insight into all 'modern' sciences from physics to human physiology. Reaction to this phase began at the beginning of the twentieth century and still persists. This reaction also has been exaggerated, based as it has necessarily been on grossly incomplete knowledge of his works. This is due to the fragmentary confusion of Leonardo's notes and their inaccessibility. Only with the publication of the *Madrid Codices* in 1974 have all Leonardo's known extant manuscripts become available for study. Many of these notes still exist only in Italian, or in French translation, and even now we possess but a fraction of his whole work – some say about one-third. The confusion of such large fragments as the 1,222 pages of the 1894 edition of the *Codex Atlanticus* and the 566 pages of the *Codex Arundel* in the British Museum presents daunting difficulties to anyone attempting to analyse and integrate them.

These difficulties would indeed have still remained insuperable except for the indefatigable studies of their chronology made throughout the first half of this century by Calvi, Brizio and Pedretti in particular. Although such studies can in most cases give only an approximate chronology, they provide sufficiently firm data for one to discern that Leonardo's concept of 'science' evolved throughout his life only after persistent, strenuous intellectual struggle. To some extent this development depended on the subject upon which he was concentrating at a given time. And the spectrum of such subjects extended from that of perspective at the beginning to the biology of man and animals at the end of his life. Thus it comes about that the task consists not

of drawing the outlines of a clear-cut pattern of a man of science but the more difficult one of tracing Leonardo's progress in creating an integrated body of knowledge of such a nature as we today call 'science'. The obvious fact that Leonardo's science was built up from the foundations of 'four elements' and 'four powers' which no longer form the alphabet of modern science necessarily means that great differences between the two exist. This greatly enhances the difficulty of presenting the shape of Leonardo's scientific creation to those whose concepts of science have been founded on the works of Newton and Einstein or statistical probability.

In order to follow the process of Leonardo's construction of that organised body of knowledge which constituted his science it is necessary to trace it approximately in the chronological sequence outlined in Chapter 1. It cannot be too strongly stressed that all stages of Leonardo's methods and concepts are visual.

THE EYE IS THE WINDOW OF THE SOUL

Leonardo's education in Verrochio's workshop initiated him into a wide field of technological skills. Here it was that he acquired the mechanic's knowledge of the five simple mechanisms which had come down from classical times through Hero of Alexandria, i.e., the wheel, lever, pulley, wedge and screw. These were held to be the components of all machinery right down to the beginning of the nineteenth century. Here too, Leonardo became skilful in the use of tools, of simple chemical or alchemical processes involving fire, but it was outside Verrochio's studio that Leonardo acquired his interest in geography, hydraulics, agriculture and the physiology of the human body.

Throughout his life Leonardo referred to himself as an 'inventor'. The essence of an inventor in his view was someone who created something that Nature had not created; painting, for instance, which can improve on nature's powers. In defending himself against 'boasters and declaimers of the works of others', Leonardo claims that as an inventor he relies on 'experience, the mistress of their masters', and an invention should be regarded and esteemed as 'something in itself rather than the reflection of that thing in a mirror, which is nothing'. 'Inventors are interpreters between Nature and Man' (CA 117 rb).

The inventor, as Leonardo sees him, surpasses nature in that 'Nature is concerned only with the production of simple things but man from these simple things produces an infinity of compounds' (W 19045 v; K/P 50 v). And 'Man does not differ from animals except in accidental things and it is in this he shows himself to be a divine thing. For where nature finishes in the production of the forms or species of things, there man begins to make with the aid of nature an infinite number of forms of natural things' (W 19030 v; K/P 72 v). In this way, by creating a new world of artifacts, man as an inventor interprets and surpasses 'simple' nature. But the process of invention can be brought about in two ways, either by unbridled imagination, or by imagination guided by 'experience'. In the first group Leonardo places fantastic, or what he calls 'sophisticated', machines which do not work in practice. In the field of painting this includes distortions and the creation of unreal 'monsters'. (He gives, incidentally, detailed instructions of how to do this by putting together the parts of different animals (BN 2038 29 or A 109 r). Such inventions belong to the realm of the 'impossible'; in contrast Leonardo defines science as concerned with things 'possible' (Triv 17 v).

As an extremely gifted technologist and painter Leonardo deliberately set out to ascertain the nature of this 'experience' which would guide his inventions along the 'possible' avenues of theoretical understanding, or science, and so to the actual creation of objects which through imitation surpass Nature by innovation. Painting, for example, is an 'imitator of Nature – a child of Nature and an instrument of philosophy [or science]' (BN 2038 20 r; A 100 r).

It is in this context that we find him raising his earliest questions about the nature of science. 'All our knowledge has its origins in our senses [*sentimenti*]' (Triv 20 r), he notes. 'Science is the knowledge of things which are possible, present or past' (Triv 17 v).

It is at this time, too, that he makes his first enquiries into the relation between the senses and the intellect. 'There are four powers, memory and intellect, lust and desire; the two first are mental and the others sensual. The three senses, sight, hearing and smell, can be but little inhibited [*proibitione*], touch and taste not at all' (Triv 33 r).

Leonardo pursues his quest with persistence and startling originality through the path of light and perspective into the field of anatomy. In a long, early (1489) passage under the title, 'How the five senses are the ministers of the soul', he enters into a detailed description of the site of all sensation and the soul [*senso comune*], discussing whether the five senses are located on the surface of the body or within the brain. He decides in favour of the latter. In the end he concludes, 'experience' depends most on the sense which he has found anatomically dominant, that of sight. This he traced via the optic nerve to the brain and *senso comune*. This part of his study of 'experience' will be described later. The psychology and physiology of experience occur within the

microcosm of the body. We will first sketch that part of visual experience occurring in the macrocosm of the world outside the body.

SCIENTIFIC PERSPECTIVE

As both an artist and an anatomist Leonardo had come by 1490 to see that the road to science is through the sense of sight. Although there had been earlier observers who used light and vision as the foundation of their scientific efforts, such as Grosseteste and Roger Bacon, it must not be imagined that the climate of opinion in Leonardo's day was favourable to such a visual approach. With the Renaissance in Florence the influence of Marsilio Ficino emphasised Platonic scepticism about the deceptive illusions of vision. Leonardo, however, brought to bear the recently developed technique of perspective (see Figure 2.1). This was as new to his age as the telescope was to the seventeenth century. Leonardo's 'perspective' was the technical perspective of Brunelleschi and Alberti. The technique as Leonardo first describes it sounds simple. 'Perspective is nothing but seeing a place directly through a completely transparent pane of glass on the surface of which the objects behind that glass are to be drawn. These can be conducted to the point of the eye through pyramids and these pyramids are intersected on the glass pane' (A 1 v) (see Figure 2.2).

This definition Leonardo gives under the heading, '*pariete di vetro*' (pane of glass). He repeats it in a number of occasions, often, however, abbreviating '*pariete di vetro*' to

Figure 2.1. Leonardo's technique for making a perspectival drawing of the sphere of the macrocosm (CA 1 ra bis).

The sphere is viewed with one eye through a small hole in a square frame steadied by the left hand whilst the right draws the projection of the sphere on the intervening vertical glass pane. Note the 'pyramidal' diminution of rays from both poles of the sphere and its central band as they pass to the eye through the vertical glass pane.

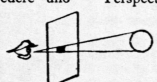

PARIETE DI VETRO	OF THE PLANE OF GLASS
Demonstration of perspective by means of a vertical glass plane (83–5).	

Demonstration of perspective by means of a vertical glass plane (83–5).

²Prospettiva · non è altro · che vedere ³uno sito dirieto uno vetro piano ⁴e ben transparēte, sulla superfitie del ⁵quale · siano · segniate · tutte le cose · che ⁶sono da esso · vetro · īdirieto: le qua⁷li si possono cōdurre per piramidi ⁸al pūto dell' ochio e esse piramidi si ⁹tagliano su detto vetro.

Perspective is nothing else than seeing a place [or objects] behind a pane of glass, quite transparent, on the surface of which the objects behind that glass are to be drawn. These can be traced in pyramids to the point in the eye, and these pyramids are intersected on the glass plane.

Figure 2.2. The principle of perspective (A 1 v). Figure from J. P. Richter, *The Literary Works of Leonardo da Vinci,* 3d edition (New York, Phaidon, 1970), vol. 1, p. 150.

just '*pariete*'. This has given rise to much confusion when '*pariete*' is translated into English simply as 'wall' or 'plane'. Leonardo always meant a solid glass pane on which he could draw.

With Leonardo *perspective* is a word which has to be interpreted in its context. He does sometimes use it in the medieval sense as covering the phenomena of 'optics' in general but far more often in the limited technical sense just described. For example, he writes, 'Perspective is nothing more than a rational demonstration applied to the consideration of how objects in front of the eye transmit their image to it by means of a pyramid of lines. The pyramid is the name applied to the lines which starting from the edges of the surface of each object converge from a distance and meet in a single point . . . A point is said to be that which cannot be divided into any parts and this point, placed in the eye, receives all the points of the pyramid' (A 31 r). Thus, Leonardo's 'pyramid' of perspective can have many sides, its base being formed by the edges of the observed object. It can have four or any number of sides, including the circular base of a cone,* in which form he sometimes draws it. Most commonly, however, he represents the perspectival pyramid or cone in vertical longitudinal section, i.e., as a triangle the base of which represents the object; its apex, the 'point', lies in the eye.

A passage in which Leonardo defines the word *pariete* when writing of a vertical glass pane also clarifies his concept of perspective. Leonardo writes, 'The confluence of the pyramids caused by bodies will show on the *pariete* the different sizes and distances of their causes. The *pariete* is a perpendicular line drawn in front of the point where a pane of glass through which you look at the different objects would perform the same function. Steadying your eye on it you can draw that which is carried there by the pyramids . . . These objects will appear so much smaller than

the originals by as many times as the resulting bases of the pyramids enter into the greater' (CA 233 rd).

The last sentence of this passage contains the reason why Leonardo attached such great importance to perspective in observation. By using this new technique he realised that he could find a quantitative relation between the 'original', or 'real' object and a measurable, proportional image by drawing or painting. The outlines of the painter's figure represent proportional linear perspective; its colour represents proportional colour perspective, and the loss of definition of its edges on the glass pane represents aerial perspective. As an object becomes more distant from the eye all three of these types of perspective will 'diminish'. Linear perspective involving both height and breadth will diminish to a point, the 'vanishing point'; colour will diminish also until it vanishes into the atmosphere, and definition of detail such as the limbs of a man will lose all shape and gradually disappear into a point on the horizon, i.e., at the vanishing point.

Perspective for Leonardo consisted of the play of proportions related to the positions of three variables: the object, the pane of glass and the eye. By moving each of these variables independently to different distances and angles he explored perspective systematically.

All these diminutions occur in proportion to the converging lines of their respective 'pyramids'. Therefore they diminish with distance quantitatively by 'degrees'* in 'pyramidal proportion'. It was noted by Leonardo that although these three kinds of perspective diminished 'pyramidally', their pyramids could be independent of one another. Aerial and colour perspectives were, for example, closely interwoven, but their pyramids might vary with 'degrees' of distance, differing from linear perspective, sometimes as in mist, modifying it deceptively.

It has often been said that Leonardo gave little attention to the study of linear perspective. This statement is based on the surprisingly sparse comments on the subject in the *Treatise on Painting* and on the supposed assertion by Luca Pacioli

---

Konos (cone) was the word used by Ptolemy in the third century A.D. In a Latin translation of the eleventh century by Eugene of Sicily 'cone' was rendered as 'pyramid'. This word was used by Roger Bacon, Pecham and other medieval writers on optics, including Leonardo. See Chapter 5, p. 153, for further references.

---

*'Degrees' are any number of sections of a pyramid made parallel to its base at equidistant intervals along its central line. Leonardo illustrates this concept clearly in Figure 4.27 (p. 113).

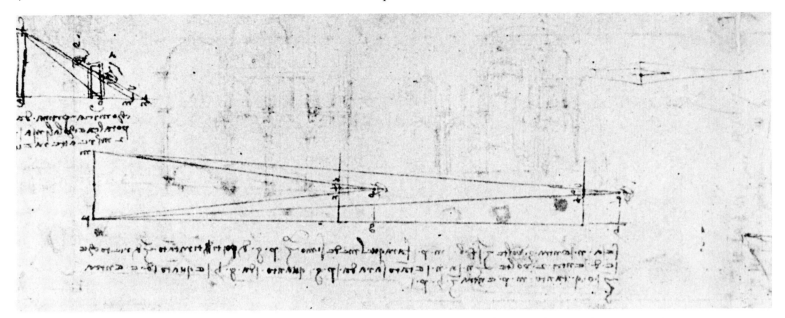

Figure 2.3.   Leonardo's diagram and text in which the ratio of size of object to distance from eye is shown to be the same as the ratio of size of projected image on the vertical glass plane, or pane to the distance from eye (CA 42 rc). This ratio holds when the distance between object and eye is doubled. See Chapter 2, Appendix, for details of the experiment.

that after seeing Piero della Francesca's treatise, Leonardo 'abandoned work on his book on the same theme'.[1] Since no such assertion has been found by Heydenreich in Luca Pacioli's works and since some of the few quantitative references to linear perspective in the *Treatise on Painting* are reported erroneously (as compared with Leonardo's notes), it would appear that Melzi, who compiled the treatise from Leonardo's notes, failed to understand their quantitative significance and therefore omitted many of them. A systematic search through Leonardo's notebooks reveals that in fact he gave a great deal of attention to the quantitative proportions revealed by the pyramidal figure of linear perspectival change. In fact, MS A folios 36 v to 42 r constitute a short treatise on perspective, richly concentrated. These studies of proportion provide a good example of his creative scientific methods.

As can be seen in the three quotations just discussed, Leonardo approached this problem of linear perspective by a series of experiments in which all three variables, the height of the object, the distance of the object from the eye and the distance of the vertical glass pane from the eye and object, were systematically compared with the 'image' painted on the vertical glass pane. In each case the ratio of the height of object to the height of the image drawn on the glass plane and the distance from the eye was observed. The whole set of data was then summarised into a 'rule' of linear perspective. It must be emphasised that these experiments on perspective apply to monocular vision only, not to binocular.

A number of these experiments have been repeated by the writer and an assistant, K. Veltman. One of the most interesting aspects of Leonardo's methods revealed by this repetition is the meticulous detail with which each variation has been quantitatively followed up. For example, when he assessed the perspectival ratios he placed his eye not only along the central line between the object and the eye but also at small intervals upwards and sideways over the surface of the glass pane, and not only with the eye at a distance of one *braccio* (about two feet) from the glass pane but at a distance of from two inches to several *braccia* away from it. Similar meticulous variations were made in varying the distance of the object from the eye and in altering the angle of oblique viewing of the object from above and below. The effects of varying the size of the '*spiracolo*' or hole through which the object is seen were also studied (see Figure 2.6, p. 48). This and the obliquity of the plane of glass led Leonardo to appreciate perspectival 'distortions' or anamorphosis. Space allows only a few simple examples of these experiments to be described here.

Leonardo's earliest experiment showing that the same object at twice the distance from the eye is projected on the glass pane at half the height was done about 1487 (Figure 2.3). For the text of this experiment see the Appendix to this chapter (pp. 77–78). Another example of perspectival measurement is found in *Codex Atlanticus* 251 va, without text. Leonardo extended this observation, stating, 'A second object as far distant from the first as the first is from the eye will appear half less [*la meta minore*] than the first, though they be of the same size'. He then continues, 'If you place the vertical glass plane [*pariete*] at one *braccio* from the eye the first object being at a distance of 4 *braccia* from your eye will diminish by 3/4 of its height in the said plane. And if it is 8 *braccia* from the eye by 7/8; and if it is 16 *braccia* distant it will diminish by 15/16 of its height and so on by degrees; as the space passed through doubles the diminution doubles' (A 8 v).

By repeating this experiment Leonardo's quantitative estimations of the height of the 'image' of the object drawn on the vertical glass pane measured under these conditions have been confirmed (see Figure 2.4).

The ancient technique of the surveyor was probably studied by Leonardo in the works of Alberti. Leonardo used it as a test for his pyramidal theory of the projection of the image of an object in Figure 2.5. In the top drawing on this page he shows 'How to measure the diminution of the pyramid through the air between the object seen and the eye'. From this it can be seen that when the surveyor's rod is held farthest from the eye it just covers the object. As this rod is drawn towards the eye it overlaps the object at both ends increasingly. 'A little further on', writes Leonardo, 'the lines will come to a point'. He is demonstrating with the surveyor's rod how the lines conveying the height of the object to the eye converge on the different positions of the surveyor's rod and come to a point in the eye.

In the three drawings at the bottom of Figure 2.5 Leonardo states his pyramidal law geometrically without reference to perspective. Leonardo expresses this law in the form of a diagram, the text to which reads thus, 'Every half-base of a pyramid has the same width as the width in the middle of the pyramid; and see how *ab* is the same as *de* and *mn* as *pq*'. By underlining this passage Leonardo indicates that he attached particular importance to these geometrical proportions within the pyramid.

Leonardo's assertions about the vanishing point being formed by the meeting of parallel lines have already been mentioned. Today we would use railway lines to illustrate this; he used the parallel lines furrowed by a plough. He also pointed out that the parallel lines on the planes of ceiling and floor seen in perspective will meet at the vanishing point. He demonstrates this by a series of lines drawn through a

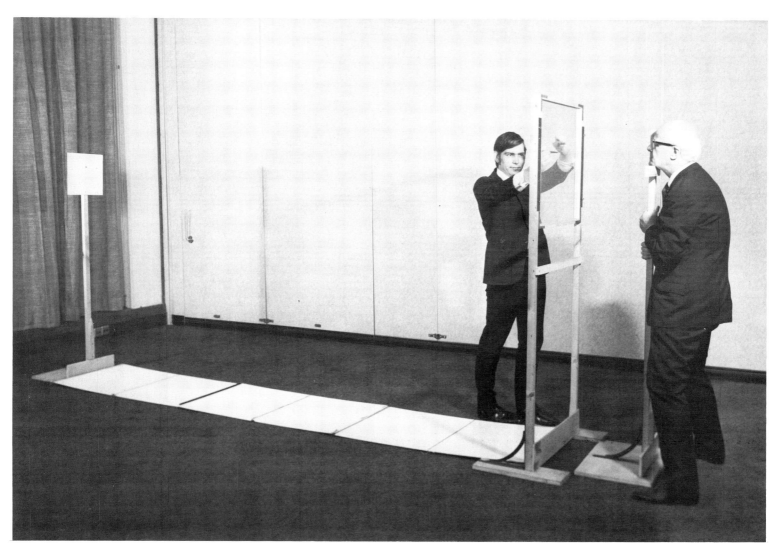

Figure 2.4. Technique used in repeating Leonardo's experiments on perspective.

The steadied eye is held 1 *braccio* (24 inches) from the vertical glass pane (*vgp*) on which the linear dimension of the projection of the object is being drawn. The distance is marked off in *braccia*.

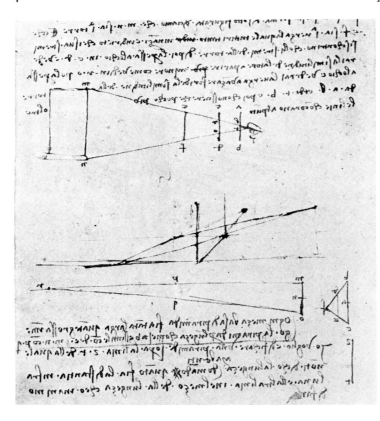

Figure 2.5. Top drawing shows how a surveying rod appears longer than the pyramidal bases as it is moved towards the eye from *ef* to *ab*. Bottom drawings have Leonardo's caption, 'Every half-base of a pyramid has the same size as the width in the middle of the pyramid; see how *ab* is the same as *de*, and *mn* as *pq*' (A 37 v). This underlined statement expresses the importance he attaches to his basic pyramidal law.

vertical glass pane to the edges of squares laid out on the ground showing how the successive lines rise on the glass pane 'until they reach the height of the eye and no further'. This meets the line of 'the flight of a bird level with the ground' which descends to eye-level at the vanishing point. Eventually he reproduces the geometrical figure of perspective described but not drawn by Alberti (Figure 2.6). Thus, Leonardo comes to make the first drawings of Alberti's perspective with the vanishing point on the horizon at the level of the eye. He has constructed all the elements of Alberti's 'costruzione legittima' from his own observations step by step, himself. In Figure 2.7 he adds an important feature. In this figure Leonardo demonstrates how parallel lines seen through a vertical glass pane *ab* will appear to meet at the vanishing point *n* with the eye at *r*. The set of perpendicular lines on the far side of the vertical glass pane

Fig. 2.6. Having explored practically every variation imaginable, Leonardo finally illustrates the validity of the 'costruzione legittima' (A 41 r).

The accompanying text reads, 'ab is the proof', and refers to the diagonal which should bisect the various diminishing squares, as well as the whole drawing. Note how the vanishing point is at eyelevel when viewed from the front as well as from the side.

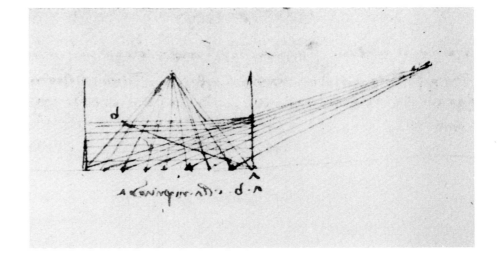

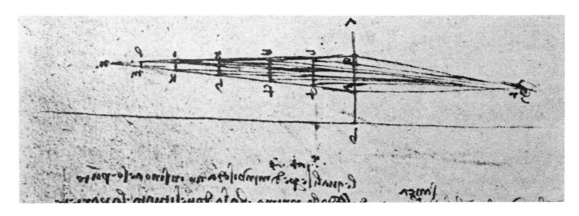

Figure 2.7. The vanishing point (A 37 r).

The vertical glass pane, *ab*, provides the base of a pyramid, *op*. The bases of more distant pyramids can be seen to diminish with distance, *cd, ef, gh, ik, lm*, until they diminish to the vanishing point, *n*.

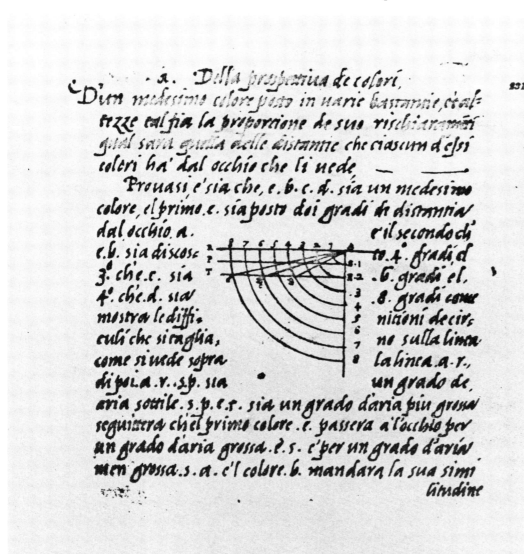

Figure 2.8. The measurement of colour perspective.

This diagram expresses in the form of a graph how a coloured surface at *EBCd*, two units of distance from *A*, the eye, is clearly seen 'in proportion to the distances which each of these has from the eye that sees them' (CU 65 r).

show the bases of smaller pyramids diminishing proportionally to their distance within the pyramid until *n*, the vanishing point, is reached. Leonardo makes it clear that this vanishing point is on the same level as the 'point' of the eye, and the visual line connecting them, *nr*, is the central line of vision. This central line becomes of great importance in his theory of vision.

Colour perspective, Leonardo saw, is intimately involved in distance judgements – intimately but not simply (see Figure 2.8). Diminution of colour is directly in (pyramidal) proportion with the distance 'as long as the air between the eye and object is of equal density' (CU 65 r). But colour perspective depends also on light intensity and therefore 'on the time of day' (TP I 224). The importance of colour perspective is that 'the distance between two objects can never be judged by the eye unless colour perspective is considered' (TP I 522). Thus both colour and aerial perspective modify linear perspective. Leonardo is as interested in

these 'causes' of perspectival conflict or illusion as he is in lineal perspectival 'truth' and endeavours to express these three perspectival factors in a geometrical figure which includes the factors of distance and colour with constant aerial 'thickness'. This figure comes very close to being a graph.

These experiments on perspective stimulated Leonardo to ask himself many questions. What is the nature of these 'rays' which carry light in pyramidal figures from the object to the eye? What is the nature of light itself and vision? Let us see how Leonardo tried to find answers to them.

THE NATURE OF LIGHT, SHADE AND PERSPECTIVE

Clearly the experiments on perspective outlined depend upon the presence of light, as does all vision. As we have already noted, Leonardo from his earliest days in Florence evinced an interest in grinding lenses and mirrors. Concave mirrors in particular brought him to a realisation of the importance of the convergence of the 'pyramidal' rays of the light and heat of the sun to a point, the focus. This was the principle of the burning glass. For many centuries such results were thought to weld light and heat together as manifestations of the natural power of the 'element' fire.

Leonardo, however, began his investigations of light from a different point of view at the same time as he was working on perspective (about 1490). He appears to have embarked on the study of light and shade after his visit to Pavia, for his plans commence with sentiments derived from John Pecham's *Perspectiva Communis* which he probably found there. 'Among the studies of natural causes and reasons light chiefly delights the observer; and among the great features of mathematics the certainty of its demonstrations is what pre-eminently elevates the mind of the investigator. Perspective therefore, must be preferred to all the discourses and systems of human learning. In this field the radiating line [of light] is explained by those methods of demonstration which form the glory, not so much of mathematics as of physics, and are graced with the flowers of both. But its axioms being laid down at great length I shall abridge them to a conclusive brevity, arranging them by the method both of their natural order and mathematical demonstration; sometimes by deduction of the effects from the causes and sometimes arguing the causes from the effects; adding also to my own conclusions some of which, though not included in them, may nevertheless be inferred from them. Thus if the Lord, who is the light of all things, vouchsafe to enlighten me I will treat of light. Wherefore I will divide the present work into 3 parts' (CA 203 ra).

Pecham's delight in the investigation of light was a reflection of the work of Grosseteste, Roger Bacon and Witelo. His work would thus bring these very congenial influences to bear on Leonardo both as regards his scientific method and the subject of light itself. Pecham's book, as mentioned at the end of Leonardo's note, was divided into three parts – on light and vision, on reflected rays and on refracted rays – to all of which Leonardo gave much attention.

Like Pecham, Leonardo at first adopted the Platonic extramission theory of vision, later turning to intramission. Under the heading, 'The nature of Light', Leonardo writes, 'The eye carries into itself infinite lines which are adherent to, or united with those coming onto it which leave the object looked at . . . Light acts in the same way, for in the effects of its lines and particularly in the operations of perspective it is very similar to the eye' (W 19148 v; K/P 22 v). On this same page he gives his first description of the principle of the camera obscura.

'The lines from the eye, sun and other luminous rays, passing through the air are obliged to keep in a straight direction . . . But if the wall [*pariete*] has in it some small perforation which opens into a dark chamber you will see the lines of the rays enter through this said foramen to intersect and to generate two pyramids with their points together and their bases opposite, carrying to the second wall the whole form of their origin both as to colour and shape, only everything will be upside down . . . It will appear clear to experimenters that every luminous body has in itself a hidden power (*una virtu recondita*) which is central, from which and to which arrive all the lines generated by a luminous surface, and from there return or leap back outwards and unless they are impeded they are dispersed through the air with equal distance' (W 19148 v; K/P 22 v).

Thus light in Leonardo's view consists of rays of power or energy radiating from the centre of any luminous body, particularly the sun. These rays move in straight lines which can intersect, as, for example, when they pass through a small hole, to produce a clear inverted transposed image of the body from which they arise. When 'impeded' these rays 'percuss' the object interrupting their straight path. In the case of an object looked at by the eye, the 'central ray' is the most powerful and produces the clearest image; peripheral 'false' rays produce blurred images. 'Light in the operation of the rule of perspective does not differ from the eye', Leonardo asserts. As usual he follows up the general statement with a series of experiments, one of which consists of the following: 'Place a light on a table and retire a certain distance away and you will see all the shadows of the objects which are between the wall and the light impressed with the shadow of the form of the objects and all the lines of length converge on the point where the light is. Afterwards bring your eye very near to the light . . . and you will see all the bodies opposite without their shadows, and the shadows on the parts of the walls will be covered as regards the eye by the bodies set in front of them'. Thus, 'the visual ray and the light ray resemble each other' (CA 204 vb).

In this connection Leonardo's concept of a point and 'line' or 'ray' is obviously relevant. Under the title, 'Definition of a line', he writes, 'The line has in itself neither matter nor substance and may be named rather a spiritual* thing than a thing of substance; and being so conditioned it occupies no space. Therefore the intersections of an infinite number of lines may be imagined to be made in a point, which has no centre, and in thickness (if thickness it may be called) it equals the thickness of one single line'. Applying this definition of a line to the all-important 'central line' of vision, Leonardo asks 'whether or not this can be intersected in the little hole [*spiracolo*] or not'. He replies, 'It is impossible that a line should be able to intersect itself; that is, that the right side of one of its faces should pass to the left side of its opposite face because such an intersection requires two lines, for the two said sides. These cannot be given motion from right to left, or from left to right unless there is a space of thickness which gives room for such motion. And if there is space it is not a line but a surface. And because a line having no centre of thickness cannot be divided therefore we must conclude it has no sides to intersect each other'. Let *afeb* in Figure 2.9 be a beam of light, with sides of *af* and *eb*, joined by lines *ab* and *ef*. 'If you move the line *ab* and the line *ef* with their front ends *a* and *e* to the position *c*

---

*Leonardo uses the term *spiritual* to signify power or energy, without corporeal substance. See Chapter 4.

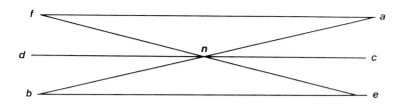

Figure 2.9. The diagram by which Leonardo shows that 'it is impossible that a line should be able to intersect itself' (W 19151 r; K/P 118 rb).

For its explanation, see text. In this drawing the diagram has been redrawn using ordinary lettering (not Leonardo's) for the sake of clarity.

you will have moved their opposite ends *f* and *b* to meet at *d*. From the two lines you will have made the straight line *cd* which resides in the centre of the intersection of these two lines at the point *n* without any intersection. If you imagine these two lines to be corporeal, necessarily through the said movement one will completely cover the other, being equal to it without any intersection at the position *cd*' (W 19151 r; K/P 118 r).

Leonardo makes it very clear that what he describes as a ray or line of light, either central or false, consists of incorporeal or spiritual power without substance. From the first he extends his concept of the ray of light to other powers. This extension is beautifully shown in Figure 2.10 in a series of nine deliberately similar diagrams in which he compares and illustrates in three drawings the actions of light and colour; in three further drawings, that of sound; the action of a magnet on iron in two drawings; and that of smell in one drawing. All these he sees as analogous processes of energy acting through 'spiritual' lines according to similar rules.

With regard to light, the top diagram in Figure 2.10 illustrates 'how light rays can pass only through diaphanous matter' from the candle *r* to the walled-in space *A*. The next drawing below shows how an object *ox* lit by *p* emits an image over the whole wall *ox*. This is reflected through the *spiracolo* (little hole) at *c* to form an inverted image *rs* on the wall of the camera obscura. The third diagram illustrates how coloured pieces of glass placed in front of candles *g* and *h* will transmit through the little holes their coloured light onto the opposite walls without interference. The fourth diagram shows how sound (created by the percussion of a hammer on an anvil) passes through a solid wall, unlike light, spreading in pyramidal shape. The fifth illustrates the spread of sound as it passes along its central line through a *spiracolo* (*b*) in that wall. The sixth illustrates the 'voice of an echo', the angle of incidence equalling the angle of its reflection.

The seventh illustrates that magnetic forces act on iron through a solid wall proportionally to the weight of the magnet and iron: 'How the lines of the magnet and that of iron pass the wall but the lighter is drawn by the heavier'. The eighth illustrates 'If the weight of the magnet and the iron is equal they will draw each other similarly'. This similarity of movement is beautifully expressed by the equal curved angulation of the cords from which the magnet and the iron are suspended. 'Among bodies which are joined together by a spiritual bond', writes Leonardo, 'that which is heavier draws the lighter' (CA 225 rc).

The last, ninth, diagram illustrates how smell spreads in the same way as the other powers. 'Smell does the same as a blow', he writes. In this series of drawings Leonardo is demonstrating that if 'Light in the operation of the rule of perspective does not differ from the eye', nor do other radiating powers such as sound, magnetism and smell – all are perspectival or 'pyramidal' in their diffusion.

From these experimental observations it is evident that Leonardo was not troubled by the question of action at a distance; the force of the magnetic 'species' is proportional to the weights of the magnetic bodies.

THE NATURE OF SHADOW

Shadow, Leonardo looks upon in many ways as a positive phenomenon which acts very similarly to light though producing opposite effects. 'Light', writes Leonardo, 'drives away darkness. Shadow is the deprivation of light' (CA 116 rb). 'Shadows appear to me to be of supreme importance in perspective because without them opaque and solid bodies will be ill-defined . . . Every opaque body is surrounded, and its whole surface is enveloped, in shadow and light. Besides this shadows have in themselves various degrees of darkness because they are caused by the absence of a variable amount of the luminous rays; and these I call primary shadows because they are first and clothe the bodies to which they are applied. From these primary shadows there result certain shaded rays which are diffused throughout the air and vary in intensity according to the varieties of primary shadows from which they are derived; and consequently I call these shadows derived shadows because they arise from other shadows [Figure 2.11]. Again these derived shadows in percussing various objects produce effects as various as the places they percuss. And since all round where the derived shadow percusses there is always a space where the light percusses and rebounds with these in a reflected stream towards its cause, it meets the primary shadow, mixes with it and becomes changed into it, so somewhat modifying its nature' (CA 250 ra).

In this context Leonardo describes the phenomenon of irradiation. 'Among objects of the same degree of darkness, size, shape and distance from the eye that will appear smallest which is seen against the background of greatest brightness or whiteness'. He illustrates this with the example of the invisibility of the shaft of a spear placed between the eye and the sun. This comes about, he asserts, because the brightest part sends the eye its image with a more vigorous ray than does the dark part (TP I 457).

Thus, Leonardo outlines the concept of shadows upon which he proposes to write seven books, much of which material is contained in MS C, in which he introduced photometry. Here he not only casts light on the principles un-

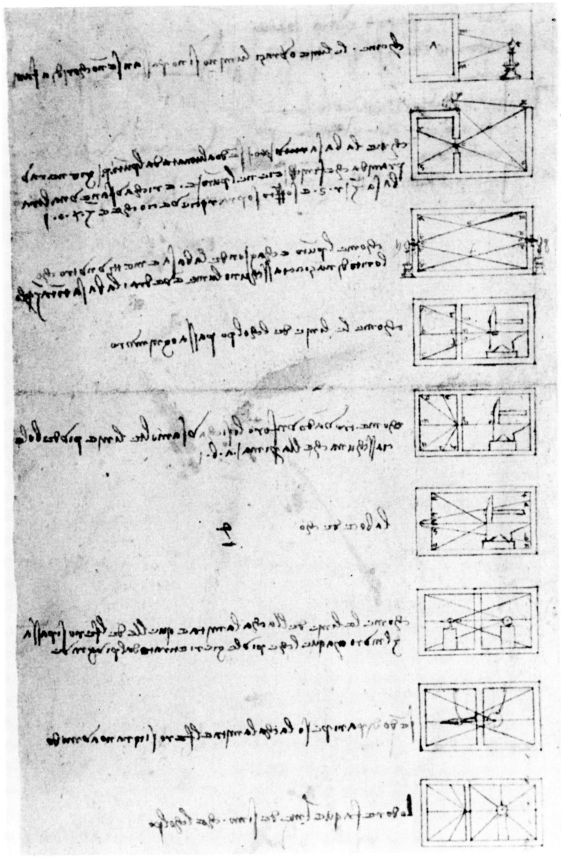

Figure 2.10. Similar patterns of transmission of the forces of light, sound, magnetism and smell, all of which percuss objects. For details see text (CA 126 ra).

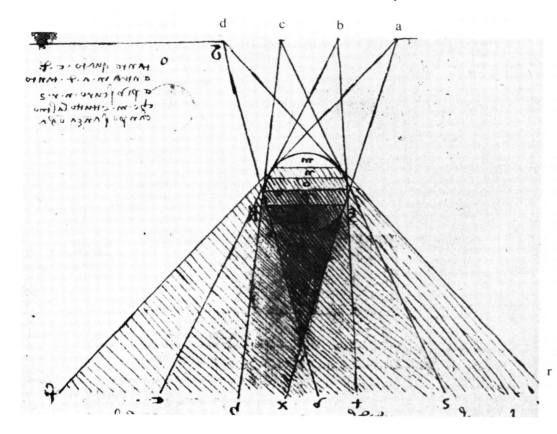

Figure 2.11. Primary and derived shadows (BN 2038 13 r; A 93 v).

In this drawing of graded shadows light falls onto the sphere from above. The shadows are graded by 'degrees' from nil at *m* through *n, o, p,* to *Q;* these are all grades of 'primary' shadow attached to the opaque body. 'Derived' shadows are also graded, their areas being designated by the letters along the base of the drawing.

derlying his chiaroscuro method of painting but also shows how light and shade contribute to the perception of distance, i.e., of perspective. Finally, and most important of all from the point of view of Leonardo's 'powers of nature', he makes it quite clear that he regarded the rays of both light and shadow as percussing the objects upon which they fall. Both therefore follow the laws of percussion.

THE RADIATION OF LIGHT

Leonardo describes the mode of spread of light (or shadow) from a luminous body in Figure 2.12 thus, 'Every body in light and shade fills the surrounding air with infinite images of itself and these by infinite pyramids infused in the air represent this body all in all and all in each part. Each pyramid that is composed of a long converging course of rays includes within itself an infinite number of pyramids, and each has the same power as all, and all as each. The equidistant circle of converging rays of the pyramid gives to their object angles of equal size; and the eye will receive the thing from the object as of equal size. The body of the air is full of infinite pyramids composed of radiating straight lines which are caused by the boundaries of the surfaces of the bodies in light and shade placed in the air; and the further they are from their cause the more acute [angled] are the pyramids, and although in their concourse they intersect and interweave nevertheless they never blend but pass through all the surrounding air independently, converging, diverging, diffused. And they are all of equal power, all

equal to each other and each equal to all. By these the images (*spetie*) of bodies are carried all in all and all in each part; and each pyramid by itself receives in each minutest part the whole form of the body which is its cause' (BN 2038 bv; A 86 v).

This passage is accompanied by Figure 2.12, which expresses Leonardo's thought more clearly than his words. From a spherical 'body' *ab* pyramidal or conical figures can be seen radiating out in all directions. It will be noticed that those pyramids with equal angles at their apices are of equal length, a fact that can be confirmed by drawing a circle through them. This is 'the equidistant circle of converging rays' to which Leonardo refers. One pyramid is prolonged to show that the longer it becomes the more acute does its apical angle become. This is demonstrated by the apices of the shorter pyramids *f, e, d* and *c*.

What does Leonardo mean by 'minutest part'? This he answers in describing the intersection of rays passing from the object to form an image in the camera obscura. At their intersection in the 'perforation' these rays form a point, and 'Every point is the termination or beginning of an infinite number of lines which diverge to form the base of a pyramid [or cone] and immediately from this same base these same lines converge to a pyramid imaging both in colour and in form. Immediately the form is created or composed an infinity of angles and lines are produced from it which, distributing themselves and intersecting each other in the air, give rise to an infinite number of angles and lines opposite each other' (W 19148 v; K/P 22 v). Thus 'the point' is

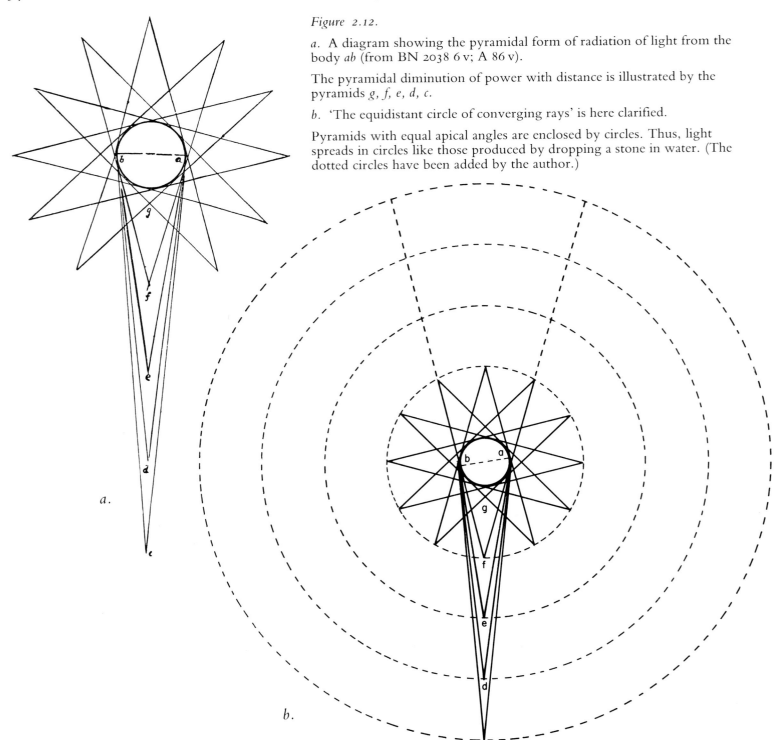

Figure 2.12.

a. A diagram showing the pyramidal form of radiation of light from the body *ab* (from BN 2038 6 v; A 86 v).

The pyramidal diminution of power with distance is illustrated by the pyramids *g, f, e, d, c*.

b. 'The equidistant circle of converging rays' is here clarified.

Pyramids with equal apical angles are enclosed by circles. Thus, light spreads in circles like those produced by dropping a stone in water. (The dotted circles have been added by the author.)

Leonardo's 'minutest part', and the image received by the eye is composed of points integrated into the form of the whole.

THE PATTERN OF REFLECTION

The apices of pyramids are points at which rays of light intersect, or, if 'impeded', percuss. If light and shade follow the laws of percussion, then it is most important to find out what happens to rays of light (or shadow) when they percuss and are reflected from a plane surface. Leonardo deals with this problem saying, 'The line of percussion and that of its rebound are placed in the middle of equal angles. Every blow that strikes an object rebounds back through the same angle as that of its percussion' (A 19 r). As is so often the case, his illustration, Figure 2.13, clarifies his words. Here he points out that 'this clearly appears if you throw a ball against a wall — it will bounce back through an angle equal to that of percussion, that is to say, the ball *b* thrown to *c* will turn back through the line *cb* making equal angles on the wall *fg*. And if you throw through the line *bd* it will turn back through the line *de*, so that the line of percussion and the line of rebound will make an angle on the wall *fg* between two equal angles, as shown by *d* in the middle between *m* and *n*'. Leonardo, having repeated this statement that the angle of incidence equals the angle of reflection in relation to the voice, adds in Figure 2.14, 'And the voice does the same as an object seen in a mirror which is all in the whole mirror and all in each part'. In a rough diagram he writes, 'Let *ab* be a mirror and the object seen in the mirror be *c*. Just as *c* sees all the parts of the mirror so all the parts of the mirror see *c*. Therefore *c* is all in the whole mirror

because it is in all its parts; and it is all in each part because it is seen in as many different parts as there are different positions of the viewer. That is, if the object *c* is in *n* it appears as much behind as it is away [from the mirror]; therefore *c* appears to be at *d*. And whoever is at *f* seeing the thing *d* sees it through a straight line, therefore *d* is on the part of the mirror *e*. And whoever is at *m* will see *d* at *t*'. Once more his diagram shows all the angles of incidence equal to the angles of reflection.

Leonardo illustrates the logical conclusion from this, how 'every base fills the air with infinite pyramids' and 'every point causes infinite bases', in Figure 2.17 (p. 57).

Leonardo considered it very significant that this law of percussion (that the angle of incidence equals the angle of

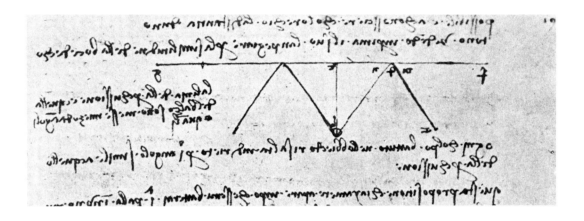

Figure 2.13. 'The line of percussion and that of its rebound are placed in the middle of equal angles' (A 19 r).

The ball shown at *b* is thrown against the wall *fg*, hitting it at *c* or at *d*. In both cases it is reflected at equal angles. At *d* the equal angles are lettered *n* and *m*.

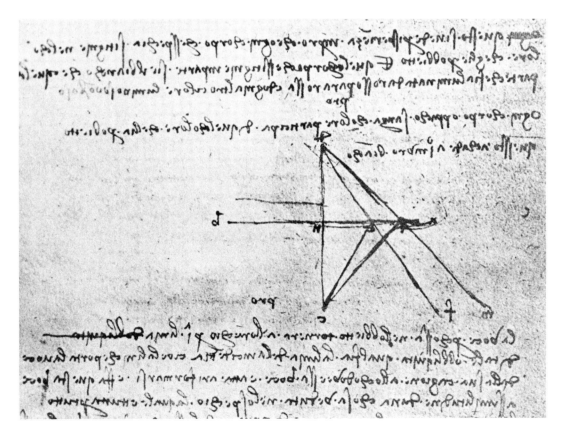

Figure 2.14. 'The voice does the same as an object seen in a mirror'; *ab* is the mirror; *c* the object (A 19 v).

If seen at *n*, its image appears to be at *d*, at an equal distance behind the mirror. If seen from *f*, it will appear at *e*, where the angle of incidence equals the angle of reflection. This applies also if the object *c* is seen from *m*, or any other place on the surface of the mirror. Sound is reflected according to the same law.

reflection) applies to the mechanical event of throwing a ball against a wall, as well as to the echo of sound and the reflection of light, for in all three cases such angles form the apices of pyramids.

EXPERIMENTS WITH THE CAMERA OBSCURA

Leonardo uses the principle of the camera obscura to 'prove' or test his concept of the distribution of the images of objects 'everywhere and all in each part'. His experiment he describes thus: 'If the front of a building or any piazza or field which is illuminated by the sun has a dwelling opposite to it, and if in the front which does not face the sun you make a small round hole, all the illuminated objects will project their images through that little hole [*spiracolo*] and be visible inside the dwelling on the opposite wall which should be made white. And there in fact they will be upside down. And if you make similar openings in several places in the same wall you will get the same effects from each. Hence the images [*spetie*] of the illuminated objects are all everywhere on the whole wall and all in each minutest part of it . . . If the bodies are of various colours and shapes the rays forming the images are of various colours and shapes, and so will be the representations on the wall' (CA 135 vb; see Figure 2.15). This experiment is repeated many times, as in Figure 2.16.

Here Leonardo takes the opportunity of further clarifying the nature of a 'ray'. Noting that if the object sends its image to the eye, the eye will reciprocally send its image to the object, he adds, 'Therefore we can rather believe it to be the nature and power of this luminous air which attracts and takes into itself the images of the objects [*che attrae e piglia in se le spetie delle cose*] which exist in it, than the nature of the objects to send their images through the air . . . From this it seems necessary to admit that it is in the nature of the air which subsists between the objects to attract the images of

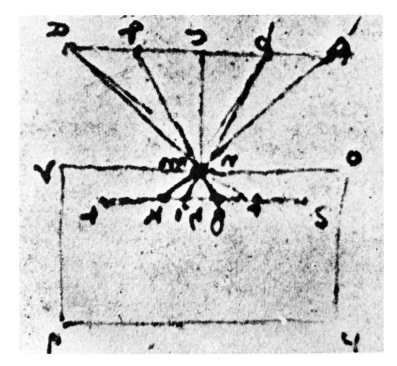

Figure 2.16. 'How the images of objects received by the eye intersect' (D 8 r).

The 'object' is *abcde*. From it the rays can be seen converging on the small hole [*spiracolo*], *mn*, whence they diverge onto a thin sheet of paper, *st*, in a camera obscura. It is noted that the image formed is turned from left to right and upside down.

Figure 2.15. The projection of images through one or more holes into a camera obscura (W 19150 v; K/P 118 v).

In the upper little sketch three 'illuminated objects', *abc*, project their images through a little hole onto the opposite wall at *iKL* in the camera obscura.

In the lower sketch the three illuminated objects, *ace*, send their images through two little holes, *n* and *p*, onto the wall of the darkened camera obscura *fi*, at *fdb* and *ghi*. All these images are 'upside down'.

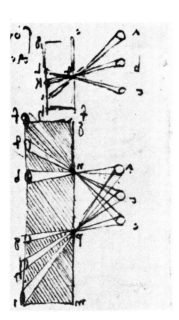

things to itself like a magnet [*calamita*] placed between them' (CA 135 vb). The power of vision is also closely related to these magnetic *spetie* – 'The air as soon as there is light is filled with innumerable images [*spetie*] to which the eye serves as a magnet' (CA 109 va, 126 ra). Thus once more Leonardo suggests that the transmission of light rays is achieved through the medium of a magnetic field. Nor does this claim for the resemblance of light rays to magnetic forces remain isolated, for he brings it into line with the other 'powers of nature', as we have already seen, in Figure 2.10, p. 52.

It may well be noticed here that from Leonardo's apparently simple description of perspective as 'nothing but seeing the image of an object conducted to the eye through a pyramid' (A 11 v), a complex and intricate structure of the physical nature of light has been built up, based on the geometrical figure of a pyramid. Leonardo elaborates his pyramidal constructions into a highly intricate web. Only a small example of this is quoted here, which is illustrated in Figure 2.17 (C 20 r). 'Every body is surrounded by an extreme surface. Every surface is full of infinite points. Every point makes a ray. The rays make infinite diverging lines. In each point of the length of any line there intersect lines proceeding from the points of surfaces of bodies, and these form pyramids. The images of the figures and colours of each body are transferred from the one to the other through pyramids. The image of each point is in the whole and the

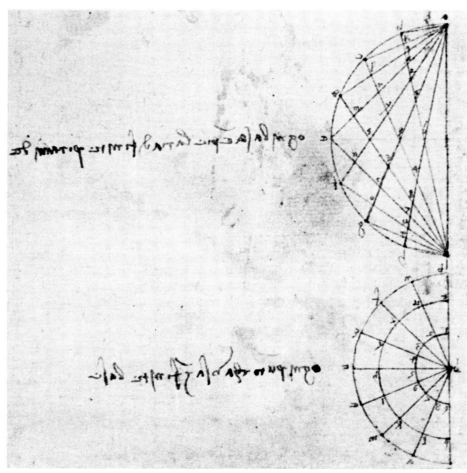

a.

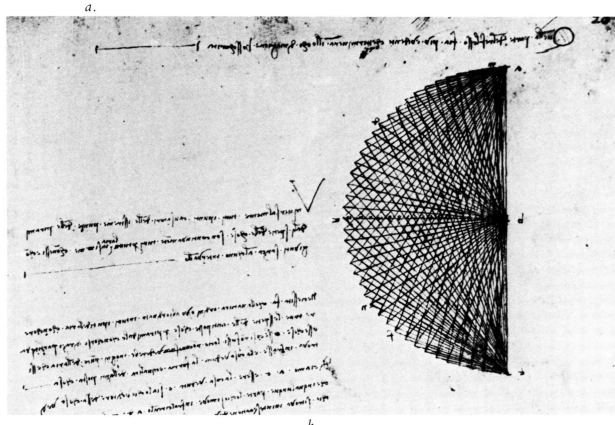

b.

Figure 2.17. How 'every point causes infinite bases' in the air (C 20 r).

a. Shows radiations from one central point, *a;* then from two points, *a* and *b* (CA 126 ra).

b. Here Leonardo develops these into the pattern of radiation from three points, *a, b,* and *c.* Radiating lines of light (or other forces) emerge from these. As they do so they spread to form bases of infinite pyramids (C 20 r).

part of the line caused by this point. Each body becomes the base of innumerable and infinite pyramids. One and the same base serves as the cause of innumerable and infinite pyramids directed in different converging courses and of different lengths. The central line of the pyramid is full of the infinite points of other pyramids. One pyramid passes through another without confusion. The pyramid with the finest point will show less of the true form and quality of the body from which it arises. The pyramid with the thickest point more than any other will dye the place where it percusses with the colour of the body from which it is derived' (BM 232 r).

THE SPREAD OF LIGHT IN CIRCULAR WAVES

At this stage Leonardo introduces into the propagation of light another geometrical figure, the circle. He broaches this in MS A 9 v. Here he writes, 'Just as a stone flung into the water becomes the centre and cause of different circles and as sound spreads itself in the air in circles, so any object placed in the luminous air spreads itself in circles and fills the surrounding parts with infinite images of itself. And it appears all in all parts and all in every smallest part. This can be proved by experiments'. And he again describes the experiment of making holes in walls all round an object like that already considered. In addition he 'proves' his statement with an experiment using mirrors placed in a circle round an illuminated object, 'which reflect each other an infinite number of times, and the image rebounds back to its cause, and in so diminishing rebounds once more to the object, and then returns once more and so it continues to infinity'. He points out that 'If at night you put a light between the walls of a room all the parts of that wall will be tinted with the image of that light . . . The same example is seen on a greater scale in the distribution of the rays of the sun which all together and each by itself convey to its object the image of its cause [the sun]' (CA 138 rb).

These experiments demonstrate the concept of Grosseteste, Bacon and Albertus Magnus that the spread or irradiation of the image of an object forms spheres or circles in all three dimensions from its central 'cause'. What is the physical nature of these spreading circles? To answer this we must return to Leonardo's experiment of throwing a stone into water. This is described in much more significant detail on A 61 r and in Figure 2.18. 'Although voices penetrating the air spread in circular motion from their causes [*cagioni*] nonetheless the circles moved from different origins meet together [*niente di meno i circuli mossi da diversi principi si scontrano insieme*] and they penetrate and pass into one another always keeping the centres which cause them, because in all cases of motion there is great likeness between water and air. I shall cite an example of the above proposition. I say, if you throw two small stones at the same time on to a sheet of motionless water at some distance from one another you will observe that around the two percussions two separate quantities of circles are caused which will meet

as they increase in size and then penetrate and intersect one another while all the time maintaining as their respective centres the places percussed by the stones. And the reason for this is that the water although apparently moving does not leave its original position because the opening made by those stones is closed again immediately. And that motion made by the quick opening and closing of the water makes in it a certain shaking [*fa in lei un cierto riscontimento*] which may be described as tremor rather than movement. In order to understand better what I mean watch the blades of straw that, because of their lightness, float upon the water and observe how they do not depart from their original position in spite of the waves underneath them caused by the arrival of the circles. Therefore the impression on the water being a tremor rather than a movement they cannot through meeting break one another, because water being of the same

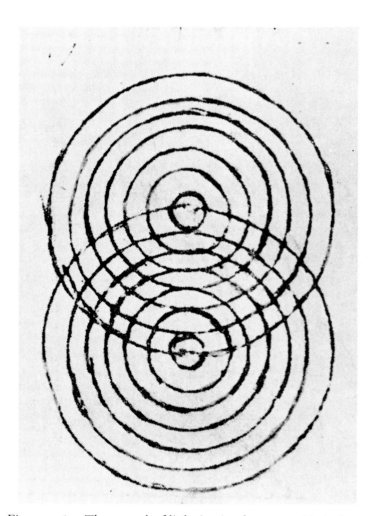

Figure 2.18.   The spread of light in circular waves (A 61 r).

Comparing the spread of light with that of sound and waves in water, Leonardo throws two stones into still water and observes the waves which spread out from these two centres of percussion. He particularly notes how these waves 'pass into one another always keeping the centres which cause them'. Neither wave-pattern appears to interfere with the other.

quality all through, its parts transmit the tremor to one another without changing their position. Thus the water, though remaining in its position, can easily take this tremor from neighbouring parts and carry it to other adjacent parts, always diminishing its power until the end'. Thus, light rays carry the images of objects by waves spreading out from their origin in percussion as divergent waves in the form of 'sectors of a circle', diminishing in pyramidal proportion in power.

This experiment performed about 1490 might well be described as crucial to Leonardo's construction of his concept of the physical universe because it introduces the irradiation of wave motion through all the 'elements', a subject which he will continue to investigate with his customary thoroughness to the end of his life. It will be noticed that the wave motion he describes here is transverse, and transverse it remains for him. All through his subsequent investigations he never considers longitudinal wave motion.

THE MOVEMENT OF ALL THE POWERS

In Leonardo's view movement arising from any source of energy such as percussion spreads out in circles. As this movement spreads in space so its power weakens in the proportion to the ever-widening base of the sector of a circle. Thus, whilst the diminishing power can be represented by a cone or pyramid arising from its base, its sides converging to its apex, the movement in space is represented by a pyramid with divergent sides, its base forming the edge of the sector of a widening circle with its apex at its source (see Figure 2.12, p. 54). At the same time as the circle enlarges and each sector widens, the waves carrying the impetus or power become of less amplitude and weaker until they ultimately disappear. Thus two pyramids, one of movement and one of power, arise from any source of power, e.g., the percussion of a stone on water or a source of light or sound. As the movement of the pyramid occupies more space its power weakens.

Leonardo was equally interested in the converse pyramidal concentration of power. For example, both light and heat travel at finite velocity. If with the aid of a concave mirror or any other means they are made to converge at a focal point, then they gain power in pyramidal proportion, i.e., by as many times as their amplitude at any section is contained in the base of the pyramid (see Figure 2.19 and Figure 4.44, p. 129). With regard to weight he gives the examples of weights on a balance being concentrated pyramidally as they approach their support and of a pyramidally pointed spear having more concentrated penetrating power than any other shape (D 1 r).

In the following notes Leonardo describes this complex geometrical pattern of the spread of movement and power. 'Any power of percussion made in water or air goes on dilating itself in the form of a sector of a circle' (CA 112 vb, see Figure 2.12, p. 54). 'Percussion makes conical forces like weight on water' (BM 37 v). 'If water enclosed in a vessel is struck at its centre a circular motion is produced

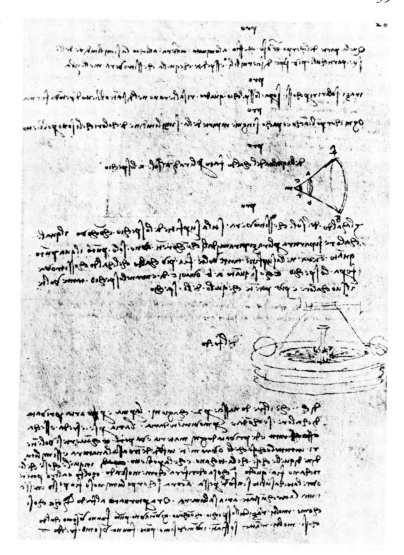

Figure 2.19. Pyramidal concentration of light and heat. (See also Figure 3.8.)

Underneath the upper drawing Leonardo writes, 'The heat of the sun which is found on the surface of a concave mirror will be reflected in converging pyramidal rays to a single point. As often as the point itself, *ab*, or *cd*, enters the base, *fg*, so many times will the heat be more powerful than that on the surface of the mirror.' In the lower figure he describes the concentration of cold thus: 'Just as the many rays of a concave mirror unite in a point making utmost heat, so many bellows blowing on the same point make utmost cold'. Air, for Leonardo, was a dry, cold element (A 20 r).

which starting from very small circles will end in very large ones' (Madrid I 126 v). 'A stone thrown into water is the centre and cause of different circles. As sound spreads in circles in air so does an object in a luminous atmosphere' (A 9 v).

It will repay us to pause here and analyse an example of the important sequence of events contained in these experiments, and to translate these events into Leonardo's terminology. In this sequence he describes a series of movements and actions which involve terms, some of which are Aristotelian, some medieval, some his own.

Stones, for Leonard, were portions of the 'element' earth; they lay at rest in their 'natural' position on the earth's surface exerting 'weight' according to 'natural gravity'. When he raised in his hand such a stone from its 'natural' place he exerted 'force', displacing it upwards; the stone being thrown passes through 'local accidental movement' exerted by 'accidental weight' or 'force'. Movement produced by a living animal or man Leonardo called 'a sensible movement' as opposed to a 'local insensible movement' such as the stone falling to the ground with 'natural movement'. For example, he states, 'Every local insensible movement is produced by its sensible mover, just as in a clock the counterweight is raised up by man who is its mover' (W 19060 r; K/P 153 r).*

When Leonardo threw a stone, its 'accidental movement' was caused by 'accidental weight' or 'force' which being imparted to it from the 'sensible movement' of the hand he called 'impetus'. This force of 'impetus' was met by the 'resistance' or 'friction' of the 'element' air. As the 'natural weight' of the stone gradually exceeded its 'impetus', the compound movement of the stone made a curve finally falling back with 'natural gravity' along the line of its 'centre of gravity' in 'natural movement'. It then 'percussed' the surface of the water. 'Percussion', Leonardo describes as 'the end or impediment of movement'. With 'percussion' the force and movement of an object may be diminished, come to rest, or 'rebound' back in such a way that the angle of incidence equals the angle of 'rebound' or reflection. In this case the stone percussing water displaced the water so rapidly that some of its parts percussed neighbouring parts making a splash by rebounding into the air which has less resistance then water. Here the drops acquired 'natural weight' and fell back into their 'natural element', water, so closing the opening. The stone still descended in the 'element' water until it rested on the bottom on its 'natural element' earth.

Waves arise from a vibration in water (or other 'elements') conveyed by neighbouring particles of water forming the waves. The force of percussion is conveyed not only by the movement of the waves but also by their impressed force, or 'impetus' a pressure or shock wave. All these 'powers', movement, weight, force, percussion and impetus, are interchangeable. Thus they may produce percussion waves in all four 'elements', earth, air, fire and water. Circular irradiation of all powers by waves is universal.

Leonardo expresses his view of the ubiquity of percussion waves thus: 'Movement of earth against earth percuss-

ing it causes a slight movement of the parts percussed. Water percussed by water creates circles at a great distance round the place where it is percussed; the voice in the air goes further, in fire further still . . . but it is finite' (H 67 r). As we have previously noted, in ancient and medieval terminology light is often referred to as a kind of fire, as in Plato's *Timaeus* (58 D). One can glimpse from this statement how Leonardo weaves the physics of light through percussion with the other 'mechanical' powers of nature. A further illustration of linking up the natural 'powers' occurs when he writes about the sun, which he considers to be the one primary source of light in the universe. 'The sun', he writes, 'sends out two kinds of species, the first light, the second heat . . . That the air attracts to itself like a magnet all the images [*spetie*] of the objects that exist in it, and not their forms only but their nature, may be clearly seen in the sun which is a hot and luminous body. All the air which is exposed to its influence is charged in all its parts with light and heat and takes into itself the shape of that which is the source of this heat and radiance, and does this in each minutest portion. The north star is shown to do the same by the needle of the compass, and the moon and each of the planets does the like without undergoing any diminution. Amongst terrestrial things musk and other perfumes do the same' (CA 138 vb).

It will be seen from this that Leonardo uses the term *species* not to describe a mere shell or visual image of objects seen but to convey their essential nature. Such are the 'species' of heat, sound or magnetic forces which travel in the same wave-pattern as light, but not at the same velocity.

Part II. Physiology: How the Eye Forms the Window of the Soul

LIGHT, VISION AND THE BRAIN

It is not to be forgotten that Leonardo's investigations of light and perspective were undertaken in the quest for a method of depicting three-dimensional 'reality' on a two-dimensional plane, the artist's panel. His excitement and diligence were continuously stimulated by the prospect of achieving by painting the illusion of reality. Half this problem was solved when he found means of reproducing proportional reductions in size, colour and definition. This one might call 'objective' observation. Leonardo was equally conscious of the effects produced by the eye and the brain. This one might call the 'subjective' side of observation. It is to be emphasised that in the first instance, since the technique of perspective was practised with one eye only, Leonardo concentrated on monocular vision. Only later does he deal with the more complex problem of binocular vision.

Leonardo's early incursions into the field of anatomy, as we have already mentioned, largely took the form of an exploration of sensory anatomy and physiology. He paid

---

*It is unfortunate that the Italian word *sensibile* cannot be translated into English by some other word than *sensible*. The word *sensitive* does not always meet the case. Leonardo's 'sensible mover' is a mover which responds actively to percussion, e.g., a bell or one possessing the capacity of feeling; 'sensible movement' is also the response of an animated (live) object to percussion.

particular attention to the connections between the eye and the brain. These studies he performed during the same years that he was developing those views on light and perspective which we have just described. In outlining his programme of anatomical investigation Leonardo pays particular attention to 'perspective through the function of the eye' (W 19037 v; K/P 81 v).

One of the earliest of Leonardo's anatomical drawings, made about 1487, is Figure 2.20, in which the human head and brain are likened to the layers of an onion. This analogy is based on the emphasis placed by past great anatomists like Avicenna, Guy de Chauliac and Mondino on the number of layers formed by the scalp, skull and membranes around the brain and eye, which is looked upon as an extension of the brain through the optic nerves. With the exception of the representation of the frontal sinus (one of Leonardo's anatomical discoveries) the whole drawing could be seen as an illustrated reproduction of the anatomy found in any one of these works. However, it will be seen that the main subject of the drawing is that of the eye and its connections with three cavities (ventricles) in the substance of the brain. In the upper half of the page, a sagittal section of the skull and brain is shown; in the lower half, a horizontal section through these parts with the top half of the section of the head imaginatively turned back like the hinge of a lid. To the left of this is a diagram in which the frontal sinus figures prominently as a triangular figure above the orbit. The orbit contains the eyeball which is coated with the two layers of the cerebral membranes carried forward. Inside this is the round lens of the eye; the break in the membranous coats of the eyeball in front of the lens represents the pupil. Running from the eyeballs are the two optic nerves which are seen converging on the foremost cerebral ventricle. Also running into that anterior ventricle are two nerves, one from each ear, the auditory nerves. Leonardo evidently believed at this time that the anterior of the three traditional ventricles of the brain was the place where these two main senses, sight and hearing, meet and enter consciousness.

This crude anatomical drawing serves well to demonstrate Leonardo's starting point in his investigation of the physiology of vision. It represents almost complete ignorance as far as 'experience' is concerned, an ignorance it should be noted which was shared by his contemporaries. Both they and Leonardo filled the gap in knowledge with imagination. Leonardo differed from his contemporaries in that he visualised and illustrated his own imaginary picture of cerebral anatomy perspectively. His drawing of the frontal sinus proves that he had obtained some direct knowledge of the human skull.

Another drawing, Figure 2.21, of about the same date (1487) shows Leonardo changing and slightly improving on the hypothetical illustration in Figure 2.20. The drawings in Figure 2.21 are almost all devoted to the anatomy of the nerves of the eye. In the lower central part of the drawing is a small illustration of the head containing the three

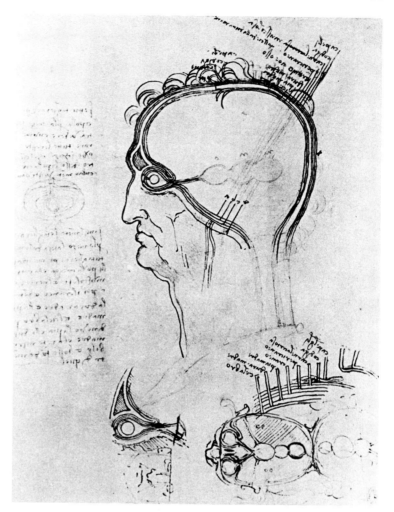

Figure 2.20.   The layers of the scalp, meninges and brain compared with an onion (W 12603 r; K/P 32 r).

In this early drawing the eye is connected to the anterior ventricle of the brain by the optic nerve. The eyeball itself appears to be composed of layers of the meninges. Its lens forms a little central round ball. The drawing is an imaginative representation of traditional anatomy realistically visualised. It shows few signs of actual observation apart from the frontal sinus depicted above the eyeball.

cerebral ventricles. The optic nerves can be seen running from the eyes at the bottom of the sketch to the anterior ventricle, as before. Alongside them are two curved lines representing the olfactory nerves of smell; these Leonardo traces to the middle cerebral ventricle. Here too the auditory nerves are made to terminate. This time Leonardo has labelled each ventricle. The anterior ventricle to which the optic nerve runs is labelled '*intelletto*' and '*imprensiva*'; the middle ventricle is labelled '*volonta*' and '*senso comune*'; and the posterior ventricle, '*memoria*'. The fact that Leonardo labels 'intellect' the place in the brain to which the nerves from the eyes run reflects the importance he attaches to vision. The term *imprensiva* is untranslatable accurately. It is

Figure 2.21. An early schema of the distribution of the optic, olfactory and auditory nerves (W 12626 r; K/P 6 r).

The optic nerves are shown going directly to the anterior ventricle, whilst the auditory nerves go to the middle ventricle. The anterior ventricle is labelled 'intellect' and '*imprensiva*'; the middle ventricle '*senso commune*' and 'will', and the posterior ventricle 'memory'.

that part of the brain in which sensory impressions are processed so that the intellect can draw understanding from them. The term was introduced by Leonardo himself. It does not appear before him, nor after him. In later stages of his progress we shall see how important his concept of the '*imprensiva*' becomes.

The '*senso comune*' corresponds to the 'common sense', a term introduced by Aristotle to indicate the place where all the senses come to a common meeting place from which they are combined to give us our unified perceptions of the outside world. Leonardo, too, uses it in this sense and later places 'judgement' with the '*senso comune*'. With this '*senso comune*' (which term I shall continue to use for this region of the brain, since 'common sense' does not reflect it adequately in English) Leonardo places '*volonta*', or 'will'. This is most important in his concept of cerebral psychophysics, for he places the origin of voluntary movement also in this middle cerebral ventricle. Here movement is linked with 'judgement'. Together with sensation and emotion these locate the soul. Both sensation and action, arising in the middle ventricle, are influenced by memory, located in the posterior ventricle, as well as intellectual impressions, located in the anterior ventricle.

The depiction of sensory and motor nerves and muscles on this page shows an accuracy and delicacy which betrays Leonardo's interest in the nervous system at this early stage of his anatomical dissections. We shall return to them later.

In another drawing of the cerebral ventricles (W 12627 r; K/P 4 r), neater and more finished than that in Figure 2.21, the nerves are similarly arranged, but the labelling of the three ventricles is different. The anterior ventricle now con-

tains the word '*imprensiva*' alone. The middle ventricle contains the words '*senso comune*', easily legible under ultraviolet light, and the posterior ventricle contains the word '*memoria*'. Leonardo is clearly changing his views about cerebral (ventricular) localisation. But one point remains constant during this period; the optic nerves go to the anterior ventricle. This is illustrated once more in a drawing which is doubly interesting in that it reflects Leonardo's concentration on perspective at the same time as he was studying the literature on the optic nerves. Figure 2.22 is a set of drawings of the human head. Three of them at the centre and upper right parts of the page show perspectival drawings of the skull and brain. From the eyeballs in each instance the optic nerves can be seen converging to the anterior of three 'ventricles' shown in perspective (middle drawing). In the other two the optic nerves converge towards the meeting place of the auditory nerves.

The common error of mistaking these views of the optic nerves for spectacles on the nose has been perpetrated by failing to detect the perspectival nature of the drawings.

Leonardo's interest in the nervous system at this time is further reflected in the performance of a vivisection experiment which involved pithing a frog. This consisted of depriving a frog of its head, heart and viscera. After which, to Leonardo's astonishment, 'the frog retains life for some hours . . . and if you prick the nerve [medulla oblongata] it suddenly twitches and dies' (W 12631 v; K/P 1 v). 'It seems', he adds, 'that here lies the foundation of motion and life' (W 12631 r; K/P 1 v). Here he is locating the soul, i.e., the *senso comune* and the centre of bodily motion (the '*volonta*') by experimental destruction of the medulla oblongata.

Leonardo's personal investigations of the anatomy of the eye and optic nerves began about 1489. They formed the central motive for his beautiful perspective demonstrations of the structure of the human skull.

In a series of anatomical studies in which he explores the skull from the front and side Leonardo exposes the orbit and its neighbouring frontal and maxillary sinuses. In each case the orbit of the eye gets special attention. In Figure 2.23, which shows the full beauty of Leonardo's method of anatomical demonstration, the whole of Leonardo's accompanying text is devoted to the eye and its orbit, and in the lower drawing the round optic foramen is especially labelled, *b*. The text to the upper drawing reads, 'I wish to lift off that part of the bone the support of the cheek which is found within the 4 lines *abcd* and to demonstrate through the exposed opening the breadth and depth of the two cavities which are hidden behind it. The eye, the instrument of vision, is hidden in the cavity above, and in that below is the humour which nourishes the roots of the teeth'.

Below the lower drawing Leonardo writes, 'The cavity of the cheek bone resembles in depth and breadth the cavity which receives the eye within itself and in capacity is very similar to it and receives veins within it through the hole *m* which descend from the brain passing through the sieve [*colatorio*, i.e., the ethmoid bone] which discharges into the nose the superfluities of the humours of the head. No other

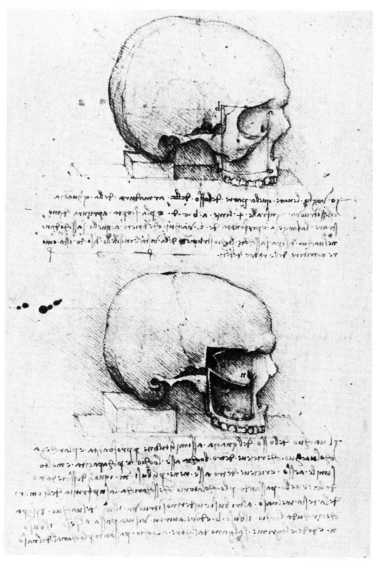

Figure 2.23. Exposure of the orbit and maxillary sinus (W 19057 v; K/P 43 v).

At the back of the orbit the optic foramen is labelled *b* (lower drawing). The maxillary sinus was one of Leonardo's anatomical discoveries.

obvious holes are found in the cavity above which surrounds the eye. The hole *b* [optic foramen] is where the visual power [*virtu visiva*] passes to the *senso comune* and the hole *n* [nasolacrimal canal] is where the tears rise from the heart to the eye, passing through the canal of the nose'.

In Figure 2.24 the orbit and maxillary sinuses are seen from the front. On the left of the drawing these are revealed by a section removing the front of the cheek, orbit and nose as far as the midline. Leonardo describes all three cavities in terms of their relation to the centre of sensation, the *senso comune*, thus, 'The cavity of the eye-socket [*cassa dell'occhio*] or orbit and the cavity of the bone that supports the cheek and that of the nose and of the mouth are of equal depth and terminate in a perpendicular line below the *senso comune*'.

Figure 2.22. Perspective drawings of the optic nerves and the cerebral ventricles (W 12603 v; K/P 32 v).

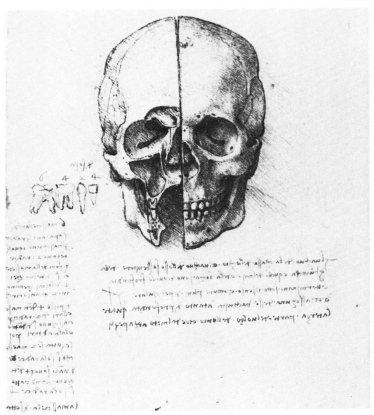

Figure 2.24. Anterior view of orbit and maxillary sinus (W 19058 v; K/P 42 v).

In the depths of the orbit the optic foramen and the orbital fissures can be seen. The different shapes of the teeth are accurately drawn, including their roots, with the exception of the premolars.

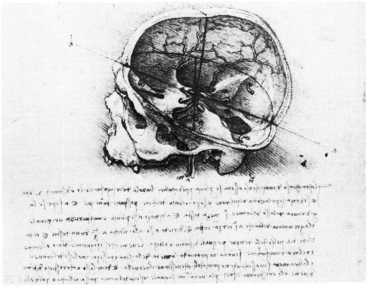

Figure 2.25. Skull divided so as to show the site of the *senso comune*, the position of which is shown by the crossing lines (W 19058 r; K/P 42 r).

The two optic nerves in front and the auditory nerves behind are shown emerging from the bones towards this site.

Two other drawings of this same group, in which the side of the skull is removed, are deliberately designed to locate the sensory centre within the skull in relation to the optic and auditory nerves approaching it, as well as other sensory nerves.

In Figure 2.25 Leonardo draws slightly oblique and vertical lines across the skull. These intersect at the site of the *senso comune*. In this figure the optic and auditory nerves can be seen emerging from their foramina in the bone, converging towards the *senso comune*. Leonardo writes, 'The confluence of all the senses has perpendicularly below it at a distance of two fingers, the uvula where food is tasted, and it lies straight above the wind-pipe and the orifice of the heart by the space of one foot. A half-head above it is the

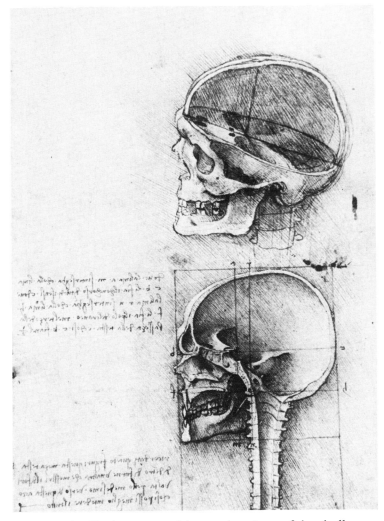

Figure 2.26. Both upper and lower drawings of the skull are designed around the intersecting lines which locate the site of the *senso comune* (W 19057 r; K/P 43 r).

In the upper drawing these lines show the distances from the shell of the skull (calvarium). In the lower drawing the skull is inscribed in a square. Intersecting perpendicular lines show the site of the *senso comune* as well as the centre of gravity, or axis around which the head moves.

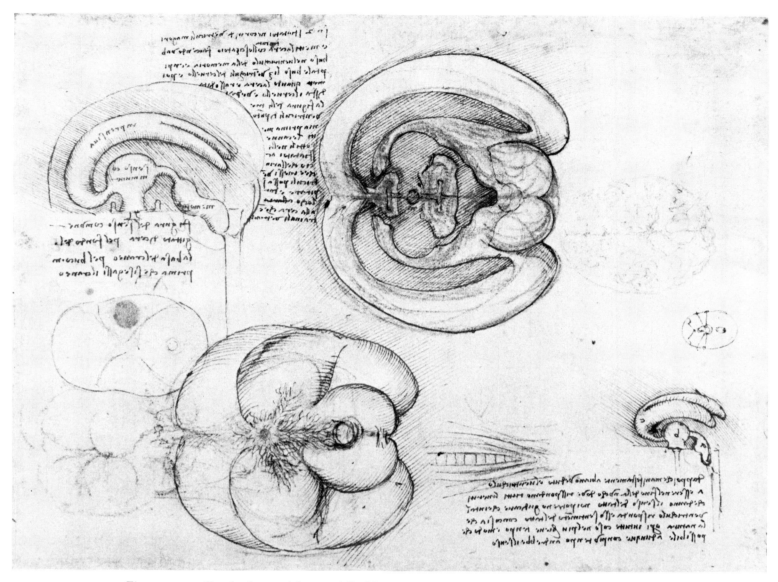

Figure 2.27. Cerebral ventricles modelled by a wax 'encephalogram' (W 19127 r; K/P 104 r).

With this discovery of the approximate shape of the cerebral ventricles Leonardo locates the *senso comune* in the third ventricle. The *imprensiva* he places in the lateral ventricles; and *memoria* in the fourth ventricle. These are labelled in the top left-hand drawing and redrawn in the lower right-hand drawing. The brain is that of an ox.

junction of the cranial bones [bregma]; and a third of a head in front of it in a horizontal line is the tear-duct of the eyes [nasolacrimal duct]; and 2/3 of a head behind it is the nape of the neck; and at the sides the 2 pulses of the temples at an equal distance and height'. The rest of his text describes (for the first time) the distribution of the anterior and middle meningeal vessels which includes clear depiction of the mastoid emissary vein.

In the upper drawing in Figure 2.26 the same intersection of horizontal and vertical lines is again to be seen locating the *senso comune*. A third perspectively drawn horizontal line between the '2 pulses at the temples' locates the *senso comune* (the third ventricle) with complete spatial accuracy in three dimensions.

The drawing below once more locates the *senso comune* where the line *am* intersects the line *cb*. The intersection of the other two lines *rn* and *hf* indicates the site of 'the axis [or fulcrum] of the cranium' in moving the head.

Thus in all these drawings Leonardo locates the *senso comune* just above the pituitary fossa, i.e., in the position of the third ventricle. Leonardo completed the logical sequence of these drawings years later. This time he used his skill as a sculptor to model the real shape of the cerebral ventricles by injecting them with wax. As a result he located the *imprensiva* in the 'anterior' ventricle, now split into two lateral horns. The *senso comune* is located in the newly found shape of the third ventricle, and memory, in the cavity of the fourth ventricle, as is shown in Figure 2.27.

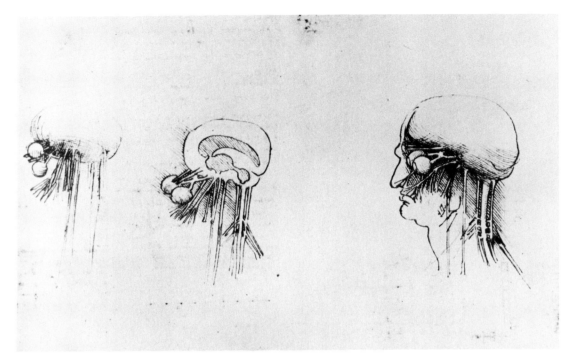

Figure 2.28. Three drawings of the optic chiasma (W 12602 r; K/P 103 r).

In all three drawings the optic nerves are seen coming from the eyeballs to cross at the optic chiasma. In the central drawing the optic tracts run to the base of the brain near the third ventricle where the *senso comune* is situated. In the right-hand drawing the surface of the brain is shown. Other cranial nerves (particularly the vagus) and the origin of the spinal cord in the neck are shown.

Figure 2.29. The base of the skull showing the optic nerves and optic chiasma (W 19052 r; K/P 55 r).

Above the optic nerve is the olfactory nerve; below it the oculomotor, the abducens, and the ophthalmic division of the trigeminal can be recognised.

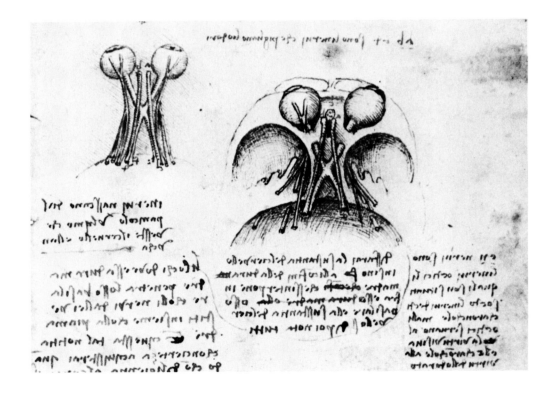

The relevance of this to his physiology of vision is demonstrated in Figure 2.28. Here in three sketches he draws the eyeballs with optic nerves emerging from them, quickly converging to form the optic chiasma where they appear to cross one another. Thence they pass to the base of the brain close to the cavity of the third ventricle where he has located the *senso comune*. The two olfactory nerves of smell are shown running to the brain just above. Below are other cranial nerves somewhat difficult to identify. Other illustrations of the optic chiasma are to be found in Figure 2.29, Figure 8.4 (p. 204) and the Weimar blatt (Figure 2.30). They represent more finished drawings of the optic and other cranial nerves. They will be described under the section on the eye (Chapter 8).

These drawings are the first known illustrations of the optic chiasma. As early as about 1500 Leonardo thought he had found the 'reason' for the optic chiasma. He approaches the problem in his typically direct way thus: 'Saw a head in two between the eyebrows in order to find out by anatomy the cause of the equal movement of the eyes. The intersection of the optic nerves is virtually confirmed as being the cause of this' (CA 305 vb). He accompanies this verdict with a simple line diagram of the optic chiasma.

HOW THE FIVE SENSES ARE THE MINISTERS OF THE SOUL

It was in 1489 when Leonardo had performed all the dissections just described (with the possible exception of the wax injection of the cerebral ventricles) that he wrote on 'How the Five Senses are the Ministers of the Soul'. 'The soul', he writes, 'apparently resides in the seat of judgement, and the judgement apparently resides in the place where all the senses meet, which is called the *senso comune;* and it is not all of it in the whole body as many have believed but it is all in this part; for if it were all in the whole and all in every part it would not have been necessary for the instruments of the senses to come together in concourse to one particular place; rather would it have sufficed for the eye to register its function of feeling sensation [*sentimento*] on its surface and not to transmit the images [*similitudine*] of the objects seen to the *senso comune* by way of the optic nerves; because the soul, for the reason already given, would comprehend them upon the surface of the eye. Similarly with the sense of hearing . . . The *senso comune* is the seat of the soul, the memory is its monitor and the *imprensiva* serves as its standard of reference . . . Thus the sense gives to the soul and not the soul to the sense; and where a sensory function of the soul is lacking the soul in such a life lacks knowledge derived from the function of this sense, as appears in the case of the dumb or one born blind' (W 19019 r; K/P 39 r).

In this passage Leonardo summarises his views on sensation in general, e.g., hearing, smell, touch. These are more appropriately described in detail in Chapters 10 and 11, on the soul and the nervous system.

As one finds so often in Leonardo's notes he develops his views elsewhere. Thus, bearing in mind his anatomical locations of the *senso comune* containing 'judgement' and the *imprensiva,* one finds enriched significance in a passage written a year or two later, probably about 1492. 'This *senso comune*', he writes, 'is that which judges of things given to it by the other senses. The ancient speculators concluded that that part of man which constitutes judgement is caused by an instrument to which the other 5 senses are referred by means of the *imprensiva,* and to this instrument they have given the name *senso comune*. And they say that this sense is situated in the centre of the head between the *imprensiva* and memory. And this name of *senso comune* is given to it solely because it is the common judge of the other 5 senses, that is, seeing, hearing, touch, taste and smell. The *senso comune* is moved through the *imprensiva* which is placed in the middle

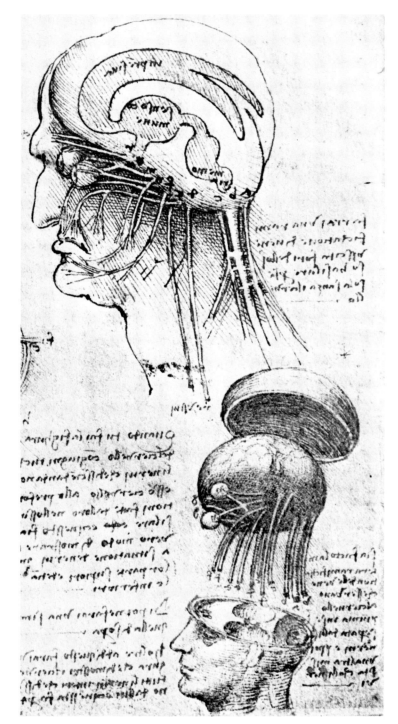

Figure 2.30. The cerebral ventricles, optic chiasma and optic tracts leading to the *senso comune* in the third ventricle. Other cranial nerves are drawn.

Below, exploded view of skull and brain (Weimar Blatt, between K/P 54 and 55).

between it and the senses. The *imprensiva* is moved through the images [*similitudine*] of objects given to it by the superficial instruments, that is, the senses, which are placed in the middle between external objects and the *imprensiva*. Similarly the senses are moved through objects. Surrounding objects send their images to the senses and the senses transfer them to the *imprensiva*. The *imprensiva* sends them to the *senso comune,* and from there they are established in the memory, and are there more or less retained according to the importance or power of the given object. That sense is most rapid in its function which is nearest to the *imprensiva* and the eye is the highest and chief of the others. Of this only will we treat, and the others we will leave in order not to make our matter too long. Experience tells us that the eye apprehends ten different natures of objects, that is, light and darkness, one being the cause of the other nine; the others are its absence [darkness], colour, and body, form and place, distance and nearness, motion and rest' (CA 90 rb). We have thus at last reached Leonardo's full interpretation of visual 'experience'.

The only further comment on this explanatory passage is to note that 'the ancient speculators' certainly did *not* refer to the *imprensiva.*

For the sake of completeness it should be emphasised here that Leonardo maintained to the end this concept of the *imprensiva* as a processing relay station for sensory nervous percussion waves. These were then passed on to the *senso comune* (and sometimes to the *memoria*) in the third and fourth ventricles. Such a relay process must be automatically controlled. In his later dissections, i.e., between 1508 and 1514, Leonardo concluded that such an instrument of automation lay in 'the worm of the brain' exerting a sphincter-like action, very much like that of the pupil in controlling the entrance of images into the eye. And whereas Leonardo located the external power of vision (*virtu visiva*) in the optic nerve head, he located the 'intrinsic or internal *virtu visiva*' (D 10 v) in the sphincter action of the choroid plexus, the 'worm of the brain', as it passes between the lateral and third ventricles − in more modern parlance, at the foramen of Monro. In a passage in which he is discussing the sphincter action of muscles in general Leonardo writes, 'On the muscle called the worm which is in one of the ventricles of the brain which lengthens and shortens itself in order to open and shut the passage of the *imprensiva* or the *senso comune* to the *memoria* (W 19117 r; K/P 115 r). And he describes 'very minute fibres of the nature and shape of the minute muscles which form the worm of the brain and of those which weave the rete mirabile' (W 19119 r; K/P 116 r).

In his investigations of the cerebral ventricles Leonardo had located the choroid plexus and noted that it passes from the lateral ventricles to the roof of the third ventricle, the site of the *senso comune,* as the tela choroidea. Here its posterior margin ends. Leonardo thus saw this choroid 'worm' as being in the position to exert a regulating sphincter action between the *imprensiva* and the *senso comune* as well as between the *senso comune* and the *memoria* located in the fourth ventricle. In this way he inserted a cybernetic or self-regulating mechanism for the sensory waves of percussion between all stages of the visual process, from the pupil to the passages from the *imprensiva* and *senso commune* back to memory.

'THE EYE, THE INSTRUMENT OF VISION'

As Leonardo emphasised, 'Surrounding objects send their images to the senses and the senses transfer them to the *imprensiva*' (CA 90 rb). Of these senses the eye is the chief; through this instrument the images of objects carried by light are transmitted to the *imprensiva.*

Having obtained some idea of Leonardo's views on the nature and properties of light on the one hand and the *imprensiva* and *senso comune* on the other, we are now in a position to examine his early investigations on the eye, the instrument of vision, as a visual bridge between the two.

As illustrated in Figure 2.20 (p. 61), Leonardo took his first ideas of the structure of the eye from the literature. Here two layers of cerebral membranes, dura mater and pia mater, surround and form the eyeball which contains a round lens. The pupil is formed by the transparent gap in the membranous coats in front of the lens. The lens with the transparent cornea in front of it appears to lie unattached, loose in some presumably homogeneous fluid. This does not correspond at all closely to the Galenic structure of the eye. It seems more representative of the simple description of the ventricles, optic nerves and eye given by Guy de Chauliac; he describes the eyes as 'made to be instruments of sight'. As such 'they are set within the bones orbital and the bones of the temple . . . and are pierced to be the way of the spirit visible'.[2] Guy also denies that the optic nerves cross in a chiasma.

Be this as it may, the two little parallel lines emerging from the round central lens to the pupil suggest extramission of visual spirit or *virtu visiva,* carried along the 'perforated' optic nerve, rather than the converging entrance of species which Leonardo later supports.

Observation and experiments on perspective, however, appear to have induced Leonardo to debate this theory of the emanation of *virtu visiva.* He writes, 'It is clearly demonstrated by experiment that visual lines and solar rays are generated from the base and terminated in the point of a pyramid, as well as from the apex of a pyramid, terminating its base' (CA 353 vb). Illustrations of this are added showing such pyramids between the eye and the object.

It was not long after this that he produced his pyramidal figure in Figure 2.3 (p. 46), showing by experiment that if the object is twice as far from the eye, it appears half as high.

By 1490 Leonardo's anatomy of the eyeball has greatly changed, as shown by the two drawings in Figure 2.31. He had previously suggested that the lens of the eye magnifies images 'like a ball of glass full of water' (CA 227 ra). On this page he combines his 'ball of glass' with the scheme of a box, representing the coats of the eye and containing a small hole (*spiracolo*) in one wall. This *spiracolo* plan of experiment we saw him using often in his analysis of light rays. The two combined together in Figure 2.31 represent the complete construction of a camera obscura. In this diagram Leonardo represents light rays approaching the *spiracolo* or pupil from broad-based pyramids at the sides, and from narrow-based pyramids, down to that narrowest of conceivable pyra-

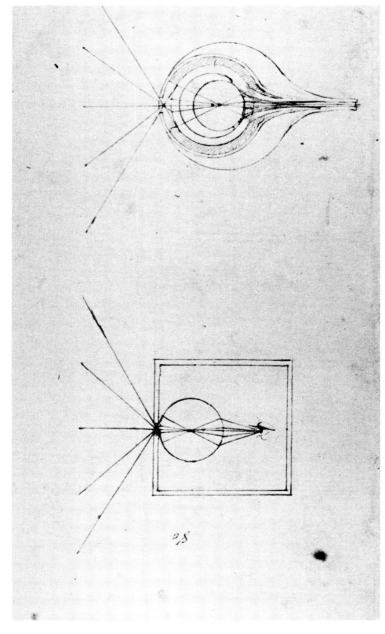

Figure 2.31. The eye (CA 337 ra).

Upper drawing: note the trajectories of light rays through the coats of the eyeball, the pupil and lens to the optic nerve.
Lower drawing: the eye is here likened to a camera obscura.
Note the two crossings of light rays at the pupil and the lens.
Leonardo represents the optic nerve by another eye.

mids, represented by the single line of the central ray. This single 'central ray' is shown passing straight through the pupil and the round 'lens' to an eye placed behind the lens in the position of the head of the optic nerve. It is the only ray that does not cross at the pupil or lens. The converging rays all cross at the pupil to enter the lens, where they cross again. After this they emerge from the lens where they are again refracted to converge on the 'sensitive' optic nerve which Leonardo places at the centre of the back of the eye-

ball. He was, of course, ignorant of the existence of the macala lutea about two millimetres away from the optic disc where central vision actually takes place.

Leonardo tries to put this schema into anatomical terms in the upper drawing in Figure 2.31. Here the central ray passes straight through pupil and lens to the centre of the optic nerve which is applied directly to the back of the lens, expanding as it were to form a restricted retina. For Leonardo the lens was suspended in surrounding 'albugineous humour'. This is represented as a clear space. It will be noticed that this albugineous humour behind the cornea appears to perform refraction before the rays enter the lens. Between it and the lens is another crescentic 'humour', possibly representing the anterior capsule of the lens or derived from the confused 'glacial humours' of Pecham and others. This does not refract the rays entering the round centrally placed lens or 'crystalline humour'. Here they cross again before making contact with the retinal optic nerve. Outside both these layers is the shaded pigmented 'uvea' interrupted by a transparent zone in front to form the pupil. Outside this in turn is a thin layer representing conjunctiva. Outermost of all is a layer expanding as it passes back towards the optic nerve, probably representing the muscles of the eye. Leonardo looks upon these two figures as so complete an exposition on the structure of the eye that it needs no further explanation, nor does he write a single word on the page. This anatomy of the eye he later modifies, but basically it remains the model upon which he builds his physiological optics.

THEORIES OF VISION: EXTRAMISSION OR INTRAMISSION

We have already noted that Leonardo's first view of vision was that it involved an emanation from the eye of visual power – a view that went back to Plato and was subsequently upheld by Euclid, Ptolemy, Galen and Roger Bacon. Only the great Arab experimental philosopher, Alhazen, supported the opposing view that visual images (*species*) enter the eye from the object.

Leonardo debated at length both points of view of the matter (CA 270 vc and rb). The debate is long and reminds one of Plato's *Dialogues* in so far as each point of view is so convincingly presented. To do Leonardo justice these pages should be quoted verbatim. The exigencies of space do not allow this, so it will be abridged – a procedure which Leonardo strongly condemned.

Leonardo begins by asserting, 'I say that the visual power [*virtu visuale*] is extended by visual rays as far as the surface of non-transparent bodies, and that the power [*virtu*] of these bodies is extended as far as the power of vision [*virtu visuale*], and that every similar body fills all the air around it with its image. Each body by itself and all together do the same, and not only do they fill it with the likeness [*similitudine*] of their shape, but also extend to it the likeness of their power'. Then under the heading 'Confutation' he cites 'those mathematicians who say that the eye does not have any spiritual power [*virtu spirituale*] extending to a distance from itself, since if it were so it would not be without great diminution in the use of the visual power [*virtu visiva*] and that though the eye were as large as the body of the earth it would necessarily be consumed in beholding the stars; for this reason they maintain that the eye receives, and does not send forth anything from itself'.

This argument he counters with examples of 'the odour of musk' which will 'permeate a thousand miles . . . without any diminution of itself'. And he asks whether the sound of a bell 'must of necessity consume the bell'. He goes on to cite the power of the magnet to attract iron. 'All these examples', says Leonardo, 'are given in order to prove how all things, or certainly many things, together with the likeness [*similitudine*] of their form send forth images [*spezie*] of their powers without any injury to themselves; and the same may happen with the power of the eye'.

Then he argues under the heading 'Contrary opinion', 'Yet if anyone wished to say that the eye, like the ear, is adapted only to receive the images [*similitudine*] of things without sending forth any power against them, they would be able to prove it with the example of the little hole

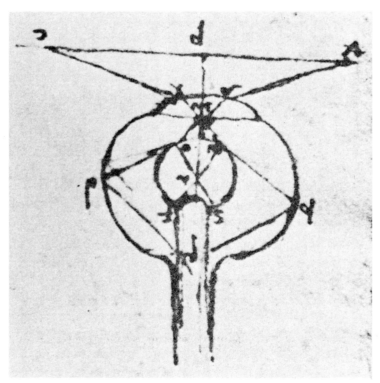

Figure 2.32. The course of central and peripheral rays through the eye (D 10 r).

Note how the central ray *bmrh* passes straight through the pupil and lens to the optic nerve. The peripheral rays from *c* and *a* are reflected from the surface of the lens to the uvea or retina, which, 'like hounds in the chase', reflects them back to the optic nerve at *h*.

[*spiracolo*] made into a window, which produces inside all the images [*similitudine*] of bodies which are objects opposite to it. Thus one could say the eye does the same'.

The refutation of this line of argument lies in the difficulty that, were the visual process 'without visual spiritual power, the images of objects would appear inverted, as they do in the dark room of the house and the eye would have to do the same, so that everything would appear there inverted'.

His 'Final proof' involves his recently acquired knowledge of the structure of the eye and its relation to the brain. 'The circle of the cornea which is in the middle of the white of the eye, is by nature suitable to apprehend objects. This has in it a point which appears black which is a perforated nerve which goes within to the intrinsic power [*virtu*] which is filled with the power of the *imprensiva* and judgement [*virtu imprensiva e guidiziale*] and which ends in the *senso comune*. Now objects placed opposite the eyes act with the rays of their images [*spezie*] in the same way as many archers who aim to shoot into the barrel of a carbine; for the one among them who finds himself in a straight line with the direction of the barrel of the carbine will be more likely to hit its bottom with his arrow. Likewise of objects opposite the eye, those will be more directly transferred to the *senso* [*comune*] which are more in line with the perforated nerve [i.e., the optic nerve].

'That water which is in the light that surrounds the black centre of the eye acts like hounds in the chase which start the quarry for the hunters to capture. So it is with the humour that is derived from the power of the *imprensiva*; it sees many things without seizing hold of them but quickly turns them thither towards the central beam which proceeds along the line to the *senso* [*comune*]; and this seizes on the images and confines such as please it within the prison of memory' (see Figure 2.32).

Leonardo's imaginative analogy between the archer shooting an arrow down the barrel of a carbine refers to the most powerful line of percussion, the 'central line of vision', which is the only line of accurate vision; the 'arrows' of 'images' shot into the aqueous or vitreous humours of the eye, lighting it up, have to be reflected by the uvea towards the central line of vision where they are picked up and transferred to the *imprensiva*; but their images are but imperfectly appreciated. He is drawing that contrast between central and peripheral vision which we now attribute to the response of the cones and rods of the retina.

Leonardo's attribution of visual power to the '*virtu imprensiva e guidiziale*' indicates that he retains the belief that vision is not just a passive process but is activated by a 'power' that flows along the optic 'nerves'. Thus 'visual power' corresponds to active nervous processes; it accepts the principle so well exemplified on our television screens, that the formation of an image which enters consciousness is processed; it involves energy or power and is not merely the imprinting of an impression onto inert wax like a seal. This power lies within the eye; it is projected outside as visual potential.

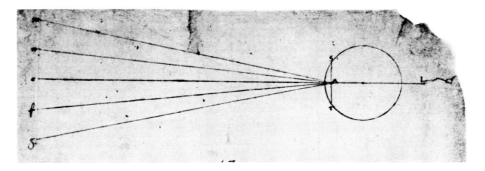

Figure 2.33. Only the central ray of light gives perfect vision (CA 85 va).

In this diagram only the central ray, *ca,* which passes through the pupil and eyeball along the line *ab* carries 'perfect' vision, as when looking at the letters of a word in a book.

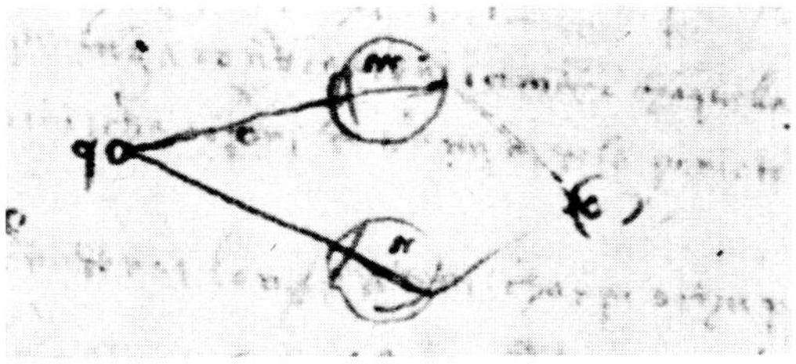

Figure 2.34. The final experiment! (CA 204 rb).

Leonardo's sketch of his experiment in which he displaces the one eyeball with his finger, so altering the position of the image of the object, *p,* which of course remains in the same position.

The reference to the 'perforated nerve' indicates Leonardo's adherence to the traditional Galenic belief that all nerves possess a central cavity along which both motor and sensory stimuli are conducted. For example, Leonardo describes the optic nerve as follows: 'The nerve which leaves the eye and goes to the brain is like the perforated nerves which with infinite branches interweave the skin of the body and through their cavities end in the *senso comune*' (BM 171 v).

'In what way the eye sees objects placed in front of it' is the title of a passage in Figure 2.33 which is accompanied by a useful illustration. 'Suppose that the ball figured above', he writes, 'is the eyeball and that the smaller part of the ball divided off by the line *st* is the cornea, and all the objects mirrored on the centre of the surface of the said cornea quickly pass on and go into the pupil passing through the crystalline humour [lens] which does not interfere in the pupil with things shown to the cornea. And the pupil having received these things from the cornea, immediately refers them and carries them to the intellect by the line *ab.* You must understand that the pupil carries nothing perfectly to the intellect or *senso comune* except when the objects presented to it from the cornea are directed along the line *ab,* such as you see is made by line *ca.* And although the lines *mnfg* are seen by the pupil, they are not taken into consideration because they are not directed along the line *ab.* And the proof is this: if the eye shown above wants to count the letters placed in front of it the eye will be obliged to turn from letter to letter because it cannot discern them unless they lie in the line *ab,* as line *ac* does. And all things seen come to the eye by the lines of a pyramid, and the point of the pyramid terminates and ends in the middle of the pupil as figured above'.

A year or two later he endorses this observation with an experiment on perspective in which he views an object through a vertical glass plane and a curved transparent plane which is 'a circular plane like the eye'. Only the central ray from the eye sees the object clearly (H 32 r).

THE 'CRUCIAL' EXPERIMENTS

What Leonardo himself describes as a crucial experiment appears on CA 204 rb and in Figure 2.34. 'The final experiment that gives a true opinion that objects come to the eye is this'. His experiment consists of displacement of one eye with his finger. 'It is clearly understood', he writes, 'how the same thing seen with two concordant eyes refer it inside to end at the same point as appears in *mnop.* But if you displace one eye with the finger you will see one seen thing converted into two. You do not touch the thing but you do touch your own eye. You do not move the thing seen but you move its image. And if the eye itself sent out visual power it could never happen that moving the eye, the source of many radiating lines, that the thing [object] should appear to be moved, because when I read, the letters seen would seem to change their places according to the changed point of view'.

Leonardo endorses this final view by a different kind of observation on the same folio, 'To prove how things come to the eye'. Here he simply states, 'If you look at the sun or other luminous body and then shut your eyes you will see it similarly inside your eye for a long space of time. This is evidence that images enter the eye' (CA 204 rb). It is also evidence of an active power or *virtu* which persists after the entering image has been cut off.

Leonardo now feels that he has disposed of the possibility that vision is performed by rays emitted from the eye to the object. He is now faced with the problems of showing how the entering images of things react on the eye. This he 'proves' by two kinds of observation or experiment. In the first group he shows how sunlight can produce pain in the eye and even destroy it. In the second group he shows how sunlight produces images in the eye.

The first group arises from what he comes to consider as his most significant discovery about the eye, the change in size of the pupil with change in illumination (Figure 2.35). He describes it thus: 'This pupil of ours increases and di-

Figure 2.35. To the left of this sketch Leonardo writes, 'Experiment on the enlargement or diminution of the pupil through the movement of the sun or other luminous body'. It is self-explanatory (I 19 v).

minishes according to the brightness or darkness of its object; and since it needs some time for it to expand and contract, it does not see all at once when emerging from the light and going into the shade, nor similarly from the shade into the light. And this thing deceived me once in painting an eye, and from that I learnt it' (Forster II 158 v).

Dilatation and contraction of the pupil were to form a very great part of Leonardo's subsequent investigations on vision. However, the significance of it as evidence of the entry of light and images into the eye was contained in the fact that bright light produces pain. 'The eye which is accustomed to darkness', he writes, 'is hurt on suddenly beholding the light, and therefore closes quickly being unable to endure the light. This is due to the fact that the pupil in order to recognise any object in the darkness to which it has become accustomed increases in size employing all its power to send to the *imprensiva* the image of things in shadow. The light, suddenly appearing, causes too great a quantity for the pupil which was in darkness; it is hurt by the brightness which supervenes, this being the exact opposite of the darkness . . . One might also say that the pain caused to the eye when in the shadow by sudden light arises from the sudden contraction of the pupil which occurs as the result of the sudden contact and friction of the sensitive parts of the eye. And if you wish to see this by experiment, observe and consider well the size of the pupil of someone who is in a dark place and then place a candle in front of him which rapidly approaches the eye; and you will see a sudden diminution of the pupil' (C 16 r). Leonardo thus explains the reason underlying his method of observing the eclipse of the sun as long ago as 26 March 1485 when he noted, 'A method of seeing the eclipse of the sun without pain to the eye. Take a piece of paper and pierce holes in it with a needle and look at the sun through these holes' (Triv 6 v). And in MS A 29 v Leonardo emphasises the converse: 'If you place or hold lenses at a certain distance from the eye and look at the sun you will lose your sight for ever'. If we ask how sunlight causes pain and damage, Leonardo answers, by percussion. 'When light percusses flat surfaces they repercuss it like the rebound of a ball' (BN 2038 14 v; A 94 v).

And percussion of light and heat may be concentrated pyramidally to produce injurious effects. Leonardo appeals to his pyramidal law below a figure of a lens concentrating sunlight to a point, saying, 'In proportion as any cut pyramid of sunlight enters into its base so many times will it be hotter than its base' (A 55 r). In Figure 2.19 (p. 59) he represents this pyramidal concentration of sunlight in the ratio of the section of pyramid *ab* into the base *fg* of a concave mirror. Later he writes of 'the sun's rays reflected from concave mirrors when their percussion will be of such brightness that the eye cannot bear it'. And on the same page, noting that the rays of the sun carry heat as well as light, he adds, 'No created thing can endure the heat of such percussion' (G 34 r).

Sunlight is for Leonardo the primary light of the universe. It is the source of daylight and therefore of the visibility of things lit up by it. Objects seen in daylight are seen only by light 'derived' from sunlight. The process of seeing the sun is thus of primary importance to him in deciding on the extramission or intramission concept of vision. In this group of observations he is decisively demonstrating that sunlight not only enters the eye but also by its pyramidally concentrated percussion can produce pain and eventually destruction of the eye. Pain and destruction are effects of percussion which Leonardo discusses at length in describing this, 'the greatest of all the powers of nature' (see Chapters 4 and 10).

However, it is only when percussion is excessive that pain and destruction are produced by light. Light from ordinary illuminated sources either during daylight or at night has its brightness and its strength of percussion suitably regulated by the size of the pupil to produce the conditions necessary for vision. Thus vision is brought about by a particular pattern of movement induced by its percussion.

THE ESSENCE OF SENSATION

This pattern is succinctly described by Leonardo in terms of the analogy of a ringing bell. 'The blow on the bell leaves after itself its similitude impressed just as the sun does in the eye' (A 22 v). This assertion he joins to that in *Codex Atlanticus* 204 ra, 'If you look at the sun or some other luminous body and then shut your eyes you will see it similarly inside your eye for a long space of time'. The after-image persisting in the eye is analogous to the persistent ringing of a bell after it is percussed.

This persistence of images he stresses time and again: 'That eye will preserve within itself more images of the sun which is looked upon a greater number of times by this sun' (CA 369 rc). 'A dark place will appear sown with spots of light; and a bright place with dark round spots when seen by the eye which has recently gazed many times and rapidly at the body of the sun' (CA 369 vd). Thus Leonardo describes positive and negative after-images.

This persistence of the image also applies to the observation of moving objects. 'Every body that moves rapidly appears to tinge its path with the likeness [*similitudine*] of its colour. This proposition is seen by experience; because lightning moves among dark clouds the speed of its serpentine flight makes its whole course resemble a luminous snake. Similarly if you move a lighted brand in a circular movement its whole course will appear like a ring of flame . . . because the *imprensiva* is quicker than the judgement'.

He dilates on this phenomenon on one of those pages which is crammed with writing without illustration, showing that Leonardo sometimes preferred expressing himself in words rather than figures, particularly when he was reaching towards a generalisation of his observations such as is contained in the following passage: 'Every impression is preserved for some time in its sensitive [*sensibile*] object and that will be more preserved in its object which was more powerful, and less so with less power. In this connection I call sensitive [*sensibile*] that object which through some impression is moved from that which was at first an insensitive [*insensibile*] object. [An insensible object] is one which though moved from its first state does not preserve in itself any impression of the thing that moved it. A sensitive impression [*impressione sensibile*] is that of a blow received by sounding a thing like a bell and similar things, and like the voice in the ear . . .

'So too the brightness of the sun or any other luminous body remains in the eye for some time after it has been seen; and the motion of a single firebrand whirled rapidly in a circle causes this circle to appear like a continuous and equal flame. Drops of rain water appear like continuous threads descending from their clouds; and so through this, one is shown how the eye preserves the impressions of the moving things that are seen by it. Insensitive objects [*li obbietti insensibili*] which do not preserve the impressions of things opposite to them are mirrors and any glittering thing which as soon as the thing impressed on it is removed becomes at once entirely deprived of that impression.

'Therefore we will conclude that it is the movement of the mover impressed on the body moved by it which moves this body through the medium whence it is moved. Further, amongst the preserved impressions in bodies one can put the wave and eddies of water' (CA 360 ra). Leonardo

continues the passage citing the movements of wind, of the voice, and the oscillations of a knife with its blade stuck into a table, as further examples of sensitive impressions (*impressioni sensibili*).

Clearly the whole passage has great significance for Leonardo's physiology of sensation, and vision in particular, since the *imprensiva* is the example within the brain of an organ which by its designated function takes and preserves the 'impressions' conveyed to it from the environment through the senses of the body. This it does by a continuous oscillatory movement like that of a percussed bell. As such it performs within the brain a function analogous to that of water percussed by a stone, or air percussed by the sound of a voice; that is to say, it initiates waves which are further relayed to the *senso comune* and the 'judgement', located in the third ventricle. Here lies what Leonardo calls 'the intrinsic power [of vision] filled by the power of the *imprensiva* and judgement, which take to the *senso comune* images of objects placed opposite the eye' (CA 270 rb).

Thus Leonardo reaches the conclusion that the pattern of movement which underlies the sensation of vision is that of oscillation or transverse wave motion. It is because these oscillations and waves persist during a finite period of time that 'The eye preserves in itself the images of luminous things which are represented there' (CA 204 v), and that 'The eye preserves in itself luminous images for some time' (Triv 43 r). Also, 'Those images are preserved more in the eye which arise from a more luminous body' (CA 250 ra). And conversely, though both light and shade percuss the eye, 'The eye will preserve in itself the images of luminous objects more than shadowed objects. The reason for this is that the eye in itself is extremely dark and since two things that are alike cannot be distinguished, therefore the night and other dark objects cannot be preserved or recognised by the eye. Light is entirely contrary and gives more distinctness, and does more damage, and differs from the usual darkness of the eye, hence it leaves the impression of its image' (C 7 v).

THE PUPIL

We have seen how Leonardo became aware of the dilatation and contraction of the pupil with its exposure to light. He relates how he first observed this whilst making a portrait, about 1490. It was not long before he asked the question, how are these movements of the pupil performed? His answer is found in *Codex Atlanticus* 125 ra: 'I find by experiment that the black, or nearly black, crinkled rough [*crespo over rasposo*] colour which appears round the pupil serves no other function than to increase or diminish the size of the pupil; to increase it when the eye is looking at a dark place, to diminish it when it is looking at a light or at a luminous thing. And you will make the experiment by holding a light close to an eye when it is looking into the darkness, and when turning the eye to this light. And you will be satisfied with this experiment' (see Figure 2.35, p. 72). This experiment with the approaching light producing contraction of the pupil is repeated on several occasions under variable circumstances. Since the principle of them all is the same, they will not be further described. The above passage is remarkable for Leonardo's detection of the 'crinkled' structure of the radial folds of the pupillary ruff. On one of the later occasions of repetition of this experiment Leonardo describes this contraction of the pupil to bright light as 'like a purse, and its closure is proportional to the brightness of the light' (F 39 v). It is almost as if he saw the purse-string structure of the sphincter of the pupil.

It will have been noticed that when Leonardo appreciated that the pupil changed greatly in size he immediately challenged the geometrical pyramidal figure of the 'perspectivists', who looked upon the pupil as the point of a perspectival pyramid. Leonardo, of course, appreciated this and struggled with the fact that he now had to regard the pupil of the eye not as a point but as the variable base of a secondary pyramid or cone of light leading into the eye, defending it from pain and damage as well as diminishing the aperture in order to allow central rays into the eyeball and to exclude the use of 'false' or oblique rays for clear vision. These attempts he returns to after fundamental revision and expansion of his mathematical ideas. Unfortunately this recognition of the change in pupil size led Leonardo to assert that 'the larger the pupil the larger the figure of the object appears. This is proved because the whole pupil is prepared with one and the same power of seeing' (D 7 r). And the 'visual power [*virtu visiva*] is infused with equal power in the whole pupil of the eye, whence the function of vision is all throughout the whole pupil and all in each part of it' (CA 237 ra). Thus, he erroneously knit together pupil size with his pyramidal law of perspective.

THE CORNEA

Leonardo was aware that the cornea, being a dense medium, must refract rays of light entering from the air. This he may well have learned from the works of Ptolemy. Leonardo first shows his experimental approach to this problem on *Forster* II 69 v, that is, about 1495. The experiment reminds one of his example of archers shooting arrows into the eye, described in *Codex Atlanticus* 270 rb. Nature, in Leonardo's view, applies the same laws for the movement of all objects, whether projectiles or light. Thus it is logical for him to apply to arrows the same rules as he applies to light. In this experiment he does just this: 'I believe that an arrow shot obliquely into water is bent, as is the visual line. And of this I will make a test [*pruova*] by fixing the bow and shooting at a frame upon which a sheet of paper is stretched, this paper being covered by water. And after you have shot at the paper, without moving the bow or sheet of paper, take away the water and you will reveal the arrow. And by means of a fine thread you will be able to discern whether the shaft of the crossbow and the centre of the hole made in the paper and the length of the arrow are in the same line or

not. And in this way you will make your general rule' (Forster II 69 v).

It is interesting that amongst the many indications of the influence of Alhazen on Leonardo's investigations of light and vision there is a demonstration of the law of the reflection of light by shooting an arrow at a mirror and observing the angle of its rebound.

Leonardo saw the cornea as a transparent membrane: 'The circle of the cornea', he writes, 'which appears in the middle of the white of the eye is of such a nature as to apprehend objects' (CA 270 rb). This 'apprehensive' function of the cornea is described and illustrated in Figure 2.36. Here he seems to be arguing against too 'mathematical' a concept of the eye, saying, 'Some of our mathematicians want to divide the eyeball into four parts as appears in *Kief*,

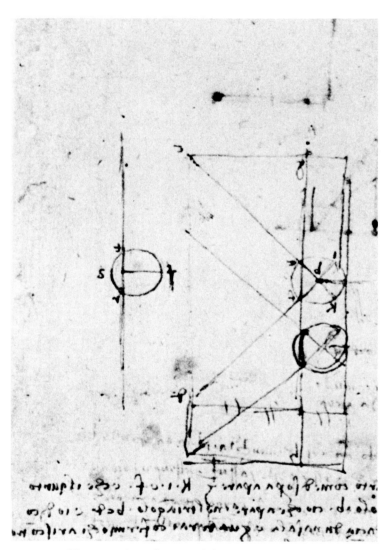

Figure 2.36.   The visual fields (CA 204 rb).

Here the visual field is shown as limited to the walls *ghcd*, i.e., 180 degrees because of the way the eyeball is divided into four equal parts. In the lower right-hand drawing there is a suggestion of a prominent cornea.

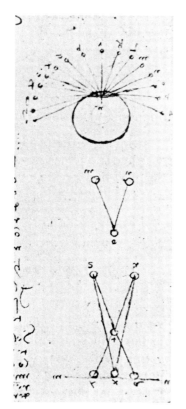

Figure 2.37. The visual field shown to be greater than 180 degrees (D 8 v).

In the lowest drawing physiological diplopia is demonstrated. The two eyes *s* and *r* if focused on object *x* see object *t* as two images at *y* and *v* on the wall *mn*.

of the humours. This, his first attempt to define the binocular visual field, is too limited. Two of the three drawings of the eyeball are strictly circular. The third (that on the lower right of the figure) suggests that it has been redrawn with the curvature of the cornea more prominent than the rest of the eyeball. Be this as it may, Leonardo in later years (about 1508) corrects the extent of the visual field, showing it to extend beyond 180 degrees, by virtue of the prominent curvature of the cornea (Figures 2.37 and 2.38).

Leonardo did, however, fully appreciate that when light enters the eye it changes its path, i.e., is refracted (Figure 2.38). 'If the eye looks in front of itself by a straight line . . . it is necessary that the line should be bent as it is changed from the rarity of the air to the density of the humours of the eye' (CA 222 ra). Later he makes experiments in an attempt to measure the angle of this change (F 30 r and Figure 2.39). And Leonardo begins to appreciate the important significance of refraction in relation to perspective (BM 220 r), pointing out here that 'the eye sees the object bigger than is given by the painter's rules of perspective' (D 2 r).

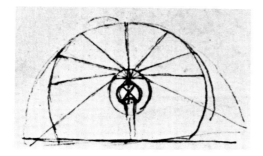

Figure 2.38. Refraction of light through the cornea (W 19152 r; K/P 118 rb).

Note how the prominence of the curvature of the cornea enables the visual field to extend backwards more than 180 degrees.

and that quarter, that is *bef,* is full of crystalline humour which refers to the angle *b* that which appears in triangle *bed* [*sic* for *f*]. I say that if you support your back against the middle of one side of a room and you look with fixed eyes at the middle of the opposite wall, that in the 3 sides, that is, left, right and opposite, there will be no movement seen except that which appears in *ghcd*. And this is a sign that the crystalline humour exists as [shown] at *tsr*'.

Leonardo's persistent ambiguity about terminology in relation to the parts of the eye leaves it possible that he is here referring to either the cornea or aqueous humour. Later, however, he fails to distinguish such a differentiation

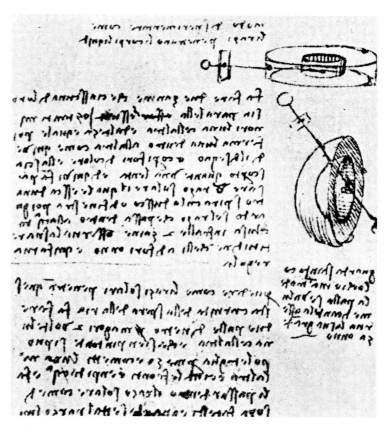

Figure 2.39. 'A method of experimenting as to how rays of light penetrate liquid bodies' (F 33 v).

This is the heading of these experiments on the refraction of light. The 'ray' is obtained by letting light through a pinhole orifice in a piece of paper. It then passes through a glass containing liquid within it, such as that contained in the eye.

THE LENS

We have previously described Leonardo's view of the lens as the site of a second crossing of the rays entering through the pupil in order to produce an upright image on the optic nerve at the back. His argument for this case is carefully worked out. He rejects the view of Galen and Alhazen that the lens is the sensitive organ whence arise the visual impulses on the following grounds: 'If the visual power [*virtu visiva*] is caused at the centre of the eye all things seen will appear of equal size because every centre is indivisible and every image by itself and many together will all be brought to this said point . . . and will appear of equal size' (CA 222 ra). He therefore refers the 'sensitive' visual power to the expanded end of the optic nerve. He accepts that the rays passing through the round lens will meet at its central point but adds, 'It is necessary that the images should make an intersection and should be turned upside down in their contact with the *imprensiva*'. This confirms his view that there is a previous intersection in the pupil in front of the spherical body (lens) which first turns the images upside down whilst the second intersection of the spherical body returns them to the upright position, 'and the *imprensiva* sees them upright'. 'If you place a ball of glass full of water in front of the eye all the images of objects that pass through it will appear upside down; and if you place 2 [glass balls full of water] one behind the other, the images of the first will appear to the eye to be re-erected into their natural direction' (CA 222 ra).

Thus Leonardo shifts the visual *virtu* from the lens to the spreading optic nerve.

THROUGH THE OPTIC NERVE TO THE *IMPRENSIVA*

The passage of images along the optic nerve to the *imprensiva* gives Leonardo the opportunity to relate his principles of perspective with vision. 'Every bodily form', he writes, 'as far as concerns the function of the eye is divided into 3 parts, that is body, shape and colour. The bodily image extends further from its source of origin than its colour or its shape; colour also extends further than shape. But this rule does not apply to luminous bodies. The above proposition is very well demonstrated and confirmed by experience, for if you see a man close by you will be able to recognise the qualities of his body, his shape and his colour likewise. But if he goes some distance away from you you will not be able to recognise who he is because his shape will lack its quality of character; and if he goes still further away you will not be able to discern his colour but he will merely appear as a dark body, and further away still he will appear a very small dark round body. He will appear round because distance diminishes the particular parts so much that they do not appear except as the greater mass. The reason for this is as follows. We know very well that all the images of objects enter the *imprensiva* through a little hole in the eye, therefore if the whole horizon *ad* enters through a similar

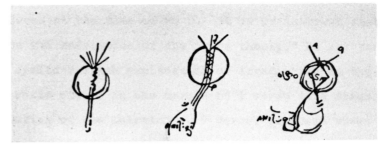

Figure 2.40. Sketches of possible routes of light through the eyeball along the optic nerve to the *imprensiva* (labelled in the central and right-hand drawings) (BM 171 v).

little hole [*spiracolo*] and the object *bc* is a very small part of that horizon it will have to occupy a very small part. And since luminous bodies have more power in darkness than any other body, it is necessary that since the little hole of vision is very dark, as is the nature of all holes, that the images of distant objects mix with the great light of the sky, and if indeed they appear at all, they will seem dark and black as does every small body seen in the brightness of the air' (BN 2038 12 v; A 92 v).

A similar physiology of vision is expressed in the *Treatise on Painting* when Leonardo explains how a man's face becomes 'obscure' at a distance. It occupies so small a part of the pupil that 'having to pass from the surface to the *imprensiva* through a dark medium, that is to say an empty nerve which seems dark, those images which are not of powerful colour are tinged with that darkness on their way to join the *imprensiva* and so they appear dark' (CU 146 r).

The little sketches drawn in Figure 2.40 in which Leonardo contrasts the straight route taken by the central line of vision with the reflected zig-zags taken by obliquely 'false' or peripheral lines of vision along the optic nerve to the *imprensiva* explain what he means by 'the line going to the *imprensiva* can be straight or oblique'. Both schemes appear in the figures. The mechanism for regulation of the visual percussion waves from the *imprensiva* to the *senso comune*, described earlier on p. 73, was suggested later after he had injected the cerebral ventricles with wax.

SUMMARY AND CONCLUSION

This abbreviated outline describes Leonardo's elementary principles of vision and perspective. It must be emphasised that so far only monocular vision has been considered, since perspective is described in terms of monocular vision. It was on this basis that Leonardo reached his first comprehension of his own meaning of terms such as *vision, experience, judgement* and *memory*. From observation of the object to the passage of its 'species' through the eye to the *senso comune*, he had constructed a consistent picture of the path of light rays by movement and percussion which connect the object with the appearance of its image in consciousness. The whole process was subject to his perspectival pyramidal law. At the *senso comune* or 'memory' this whole movement comes to rest. Here the conscious image, 'bridled by the imagination' and 'monitored by memory' is experienced and judged by the soul located in the *senso comune*. From here, too, the 'will', producing the energy of all the movements of the body, arises. In this way the sensory and motor aspects of the nervous system are linked.

This whole scheme of the eye as 'the window of the soul' is geometrical in nature, thus conveying for Leonardo a rational, mathematical relationship between the 'real' outside object and the subjective image seen. This 'reality' was

what he sought for as the science on which to base his art of painting. This is how 'the observer's mind must transform itself into nature's mind (TP I 55). And this is how 'Experience interprets between formative nature and the human race' (CA 86 ra).

Leonardo constructed this physiology of vision from the physics of his 'four powers' of nature which control all the movements of the 'four elements' in the macrocosm of the world *and* the microcosm of man. We will therefore turn now to his analysis of these 'elements' and 'powers'.

Appendix

An experiment to show how Leonardo investigated the effects on linear perspective of the distance between the eye and the observed object.

LEONARDO'S EXPERIMENT

In Figure 2.3 (p. 46) Leonardo gives a detailed account of an experiment on perspective, supplemented by a clear drawing. Since this folio dates from about 1487–1490, it provides a chronological base-line with which to compare other experiments on linear perspective.

Before marrying the text with the figure, however, errors of transcription and translation which hitherto have made it impossible to appreciate its significance have to be corrected.

It will be seen from the figure that Leonardo has written an illegible letter against the distant eye. Richter in his transcription of 1883 (101) renders the phrase referring to this labelled eye as follows: '*quanto dc entra og tanto mp entra in hp*'. A glance at the diagram reveals that this makes no sense. In the third edition (101) *og* has been changed to *op*, probably now taking Leonardo's mark near the eye as a *p*. Unfortunately Leonardo has already used the letter *p* in labelling the object observed *mp*. In other diagrams Leonardo often uses the same letter twice and there can be little doubt that he has done so here, thus creating the original confusion felt by Richter. Accepting this letter as *p*, it may be called for clarity's sake *p₂*. The passage now reads, 'If *an* goes 3 times into *fb*, *mp* will do the same into *pg*. Then go backwards so far that *cd* goes twice into *an*, and *pg* will equal *gh*. And *mp* will go into *hp* as often as *dc* into *op₂*'. This statement can be tested experimentally. It will be noticed that Leonardo does not measure the object *mp* or its images on the vertical glass pane. His experimental proportions can therefore be tested against actual measurements.

DEMONSTRATION OF LEONARDO'S EXPERIMENT
(See Figure 2.4, p. 47)

On a vertical glass pane of 24 inches square an arbitrary image of an object (*mp*) was taken, measuring 8 inches (*an*), the eye being placed 24 inches (1 *braccio*) from the glass pane (*fb*) so that '*an* goes 3 times into *fb*'. The height of an object

on a pole (*mp*) 5 *braccia* (10 feet) distant was then measured as viewed through the vertical glass pane. This measurement, 38.5 inches, was later shown to be in error by 1.5 inches, the correct measurement of the object being 40 inches – a good fit considering necessarily approximate units of measurement.

Using Leonardo's statement, one finds that if *an* (8 inches) goes three times into *fb* (24 inches), *mp* (38.5 inches) goes 3.12 times into *pg* (120 inches) or 5 *braccia*. This confirms Leonardo's statement, allowing for experimental error.

In the second part of the experiment Leonardo withdraws the observer's eye, keeping the vertical glass pane at a distance of 1 *braccio* until the projection of the image of *mp* on the vertical glass pane measures 4 inches. To quote: 'Go backwards so far that *cd* [4 inches] goes twice into *an* [8 inches]'. It is now found that the eye was 10 *braccia* (20 feet) from the object (*mp*); i.e., '*pg* equals *gh*', as stated by Leonardo.

The third part of the experiment states that '*mp* [38.5 inches] will go into *hp* [240 inches] as often as *dc* [4 inches] goes into *op₂* [24 inches]'. Here experimental error was the same as in the first part of the experiment, i.e., between the projected and actual measurements of the object *mp*. This error of 3.7% is thus accounted for.

COMMENT

Thus the experiment described in detail by Leonardo has been repeated, confirming his results at each stage. The details he gives strongly confirm that he devised and carried out the whole experimental procedure himself, achieving valid results. It would appear that this is one of the first of similar experiments from which Leonardo concluded that observations on perspective (and also vision) were to be based on 'pyramids', the linear height of the object (represented by the base of the pyramid) being inversely proportional to its distance from the eye.

That his proportion of base to perpendicular distance to the apex of the 'pyramid' was reached by experiment is further supported by the fact that in earlier manuscripts Leonardo did *not* find this relationship. On MS B 17 r (circa 1487) Leonardo specifically relates the length of the base of a 'pyramid' to the length of its *side,* not its perpendicular distance from the apex. From this circumstance one is justified in suggesting that the reason for the meticulous description of this present experiment lies in his awareness that it refutes his previous view, and at the same time opens up the existence of that 'central line', so important to him in his later concept of the 'vanishing point' and physiology of vision.

Leonardo describes a variant of this Experiment 1; he writes, 'Linear perspective deals with the action of the lines of sight proving [or testing] by measurement [*a provare per misura*] how much smaller a second object is than the first . . . I find by experience that if a second object is as far beyond the first as the first is from the eye, although they are of the same size, the second will seem half the size of the first' (BN 2038 23 n; A 103 r).

In demonstrating this variant one soon appreciates that the only difference between this and that already demonstrated is that the object is moved away an equal distance from the stationary eye instead of the eye and vertical glass pane away from the stationary object. Thus the ultimate spatial relationship between the three variables, eye, vertical glass pane and object, are the same. These results thus again confirmed Leonardo's statement that 'a second object as far beyond the first as the first is from the eye seems half the size of the first'.

So far for this particular set of spatial relationships Leonardo has demonstrated that the linear height of the observed object is proportional to the distance of the object from the eye when this distance is doubled. In MS A 8 v he extends his experiments under the heading, 'On diminution at different distances'.

REFERENCES

1. Heydenreich, Ludwig H. 'Introduction', in Leonardo da Vinci, *Treatise on Painting (Codex Urbinas Latinas 1270),* trans. and ed. by A. Philip McMahon. Princeton, N.J., Princeton University Press, 1956. Vol. 1, p. xxx.
2. Guido de Cauliaco. *Guydos Questions Newly Corrected.* London, Thomas East, 1579. Reproduced by Theatrum Orbis Terraram Ltd., Amsterdam, and Da Capo Press, New York, 1968, p. 15 v.

Chapter 3

The Elements of Leonardo's World

It is not easy to put ourselves into Leonardo's world. The accepted medieval view of the physical world into which he was born was very different from our own twentieth-century view of it. To all intents and purposes Leonardo found himself in a world composed of the ideas of Aristotle and Plato, modified by later Moslem and Christian philosophers. Aristotle's was the common-sense point of view that the world and the universe were largely what our senses told us about them. The whole was built up from a basic substance which took its form and shapes according to 'natural qualities', those of earth, air, fire and water. These 'elements' were arranged in spheres with the earth at the centre surrounded by water, air and fire concentrically. Outside this were spheres of planets; beyond this, the sphere of the stars; and beyond that, nothingness. Solid, liquid, gaseous and fiery substances occupied their own respective spheres according to their 'natures' or 'qualities', i.e., dry, moist, cold and hot.

Movement 'upwards' meant away from the centre of the earth; movement 'downwards' was towards its centre. When one element was displaced into another it 'desired' to return to its own sphere by 'natural' as opposed to 'violent' movement or force.

The moon divided this universe into 'terrestrial' and 'heavenly' or 'celestial' parts. Below the moon 'natural movement' was in a straight line; above the moon movement was circular, eternal and not subject to the 'terrestrial' laws of nature. For Plato the stars and planets were spiritual entities exerting influence on terrestrial events by virtue of powers derived from their divine creator.

In the ninth century John Scot, or Erigena, propounded the view that all nature is the manifestation of God; God alone, the creator of all, has true being. The concept is still potent in the twentieth century. Erigena blended neo-Platonic ideas with Christian concepts, incorporating also some aspects of Aristotelian thought. Thus, true religion became philosophy and true philosophy, religion. Theology came to be crowned queen of the sciences. However, with the later scholars Platonism and Aristotelianism crystallised into conflicting schools. For the Platonists only ideas were real; the world of the senses, an illusory shadow of reality. For the Aristotelians the senses provided the real-

ities, classification and general ideas being stages of abstraction towards mental concepts.

This debate was very real to Leonardo. It came down to him through the works of a number of medieval scholars, whom he names in his manuscripts. Plato's *Timaeus* was well known to him, being one of the few works of the ancient Greeks to remain alive throughout the Dark and Middle Ages. Leonardo possessed some of Aristotle's works himself, as he notes in his library list.

Like any good scientist, Leonardo 'read the literature' as far as he could. He was limited by the fact that most of the works he wanted were not printed since this invention dates from about the time of his birth in 1452. He was limited also by his ignorance of Latin, an obstacle which, we know from his manuscripts, he made great efforts to overcome.

The influence of the medieval outlook is therefore inevitably to be found in Leonardo's work. For example, it was a feature of medieval thought that the universe was composed of a fusion of natural, moral and spiritual events constituting ultimate reality. This may indeed be true, but the whole analytical trend of scientific work in subsequent centuries has been to divide existence into different fields of investigation. Leonardo habitually fused them. This is particularly evident in his 'fables' and 'prophecies'.

One of the clearest examples of Leonardo's medieval outlook, and one that greatly influenced the pattern of his researches, was the identification of the macrocosm of the Earth with the microcosm of Man. This idea, which he might well have found in Plato's *Timaeus,* has much deeper roots. It is to be found amongst the Pythagoreans, who in turn may have derived it from the ancient Egyptian gods. It appears, too, apparently quite independently as a potent concept in ancient Chinese thought. We shall try to show how Leonardo integrated some medieval views of the macrocosm into his own picture of the world, particularly that of Man.

Leonardo gives his own neat summary of this ancient doctrine thus, 'Man has been called by the ancients a lesser world and indeed the term is rightly applied, seeing that if man is compounded of earth, water, air and fire, this body of the earth is the same. As man has within himself bones as a support and framework for his flesh so the world has the rocks which are the supports of the earth. And as man has within him a pool of blood wherein the lungs as he breathes increase and decrease, so the body of the earth has its ocean which also increases and decreases every six hours with the breathing of the world. As from the said pool of blood proceed the veins which spread their branches through the human body, in just the same manner the ocean fills the body of the earth with an infinite number of veins of water. In this body of the earth, however, there is lacking the sinews, and these are absent because sinews are created for the purpose of movement, and as the world is perpetually stable no movement takes place, and in the absence of any movement sinews are not necessary; but in all other things man and the world are very similar' (A 55 v).

Two characteristic points in Leonardo's version of the analogy are worth noting. His comparision between the world and man is framed entirely in mechanical terms. He ignores any astrological influence from the stars in either the greater or lesser world. On the contrary, time and again he uses such phrases as 'This earth is the world machine' (CA 252 rb), and he refers to water as 'the vital humour of the terrestrial machine'.

Leonardo's emphasis on mechanism must not, however, be allowed to mislead one into believing that he failed to acknowledge the existence of 'soul' or 'spirit' as active presences in both greater and lesser worlds. Indeed, he uses these terms to describe the four 'powers' which make both the greater and lesser machines work.

WHAT ARE ELEMENTS?

Though Leonardo accepted the four elements earth, air, fire and water as the basic materials upon which he built up his concepts of the organisation of the world and man, he had his own views as to their interrelationships. Most of these views will emerge when considering the four powers of movement, weight, force and percussion. But some salient points about the elements need describing first to render these 'powers' intelligible.

Leonardo early rejected the alchemist's views of the elements. He writes about 1490, 'The false interpreters of nature declare that quicksilver is the common seed of all metals, not remembering that nature varies the seed according to the diversity of things she desires to produce in the world' (CA 76 va). He persists in this attitude towards alchemists. In 1508 he elaborates it (somewhat less unsympathetically) in words which still apply to the twentieth century. Man, he calls, 'the greatest instrument of nature . . . Because nature is concerned only in the production of simple things but man from these simple things produces an infinity of compounds . . . And of this the alchemists of old will bear witness, who never by chance or planned experiment happened to create the smallest thing which can be created by nature. And this generation deserves infinite praises for the usefulness of the things which they have discovered for the use of men, and they would deserve them more if they had not been the inventors of noxious things

like poisons and other similar things which destroy life or mind from which they are not exempt. For by much study and effort they are seeking to create, not the least noble but the most excellent production of nature, that is, gold'. Having attributed this desire to 'stupid avarice', Leonardo declares its futility. 'There is no quick-silver, no sulphur of any kind, no fire nor other heat than that of nature, vivifier of our world' (W 19045 v; K/P 50 v).

Such emphatic rejection of alchemy did not, however, prevent Leonardo himself practising his own chemistry. Curiously enough, one of his earliest recorded examples of practical chemistry concerned the production of 'salts' by the distillation of human faeces. The experiment is described thus, 'Salt may be made from human stool, burnt and calcined and made into lees and dried slowly by fire. And all stools produce salt in a similar way, and these salts when distilled are very penetrating' (W 12351 v).

Later, in discussing how smoke takes up with it heavy earthy particles, Leonardo draws an analogy with the sublimation of mercury: 'This is seen in distilling quick-silver in an alembic; you will see that when this very heavy quick-silver is mixed with the heat of the fire it is lifted up in a smoke, falling back into the second receptacle in its first natural state' (A 57 r).

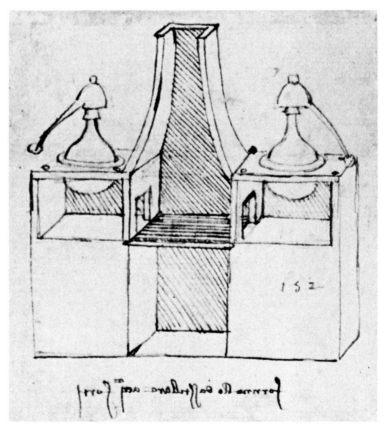

Figure 3.1. 'Furnace for distilling acqua forte [nitric acid]', writes Leonardo beneath this drawing (CA 335 rb).

The chief defect of this apparatus is its lack of a satisfactory method of cooling the distillate.

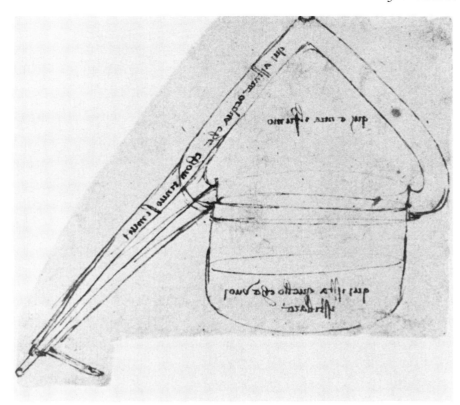

Figure 3.2. Leonardo's continuously cooled still-head.

The head of the still, as compared with that in Figure 3.1, as well as the tube leading to the receiver, is cooled by a continuous stream of cold water which surrounds the head and neck of the tube and flows out through the hole at the top. Within this involucrum Leonardo writes, 'here there is water which is continuously changed' (CA 400 vc).

It is not surprising to learn that in order to perform such chemical experiments Leonardo invented new chemical apparatus. For example, he produced new designs for several types of chemical furnaces and various modifications of the traditional Arabian type of distillation apparatus[1] (see Figure 3.1). One of these, Figure 3.2, achieved far more efficient condensation of the distilled vapour by a jacket containing continuously circulating cold water which extended along the neck of the alembic to the 'second receptacle'.

Leonardo prepared nitric acid and copper nitrate and red ferrous oxide (Forster III 59 v). Moreover, he noted that a copper salt solution when soaked up by felt (which he used as filter paper) burnt with a green flame (B 51 r). A number of 'recipes' are to be found for the manufacture of gunpowder and Greek Fire, as well as medicinal substances. Much of this chemical work was carried out as part of his perennial search for new pigments, varnishes and gums. He even invented an 'extrusion' apparatus for the production of plastic materials in threads which he called 'vermicelli' (K 116 r).[2]

Leonardo's work in chemistry has been masked by his forthright condemnation of alchemy. Many people believe that *chemistry* and *alchemy* were synonymous terms at the end of the fifteenth century. Leonardo himself made the distinction clear. The line between the two was drawn by both motive and belief or disbelief in the triad of mercury, sulphur and salt, as basic elements, rather than in chemical procedure. He expresses his views on the accomplishments and shortcomings of alchemy (and chemistry) thus, 'alchemy which consists of the manipulation of the simple products of nature, which function cannot be exercised by nature herself because in her there are no organic instruments with which she might be able to do that which man performs with his hands, which when so employed have made glass, etc.' (W 19048 v; K/P 49 v). In Leonardo's view, since the human body contained no such crude apparatus as the alchemists used, i.e., such 'instruments' as furnaces and alembics, similar chemical processes could not be produced in the human body. Therefore, since he was not acquainted with catalysts or enzymes, he concluded there could be no biochemistry.

He did not, however, feel the same about the mechanical structure of the body; he certainly did believe in the existence of biomechanics. For this reason he confined most of his investigations of both the macrocosm and microcosm of man to the physical changes occurring in the four elements of earth, air, fire and water, which had been traditionally accepted by all (except the alchemists) since the days of Empedocles. 'Man', said Leonardo, 'is composed of earth, air, water and fire, in which thing man and the world are very similar' (A 55 v).

THE STUFF OF THE FOUR ELEMENTS

It was Leonardo's view that certain fundamental properties were shared by all four elements. For example, they all consisted of 'atoms' or corpuscular particles. 'Atoms' for Leonardo were 'natural points'. And under the heading *Natural Point* he writes, 'The smallest natural point is larger than all mathematical points, and this is proved because the natural point is a continuous quantity and anything that is continuous is divisible to infinity; and the mathematical point is indivisible because it is not a quantity . . . The natural point is in infinite proportion to the mathematical point', he reiterates (BM 204 v). Elsewhere he states, 'An atom exists in nature but it can never be divided into any parts' (BM 176 v). With these statements Leonardo tries to reconcile 'discontinuity' with infinitely divisible 'continuity', entering into the very fundamental subject of the relationship between mathematics and physical events. To this we shall return when describing the development of his mathematical concepts. Suffice it here to say that Leonardo does not assume that the mathematical point is outside natural phenomena. We have already described, for example, the 'point' through which infinite numbers of light rays pass into a camera obscura or the pupil of the eye. Such a point was by no means to be looked upon as merely mathematical; it was a physical point of energy in Leonardo's opinion, occupying no space. In the division of the elements into their atoms, that is, 'towards infinity', Leonardo distinguishes between division of substance and division of power. Later in life he summarises this distinction thus: 'Divisions made by power towards the infinite, change the substance of the matter divided. These divisions will return to the composition of the whole, the parts reuniting in the same degrees in which they were divided. For example, let us take ice and divide it towards infinity; it will become changed into water and from water into air and from air into fire, and if the air should become thickened again, it will change itself into water and from water into hail, etc.' (E 60 r).

The four elements were thus for Leonardo four states of increasing particulate density from fire to air, water and finally to earth, produced by the agglomeration of 'natural atoms' constituting 'substance'. This concept of the relation of the four elements reflects his attempt to reconcile Aristotelian ideas of continuity of infinitely divisible substance with the discontinuity of the atoms of Democritus.

Leonardo's version is not in accord with Aristotle's conception of the transformation of the four elements. In other places Leonardo speaks of the movements of the particles of elements, which is related to the 'size of the granular parts of all elements . . . I maintain that air is full of parts which have dissimilar movement and therefore there is no uniformity of parts moved by contact with air' (M 41 r). And in referring to water vapour (which he clearly distinguishes from air) Leonardo states, 'No element has gravity or levity in its own sphere; vaporised elements stay for some time in one another, but in unit quantities' (BM 164 v).

Aware as he was of the 'continuity' of the elements, Leonardo also takes the view (not without considerable discussion on it himself) that the elements are particulate, using such terms as *atoms, granules* and *unit quantities* to express this. In an experiment on the flow of water he discusses at length how legitimate it is to use a 'discontinuous' marker of millet seeds to analyse the pattern of movement of a 'continuous' element. He defends his procedure on the ground that mathematicians consistently treat continuous quantities such as time as if they consisted of

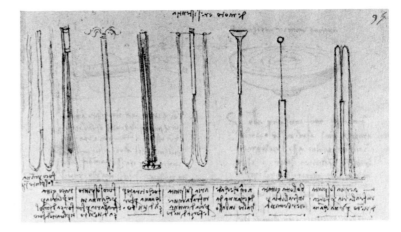

Figure 3.3. This modest-looking set of eight drawings is headed, 'On movement and resistance' (Madrid I 95 r).

The drawings are designed to show how resistance splits fluid jets of any sort whether they consist of water, air or fire. From right to left (as Leonardo lays them out) the captions describe (1) 'water pushed into the air through a tube by the force of wind'; (2) 'a cannon ball thrown into the air from a gun' (being solid, this does not split); (3) 'water falling down from a tube' (this is split gradually into a pyramidal shape); (4) 'air forced under water through a tube' (this splits and rises up to rejoin its own element air); (5) 'fire drawn back through a tube by suction'; (6) 'fire forced into the air through a tube' (this splits like water falling through air in 3); (7) 'a rocket thrown through the air by the force of the multiplication of fire' (the tail of fire spreads in the same way as does water in 3); (8) 'fire forced down through a tube in air' (this is split and rises up to its element fire, just as air goes back to air in 4 and water returns to water in 1. All these shapes are produced by resistance to movement.

discrete and separate parts (CA 126 va). His later reference to the existence of 'natural' as opposed to 'mathematical' points indicates that he was prepared to treat the elements from some points of view as 'discrete' or 'discontinuous', from other points of view as 'continuous'. This distinction is particularly clearly applied to the element of water. For example, drops of water he treated as discontinuous; streams of water, as continuous. Other features which Leonardo finds common to the elements are 'movements'. These 'are of the same pattern in all three elements, fire, air and water, but not earth' (BM 145 v). Percussion also made by all four elements has similar results; 'Water, air and fire make similar effects in their rebound on objects which are placed across their course' (C 22 v). He makes a long list of elemental 'qualities' and 'quantities' (BM 189 r). Like many of Leonardo's assiduously compiled lists written to clarify his concepts, this makes somewhat tedious reading. It begins thus: 'Two are the qualities of the elements, i.e., rare and dense; fire, air and water are called rare, earth alone can be said to be dense. Two are the quantities of each element, i.e., continuous and discontinuous. Two are the natural motions of the continuous and discontinuous elements, that is movement upwards and movement downwards . . . No part of an element weighs in its own element'. This last statement is repeated many times on this page about elemental properties. Many of these dicta are elaborated further in considering each element and power. Perhaps one of the most relevant to the science of man is the statement that 'liquid in liquid has no weight' (W 19062 r; K/P 155 r).

Two other general features of elements help in defining Leonardo's outlook. They are the importance he attaches to 'resistance' and his insistence on the existence of a fifth quintessence.

As his ideas develop Leonardo puts increasing emphasis on the factor of resistance or friction between moving elements. Resistance increases with density of the medium and diminishes with 'rareness'. The whole subject of the movements of each element within another is illustrated in Figure 3.3, eight drawings, each of which is meticulously and lengthily discussed in the text on Madrid I 95 r. The illustrations deal with the movements of water in air, a solid earthen ball in the air, water and air in each other, fire in air, and they finish with an illustration of 'fire forced down through a pipe', i.e., a rocket raised into the air 'by multiplied fire'. Explaining a rocket, Leonardo says, 'The flame does not flee from the rocket but the rocket flees from the flame' (CA 227 rc). All these movements are shown to be modified by the resistance of the medium or element in which they take place. However, Leonardo's appreciation of 'resistance' in movement within the elements is more simply shown from the details of a neat experiment in *Codex Atlanticus* 284 v. Here Leonardo states, 'The proportion of resistance of air in fire is the same as the resistance of earth in water'. He demonstrates this by an experiment with the balance, one of the earthen weights being immersed in water. 'Observe', he writes, 'how much extra weight is necessary for that in water to equal the weight in air'. Here one can see Leonardo's 'resistance' beginning to take on the features of specific gravity. This experiment is repeated on Madrid I 181 r (Figure 3.4) where a lump of iron is weighed in air and water, with the result as reported by Leonardo, 'Iron placed in water loses 2/11 of its weight in air; wood loses its whole weight'. The specific gravity of iron is 7.9, not 9, as Leonardo's experiment suggests. (Compare Leonardo's experiments on specific gravity, p. 115.)

The 'fifth essence', or quintessence, Leonardo considers necessary for the formation of each element from primary 'substance'. Thus he says, 'The fifth essence is infused throughout the air as is the element of fire . . . and through it each particle is given nutritive material and acquires growing form and increase in size; and if the nutriment is taken away they immediately abandon such a body and return to their primary nature' (CA 393 va). Thus this fifth, spiritual, essence is as necessary for the formation and growth of the elements of the world as is the soul for the formation and growth of the human body – a theme which reminds one of his emphatic assertion that 'man is a model of the world'.

This quintessence permeates but does not occupy space. It exerts power; it is the soul of the universe, the source of all the 'powers', shaping the growth and movement in the world composed of the four elements – macrocosm as well

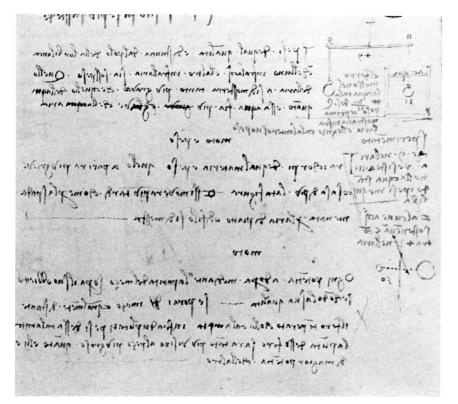

Figure 3.4. The resistance of air and water to a lump of iron is compared in the upper drawing.

The iron in the vessel labelled 'water' weighs eleven units (as marked) in air, but in water only nine as marked on the other end of the balance. Therefore, as he writes below, 'Iron placed in water loses 2/11 of its weight in air'. Compare Figure 4.30.

The lower drawing illustrates the effect of two magnets on a rod of iron. 'If you place tempered iron at an equal distance from two magnets, one twice the weight of the other, the point of the iron will be as much nearer to the heavier one as its power is greater' (Madrid I 181 r).

Compare with Figure 2.10 (p. 52). The two forces of gravity and magnetism obeyed the same laws in Leonardo's view.

as microcosm. In discussing man's 'hope and desire of going back to his own country or returning to primal chaos', Leonardo claims that 'this longing is in its quintessence the spirit of the elements which finding itself imprisoned within the life of the human body desires continually to return to its source. And I would have you know that this same longing is in its quintessence inherent in nature, and that man is a model of the world' (BM 156 v).

On page 146 r of the *Madrid Codex* I, devoted to studies of gravity and levity, Leonardo rewrites one paragraph three times in an endeavour to express himself clearly with regard to the nature of the centre of gravity. In the end he says, 'Of the second kind [*spetie*] of centre we find only one in each body. This is that true centre which resides at the centre of the natural gravity of this body without occupying space. This seems truly to be the soul of weights because always where it resides it becomes the guide of the bodies' motion joining its centre with the path made through the air. If someone were to take the earth away from its site and let its body free in the air this would be the centre of the universe'. Thus Leonardo expresses his intuition of weightlessness in space.

Thus the centre of gravity is a force which does not occupy space but does exert power. Such are the consistent characteristics of the spirit or soul whenever Leonardo describes it.

Leonardo's analysis of the stuff or substance of the elements goes deeper. It takes the form so common with him of a long debate with an 'adversary' as to whether a geometrical 'point' exists in nature or not. This argument has been beautifully presented by Augusto Marinoni in his Leonardo da Vinci Lecture of 1960.[3] Only its conclusion can be summarised here. This is expressed by Leonardo thus: 'The point is said to have no parts, and it follows from this that it is indivisible, and an indivisible thing has no centre, and that which has no centre is resolved into nothing. Therefore the point is nothing. And upon nothing one cannot begin any science. And in order to avoid such a beginning I shall say the point is that than which nothing can be less, and a line is created by the movement of a point, nothing can be narrower or finer, and its ends are two points. A surface is generated from transverse movement of a line, and nothing can be thinner, and its edges are lines. *And a body is made of movement*' (BM 159 r; my italics). Thus Leonardo proceeds to assert that a geometrical point does exist in nature by virtue of its movement. It is by *movement* of this point that lines, surfaces and physical bodies are built up.

He is still faced with the difficult problem of defining the nature of the geometrical point itself. His adversary on CA 289 ra says that a point must have a place (*sito*); if it has not it does not exist in nature. If it does exist, it must be as one or many points; and if it is one, it must be mobile or immobile. Leonardo replies, 'A point is in a place [*sito*] without occupation of that place [*sito*] . . . the point does exist in nature; points are infinite. The point is mobile together with the place [*sito*] in which it resides. The move-

ment of the point describes an inanimate [*insensibile*] line which in itself is divisible to infinity'.

Leonardo's geometrical point exists as non-spatial energy; it is the source of the movement of the spatial place (natural point) in which it resides. Movement of a 'natural point' can build up substance and bodies. Movement of a geometrical point can also exist incorporeally or spiritually in the form of the 'powers' of nature, as mathematical lines, for example, or as rays of light which intersect in points. These have incorporeal but 'real' existence. It is very significant that Leonardo uses the same phrase, 'resides in a place', to describe the location of his geometrical point, the centre of gravity of a body, and the location of the human soul in the *senso comune*. All these he sees purely as sources of energy, incorporeal and 'spiritual', residing in their different sites.

Perhaps one of the most significant results of Leonardo's dynamic concept of a body, particularly a live body, as ultimately composed of movement was his appreciation of the transient nature not only of the patterns of 'powers' but of 'bodies' also. This leads him in the case of the human body to his appreciation of the transformations of movement, growth and metabolic equilibrium, ageing and death.

THE FOUR ELEMENTS

EARTH

In contrast with the other elements, water, air and fire, Leonardo gave relatively little attention to the earth as an element, as opposed to the Earth as the world. He, of course, appreciated that there are many different kinds of earth as represented in the different rocks, just as he appreciated that there are many different tissues of the human body. But he makes no attempt to analyse those different earths further. This probably arises from his relative indifference to chemistry.

Leonardo was concerned mostly with the movements of the element earth in relation to the other elements and the four powers. Earth is the heaviest of the elements, the only 'dense' one; it weighs most, and its properties were therefore studied most intensively in the hundreds of experiments he performed on the balance (see Figure 3.5). However, through all these experiments on weight, one theme is constant from the beginning: 'No element has weight in its own element (Triv 6 v). Therefore, 'earth does not weigh in earth, nor air in air', etc. Earth begins to acquire weight only when it is displaced from its own element upwards, for example, into water. Even then its weight varies, for there are all degrees of such mixtures; 'That earth which is mixed with most water will offer least resistance to weights placed upon it. That water which is most mixed with earth will offer the greatest resistance to weight placed upon it' (Triv 33 r). As 'earth' is displaced upwards further from its natural place, for example, from water into air, it acquires more weight, since the 'resistance' of air is less than water. More-

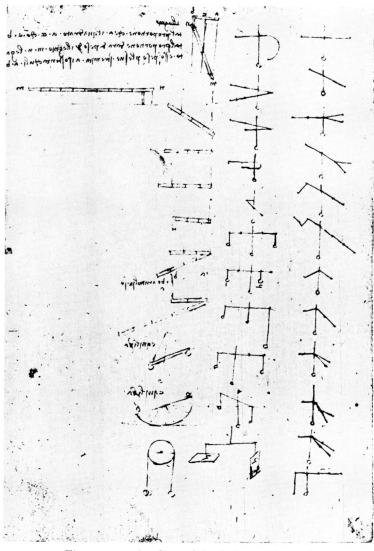

Figure 3.5. Studies of the balance (A 87 v).

This page contains thirty-two plans for the study of different structures of the balance, the great majority of which Leonardo confirmed experimentally. They begin at the upper right corner of the page, increasing in complexity as they progress. Suspension of the beam from above is studied as well as its support by a fulcrum from below.

This visualisation of problems is typical of Leonardo; it may be compared with his presentation of percussion.

over, the upper air is 'rarer' than lower air, which is 'denser'. Thus earth weighs more as it is displaced higher, until it pierces the outermost sphere of fire, outside which 'nothingness exists' and gravity is not exerted.

'Every part of an element separated from its mass desires to return to that mass by the shortest path' (CA 273 rb). The shortest path is the straight line, and this path is pursued by the 'centre of gravity' of the displaced earthen object to the centre of gravity of the world. Moreover, if elements are displaced, 'that which has the greater mass will draw to itself fellow-elements by the same line that separates

them . . . like earth to water' (CA 273 rb). This principle Leonardo applies to 'levity' as well as 'gravity', to the movement upwards of air or fire in water, as well as earth in water downwards, towards the centre of gravity of the world. Thus earth and water he called 'heavy elements', air and fire, 'light elements' (CA 268 vb).

The centre of gravity of the world is the goal of all 'heavy' objects. 'The desire of every heavy body is that its centre should be the centre of the earth. Every heavy body when falling is directed to the centre, and that part which weighs most will be nearest to the centre of the world' (Forster III 55 r and 51 r). So far so good; the centre of gravity of any heavy body is directed along a straight line to the centre of the world. But where is the centre of the world? This is as elusive as the centre of gravity of any other heavy body, as described on Madrid I 146 r. 'The centre of the world is indivisible', writes Leonardo, 'and since nothingness alone is indivisible, this centre is equal to nothing. And if one were to make a hole to the centre along the diameter of the world and throw a weight there, the more it moved the greater would its weight become. So when it reached the centre of the world (which is only a name being equal to nothing) the weight thrown would find no resistance at this centre but would pass it and return' (Forster II 59 v). The centre of Leonardo's world is like the geometrical point, nothing, in spatial terms; it 'resides at the centre of the body of the body [the world] without occupying place' (compare Madrid I 146 r). Moreover, since heavy objects composed of the 'heavy' elements earth and water are continually moving within their spheres, the earth's centre of gravity must be continually shifting. Thus it is that Leonardo comes to realise that 'The earth is moved from its position by the weight of a tiny bird alighting upon it' (BM 19 r). Often cited as a literary or poetic statement, this is for Leonardo a scientific truth.

The detailed study of the growth of the earth, and its movements in waves or tremors, within the other elements comprises Leonardo's view of the earth as a living organism (see Figure 3.6). 'The element of earth enlarges', writes Leonardo, 'whilst water stays at its level and fire does the same' (CA 320 vb). He even goes so far as to propose an experimental method of measuring the growth of the earth by the increase of soil over a period of ten years (CA 265 ra), and in one of his early notes he envisaged the earth growing through the elements of water and air to be 'burnt to a cinder' in the element of fire. This would be the 'end of terrestrial nature' (BM 155 v).

WATER

Water was the element which fascinated Leonardo most because it was visible, unlike air, and with the aid of markers its movements could be traced in geometrical patterns. Its oceans form an almost perfect sphere, broken by land masses. 'Of the four elements water is the second in weight and second in respect of mobility. It is never at rest until it unites with the sea where . . . it is stabilised and rests with its surface equidistant from the centre of the world. It is because the minute parts of the cloud fasten together to form drops' (A 26 r).

'No surface of water can be lower than the surface of the sea' (A 60 r). At this level water (as in the lagoons of Venice) is 'dead' and 'cannot move itself or anything else'. On the other hand, all water raised above the level of the sea is impelled to move down to the level of the sea. Leonardo observes that water often rises to springs at the tops of the mountains. How has it got there? At first he answers, by the force of the heat of the sun which raises vapour up into the sky. 'I say that just as the natural heat of the blood in the vessels keeps it in the head of man . . . When the sun warms a man's head, the blood increases and rises so much with other humours, that by pressure in the veins headaches are often caused. In the same way veins ramify through the body of the earth and by the natural heat which is distributed throughout the containing body, the water is raised through the veins to the tops of the mountains. And this water being dead will not pass through a closed channel in the mountain unless it is heated by the vital heat of the first vessel [compare heat of the heart]. Moreover, the heat of fire, and sun by day have power to raise up the moisture of the low places in the mountains and draw it up high in the same way as it draws the clouds and raises up their moisture from the bed of the sea' (A 56 r) (see Figure 3.6).

Having reached the tops of mountains (and man) the water or blood 'pours out through the broken vessels of the forehead. As water rises from the lowest part of a vine to its cut branches, so from the lowest depths of the sea water is raised to the summits of the mountains, where finding the vessels broken, it pours down and returns to the bottom of the sea. Thus the movement of the water inside and outside varies in turn; now in accidental motion it is forced to rise up, then with natural motion it freely descends. Thus joined together it goes round and round in continuous rotation, flowing hither and thither, from above and below, never resting in quiet' (CA 171 ra).

Thus, Leonardo depicts a circulation of 'humours' in the earth, in plants and in man. This concept of circulation was not modified until his later anatomical dissections. His last analysis of the movement of blood in the heart and aorta draws heavily on his vast experience of observations and experiments of the patterns of flow-waves and eddies reported in MSS F and K.

To return to the part played by water as an element around the earth, Leonardo sees water as sculpturing the shape of the earth's surface into its mountains and valleys during its flow down from the heights back to the level of the sea. It was his conviction that originally the whole of the earth was covered by the sea (CA 126 vb). Then the earth grew through it so that 'the mountains, the bones of the earth, with their wide bases penetrated and towered up amid the air, covered over and clad with much high-lying soil. Subsequently incessant rains have caused the rivers to increase, and by repeated washing, stripped bare part of the

Figure 3.6. How the heat of the sun raises water vapour, and how the earth grows through the seas (A 56 r).

The heat of the sun is simulated by a hot coal in a vase inverted in water (top drawing). Its effect on the 'veins' going to the tops of mountains is shown below. This happens also to blood which rises to a man's head, as described in the text alongside.

Below, the earth is shown growing through the sphere of water.

readily raised by heat in thin vapour through the air. Cold causes it to freeze and lack of motion corrupts it. It is the expansion and humour of all vital bodies. Without it nothing retains its form. By its increase it unites and augments bodies' (C 26 v).

The circulation of water impresses Leonardo as the basis of the 'life' of the world. He approached, but did not solve, the problem of a similar circulation in the bodies of men. 'It is its [water's] nature to search always for low places when unobstructed. Readily it rises up in vapour and mist, and converted into cloud [by cold], it falls back again as rain lofty summits of these mountains so that the rock finds itself exposed to the air, and the earth has departed from these places. And this earth from off the slopes and lofty summits of the mountains has already descended to their bases and raised the beds of the seas which encircle these bases and caused plains to be uncovered and in some parts driven away the seas over a great distance' (CA 126 vb).

The soil thus washed down falls into 'valleys formerly in great part covered by lakes because this soil always forms banks of rivers and seas which afterwards, through the per-sistent action of rivers, cut through the mountains; and the rivers in their wandering courses have carried away the wide open plains enclosed by the mountains. And the cut-tings of the mountains are shown by the strata in the rocks which correspond to the cuttings made by the said courses of the rivers' (Leic 1 v).

This general description Leonardo founds on two sets of data, literary and observational; one he may well have based on Aristotle's account in his *Meteorology*, Book I, 14. Here Leonardo describes the Danube flowing into the Black Sea which eventually broke through the cutting of the Bosporus and the Dardanelles to the Aegean Sea. This raised the level of the Mediterranean 'river' until in its turn it broke through the cutting of the straits of Gibraltar to reach its real basic 'sea-level' in the Atlantic Ocean (Leic 8 v).

The other description was undoubtedly the result of his own personal observations of the Arno watershed of which he made a uniquely accurate map, Figure 3.7. 'In this upper part of the Val d'Arno [from Perugia] as far as Arezzo a lake was formed [see Figure 1.24] which emptied its waters into the lake where we now see the city of Florence flourish together with Prato and Pistoia; and Monte Albano fol-lowed the rest of the bank down to where now stands Ser-

Figure 3.7. Map of the Arno Valley.

The river Arno runs down the valley to the right. It passes between Monte Lupo and Monte Albano to Pisa before it reaches the sea. The narrow space near Monte Albano illustrates the site of the water breaking through from the inland 'sea' around Florence towards the coast, an example of Leonardo's geological theory of the formation of the earth's surface (W 12683).

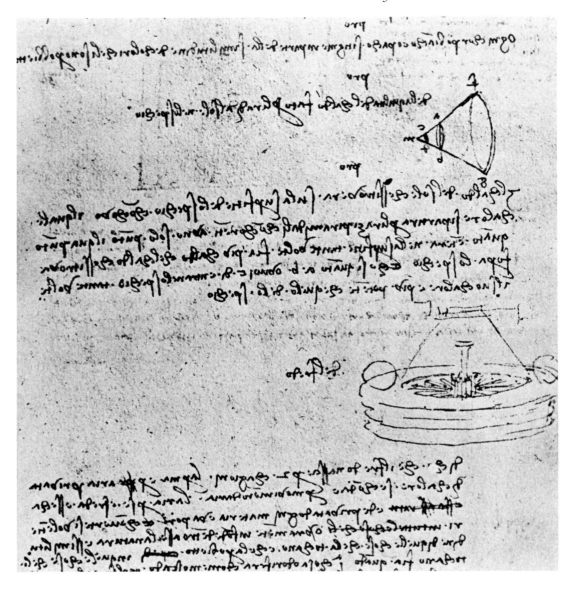

Figure 3.8. Air; the concentration of heat and cold in the air.

In the upper drawing rays of the sun are shown as 'pyramidal rays converging to a point' so that as many times as *ab* or *cd* enters *fg*, 'so many times is the heat more powerful'.

Below, he attempts to represent a similar concentration of cold air, saying, 'Just as many rays in a concave mirror joined in a point make the utmost heat, so many bellows blowing on to the same point make the utmost cold' (A 20 r).

ravalle'. The Arno is shown on his map breaking through another gorge near Vinci, at the base of Monte Albano, and so finding its way to the sea near Pisa. His maps show beautifully the shapes of all these inland 'seas' and their narrow gorges of outflow. But the most striking illustration of this geological process of aqueous earth-sculpture is to be found in the background of the portrait of the *Mona Lisa* (see Figure 1.27, p. 35). Leonardo forecasts that the 'raising of the sea levels' by alluvial deposits will eventually fill the Mediterranean 'laying bare its depths to the air' (Leic 20 r) so that the African land mass will join Europe – his explanation of 'continental shift'.

AIR

Outside the sphere of water in Leonardo's world was the sphere of the air. 'Air,' he writes, 'is a liquid body clothed in a spherical surface, and it is penetrated by the rays of the sun which converge in their course, and they are so much the hotter as they are narrowed to their point of meeting' (BM 204 r). Thus air for Leonardo is a fluid spherical medium which refracts the rays of the sun like a lens.

In itself Leonardo insists that air is cold and dry. These properties he tries to illustrate by concentrating the projection of air from a circle of bellows with nozzles pointing to the centre. On this same page (A 20 r, Figure 3.8) he illustrated the 'pyramidal convergence' and concentration of the light and heat of the sun's rays (Figure 3.8, upper drawing).

He is keenly aware of both the similarities and differences between air and water. 'Air moves like a river and carries the clouds with it just as running water carries all the things that float on it' (G 10 r). He reaches the conclusion (late in life) that 'In order to give the true science of birds in the air it is necessary first to give the science of the winds and this we shall test through the movement of water' (E 54 r). Water, being visible, will serve as 'a ladder to arrive at perception of flying things in air'.

Air differed from water, too, in being 'infinitely compressible', capable therefore of being infinitely 'dense' or heavy, and 'the greater the velocity of a moving thing in air the more the air is condensed' (CA 108 r). Water on the other hand is incompressible, which is why 'The surface of the sphere of water is moved by a tiny drop of water falling on it' (BM 19 r). Since 'no element weighs in itself', air does not weigh in air, but it does do so in fire, and in water it rises by the force of levity as bubbles. Moreover, since air increases in weight the more it is condensed, a bladder blown up with air weighs more than an empty bladder because it contains condensed air. Air being 'grosser' as it is lower, and 'finer' or clearer as it is higher, weighs more near the earth or sea. In nature air contains vaporised water particles. These rise with heat, condense as they cool at a height roughly halfway between the elements of water and fire, to form clouds. These condense further by virtue of increasing cold and pressure, falling as rain drops which, by friction as they fall, may vaporise again into minute invisible particles. These minute particles of moisture catch the sunlight and reflect it all around, so producing the varying greys and blues of the sky, blue being a dilution of black by white light. Leonardo states his rule that action equals reaction for air as well as the other elements: 'A thing makes as much resistance against air as air against the thing' (CA 381 va).

Leonardo was most interested in the effects of heat and cold on air. One can see how he forges a link between the part played by air in the macrocosm and microcosm of man in the following passage: 'As the water flows in different ways out of a squeezed sponge or air from a pair of bellows, so it is with the thin transparent clouds that have been driven up to a height through the reflection occasioned by the heat. The part that finds itself uppermost is that which first reaches the cold region and there, halted by the cold and dryness, awaits its companion. The part below as it ascends towards the stationary part treats the air in between as though it were a syringe; this then escapes transversely and downwards [forming wind] . . . Similarly the natural warmth spread through the human limbs is driven back by the surrounding cold which is its opposite and enemy, flowing back to the lake of the heart; and there in the liver it is protected or fortified, making of these its fortress and defence. So clouds being made up of warmth and moisture, finding themselves in a cold region, act like certain flowers and leaves which when attacked by a cold hoar-frost press themselves close together and offer greater resistance' (CA 212 va).

Thus heat expands and rarefies elements infused by it; water is expanded and vaporized and blood spreads by its warmth into the periphery of the body from the heart. Cold contracts it, bringing water back to the earth's body as rain drops or pressing blood back to the heart and liver.

The 'levity' of air in water makes it rise as bubbles. Bubbles normally have thin 'skins', which is why they often remain as hemispheres on the surface of water. Leonardo gives such 'bubbles' of air real skins which he calls 'bags' and fits them into cylindrical vessels and notes that they raise water and spill it over the rim of the cylinder (A 25 v). This principle he applies to raising 'heavy weights and ships' from beneath the sea (Madrid I 82 a) – a principle used in the modern 'camel'. Regulating the rate of rise by screw-action, he makes a water clock. On this very page he applies the same principle to the bringing up of air and water through the human gullet (oesophagus) and in this way explains belching up wind and water-brash (W 12282 v; K/P 125 r).

Wind is nothing but 'moving air'. Thus when Leonardo comes to examine the *pneuma* of Greek philosophy as a 'power' within the body, he sees it as the movement of air (wind) in the ventricles of the brain and along the 'hollow nerves' to the muscles (see Chapter 13). And on the grounds of the impossibility of the nerves, even 'the whole body', containing a sufficient volume of air, he rejects the pneumatic theory of muscular action (W 19017 r; K/P 151 r). At one period he also sees air as being burnt up as a flame in the heart (see Chapter 15).

FIRE

Fire forms the outermost sphere of the elements. The centre of the element of fire does not move; the centres of the other three elements do continuously 'because of shifts of earth, water and air' (CA 153 va). It is the 'rarest' of the elements, of greatest 'levity' or lightness. Therefore when displaced downwards it always attempts to rise to its natural place. This is seen in the piercing, pyramidal shape of great fires. (Leonardo draws such fires near the city of Milan, in Figure 3.9.) It is because fire strongly desires to re-enter its own sphere that its power carries with it a rocket to ascend into the sky (Madrid I 81 v and 95 r) (see Figure 3.3). He asks on *Madrid* I 178 v, 'Why does fire rise up in the air like a cone?' The reason, he finds, is the same as that for water falling down in air from a tap. This also takes a conical or pyramidal shape (CA 151 ra). This shape pierces the elements most quickly and efficiently, just as a pointed spear pierces solid earth more quickly and efficiently than any other shape, and nature works in the shortest way possible (BM 85 v and CA 273 rb).

Leonardo from his study of candle flames concludes that pure fire is blue; only incandescent particles incompletely burnt give the light of a candle flame before it 'dies'; when further cooled, it changes into dark smoke (CA 270 va). In this analysis Leonardo claims that the blue part of a candle flame is primarily spherical and does not move of itself, nor does the nutriment it takes up from the candle; 'this movement is generated by air which furiously moves in to fill up the vacuum left by that which is consumed by the flame'. Here Leonardo recognises clearly that fire 'consumes' a part of the air. Moreover, it is the pressure exerted by eddies of surrounding air that compress the candle flame into a pyramidal shape.

'Where the air received is not proportional to the flame, no flame can live, nor can any animal, terrestrial or aerial', writes Leonardo (CA 270 ra). And in the course of a very lengthy and interesting essay on the nature of flame, he

Figure 3.9. Pyramids of fire (W 12416).

Two separate fires at Desio, near Milan, were started by the Swiss in December 1511. The first, on the right, shows a fire in calm air making an almost typically pyramidal ascent through the air to its own element. The dark line at about a third of its height corresponds to an arrest of the normal diminution of temperature with height, an 'inversion', from which the pyramid rises anew. On the left a wind blows the smoke leeward, bending the normal pyramidal shape into a 'falcate', i.e., a pyramid with bent or curved sides. Eddies are seen at the edges.

These shapes may be compared with those of a candle flame. See Figure 15.8b.

adds, 'Where flame does not live an animal that breathes cannot live'. The analogy between life and the burning of a flame is developed particularly in relation to events occurring within the heart which are likened to a candle flame and so to the source of heat or vital spirit, and the smoke to exhalation of burnt vapours through the lung (see Chapter 15). Thus fire enters into Leonardo's anatomical studies of the microcosm of man.

THE ELEMENTS OF THE MOON AND STARS

The greater and lesser worlds of the earth and man being composed of the same elements organised by their 'souls' and 'vital spirits', are alive. What of the moon, planets and stars? Leonardo's source of interest is revealed in the question which he asks, 'How does one test whether the moon is another world like our own' (A 96 v). The 'test' by experiment does not appear until some sixteen years later when he puts a burnished ball of gold in darkness and shines a light on it. 'Although this illuminates about half the ball, the eye sees the light reflected from a small part of its surface; all the

rest of the surface reflects the darkness that surrounds it . . . The same thing would happen to the surface of the moon if it were polished, bright and solid' (F 93 r). Having decided that the moon is not itself luminous but reflects the light of the sun, Leonardo comments that its outer surface cannot be 'uniformly spherical' since it does not reflect light like the burnished ball in the experiment. Therefore the moon must reflect light by a 'rugged surface like mulberries' (see Figure 3.10). Such a surface could in his view be produced only by waves of water. 'The surface of the waters forming the seas of the moon and of our globe is always more or less ruffled; this rugosity causes an extension of the innumerable images of the sun which are mirrored in the ridges and hollows, sides and fronts of the innumerable waves' (BM 94 v).

This reflection of sunlight from waves also travels in the reverse direction so that 'Anyone standing on the moon . . . would see this our earth with its element of water just as we see the moon, and the earth would light it as the moon lights us' (F 41 v). And describing the 'glimmer of light visible between the horns of the new moon', he denies that this arises from the moon itself, but 'it is derived from our oceans, for they are at that time illumined by the sun which is then at the point of setting in such a way that the sea then performs the same function for the dark side of the moon as the full moon does for us when the sun is set' (Leic 2 r). He then suggests a technique for assessing the two 'brightnesses' of the light from the new moon. Leonardo's astronomy is essentially limited to an extension of his investigations of perspective, light and shade. 'Astronomy is a product of visual lines and perspective without which the art of geometry is blind' (TP I 15).

Having 'proved' that the light part of the moon is reflected from its waves, and that the dark parts consist of earth, Leonardo points out that there would be no waves unless air ruffled the surface of the waters. Thus he concludes that the moon, like the earth, has its set of 'elements' within which it has its own centre of gravity, and since all this occurs beyond earth's element of fire, the earth's gravity is not exerted on the moon. Therefore the moon does not fall onto the earth. Applying similar reasoning to the light of the planets and stars, Leonardo concludes, 'All your

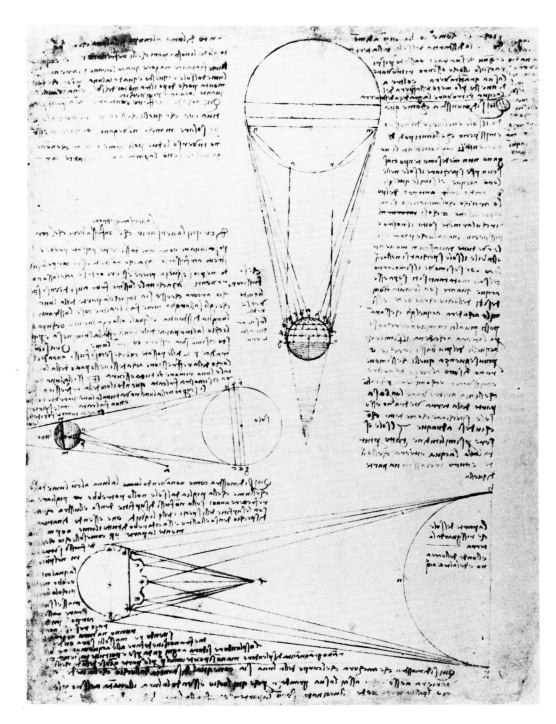

Figure 3.10. How the light of the moon is reflected from the sun (Leic 1 r).

In the top figure the sun is shown shining down on the earth.

The upper drawing on the left shows how the light of the sun shining on a spherical moon would be reflected.

The lower drawing depicts a wavy surface which Leonardo describes thus, 'If the lustre of the moon arises from a liquid body . . . which is full of waves such as we see made by the waters of the sea then its brightness will be given out by each wave on its own, therefore all together . . . they will make a great quantity of brightness'. Here the sun casts its rays on a mulberry-shaped moon. Each bulge represents a wave which reflects sunlight to the observer's eye placed in position on the earth between the sun and the moon. Thus the moon must have seas with waves of water on its surface, like the earth.

discourse points to the conclusion that the earth is a star much like the moon, and thus you will prove the nobility of our world' (F 56 r). Accepting stars of the cosmos as equally 'noble' entities, Leonardo concludes that the four elements are organised in an infinity of living 'worlds' through space. For him presumably each macrocosmic living world is capable of bearing on its surface innumerable 'lesser' microcosms, like living man.

REFERENCES

1. Reti, Ladislao. 'Le Arti Chimiche di Leonardo da Vinci', *La Chimicha e l'Industria,* Milano, Soc. An. Editrici Chimica, 1952. Anno xxxiv, pp. 655–721.
2. *Ibid.*
3. Marinoni, Augusto. 'L'essere del nulla', *Lettura Vinciana* 1960, ed. by Paolo Galluzzi. Florence, Barbèra, 1970.

Chapter 4

The Four Powers

All powers . . . are pyramidal (Madrid I 128 v).

From a vast number of observations on perspective and experiments on light, heat, sound, levers, pulleys, etc., Leonardo constructed a general law, 'We will be telling the truth by affirming that it is possible to imagine all powers capable of infinite augmentation or diminution. Consequently all powers are pyramidal because they can grow from nothing to infinite greatness by equal degrees. And by similar degrees they decrease to infinity by diminution ending in nothing. Therefore nothingness borders on infinity' (Madrid I 128 v).

Leonardo made this assertion about 1495. It appears in the context of a 'demonstration that the power of the arms of a balance has a pyramidal nature'. It is repeated and amplified some ten years later in the *Codex Atlanticus* (Figure 4.1). Here Leonardo heads the page with the words 'All natural powers have or are to be called pyramidal inasmuch as they have degrees in continuous proportion towards their diminution as towards their increase. Observe the weight which in each degree of its free descent is in continuous [arithmetical] geometrical proportion, and similarly for the force of levity'. The word *arithmetical* is erased; *geometrical* refers to the triangular or 'pyramidal' geometrical figure which accompanies the statement.

Leonardo's momentous generalisation is reinforced by five drawings on the same page (Figure 4.1), each illustrating a different aspect of the statement. On the left border of the page Leonardo illustrates the statement expressed in *Madrid Codex* I 128 v depicting a balance suspended on a fulcrum. To the left of the fulcrum a triangular section of a pyramid is shown, increasing in width with each of eight degrees marked out along its beam. This is balanced by a weight suspended at one degree of length of the right-hand beam. Against this Leonardo has noted, 'That weight which descends more goes the faster and becomes heavier'. The next diagram illustrates the 'diminution by degrees in continuous proportion' of the movement of a weight thrown upwards. Leonardo labels the drawing 'accidental movement'. Next to it he illustrates the 'pyramidal' increase in velocity of a 'falling weight', and further to the right, 'pyramidal' increase and decrease of a horizontal movement, also labelled 'accidental movement'. The upper right border of the page in Figure 4.1 is devoted to an

attempt to weigh the force of percussion of falling water. Below this Leonardo illustrates the 'pyramidal' shape 'of a falling stream of water'. He draws it narrowing pyramidally 'by degrees' as it pierces the air.

These diagrams of the balance and the falling weight immediately and clearly illustrate Leonardo's meaning of the term *pyramidal degrees*. For it will at once be seen that they are of similar shape to the figures which he uses to illustrate perspective. Herein lies the principle of Leonardo's approach to the 'four powers'.

For many years it was appreciated by artists that Leonardo not only worked out the laws of linear perspective geometrically, using the device of the interposed glass pane, but also extended the linear perspectival laws of the pyramidal figure to colour perspective and the perspective of distinguishing detail, sometimes called aerial perspective. Leonardo saw all these forms of perspective, as he showed in his diagrams in his *Treatise on Painting,* as inversely proportional to the horizontal distance of the eye from the object. This was an application and generalisation of the surveyor's technique.

Leonardo also saw painting, governed by this perspective of the interposed plane, as capable of presenting the 'real' proportional relation of objects to distance. In this he found its potentiality for science.

As we have shown in our brief outline of Leonardo's descriptive analysis of visual experience, he considered light as a 'power' bringing images of illuminated objects to the eye. This power diminishes in 'pyramidal' proportion according to the distance it travels in spreading waves from the object seen. These waves, carrying their 'images', percuss the eye and travel up the the optic nerve to the *imprensiva* and so to the *senso comune,* where they enter consciousness.

The perspectival power of light and colour diminishes 'pyramidally', and the power of the waves from the percussion of a stone in water also diminishes 'pyramidally'. For Leonardo the term *perspective* therefore applies not only to the transmission of light, producing vision, but also to other mechanical powers. In this way Leonardo's studies of perspective opened up to him the hypothesis that *all* powers were perspectival or, as he calls it, 'pyramidal'.

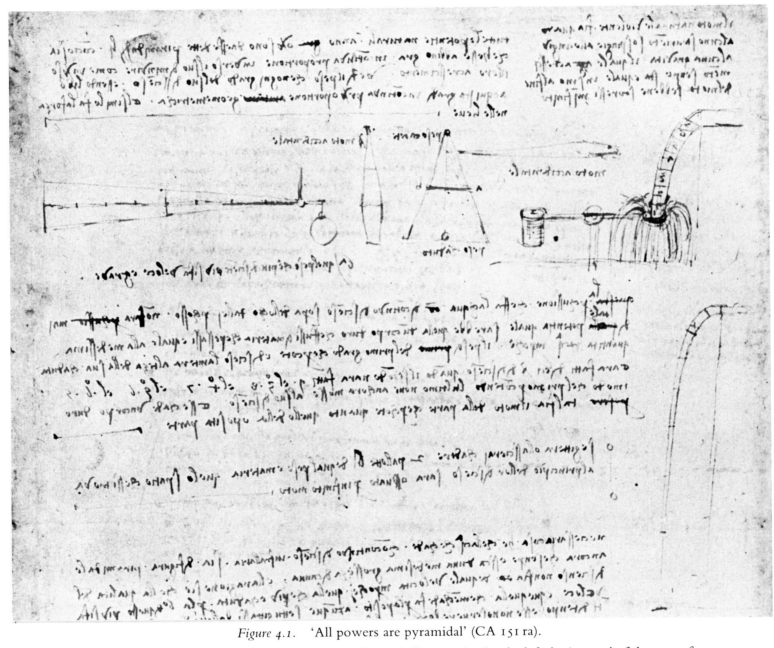

Figure 4.1. 'All powers are pyramidal' (CA 151 ra).

This statement heads the page. All the drawings beneath illustrate it. On the left the 'power' of the arm of a balance is depicted in pyramidal form; next to it, the deceleration of an object thrown upwards; next to that, the acceleration of an object accelerated by gravity. On the right an attempt to weigh the percussion of water is drawn; below this the pyramidal narrowing of a stream of water as it drops is drawn.

During the same years that he was working out the details of the laws of visual perspective on the glass pane interposed between the eye and the object, Leonardo was inventing, modifying and studying the principles of power transmission in machines. He brought his perspectival laws to bear on these mechanical problems. This enabled him to draw machines in measured proportions so successfully that working models of them have been constructed from his drawings nearly 500 years later. Only by applying his perspectival rules could he represent quantitatively proportioned drawings of levers, etc., particularly those of gear wheels and pinions, in different planes with correctly measured and numbered teeth, for the creation of their mechanical interplay in real machines. Thus his machines took shape in his mind in transparent perspective and were represented by the art of his hand perspectivally. The result was a drawing of a machine shaped according to its function. Leonardo expresses this beautifully in one sentence: 'Once the instrument is created, its operational requirements shape the form of its parts [*membra*]' (Madrid I 96 v). The word *membra* Leonardo uses for the parts of a machine, for the branches of trees and for the limbs of man and animals.

Here he means all of these; for as we have emphasised, he saw the earth and man as machines. Leonardo's drawings of plants, animals and man were therefore seen by him as perspectival studies of the proportions of their parts. Like those of his inorganic machines, they were designed to express the geometry of the power-patterns, the 'operational requirements', of these 'machines'. And all these power-patterns were for him variations on a basically 'pyramidal' theme.

THE ELEMENTS OF MACHINES

At the same time as he was working out the significance of perspective, Leonardo was analysing and constructing machines from their basic constituent mechanisms, what he called the 'Elements of machines'. He proposed to write four volumes on this subject. Although such a work, if ever written, has been lost, the *Madrid Codex* I contains a number of such studies. One example must suffice.

Clocks were a persistent challenge to Leonardo's mechanical ingenuity. Clockwork involved the use of springs. Springs lose force and movement as they unwind. Leonardo's first effort at making a clock is drawn on B 50 v (Figure 4.2). Here a series of springs is shown with their diminishing rate of movement compensated by their threads being taken on to a series of inverted cones or 'pyramids' so placed that each screws into the one above and so continues the movement until all four springs are unwound.

A series of more complicated clockwork schemes are produced in Figure 4.3 and *Madrid Codex* I 16 r and 45 r, all of which clearly show different kinds of 'pyramidal' mechanisms in the form of either their pinion wheel (16 r), spiral gear (Figure 4.3) or a combination of both (45 r). All of them have the same purpose, i.e., that of compensating for the 'pyramidal' loss of power and movement from the unwinding mainspring (enclosed in these drawings within a cylindrical casing) by a conversely pyramidal movement. Together, therefore, they demonstrate one of Leonardo's elements of mechanics based on the principle that the power and movement of a spring diminishes pyramidally. The variable-speed gearing device in Figure 4.4 shows a similar pyramidal principle.

Perhaps Leonardo's clearest illustration of the pyramidal forces in a spring is that on Madrid I 85 r (Figure 4.5). Here the spring or beam is represented by a series of suspended bows, the lowest string of which bears loads of ten pounds. The series is likened to the coils of the spring drawn next to it; both have their central lines of gravity indicated. They illustrate that 'The spring with every degree of motion gives degrees of force or weakness', i.e., the force exerted by a spring is directly proportional to its movement.

Below this in Figure 4.5 is another row of drawings showing a clock-spring being wound up by a central key. The text discusses the forces which are exerted in winding up the spring. Leonardo ends by saying, 'The power of the spring diminishes pyramidally, and this is the reason why

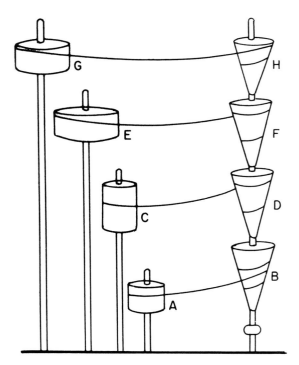

Figure 4.2. Design for long-lasting clock springs.

In explanation Leonardo writes, 'Four springs for a clock made in such a way that when one has completed its run another begins. During the revolution of the first the second remains fixed. The first fastens itself upon the second by means of a screw, and when it is quite firmly screwed, the second spring takes up the same movement; and so on with the remainder' (B 50 v).

ACEG are a series of four springs connected by a thread with conical fusees working on the screws, *BDFH,* also in series. Pull by spring *A* rotates the cone *B,* so screwing it into cone *D.* The pyramidal decrease in movement of spring *A* is compensated by pyramidal increase in the diameter of the cone *B.* As soon as the cone *B* is screwed into cone *D* the spring in *C* comes into action turning the cone *D* until it is firmly screwed into *F.* This process can be designed to continue for as long as required; theoretically, indefinitely.

The original drawing being rough, this diagram has been taken from Ivor B. Hart, *The World of Leonardo da Vinci* (London, Macdonald, 1961), p. 304.

clockmakers counterpoise it with a pyramid, as represented in the drawing beneath'. Here he shows a spring with its thread attached to the conical fusee mechanism used by clockmakers. Beneath this he writes, 'If the spring is of uniform thickness its power diminishes gradually as it is unwound. We shall say therefore that such power is in the nature of a pyramid, since its beginning is great and it diminishes to nothing. For this reason it is necessary to oppose such a pyramidal power with another pyramidal power that has an opposite diminution of resistance'.

The same 'pyramidal' rule is illustrated even more obviously on Codex Madird I 84 v (Figure 4.6). Here Leonardo illustrates how bending a straight spring or elastic

Figure 4.3. Method of equalising the force of a
clock spring.

Leonardo writes here, 'When you wish to wind
up the spring turn the crank *m* contrariwise,
and the wheel *s* will come to a stop together
with its axle; and the pinion *n* will rotate on
this axle together with the spring, the endless
screw, the pinion and its cog-wheel'
(Madrid I 14 r). The conical fusee has
here been modified into a simple form
of gear wheel.

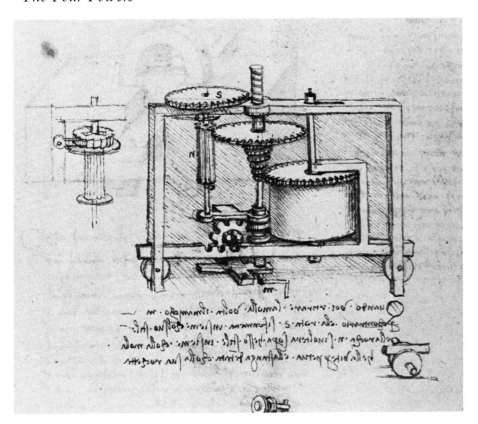

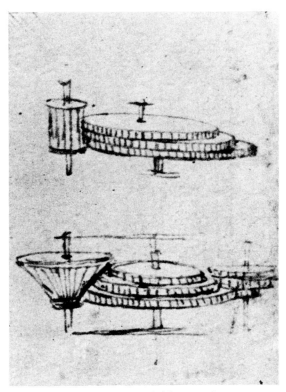

Figure 4.4. Variable-speed gears.

The transition from the clock-spring mechanism
in Figure 4.3 to the variable-speed gears is clearly
illustrated by the upper and lower drawings. It
will be noticed that this variability is obtained by
confronting a pyramidally shaped pinion wheel
with an inversely pyramidal set of cog wheels
(CA 27 va).

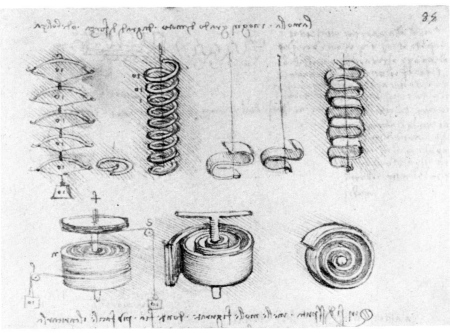

Figure 4.5. The pyramidal forces of springs
(Madrid I 85 r).

A weight of 10 pounds attached to the suspended
series of five bows is compared with the effect of
the same weight attached to a spring drawn alongside. Below these Leonardo describes the effect of
a weight of 10 pounds in winding up a spring
from the outside, *l,* or inside, *s,* with a key at *f.*
The page ends with a sketch and description of the
fusee showing the conical or pyramidal shape
necessary to compensate for the pyramidal weak-
ening of the spring as it unwinds.

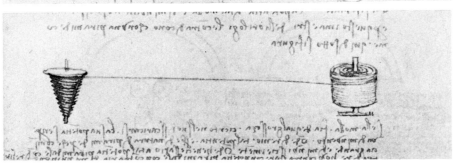

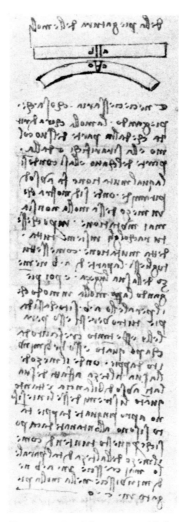

Figure 4.6. The pyramidal-
shaped gap in a bent beam or
bar (Madrid I 84 v).

It will be noticed that the
parallel lines at *a* and *b* in the
upper drawing of a straight
bar become converging 'pyramidal' lines when the bar is
bent. The text below explains
this.

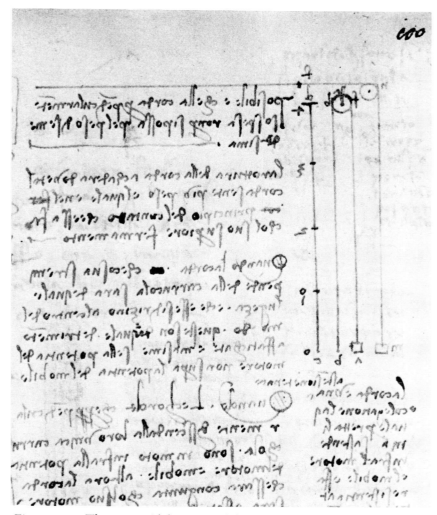

Figure 4.7. The pyramidal nature of the power of a rope.

Experimenting with weights on ropes Leonardo notes the stress on the
rope *c* in this drawing in these words, 'If the extension is straight its
power is pyramidal, that is, in each degree of its height it acquires a degree
of weakness; since if the length is 4 degrees at its upper end it bears 4; at
the second 3, and at the third 2, and at the fourth nothing' (Madrid II
100 r). It will be seen that this pyramidal proportion is enumerated on the
rope *c*.

beam produces a 'pyramidal mutation'. He writes, 'If a
straight spring is bent it is necessary that its convex part
becomes rarefied or thinner and its concave part denser or
thicker. This mutation is pyramidal, whence it is demonstrated that there will never be any change in the middle of
the spring [i.e., the neutral axis]'. This 'pyramidal' nature
of stress or strain he applies widely, to the weights breaking
ropes round pulleys, the bending of iron bars and wooden
beams (see Figure 4.7) and even to the action of muscles (see
Figure 13.8, p. 277).

On *Madrid Codex* I 82 r Leonardo describes his plan for
this codex: 'We shall discuss here the nature of the screw
and of its lever, and how it shall be used for lifting rather

than thrusting . . . how endless screws shall be combined
with toothed wheels . . . we shall examine the nature of
nuts', and he continues with plans to 'examine the nature' of
levers, wheels, the ratchet and pawl, axles, ropes and
pulleys, capstans, rollers, etc. The whole *Madrid Codex* I
consists of studies of all these mechanical instruments on the
one hand, composed into formidable-looking machines on
the other. These are analysed into their simplest elements of
movement, weight, force and percussion—'the four
powers'. Similar studies form the main parts of *Codex Atlanticus, Forster* II and MS M. Through these extensive investigations Leonardo stepped out of the field of medieval
empirical technology into that of science.

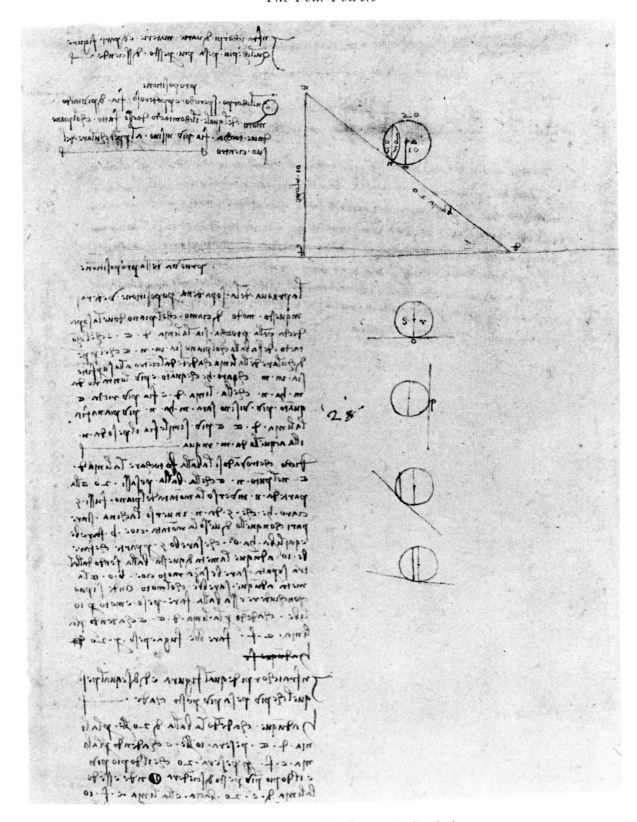

Figure 4.8.   Movement of a ball on an inclined plane.

In the top drawing the movement is shown to depend on the distance between the centre of gravity, *am,* and the point of contact, *n,* with the inclined surface, *ed,* and its rate is proportional to the ratio of the distances *ef* to *ed,* that is the sine of the angle of inclination. In the second drawing below the centre of gravity and the line of contact coincide; the part *s* equals the part *r.* Therefore, this ball on a level plane does not move. The three lower drawings vary the position of the point of contact and the centre of gravity with the inclination of the plane. At *q* (third drawing) the ball makes contact with a perpendicular plane (CA 338 rb).

THE FOUR POWERS

Leonardo's reduction of the manifold powers of nature to four, achieved by his analysis of machines, was applied to the macrocosm and microcosm consisting of four elements. As we have already seen, the most fundamental of these powers was movement. We shall follow Leonardo's own line of thought in dealing with these four powers. 'Speak first', he writes, 'of movement, then of weight because it originates from movement, then of force which arises from weight and movement, then of percussion which springs from weight, movement and often from force' (CA 155 vb).

MOVEMENT

The most important features of Leonardo's conception of movement are that it is born of the point in creating a line (see p. 84) and that it can be the source of the other three powers. 'Gravity, force and percussion are of such a nature that each by itself alone can arise from each of the others . . . And all together and each by itself can create movement and arise from it' (BM 37 v).

'Natural' movement of any element is towards its natural place of rest in its own element. In the case of earth, for example, this is down through the other three elements. As it descends in air it accelerates in 'pyramidal' proportion (as illustrated in Figures 4.27a and b, p. 113). Other movements produced by force or percussion are 'accidental' movements. These either displace an element upwards out of its 'natural' place, or they produce movement within an element such as currents or eddies of water within water.

Movement in Leonardo's view results from inequality of force or weight about the centre of gravity. Equilibrium or balance results from equality of forces around the centre of gravity. If the centre of gravity shifts, then local movement occurs. For example, a ball on a level surface with its centre of gravity over its point of contact with the ground is at rest (CA 246 ra), but as soon as the centre of gravity shifts from this perpendicular line the ball moves (see Figure 4.8, CA 338 rb). 'Motion is due to inequality of weight distribution around the centre of gravity', writes Leonardo, and 'loss of motion is due to loss of inequality of weight distribution'. Further, 'That object moves most rapidly which is farthest from its balance' (TP I 346). 'Man or any other animal', writes Leonardo, 'will move from a position with more or less velocity in proportion as the centre of gravity is further away from, or nearer to, the supporting foot' (TP I 350) (see Figure 4.9b). Thus the velocity of movement is 'pyramidally' proportional to the distance of the point of contact with the ground from the centre of gravity both in a ball and in man. With movement a body acquires 'impetus', which is a force infused by the mover into a moved body. Impetus is increased in proportion to the weight of the body and its velocity, but it is never enough to overcome permanently the resistance from air or any other medium; the body will thus always be brought to rest at some distance, even if that distance is 'infinity'.

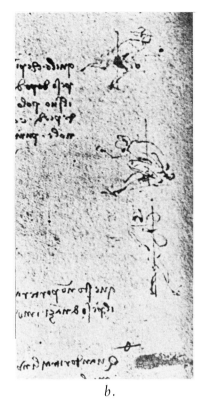

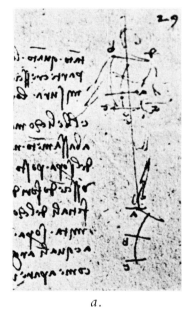

a.

b.

Figure 4.9.

a. The centre of gravity in a standing and moving man (A 29 r).

In this sketch of a standing man his centre of gravity passes through the left foot. So does his line of support or resistance.

b. The centre of gravity in a moving man (A 28 v).

Here the line of support or resistance does not correspond to the centre of gravity. The line of support is drawn in all three human figures. In the running man, the centre of gravity is in front of the drawn line of support; it is also in front of the supporting point of contact in a man rising up from a sitting position on the ground. But the centre of gravity is behind the supporting foot in a man carrying a load in front of him.

In these sketches only the lines of support are drawn. Those of the centre of gravity are omitted. Both are drawn in the examples of movement of a sphere in Figure 4.8, where they correspond to the lines of 'resistance' and centre of gravity drawn in Figure 4.19b (p. 107); here they illustrate the flight of a bird.

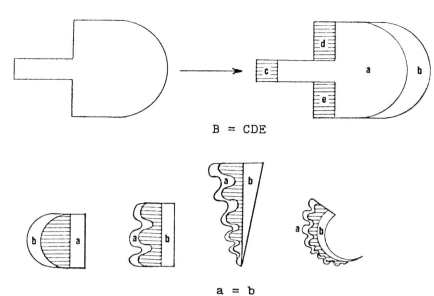

B = CDE

a = b

Figure 4.10. 'Of everything that moves, the space which it acquires is as great as that which it leaves' (CA 152 va and 166 ra).

This statement is the physical counterpart of Leonardo's geometrical transmutations, particularly of curved figures into straight-sided figures (quadrature). In the top drawing it will be seen that by moving the bullet-headed figure to the right the curved figure *b* equals the spaces left *cde*. In the lower row the curved forms are varied so that *a* always equals *b*. This realisation led Leonardo to appreciate the principle of continuity in relation to the flow of water in rivers.

I am indebted to Dr. James E. McCabe for these illustrations from his work, *Leonardo da Vinci's De Ludo Geometrico* (Los Angeles, University Microfilms Library Services, Xerox Corporation, Ann Arbor, Mich., 1972).

When a man wants to stop himself when running he throws his centre of gravity behind his feet. When he stands still his centre of gravity is between or on his feet. When he walks or runs forward it is in front of his feet. By such studies of movement of the centre of gravity with movements of the body, particularly the spine and limbs, Leonardo transfers his investigation of movement of inanimate objects to that of animals and man in different media, i.e., on the earth, in or on water and in the air. These movements of walking, running, swimming and flying comprise a vast section of Leonardo's work, to which we shall return.

All bodies, as we have seen, are composed of persistent patterns of energy or movement of variable duration. Usually they are defined by the shapes of their surfaces, but 'Bodies are of two kinds, the first is without shape or any distinct or definite edges, which though present are imperceptible . . . The second kind of body is that whose surface defines and distinguishes the shape. The first kind is that of fluid bodies like mud and water, mist or smoke with air . . . whence by this intermingling their boundaries become confused and imperceptible, for which reason they are found without surface since they enter into each others' bodies (CA 132 rb).

'Of everything that moves, the space which it acquires is as great as that which it leaves' (M 66 v; see Figure 4.10). This may seem obvious enough for solid bodies with definite edges but not for the 'imperceptible' bodies which mingle or change their shape, as applies, of course, to air, water and fire. Leonardo pays particular attention to the fact that though a thing that moves (such as a river) acquires as much space as it loses, it by no means necessarily preserves the same shape. In *Codex Atlanticus* 152 va he gives an example, showing how a curved narrow form may by movement change into rectilinear shapes of equal area or volume. Observation of such transformations in shape lead him to say of rivers, 'Although it may be of different tortuosities, breadths and depths, the water will pass in equal quantities in equal times through every degree of the length of the river' (CA 287 rb).

Figure 4.11. The movement of water: eddies.

a. An eddy seen from the side diminishing pyramidally towards its apex, to form a cone.

b. The same eddy seen from above diminishing to a point.

To the right of this the movement of the water in the wall of the eddy is drawn. At its apex it continues with reversed movement to form a secondary eddy.

Both figures are taken from E. Carusi and A. Favaro, eds., *Del Moto e Misura dell'Acqua: Libri nove ordinati da Luigi Maria Arconati* (Venice, n.d.), Lib. 4, Caps. 41, 53. This work consists of extracts of Leonardo's notes on water, many of which exist nowhere else.

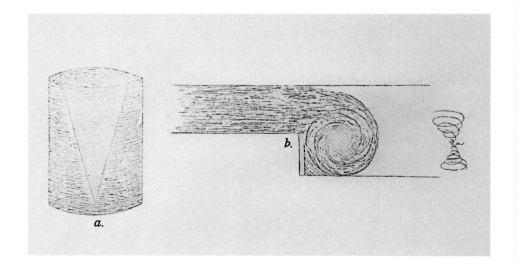

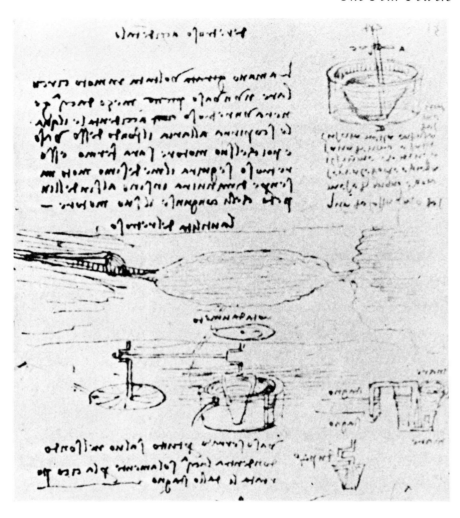

Figure 4.12.   The centrifugal pump.

Sketches for creating an artificial eddy or vortex for draining swamps (F 13 r).

We shall see Leonardo applying this same principle of continuity later to the movements of blood in the heart and blood vessels and the flow of bile and urine.

By using his geometrical principle of transformation Leonardo from about 1505 onwards comes to deal with pyramids having curved sides equivalent to those with straight sides. He calls such figures 'falcates'. In this way he constructs screws or spirals, seeing them simply as 'curved pyramids' (CA 235 vb). Eddies in water or air are examples in nature of such figures (see Figures 4.11 and 4.12). As an eddy in water deepens it becomes more pointed, its sides constituting a cone. But within the curved wall of the eddy, water is moving round in circles which 'diminish by degrees', i.e., 'pyramidally', towards the apex of the eddy. Leonardo asks, 'Why do not the sides of the eddy fall in'. He finds the answer in the circular movement round the side representing the constant force of 'accidental weight' of water in water. Since the shape of the eddy will persist only as long as the accidental weight of the water persists, and since this 'accidental weight' will remain constant only so long as its velocity and weight together (impetus) remain constant, so, as the point of the eddy cone is reached the velocity of the circulating water increases, or as Leonardo puts it, 'The helical or rather rotatory movement of every

liquid is swifter in proportion as it is nearer the centre of its revolution. This is a fact worthy of note since movement in a wheel is that much slower as it is nearer the centre of the rotating object' (CA 296 vb). This type of movement is later studied particularly intensively in relation to the movement of blood in the heart and aorta.

The main features of wave motion have been described in Chapter 2. Leonardo's deep understanding of it is reflected in studies to be found on Madrid II 126 r (Figure 4.13a). Here comparison with Huygen's figure of the wave motion of light is shown (Figure 4.13b). In two small red chalk drawings Leonardo describes the rebounding wave-fronts so showing his awareness of the continual creation of waves from all points of contact with one another: 'The reticulation of the water surface will have as many lines as there are lines that rebound from the points percussed by the incident motion of the water'. Similar waves are drawn by Leonardo in describing the movement of blood in the aorta.

The movement of 'rebound' is primarily a manifestation of percussion on a resistant object. A single 'simple rebound' is distinguished by repetition from Leonardo's '*moto ventilante*' which describes oscillation such as that of a pendulum or that of a knife whose blade is stuck into a wooden surface (A 33 r and BM 2 r). In *Codex Trivulziano* 33 v he

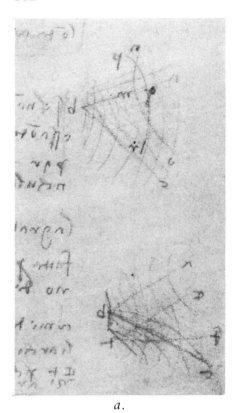

Figure 4.13.

a. Pyramidal and wave movements combined.

'The reticulation of the surface of the water will make as many lines as there are lines which rebound from the points percussed by the incident motion of the water' (Madrid II 126 r).

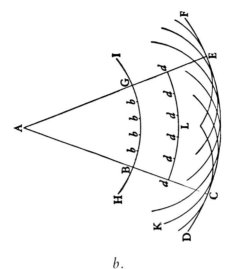

a.   *b.*

b. Huygens's diagram of the spread of light from a luminous point A.

He, too, describes the spread of light by straight lines combined with waves created by each particle, which makes a wave of which it is the centre.

Figure from Christian Huygens, *Treatise on Light* (1690), trans. by Silvanus P. Thompson (Chicago, University of Chicago Press, 1952), chap. 1.

describes this oscillatory movement as 'standing still by diminishing percussions'. In MS A 26 v and Madrid I 147 r (Figure 4.14) he applies it to pendular movement, noting that the 'accidental' movement of the rising pendular bob never equals the 'natural' movement of descent; in clocks this demands regulation by escapement mechanisms. As the pendulum makes smaller angular movements to each side, so these movements become more and more nearly equal (Madrid I 183 r). Later, Leonardo analyses the different types of *moto ventilante*. He defines it as 'a flexible movement which cannot follow its impetus in the direction in which it was created', so that the incident movement is converted into a reflected one, and this continues until 'the impetus dies out' (BM 2 r). It will be noticed that this description corresponds precisely to the oscillations of an object thrown down a hole to the centre of gravity of the world, as described in *Forster* II 59 v. He illustrates such movements with figures of a pendulum and a knife stuck in a board. More significantly, he extends it to the reflected movement of water or air inside curved bowls and so to expanses of water within curved bays. In this way he reaches the conception (which he never relinquishes) that the tides of the sea are manifestations of such oscillatory movements. By such steps he reaches the idea that the ventilatory movements of breathing, of the world and man (macrocosm and microcosm) are tidal and reciprocal in nature, as are those of contraction and dilatation of the heart (systole and diastole). Moreover, such oscillatory movements are often 'sensible', i.e., occur in 'sensitive' bodies such as animals and man (CA 360 ra).

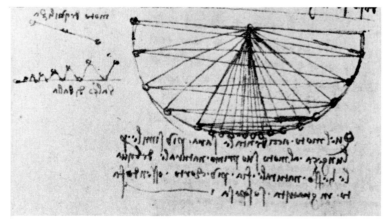

Figure 4.14.   '*Moto ventilante*' illustrated by pendular movement.

A straight cord attached to a fixed point, if released (top left), will always fail to rise to the same height on the opposite side. This inequality of movement is drawn occurring progressively in smaller arcs until the pendulum bob comes to rest below its point of fixation. Below this drawing Leonardo notes, 'The smaller the first natural movement (downwards) of a suspended weight the more equal will be the following accidental movement upwards' (Madrid I 147 r).

It will be noticed that he here forestalls Galileo's famous observation of pendular motion in the cathedral at Pisa.

In the little drawings on the left labelled 'motion of a balance' (upper) and 'bounce of a ball' Leonardo indicates that these, too, are examples of '*moto ventilante*'. Such little sketches are reminders to make further studies on these points.

The evolution of Leonardo's ideas on circular, or 'circumvoluble' movement, as he calls it, is ultimately significantly applied to human physiology, e.g., the movement of blood in the aorta.

Using his perspectival or pyramidal view, Leonardo finds that the angular or circular movement of a beam at one-third of its length enters three times into the movement of its end (A 7 r). In *Codex Atlanticus* 396 ra Leonardo makes his first observations on the movements of a spinning top, a problem which some years later he links with the larger problem of wheels rotating on perpendicular and inclined axles (CA 305 vb) (see Figure 4.15a). It is in this context that he notes that 'that part of a wheel revolves with least movement which is nearest to its centre of revolution' (E 36 r) and as we have seen, contrasts this with the spiral movements in the wall of an eddy (CA 296 vb). He finally comes to the conclusion that 'the revolving movement of a weight round a fixed axis such as a spinning top will persist as long as the impetus of the revolving movement is not less than the power of inequality' (E 50 v), i.e., the movement which shifts its centre of gravity (see Figure 4.15b). When this happens the top tilts, its point of contact with the ground describes a spiral course as illustrated on MS E 34 r, and it finally falls on its side, the centre of gravity of the whole mass now having shifted to lie between the two points of contact with the ground.

Leonardo also points out that the more distant a circular movement is from its centre (i.e., the longer its radius), the more such a movement approaches a straight line, so that if separated from its mover it goes off in a straight line, i.e., tangentially. He recognises centrifugal force created by circumvoluble movement, describing it in relation to a man's arm at wrist and elbow, or the use of a sling. However, as mentioned in MS A 7 r, he thought that the centrifugal force increased 'pyramidally' with the radius and so the force of the projectile was in simple proportional relation to the force created at the centre. For example, the force of the stone from a sling is proportional to the force at the shoulder of the throwing or sling arm (W 19117 v; K/P 115 v). This proportional relation was reached partly by experiment which he performed on the movements of rotating circular plates described on *Forster* II 68 v. Since Leonardo early came to realise that many joints of the human body are constructed for circumvoluble movement, particularly the 'universal' ball and socket joints of the hip and shoulder, he applies these principles of circumvoluble movement to the movements of arms and legs. Needless to say, Leonardo first describes and draws on Madrid I 100 v such 'universal joints' as part of his investigations of the movements of axles (see Figure 4.16). Circumvoluble movement also plays a big part in the formation of eddies in the aorta by which the aortic valves are closed. (See Figure 15.18.)

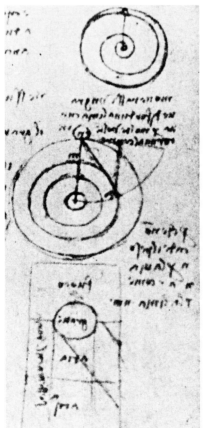

a.

b.

Figure 4.15.

a. 'Circumvoluble' or revolving movement (BM 175 v).

Below the top drawing Leonardo writes, 'A weight will be moved away along a curved line many times around its centre'. This curve shown for a top is drawn here. In the drawing below he deals with a weight falling to earth through the revolving elements of fire, air and water labelled in the bottom drawing.

b. The curved route of a hemisphere rolled on its side as it comes to rest is drawn here.

This particular experiment Leonardo probably derived from Nicolaus Cusanus (E 34 v).

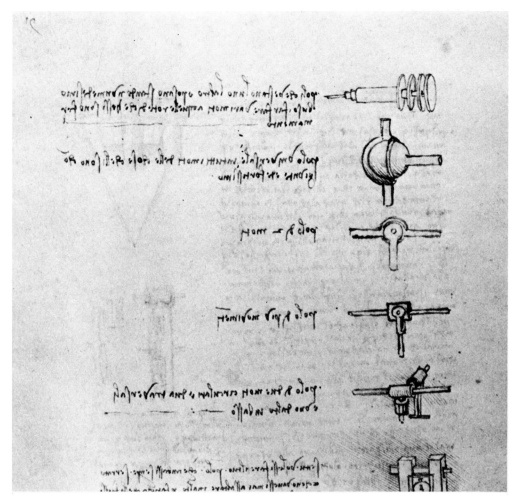

Figure 4.16. The universal joint.

Beside the top drawing Leonardo writes, 'Axles or joints which enclose one another or are placed in the same hole to make different movements of those wheels to which they are fixed'. The drawing below he describes as 'A universal joint permitting all movements to objects joined to it; it is very strong'. The other joints below are capable of two or more movements, vertical or horizontal (Madrid I 100 v).

The universal joint has been generally attributed to Girolamo Cardano and called the Cardan shaft. Cardan's father was a friend of Leonardo, and Girolamo had intimate knowledge of Leonardo's notebooks.

CLASSIFICATION OF MOVEMENT

In the course of his investigations of movement Leonardo made a number of different classifications to suit different contexts. For example, he differentiated between the 'sensible' movement of live men and animals and the 'insensitive' movements of inorganic bodies. The movements of both man and inorganic bodies he finally classified into three types: 'Movement is of three kinds, i.e., local, simple action, and a third motion compounded of action and local [movement]' (*Li moto sono di tre spetie cioe lochale, actionale semplice et il terzo e'moto composto d'actionale colochale*) (CU 111 r). These he describes thus: 'Local movement is when the animal moves from place to place and the movement of action is movement which an animal makes in itself without change of place' (*Il moto lochale e quando l'animale si move da locho a locho el moto actionale e'l moto che fa l'animale in se medesimo senza mutation di locho*) (CU 111 r). 'That is called simple movement in man when he simply bends forward or backwards or to the side . . . That is called compound movement in man when some purpose requires bending down and to the side at the same time' (TP I 356, 357).

To simple and compound motion he finally adds '*moto deconposto*' (*Madrid* II 53 v), i.e., doubly compounded motion. This he illustrates in Figure 4.17, where he demonstrates a compound motion further compounded. He uses the same term to describe the complex movements of the neck which may consist of simple movements such as nodding or bending of the head; compound, consisting of a mixture of the two; and decomposite (*deconposto*), when these compound movements are performed with 'one eye above the other' (W 19003 v; K/P 137 v).

THE MOVEMENT OF TIME

All forms of movement are inseparably connected with time by their velocity. Leonardo gives his attention to time from his earliest years in Florence. In Book XV of Ovid's *Metamorphoses* Leonardo read of Pythagoras's views that 'All things are ever changing, nothing perishes . . . All things are flowing onward . . . Even time glides onwards with constant progress no otherwise than a river'. All these views are to be found assimilated into Leonardo's thoughts on time and movement, whether of inanimate inorganic things or living bodies. From such beginnings he came to his far-reaching concepts of perpetual geometrical transformation of the macrocosm of nature, and the microcosm of man's body.

We have already seen (p. 95) how in Leonardo's hands the measurement of time soon took shape in the invention of clocks of various types: water-clocks, clocks on the sand-glass principle, striking clocks (CA 288 ra; W 12688 and 12716), alarm clocks (B 20 v), sundials (H 97 v), a siphon clock (G 44 v), a piston-weighted clock (CA 12 va) in which 'the descent will show you the hour', and a mercury clock, 'divided into hours, minutes and seconds' (BM 191 v). These are but a few examples of Leonardo's endeavours to measure time down to the smallest interval possible. His most systematic attack on the problem of clock construction is to be found on many pages of *Madrid Codex* I (see Figures 4.3 and 4.5, p. 96). Here the use of the spring as driving force, of the fusee for its regulation and of various forms of escapement mechanisms are tried out and discussed in detail. Here, too, these regulatory devices are applied to a pendulum clock. The whole progressive pattern of Leonardo's work on the measurement of time is explained in detail by Ladislao Reti and S. A. Bedini in *The Unknown Leonardo* in the section entitled 'Horology'.[1]

Leonardo attempted to produce a clock which would 'Divide an hour into 3,000 parts; and this you will do by a clock, lightening or weighting the counterweight' (BM 191 r). However, in his practical work he uses the 'harmonic tempo' of 1,080 beats to the hour, as for example in counting the pulse rate (W 19081 r; K/P 164 r), and for a time and motion study of men shovelling earth (BM 207 r).

However, Leonardo was not satisfied with merely measuring time. In a set of notes which convey the impression of an almost passionate attempt to define such terms as 'the point, angle, surface', etc., he notes, 'The instant does not have time; and time is made from the movement of the

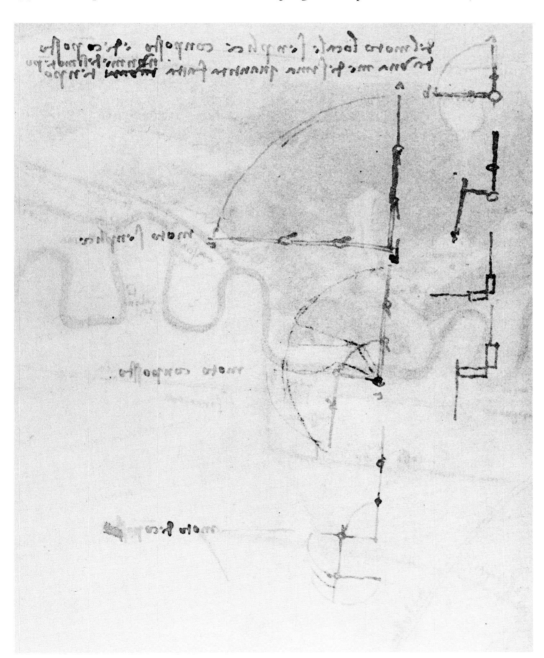

Figure 4.17. 'On local motion, simple, composite and decomposite motion' (Madrid II 53 v).

Beneath this heading Leonardo makes three diagrams, the top one labelled 'simple motion', the next 'composite motion', the bottom one 'decomposite motion'. The ranges of the different movements are indicated. However, the meaning of these terms is perhaps more clearly shown by the series of diagrams on the right margin, particularly if one sees them as the leverage of upper arm, forearm and hand. These diagrams are somewhat obscured by the drawing of a map of the Arno Valley coming through from the opposite side.

instant; and the instant is the end of time' (BM 176 r). This definition comes alongside his definition of a point: 'The point has no parts; the line is the transit of a point; and points are the ends of lines'. The analogy between the two is unmistakeable. He then adjures himself to 'Write on the quality of time apart from its geometry'. He explains this in *Codex Arundel* 173 v, where he accompanies similar definitions of the point and the instant of time with the observation that 'Although time is numbered amongst continuous quantities, yet through being invisible and without body it does not fall wholly under geometrical power which is divided into shapes and bodies of infinite variety as may be seen constantly in visible and corporeal objects. But only in first principles do they come together in point and line'. And so he is driven back to defining the instant of time as 'nothing'. 'That which is called nothingness [*niente*] is found only in time and speech. In time it is found between the past and the future and retains nothing of the present . . . Amid the vastness of the things among which we exist the existence of nothingness holds the first place; its function extends over all things that have no existence, and its essence resides between the past and the future and possesses nothing of the present . . . and in nature it is included amongst things impossible; hence it has no existence' (BM 131 r). This abstract reasoning may be clarified by Leonardo's concrete example: 'In rivers the water that you touch is the last of what has passed and the first of that which is coming; so with time present' (Triv 34 r). The close relation between movement in space, time and light finds expression in Leonardo's statement, 'Observe the light and consider its beauty. Blink your eye and look at it again. That which you see was not there at first, and that which was there is no more' (F 49 v). But however fast light moves, its velocity is not infinite; for in discussing movement within the elements, Leonardo notes that the fastest movement of all, even faster than that of fire or light, is that of the mind. 'Mind ranges over the universe but being finite it does not extend to infinity' (H 67 r).

The movement of time is more basic than the movement of any other power of nature. 'Movements are of 3 natures of which the first is called temporal because it is concerned solely with the movement of time; and this embraces within itself all the others . . . But among these [other] movements that of the images of things in the air is swiftest because it covers a very long distance in the same time as a very short distance, and this [movement] is lost through distance because the air thickens' (CA 203 va).

On most occasions Leonardo subscribes to the Aristotelian relationship between time and movement. 'If a power moves a weight a certain distance in a certain time the same power will move half the weight double the distance in the same time' (CA 212 vb). But he breaks away from this relationship later, saying 'It would follow from this that an atom would be almost as rapid as thought itself, or as the eye which ranges in an instant to the height of the stars . . . This opinion is therefore condemned both by reason and experience'. He continues on the same theme, likening the 'atom' to smoke from a gun and stressing the absurdity of

such a 'rule'. He cannot resist exclaiming, 'You investigators therefore should not trust yourselves to authors who, by employing only imagination have wished to make themselves interpreters between nature and man, but [trust] only those who have exercised their intellects not only with the signs of nature but with the results of their experiments' (I 102 rv).

On many occasions Leonardo pushes back geological or macrocosmic time regardless of the biblical boundary of some 7,000 years. He draws his own vivid word picture of himself looking into a cavern, 'my back bent and arched, my left hand clutching my knee while with the right I made a shade for my lowered and contracted eyebrows' (BM 155 r). And there he saw the fossil fish which he apostrophised, 'O powerful and once animate instrument of constructive Nature, thy great strength not availing thee, thou camest to abandon thy tranquil life to obey the law which God and time gave to generative nature . . . O time, consumer of all things, changing them within thyself, thou givest to the taken lives new and different habitations. O time, swift despoiler of created things, how many kings, how many peoples hast thou undone, and how many mutations of state and circumstance have followed since the wondrous form of this fish died here in this cavernous and winding recess. Now destroyed by time thou liest patiently in this confined space with bones stripped and bare serving as armour and support for the mountain placed above thee' (BM 156 r).

And in proving that 'the shells at a height of a thousand *braccia* were not carried there by the Deluge which happened at the time of Noah' (Leic 3 r and CA 155 rb), Leonardo conceives of the sea, 'where we now see the flourishing city of Florence' (Leic 9 r), 'and the peaks of the Appenines stood up in this sea in the form of islands surrounded by salt water' (Leic 10 v). 'Since nothing is born in a place where there is neither sentient, vegetative nor rational life . . . we can say that the earth has a vegetative soul, and that its flesh is the soil, its bones the arrangement and connections of the rocks of which the mountains are composed . . . its blood the veins of water . . . and the heat of the world soul is the fire that pervades the earth' (Leic 34 r). Therefore, the 'Earth has lungs, nerves, muscles and cartilage within itself' (CA 260 rb).

Thus, geological time for Leonardo began with the formation of the living Earth, and it will continue until the death and destruction of that Earth. This he vividly depicted in the so-called Deluge drawings.

This intense awareness of the changes wrought by time Leonardo focuses on the anatomy and physiology of the human body—the microcosm. One of the many plans for the arrangement of his book on anatomy begins, 'This work should commence with the conception of man, and should describe the form of the womb, and how the child lives in it, and to what stage it resides in it and in what way it is given life and food. Also its growth, and what interval there is between one degree of growth and another; and what it is that pushes it out of the body of the mother . . . Then you will describe which parts grow more than others

Figure 4.18.

A sketch of Helen moulded
by the hand of time (Forster III 72 r).

after the infant is born; and give the measurements of a child
of one year. Then describe the grown-up man and woman
and their measurements' (W 19037 v; K/P 81 v).

Thus, he focusses his attention on the daily changes in the
body brought about by breathing, feeding and excretion,
even to the extent of realising that balance of intake and
output which is summed up in the term *metabolism*.

Thus, too, Leonardo is keenly aware of the external and
internal changes in the human body brought about by ageing, for example, in the beauty of women. And he ends a
paraphrase of Ovid's *Metamorphoses*: 'O envious age, you
destroy all things and consume all things with the hard teeth
of old age, little by little with slow death. Helen when she

looked in her mirror and saw the withered wrinkles which
old age had made in her face, wept, and wondered to herself
why ever she had twice been carried away' (CA 71 ra). See
Figure 4.18.

For Leonardo Time is the great transmuter.

WEIGHT

The nature of the centre of gravity troubled Leonardo
over many years during which he attempted definitions on
dozens of occasions. His final view has been summarised in
Chapter 3, describing his concept of the element earth. Here
we shall try to show how he viewed weight as one of the
natural powers. It was fundamental to him that none of the
four elements weighed at all when in their own element;
that is, air does not weigh in air, water does not weigh in
water, etc., except by virtue of movement. This he demonstrated experimentally in *Codex Arundel* 211 r. One of his
early experiments consisted of defining and locating the
centre of gravity of a body by suspending it. He shows in
Figure 4.19a how 'The centre of all suspended weights is
established under its point of suspension'. This same principle he applies to bodies of any size and shape, alive or
dead. For example, some fifteen years later he draws a simi-

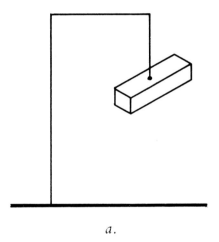

a.

Figure 4.19.

a. Centre of gravity of a beam.

Here Leonardo writes, 'The centre of all suspended weights is
established under its point of suspension' (B 18 v). See also
Figure 3.5.

b. Centre of gravity of a bird.

Opposite the top drawing Leonardo writes, 'This is done to
test for the centre of gravity of the bird without which
instrument the machine would have little value'. Below, he
draws a bird falling downwards; the line of its centre of gravity
is in front of its line of resistance. Leonardo explains, 'When
the bird drops down its centre of gravity is in front of its centre
of resistance; thus if the centre of gravity is on line *ab,* the
centre of resistance is on line *cd*'. In the sketch below he draws
a bird rising upwards. Here the converse applies. He writes,
'And if the bird wants to rise then the centre of its gravity is
behind the centre of its resistance. Thus let the centre of gravity
be *fg* and the centre of resistance is at *eh*' (Sul Vol 15 v).

b.

lar figure (Figure 4.19b), only the block of wood is replaced by the figure of a bird. Against it he writes, 'This is done to find the centre of gravity of the bird'. The core of his investigation into the flight of birds depended on the spatial relation between the movements of the centre of gravity of the bird and its centre of 'resistance'. It needs only the most superficial appreciation of Leonardo's approach to mechanical problems to anticipate that he would apply similar investigations to the moving centres of gravity and resistance in man (compare Figures 4.8 and 4.9).

However, the main instrument through which Leonardo investigated the 'power' of weight was the balance (see Figure 3.5, p. 85). In these studies it has been clearly demonstrated by Marshall Clagett[2] how much Leonardo derived from previous workers on levers and the balance, particularly Archimedes and Jordanus Nemorarius, ex-

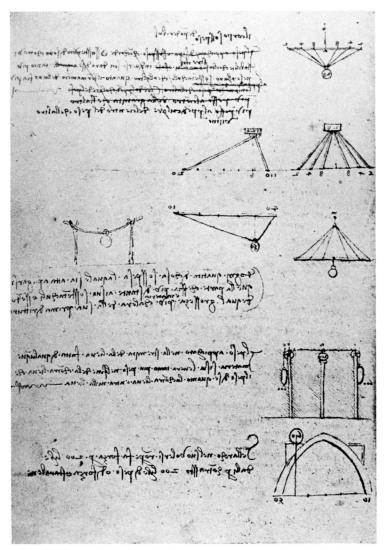

Figure 4.20. Studies of weights on beams (A 47 v).

The proportions between a weight and its distance from the fulcrum (above) are here developed and applied to the bending of a beam (left) and the rupturing of an arch (below) (A 47 v) (Pubblicati della Commissione Vinciana, Rome, 1936).

tracts from whose works are to be found in Leonardo's notes written about 1500. However, Leonardo's investigations commenced about 1487. Their beginnings are to be found in the *Codex Trivulziano* 26 r, where Leonardo states, 'That part of a beam which is furthest from the fulcrum will be less sustained by the fulcrum'. Conversely he asserts, 'All weight makes its maximal load by the perpendicular line of its resistance under it' (Triv 3 r). Later he states that all powers, including weight, 'increase the more they approach the centre' (Figure 4.20) (A 47 v and D 1 r). A modern example of this is the damage caused by the stiletto heels of ladies' shoes.

If the 'power' of weight is maximal when most concentrated, it is minimal when most 'expanded'. Leonardo gives a simple example. 'That body weighs less upon the air which rests upon a greater expanse of air. We may take as an example the gold from which money is made, which is extremely heavy but which when spread out in fine leaf for gilding maintains itself upon the air with each slightest movement of the air' (CA 395 rb). In a similar way, 'a bird acquires more lightness the more it expands its wings and tail . . . thus one infers that by means of a wide expanse of wings a man's weight can be supported in the air' (E 39 r).

On a page (A 5 r) devoted to studies on weights (Figure 4.21) Leonardo describes an experiment worth quoting: 'I ask if the two arms of a balance are divided into equal parts, and if one pound be placed at each of the points, *a, b, c, d, e,* what weight would resist their combined effect at *f*? You do it thus: *a* gives a resistance of one pound; *b* makes a resistance of two; *c* of 3; *d* of 4; *e* of 5; so that the sum gives a resistance of 15 pounds placed at *f*'. This passage illustrates Leonardo's performance of a quantitative experiment revealing that with each equal 'degree' of distance from the fulcrum each unit of weight exerts a force which increases in arithmetical or pyramidal proportion. Leonardo's recognition of the concepts of moments about a fulcrum as measured by the product of a weight and the distance of the line of its action from the fulcrum is repeatedly expressed. He applies it to his calculation of the distribution of weight 'of a man standing on one foot' (W 19008 r; K/P 140 r). Noteworthy is Leonardo's statement that weight makes its 'maximal load' along the 'central line' of gravity through the support of the balance at the fulcrum. His 'central line' of maximal weight power corresponds to the 'central line' of visual power in his figures of linear perspective.

One of Leonardo's neatest and most typical experiments demonstrating the pyramidal power of the arms of a balance is found on *Madrid* I 128 v (Figure 4.22). Here, beneath a very simple drawing of a balance Leonardo writes, 'It is demonstrated here how the power of the arms of a balance is pyramidal, that is, when arm *rh* is straight it counterbalances or resists exactly the arm *hs*. And if you bend one arm by the smallest part possible at one of the ends, coming by the smallest equal degrees to the bending of the whole, you will find at each degree of motion it acquires degrees of weakness against the power of the weight of its opposite arm *hs*. And as every continuous quantity is divisible to

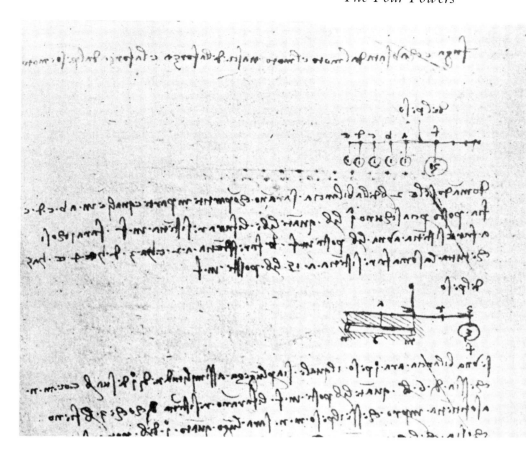

Figure 4.21. The rule of lever and counter-lever.

The top drawing describes the experiment quoted in the text. The two arms of a balance Leonardo describes as lever and counter-lever.

In the drawing below he demonstrates that if a weighted beam, *mn,* is attached to the lever it acts on the centre point of its length, so that its weight of 6 acting at 1 unit of distance (*a*) is balanced by a weight of 3 acting at a distance of 2 units of distance (*f*) (A 5 r).

Figure 4.22. Demonstration that the law of the balance is pyramidal (Madrid I 128 v).

The weight of the arm *hr* is progressively diminished by bending portions down from the end towards the fulcrum, *h*. From the results Leonardo infers that 'every power is pyramidal'.

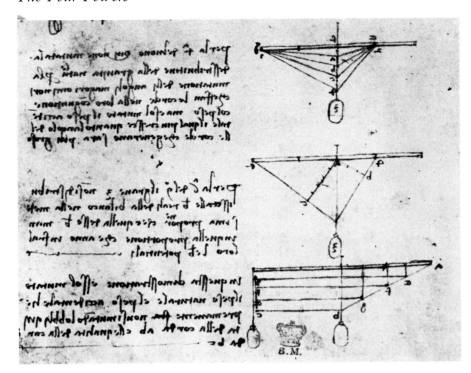

Figure 4.23.   The 'real' and 'potential' arms of the balance.

The top drawing depicts the distribution of a weight in proportion to the angle of suspension from the balance. The middle sketch draws the 'potential' arms. Leonardo writes that 'the weight 3 is not distributed to the real arms of the balance in the same (inverse) ratio of these arms but in the ratio of the potential arms', i.e., *ab* and *ac* (BM 1 v).

These drawings show his logical progress from comparison with Figures 4.20 and 4.22 (pp. 108, 109).

infinity the arm *hs* will acquire infinite power against the diminution of the arm *rh*. Such power grows in proportion to the diminution of the length of the opposite arm. Therefore we will be telling the truth by affirming that it is possible to imagine every power as capable of infinite augmentation or diminution. And consequently every power is pyramidal'.

Between 1490 and 1500 Leonardo made systematic and detailed investigations of all the separate parts of a balance, studies that remind one of his similarly systematic investigations of perspective. In what he calls his 'definition of the balance' he describes the characteristics of a material beam as opposed to a 'mathematical' beam, i.e., a geometrical line. He discusses the effects of different modes and siting of suspension or support of the beam, different weights and their 'appendicoli', or attachments. He systematically alters each of these variables within the context of other constants in this system, so coming to consider the angular balance, the bent-arm balance, etc.. In this way he reaches the concept of the 'potential' arm as opposed to the 'real' or material arm of the balance (see Figure 4.23).

These experiments led Leonardo later to write on 'Where the science of weights is led into error by the practice'. He explains this as follows: 'The science of weights is led into error by its practice which in many instances is not in agreement with this science, nor is it possible to get agreement. This arises from the beams of the balances through which such weights make science. These beams according to the ancient philosophers were given the nature of mathematical lines [i.e., central lines], and a particular place was indicated by a mathematical point. Now these points and lines are incorporeal, whereas practice deals with them as corporeal since necessity demands it for supporting the weight of the

balances together with the weights on them which are to be judged . . . which errors I set down here'. Leonardo then draws (see Figure 4.24) material beams of the balance with three different kinds of axis or central line. 'The central line can run below, through, or above the fulcrum – the perfect one is through the middle, the worst when it is above; that below is less bad'. Finally he concludes that 'Any inequality of weight and length which the arms of a balance acquire in being made oblique is the cause of titubation [i.e., oscillation or *moto ventilante*] of the balance. Through this impetuous motion is generated that which exchanges gravity from the upper to the lower arm' (CA 259 ra). Thus, the 'perfect' balance beam will remain in an oblique position if placed there and not return to the horizontal.

Leonardo's analysis of the 'supports' of a balance leads him to conclude that 'Many small supports bound together can bear much more weight than the sum supported by each added together' (A 3 v). On A 48 v he finds the strength of a support proportional to the area of its base and inversely proportional to its height (see Figure 4.25). As usual, he varies the 'size' of his supports until he is dealing with a great variety of combinations of beams and cords supporting or suspending weights.

With regard to weight on an inclined plane, Leonardo makes a most interesting error. From his authority, Jordanus Nemorarius, Leonardo learned that the weight distribution was proportional to the length of the oblique line over the perpendicular. But he does not always accept this correct solution. Instead he 'solves' the problem perspectivally (and wrongly) by his pyramidal law and measures the distribution of weight according to the obliquity from a perpendicular line starting at the apex of two planes, increasing pyramidally according to each degree of the apical angles. Thus, he believed the weights on the inclines to be inversely proportional to the tangents of these two apical angles of obliquity as represented by the bases of the triangles subtended. (The angle of obliquity for Leonardo is the angle of divergence from the perpendicular from any place on the circumference to the centre of a circle, e.g., the sphere of the earth.) Leonardo continues to make this error to the end (e.g., G 77 v) in relation to weights and velocities on the inclined plane. Compare Figure 4.8 (p. 98).

When, however, he applies the concept of the potential lever arm to the problem of the flight of the bird he obtains a 'correct' answer. Leonardo applies both methods to two similar problems of weights on inclined planes: first, that of the potential lever, secondly, the surveyor's perspectival or pyramidal method, thus obtaining one correct and one incorrect answer (see Figure 4.29, p. 114). This has been beautifully demonstrated by Marshall Clagett.[3] Indeed both can be looked upon as 'pyramidal' or perspectival solutions depending whether the eye is placed at the angles of attachment of cords, for example, or at the perpendicular attachment of the weight to the cords (see Figure 4.23, p. 110). Leonardo uses both methods of observation. It would appear that he did not realise the significance of their differences.

On BM 193 v it would appear that Leonardo discovered the centre of gravity of a tetrahedron by applying his perspectival approach. As illustrated in Figure 4.26, a series of pyramidal figures shows how Leonardo following Archimedes has found the centre of gravity of a triangle at the point of intersection of three perspectival central lines drawn from each angle to the centre of the opposite side at right angles. On one side of this basic triangle he has erected another equal triangle to form one surface of the tetrahedron. This too can be seen to be divided by three 'perspectival [central] lines', i.e., from each angle to the centre of the opposite side 'between equal angles', i.e., at right angles. Finally he has drawn a perspectival line from the right lower angle of the tetrahedron to the mid-point of the opposite surface of the tetrahedron. This cuts the line from the apex to the mid-point of the basal triangle at *a*, the letter *a* coinciding with the junction of the third and fourth quarters of this line from apex to base. Underneath this drawing Leonardo writes, '*a* is between the third and fourth

Figure 4.25. The strength of a support in relation to 'size' and height.

Leonardo writes, 'If the reed *ef* bears one ounce and its height is 100 diameters of its size, *ab* also being composed of 100 diameters of 100 reeds will support 100 ounces because of its proportion to the reed *ef*. If you take 100 reeds like *ef* and bind them together firmly and continuously, since *ef* is of 100 times the size [*grosseze*] and *cd* of 5 times the size, 5 goes 20 times into 100, and so each of the reeds bound together will support 20 times as much as when single; so that if *ef* bears one ounce *cd* will bear 2000' (A 48 v).

Similar considerations are applied to the 'sizes' of the vertebrae of the spinal column on which the skull balances.

[parts] of the axis of this pyramid'. Similar explanatory sketches are drawn on the left of Figure 4.26. This location of the centre of gravity of the tetrahedron is confirmed in the text, which says, 'The centre of gravity of a body consisting of four triangular bases is in the intersection of its axis, and in the 4th part of its length'. The statement is demonstrated by experiment on several occasions, e.g., Forster II 80 r, CA 100 vc and *Codex Arundel* 31 r and 67 r. In each experiment a pyramid is suspended at a point between the first and second quarter from its base. The unequal arms of such pyramidally shaped scales balance the whole pyramid. In one case (BM 67 r) the pyramid is described as 'round', i.e., a cone.

Leonardo's view of the nature of gravity changes with the years. After about 1505 he habitually speaks of a body 'having three centres', a centre of magnitude, a centre of natural gravity and a centre of accidental gravity (E 68 v). We have seen how he interprets the centre of natural gravity by suspension and by geometrical means. Natural gravity acts at every point (BM 187 r) along the 'intercentric line' – the line between the centre of gravity of the body and that of the world (CA 115 rab). Accidental gravity is the centre of gravity of a moving object, its centre of 'resistance'. Movement shifts the centre of gravity, as we have seen, in the case

Figure 4.24. Central lines of the arms of a balance.

The three drawings are taken from Leonardo's diagram describing central lines which can run 'below, through or above the fulcrum' (CA 93 vb).

of a moving man or bird, and moving the 'elements' shifts the centre of gravity of the world. It does the same in a moving pyramid, for example, so that if thrown into the air with its point forwards it turns round so that its base comes to the front (G 51 r). Its centre of accidental gravity is now placed at a third of its length from the base, not a quarter (CA 100 vc). Similar changes in the location of the centre of accidental gravity occur, of course, in the movements of the bodies of man and animals and in parts of the body, particularly the arms. The faster a man or a horse runs, the more does accidental gravity dominate over natural gravity, therefore the less he weighs on the ground. Thus accidental gravity can be looked upon as created by the 'impetus', i.e., the momentum of a moving body. Such impetus will, of course, act not only on a body travelling horizontally but also on one falling to the ground. This gives rise to the acceleration due to gravity which Leonardo recognises as a 'pyramidal' form of acceleration, i.e., with 'every degree' of time there is an increase of a degree of velocity.

'Gravity that descends freely', writes Leonardo, 'in every degree of time acquires a degree of movement' (M 45 r) (see Figure 4.27). This he illustrates with a pyramidal figure on M 44 v explaining the use of this 'pyramid' in now familiar terms: 'Prove the proportion of time and movement together with the speed made in the descent of heavy bodies by the shape of the pyramid, because the aforesaid powers are all pyramidal since they commence in nothing and proceed to increase in arithmetical proportion' (M 44 r).

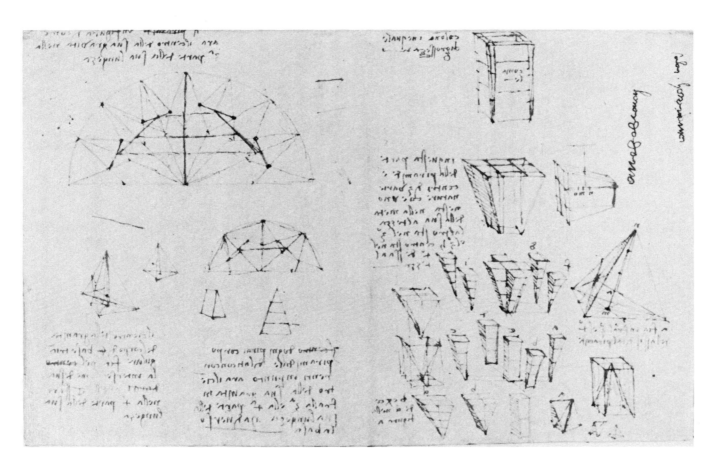

Figure 4.26.   The centre of gravity of a pyramid.

In drawings in the middle of the right and left margins two tetragons are shown. Their triangular bases are divided by central lines from each angle to the centre of the opposite edge. These intersect at *m* (right drawing). From *m* a perpendicular line is raised, *mn*; *m* forms the apex of a triangle on one side of the basic triangle. Its centre is marked in a similar manner, and a perpendicular line from the opposite angle is drawn to this central point. It cuts the line *mn* at *a*. Thus Leonardo writes, '*a* is between the 3rd and 4th parts of the axis of the pyramid'. This point is its centre of gravity.

Below the similar drawing on the left margin Leonardo writes, 'The centre of gravity of a body consisting of 4 triangular bases (a tetragon) is in the intersection of its axis, and will be at the 4th part of its length' (BM 193 v).

These geometrical findings he confirmed by weighing.

a. *b.*

Figure 4.27.

a. The acceleration due to gravity.

This is here described by the 'pyramidal' figure (as explained in Figure 2.5, p. 48), realising that Leonardo took a two-dimensional section of a pyramid, i.e., an isosceles triangle, to express this proportion. Alongside this drawing he writes, 'These units are marked so as to show that the excesses of each degree are equal'. The page is headed: 'Of local motion, which being natural is of finite pyramidal power'. 'The weight which descends by natural motion in each degree of motion acquires a degree of velocity. And always when it has doubled its motion it will have doubled its velocity; and thus it will continue successively to the centre of the world' (Madrid I 88 v).

b. Acceleration due to gravity is again expressed by a similar pyramidal drawing.

Leonardo explains, 'The natural motion of heavy things at each degree of its descent acquires a degree of velocity. And for this reason such motion as it acquires power is pyramidal because the pyramid acquires similarly in each degree of its length a degree of breadth. Such a proportion of acquisition is found in arithmetical proportion, because the parts in excess are always equal' (M 59 v).

For further comment on Madrid I 88 v, see p. 143.

He has now learnt to equate pyramidal with arithmetical proportion. He then explains the pyramidal proportions marked in the accompanying figure. In MS M 44 v Leonardo couples distance with time, saying that with 'doubled quantity of time the length of the descent is doubled'. He has failed to detect the difference between the time and distance relations of the acceleration of gravity by seeing the problem only in perspectival pyramidal terms.

On M 42 v (Figure 4.28) Leonardo repeats his experiment on the acceleration due to gravity, using the inclined plane. The illustration brilliantly illustrates his thought, for the inclined plane is represented by an oblique 'pyramid' on

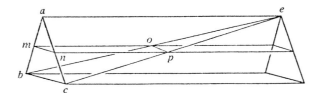

Figure 4.28.   The acceleration of gravity on an inclined plane.

Here too Leonardo adheres to the view that acceleration is in 'pyramidal' or arithmetic proportion. He writes, 'Although the motion is oblique it observes in its every degree the increase of the motion and its velocity in arithmetic proportion' (M 42 v).

The inclined plane is represented by the triangle *ebc*, vertical fall by the 'pyramid' in perspective *abc*. Both being on the same base and between the same parallel lines are 'equal'. Horizontal lines drawn at half their heights *mn* and *op* are equal. Velocities here and at their common base, *bc*, are therefore equal, but their times are unequal, in proportion to the deviation or obliquity of the plane *eopbc* from the vertical plane *amnbc*.

Figure from Marshall Clagett, "Leonardo da Vinci: Mechanics," in *Dictionary of Scientific Biography*, ed. by C. C. Gillispie (New York, Scribner's, 1973), vol. 8, p. 215.

the same base as a vertical 'pyramid' and of the same height. Such 'pyramids' are equal and their bases at half the height are also equal; this too is illustrated. Once more Leonardo describes 'the motion and the speed' as increasing in arithmetical proportion. In these circumstances one might expect Leonardo to have measured the distance covered by a descending ball and so to have exposed his error, as did Galileo.

'Levity' was as real to Leonardo as 'gravity'. Both are relative, created by the movement of elements displaced out of their 'natural' place. If an element is displaced upwards, 'gravity' is created; if displaced downwards into dense elements, 'levity' is created. Many experiments are done which consist of 'pushing' air under water to observe the phenomena of 'levity' in rising bubbles of air. Fire in its turn rises in pyramidal form in air because of its 'levity', as we have noted with the candle flame (see also Figure 3.8, p. 88).

In a remarkable series of experiments described on Madrid I 145 v (Figure 4.31, p. 116) and *Codex Arundel* 181 r Leonardo measures the 'weight' of air in water and water in air, using the surface of the water as a fulcrum. From these experiments and many observations that no element possesses weight or levity in its own element, he comes to the conclusion that heaviness and lightness are relative only, not naturally absolute phenomena. Their expression is written in movement. For example, the weight of the element earth in water is expressed by its movement downward; the lightness or 'levity' of air in water is expressed by its movement upwards. An 'element' can 'weigh' in itself if it moves, i.e., water moving within itself can exert weight downwards. But the power of movement in any direction comes into the category of a 'force'. Therefore, says

Leonardo, there is no such thing as 'natural' weight in the Aristotelian sense. All weight is 'accidental', i.e., exerted by force (W 19060 r; K/P 153 r).

Outside the spheres of the four elements objects are weightless. For this reason the stars remain in their places and the moon does not fall into the earth. Moreover, the moon and other celestial bodies are, in his view, surrounded by their own coats of the four elements of which they are the centre (BM 94 r). Thus the moon has its oceans of water like the earth, and these ocean waves reflect the light of the sun to the earth, just as our oceans reflect the light of the sun, so producing the faint rounded image between the 'horns' of the new moon (BM 94 v).

Such was one train of thought initiated by Leonardo's view of the relativity of heaviness and lightness. From another point of view it can be seen as an extension of the concept of specific gravity. Leonardo gained some idea of specific gravity from his acquaintance with the work of Archimedes. On one of his drawings for a canal he refers to the fact that 'the great weight of a barge which passes along the river supported by the arch of the bridge does not add weight to this bridge because the barge weighs exactly as much as the weight of the water it displaces' (CA 211 va). This practical observation is generalised in MS H 92 r, 'as much weight of water will be displaced as the total weight that displaces the water' (see Figure 3.4). He then proceeds to compare the weights of other objects in air and water. On

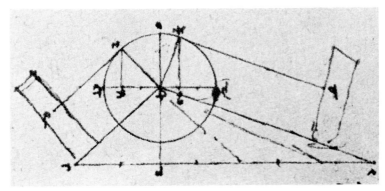

Figure 4.29.   The distribution of weight on inclined planes (Sul Vol 4 r).

In this drawing Leonardo reaches a correct solution when using the concept of the potential lever arms *dn* and *dm* round the fulcrum *d*, where distribution is shown to be proportional to the sines of the angles *dca* and *dac*. But when he uses the angles of obliquity *bdc* and *bda* he distributes the weight as inversely proportional to the tangents of these angles, i.e., in accordance with pyramidal, perspectival or surveying principles. This error therefore recurs.

It may be noted that this same principle applied to the problem of acceleration of gravity is correct, as in Figure 4.28.

Figure from Marshall Clagett, "Leonardo da Vinci: Mechanics," in *Dictionary of Scientific Biography*, ed. by C. C. Gillispie (New York, Scribner's, 1973), vol. 8, p. 215.

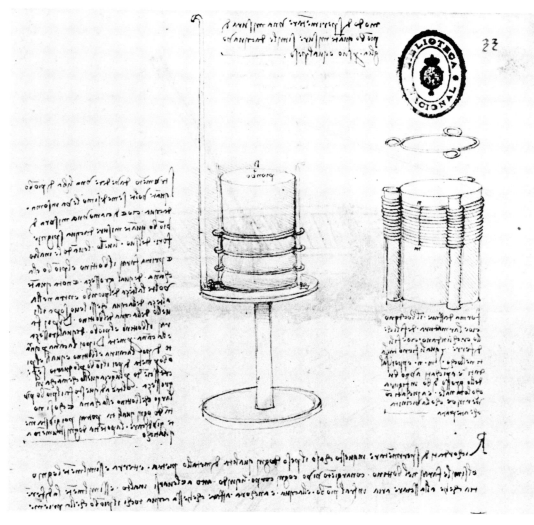

Figure 4.30. Specific gravity (Madrid I 33 r).

This page is headed, 'An experimental method of finding out how many measures of water are needed to counterbalance one measure of lead'. On the right are instructions for surrounding a piece of thick oxen intestine with a framework of strands of iron wire. Thus he constructed the cylinder drawn on the left. From its base emerges a pipe up which water is displaced by the lead (labelled 'pionbo') fitted onto the surface of the water in the cylindrical 'vessel'.

a page in Madrid I 33 r containing remarkable experiments on this subject, Figure 4.30, Leonardo constructs a cylindrical 'instrument' for ascertaining 'how many pounds of water are carried [balanced] by one pound of lead', the water being displaced through an attached pipe arising from the bottom of the cylinder. He continues, 'Remember to experiment in the same way with every type of metal, stone, earth and also wood . . . You will likewise proceed by placing quicksilver and other liquid bodies capable of rising up [into the attached pipe] . . . Should you wish to find out how many measures of water are needed to counterbalance a measure of lead proceed this way'. He then describes how to make a lead cast in wax: 'empty out the lead and fill the wax cast with water, and compare the weights of the enclosed lead and water'.

In *Madrid Codex* I 150 v this experiment is elaborated by lever systems designed to assist in collecting the water displaced by the lead in the cylinder. Leonardo's measurements tell him that 'Every jar of lead requires 12 jars of water and it will have the same weight . . . the water in the pipe will never be raised more than 12 times the height of the lead, but never less'. Thus he measures the specific

gravity of lead as twelve times that of water.* 'Wood' he considers 'loses all its weight' and is therefore of the same specific gravity.

When he comes to 'weighing' air in water Leonardo finds that it has negative weight, or 'levity'; 'it wishes to rise and the water is poured out at the top' (CA 109 va). 'Levity' is therefore the opposite of 'gravity' in that the air wishes to rise from water into its own element with the same 'weight' as water wishes to fall through air into its own natural element. Leonardo tries to demonstrate this experimentally on several occasions, in *Codex Arundel* 37 v and 181 r and *Madrid* I 145 v (Figure 4.31). This figure contains the clearest illustration of his view of the problem. Here in the uppermost of the four drawings on the right-hand side of the page, two bags of air are seen attached to the arms of a balance, the whole immersed in water. One bag, labelled 8, is clearly twice the size of the other, labelled 4. The levity of the air in water is symbolised at the left border of the drawing next to that of gravity. Below is a drawing showing a container of water above air connected to a container of air

---

*The specific gravity of lead is 11.35.

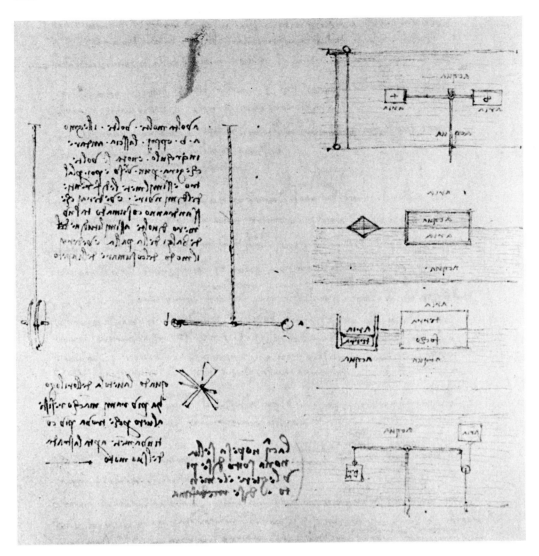

Figure 4.31. The relativity of gravity.

On the right side of this page are four drawings in which the gravity or levity of one 'element' within another is considered. In the top drawing a balance under water has at one end a load, 4, of air, at the other a 'weight', 8, of air. Clearly the air goes upwards and rotates the arms of the balance accordingly.

In the drawing below, water and air are put into compartments, using the surface of the water as the fulcrum; the air rotates upwards. Below this, to the left, air is shown separate from earth whilst to the right, earth is placed above fire. Both use the surface of the water as a fulcrum. The earth descends in both cases.

In the lowest drawing earth is suspended on one arm of a balance in water, air exerts levity round a pulley on the other arm. Leonardo writes, 'Water does not have weight unless it has beneath it an element lighter than itself' (Madrid I 145 v).

below water level. The bottom drawing shows how a volume of air by means of a pulley 'weighs' upwards whilst an equal volume of earth on the opposite arm of the balance weighs downwards. Leonardo's comment on this is, 'Water does not have weight unless it has beneath it an element lighter that itself'. Here he demonstrates the reason why 'male dead bodies float on their backs' (H 31 v; CA 119 va).

These experiments are continued in *Codex Arundel* 181 r with somewhat similar figures except that the supports of the balances are drawn separately in both air and water. Lower on the page they are replaced by his imaginative notion of using the surface of water itself as the support and fulcrum around which water contained in the air above and air contained in the water below can move. Leonardo describes this ituation thus on BM 18 r: 'The balance beneath the water does the same with its weight of air as does the balance with the weight of water in air. Left at liberty, these two balances would touch one another at the same time at the surface of the water. If this instrument is turned round,

at once the two weights are abolished. *ab* is the surface of the water and the fulcrum of the two weights . . . Gravity is a certain action which arises when one element is drawn into another and, not being able to be received there with continuous combat, attempts to return to its own place. Therefore a bladder full of air weighs as much in deep water as a bag of water of similar size in air. This is proved by two bags tied together, one full of water, the other full of air; and if you put the one full of air under water and that full of water in the air in such a way that they cannot be turned round you will see that the surface of the water is the fulcrum of the two weights. Therefore if you turn both of them upside down both weights will die. Gravity is an accidental action made by one element being drawn into another; it has as much life as the desire of the elements for repatriation'. This remains Leonardo's final view on gravity, for he repeats it in a passage 'On the Nature of Gravity' written in 1518, within a year of his death. 'Gravity is an accident joined to bodies removed from their natural place'. And 'On the Nature of Levity' he writes, 'Levity joined to gravity are like unequal weights joined together on a balance, or light liquids placed under dense liquids' (BM 264 r). Thus, levity for Leonardo is negative gravity.

Leonardo's view of gravity in the world may be summarised thus, the centre of gravity of the world is at the centre of the sphere of water. This centre is at the termination of an infinite number of intercentric lines of the force of gravity arising from the four elements and converging to a central point. Around this centre are the concentric spheres of the four elements. These are of relative gravity or levity, at rest in their own spheres, unless moved by outside forces. They exert gravity or levity only if displaced upwards or downwards from their natural positions. Water is the most obviously spherical of the elements. 'All parts of the surface of the water desire to be equally situated from the centre of the elements' (H 37 r). Here then lies the point nearest to the centre of gravity.

GRAVITY OF A DROP OF WATER

But what is the nature of this gravitational force? This question may be answered by Leonardo's concept of a drop of water. Small drops of water are round – the smaller, the rounder. Why? 'The centres of sphericity of water are two', writes Leonardo, 'the universal, which serves the whole of the watery sphere . . . and that which occurs in the tiniest particles of dew which are seen in perfect roundness on the leaves of plants . . . Its surface is drawn to itself equally from every side . . . each part runs to meet the other with equal force and they become magnets one of another with the result that of necessity each becomes of perfect roundness' (Leic 34 v). He describes the formation of raindrops from mist particles in similar terms. This coalescence to form round drops Leonardo ascribes to 'two gravities', one for the centre of all the 'elements', and a second for the centre of the drop which he likens to 'magnetic action when it draws iron' (CA 75 va). In Figure 4.32 he performs an exquisitely delicate experiment on a spherical drop of water to find out 'How it [water] contains within itself the body of the earth without the destruction of the sphericity of its

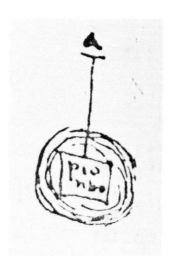

Figure 4.32. How water contains the body of the earth (F 62 v).

For the description of this experiment, see text.

surface. Take a cube of lead [labelled 'pionbo' in the drawing] of the size of a grain of panic grass and by means of a very fine thread attached to it submerge it in the drop. You will perceive that the drop will not lose any of its roundness although it is increased in size equal to that of the cube enclosed in it' (F 62 v). The cube, it should be noted, symbolised Plato's regular solid shape for the world. Thus, in miniature, Leonardo experimentally immersed the world in a dewdrop and found the sphere of water still round. This force of gravity he likened to 'magnetic action'. In relation to magnetic action Leonardo illustrates in Figure 2.10 (p. 52) how its force is related to the mass of magnetised iron. In *Codex Atlanticus* 225 rc he generalises his assertion: 'Among bodies which through a spiritual bond are joined together that which is heavier draws the lighter'. This is the nearest he gets to anticipating Newton's concept of gravity.

FORCE

Leonardo's many efforts to define force have been praised by literary men as poetry and scorned by scientists as speculations, both groups being unable to sympathise with his struggles to give geometrical form to the complex chaos of 'powers' which he observed around him. Clearly 'force' overlapped his other 'powers'. For example, 'accidental weight', as we have seen, came almost entirely to replace 'natural weight'. In the end Leonardo sees all weight as 'accidental' (W 19060 r; K/P 153 r).

One of his earlier attempts at definition of force is the clearest: 'Force I define as a spiritual power, incorporeal and invisible, which with brief life is produced in those bodies which as the result of accidental violence are brought out of their natural state. I have said spiritual because in this force there is active incorporeal life; and I call it invisible because the body in which it is created does not increase in either weight nor size; and of brief duration because it desires perpetually to subdue its cause, and when this is subdued it kills itself' (B 63 r).

This definition, as we have previously noted, describes force in both organic and inorganic worlds, i.e., force arising from the soul, which he calls *sensibile,* as well as the inorganic forces arising in the natural world around us. Leonardo defines force in animals thus: 'I affirm that the said force of movement is based on different poles or fulcra [*poli*]. Force is produced by the diminution and contraction of muscles which draw back, and by the nerves which reach as far as the stimulus [*sentimento*] communicated by the hollow nerve dictates' (B 3 v). Later he speaks of 'two kinds of moving forces', the 'sensitive' (*sensibile*) which have life and the 'insensitive' which have no life (E 52 r).

The primary source of 'spiritual power' which disturbs the 'natural' restful state of the elements is the heat of the sun. 'The sun has substance, shape, movement, brightness, heat and generative power; and all these qualities emanate from itself without its diminution' (CA 270 vb). Heat, light and generative power are the primary causes disturbing the equilibrium of the world and the equanimity of living

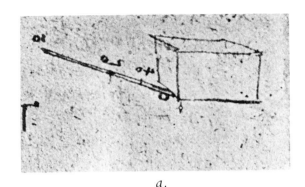

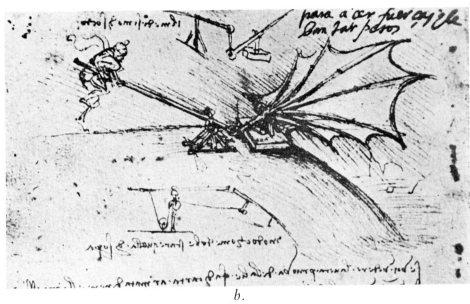

a.

b.

Figure 4.33.   Studies for flight using the lever and balance.

a. '10 pounds at the end of a lever will do the same as 20 pounds at the middle, and as 40 pounds at the fourth part' (A 33 v).

b. This principle applied to rapid lowering of a wing so as to raise a block weighing 200 (florentine) pounds, the weight of a man (B 88 v).

c. Experiment to ascertain the rate and force with which the wing must be lowered in order to raise the weight of a man. Above and to the right, rough sketches of ornithopter power (compare Figure 1.25, p. 33). Right upper margin, figure of a man descending by a parachute 'without sustaining any injury' (CA 381 va).

c.

beings. The effects of heat producing 'circulation' of the elements are described several times. 'Heat and cold', he writes, 'come from the propinquity and remoteness of the sun. Heat and cold produce the movements of the elements. No element has in itself gravity or levity. Gravity and levity without augmentation arise from the movement of an element in itself, in its rarefaction and condensation, as is seen in the air in the generation of clouds through the moisture infused in them' (BM 204 v). Thus, he leads up to his view that 'The cause which moves the humours in all kinds of living bodies contrary to the natural law of their gravity is really that which moves the water pent up within the veins of the earth . . . in continual revolution . . . suffering itself to be sucked up by the heat' (CA 171 ra).

Before 1490 Leonardo had declared one of the most important laws of force which he continued to reiterate throughout his life. As with many of his earlier general statements, it accompanies a particular observation, this time the flight of an eagle (see Figure 4.33). Leonardo writes, 'There is as much force exerted by an object against the air as the air exerts against the object' (CA 381 va). He then continues with his example: 'Observe how the wings percussed against the air sustain the heavy eagle in the upper rarefied air'. It is from this observation and this general 'rule' that Leonardo reveals his more practical objective: 'From these things demonstrated and adduced one should be able to know how man with great wings attached to him making a force against the resistant air and overcoming it, is able to subdue it and raise himself up on it' (Figure 4.33b).

This principle is turned into practical experiment on the same page. Here not only does Leonardo describe the parachute, but he describes and draws a method of ascertaining 'what weight this wing will support' by placing yourself on one side of a balance, fastening yourself to a lever attached to a wing in such a way as to move it down quickly enough to raise your own body up. Sketches of ornithopters in Figure 4.33c show an alternative mechanism.

Repetition of the same law occurs some twenty years later in 1508 in MS F 37 v and in *Codex Atlanticus* 214 rd, where it is stated in relation to the force of water or air against an object; this equals the force of the object against the air or water. 'The amount of movement made by an oar against immobile water equals the amount of movement made by water against an immobile oar' (CA 175 rc). For Leonardo action consistently equals reaction, in all four elements.

POWER TRANSMISSION AND TRANSFORMATION

PERPETUAL MOTION

Leonardo's concern with power transmission led him to investigate the prevailing medieval conviction that perpetual power could be obtained through perpetual motion machines. Undoubtedly at first he accepted this idea. In his early notes (about 1480–1489) he draws up such schemes using the well-known Archimedes screw (see Figure 4.34). He uses all his ingenuity in an attempt to keep water in perpetual motion by means of a feed-back of water carried upwards by one screw, down by another, to be picked up once more by the first from water level (Forster I 41 r and v, and 43 v). Some three years later he denies the possibility of perpetual motion (A 22 v), and a year or so later still he scoffs at such attempts. 'Among the superfluous and im-

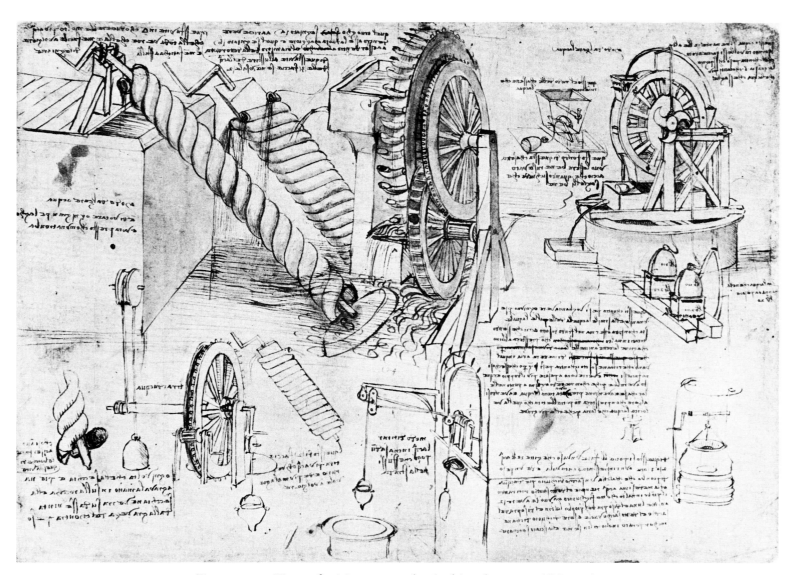

Figure 4.34. Ways of raising water; the Archimedes screw (CA 7 va).

Through the mechanism of the Archimedes screw Leonardo at one time hoped to achieve perpetual motion in water by collecting the downflow and then elevating it again, continuing the cycle ad infinitum.

possible delusions of man', he writes, 'there is the search for continuous motion, called by some the perpetual wheel. For many centuries almost all those who worked on hydraulic and war machines and other subtle engines, dedicated much research and experimentation to this problem, incurring great expense. But finally the same befell them always as befell the alchemists; for a small detail everything was lost . . . I remember that many people from different countries led by their childish credulity went to the city of Venice with great expectation of gain to make mills in dead water; but after much expense and effort, unable to set the machine in motion, they were obliged to move themselves away in great haste' (Madrid I cover). Leonardo had come to realise that 'dead' water, such as still sea water, cannot make power; or as he puts it one cannot 'lift still water out of the sea or lake by instruments, and once lifted in its descent put it to work on mills as if it would work like other water. O fools', he scoffs, as if he had never tried to do it himself.

Leonardo found that water in communicating vessels, whether under the earth or in U-tubes, settled so that 'their surfaces are of equal height' (CA 219 va; see Figure 4.35). It will never lift another quantity of water above itself because water at rest always comes to the surface of a sphere equidistant from the centre of the earth. He also devotes a number of pages to showing why perpetual motion of any wheel is impossible, since any weight will come to rest under the centre of the wheel, even ignoring the factor of friction.

On the other hand, as his experiments on the specific gravity of lead showed, he was aware of the power of the piston and formulated the quantitative relation between piston pressure and area on one side, and area and height of the water column on the other. Leonardo approaches Pascal's Law in relation to the pressures of water on all sides within a cask or a flexible bag in drawings in MS C 7 r and MS E 74 v and states it fairly clearly in Figure 4.36.

In spite of this evidence he can be said to have only partially anticipated Pascal's Law, because Leonardo curiously enough believed that the sphere of water, being lighter than earth, does *not* weigh on the earth. After much debate he

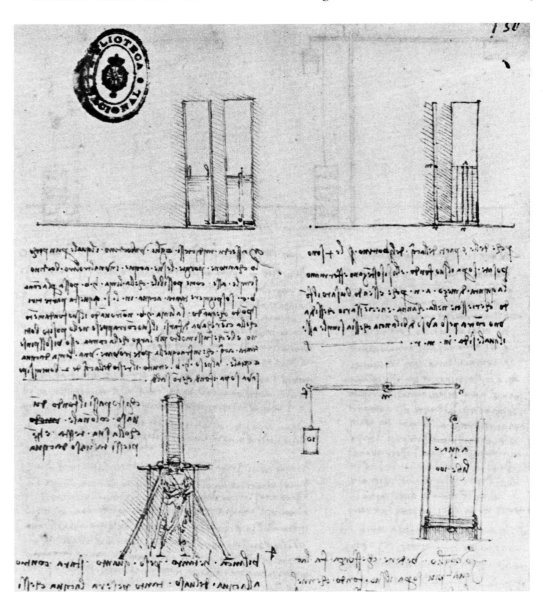

Figure 4.35. The balancing of columns of liquid (Madrid I 150 r).

In the two upper drawings Leonardo shows that liquid connected through U-tubes rises to an equal level whatever the relative sizes of those columns, i.e., narrow or wide.

In the lower drawings he experiments on the distribution of 'the force on the bottom and the sides'. In the right-hand drawing he inserts a leather piston at the bottom of a column of water attached to the arm of a balance in order to weigh this 'force'. On the left he arranges a support for the apparatus and inserts a man's head at the bottom instead of the leather piston. The man is of known weight before the experiment; therefore, presumably he is to be weighed again with the column of liquid on his head, his feet being on the pan of a balance, as in Figure 4.33c (p. 118).

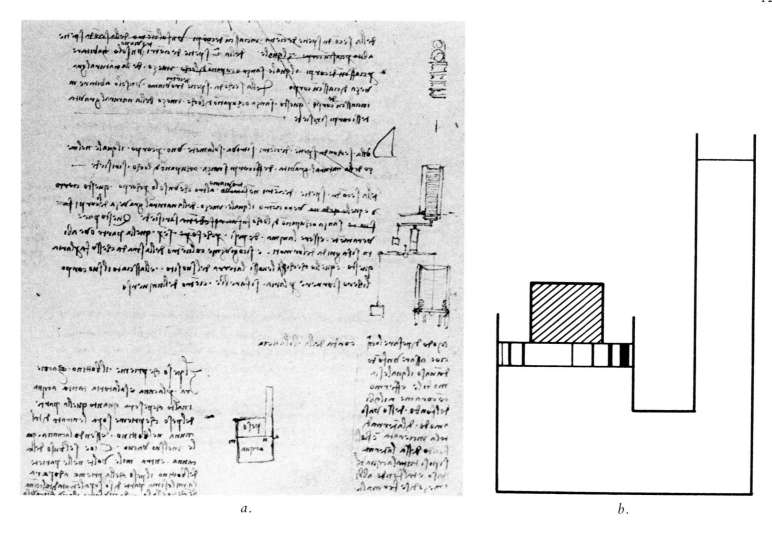

Figure 4.36.   The principle of the hydraulic press.

a. The little diagram labelled 'weight' and 'water' explains the principle upon which Figure 4.36b is based. Leonardo writes, 'The weight which presses the vessel will drive up through the exhalation pipe as much water as that part of the weight which presses on the entrance of the big pipe into the vessel'. *mn* is the dimension of the big pipe, *na* of the 'exhalation' pipe (Madrid I 146 r). The drawings in the right margin are preliminary sketches for the more finished right, lower drawing in Figure 4.35.

b. Here Leonardo shows that the ratio of the pressure of the piston to the weight of water raised in the 'exhalation pipe' is proportional to their two surface areas (Leic 11 r).

was led to this conclusion by two considerations: 'very delicate plants on the bottom wave about in the water', and 'light mud floats in the water making it turbid. If water weighed on these, the plants would be crushed and the mud would be petrified' (Leic 6 r). Thus he failed to appreciate water pressure on earth, let alone the existence of atmospheric pressure.

PULLEY SYSTEMS

Leonardo developed a thorough understanding of pulley systems and their mode of transmitting power. He looked upon a pulley as a circular balance, the fulcrum of which is at the centre with its two beams formed by the radii of the pulley, the pulley-rope acting tangentially. With typical methodical care he assesses the results of adding together fixed and movable pulleys. Some idea of the complex systems he evolved can be obtained from *Codex Atlanticus* 153 ra. Leonardo works out correctly the link between velocity and mechanical advantage in all pulley systems, often enumerating the weights involved by each pulley-rope, as in Figure 4.37a. Finally, he sums up this whole field of mechanical knowledge in terms of his pyramidal law. 'The ropes of a pulley tackle share in equal parts the weight that they support. The power that moves the tackle is pyramidal since it proceeds to delay with uniform difformity to the last rope. And the movement of the ropes of this tackle is pyramidal because it slows down with uniform difformity from the first cord to the last' (G 87 v). The term 'uniform difformity' indicates change by equally proportional degrees, i.e., pyramidal proportion.

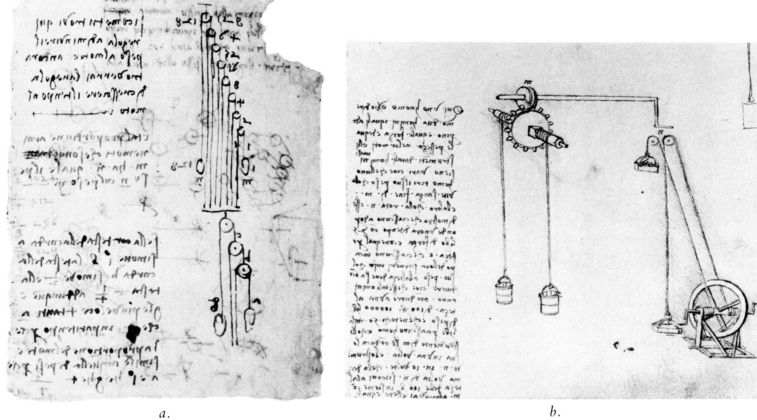

a.   *b.*

Figure 4.37. Pulleys.

a. A pulley system in which mechanical advantage and velocity ratio are expressed. Leonardo writes, 'Just as you can find here the law of diminishing weight applied by the mover you will further find the law of the increased time of the movement. The movement of *m* stands in proportion to that of *n* as the weight *n* is to the weight *m*' (CA 120 vc).

b. Here he contrasts the velocity between raising a load by means of a simple pulley, *n*, and a man working through an 'endless screw' or worm gear, *m* (Madrid I 35 r).

This mechanical composition of a worm gear and toothed wheel was not reinvented until about 1740 by Hindley, an English engineer. See Ladislao Reti, 'Leonardo da Vinci: Technology', in *Dictionary of Scientific Biography,* ed. by C. C. Gillispie (New York, Scribner's, 1973), vol. 8, p. 210.

By his investigations of perpetual motion and pulleys, amongst others, Leonardo found himself able to dispel the prevalent medieval tradition that mechanical machines could actually increase the power of a man. His analysis of the elements of such machines showed him that they only transform power, they do not make it. He gives a good example of this on Madrid I 35 r (Figure 4.37b). Here he describes and illustrates two machines, one raising a weight by a crank and simple pulley, another by a crank working through a worm gear. He explains, 'If . . . two men apply equal force and velocity in their motions, though the instruments moved by men are different the load is raised to the same height at the end of an hour, the simple pulley *n* raising its total load in ten journeys, the worm gear *m* much more slowly, in one journey'. Leonardo concludes, 'It is impossible to increase the power of instruments used for weight-lifting if the quantity of force and motion is given' (Madrid I 175 v).

Leonardo uses the action of pulleys to explain the elevation of the ribs described in Chapter 14.

IMPETUS

How does a ball preserve its movement after it has left the hand that throws it? This question severely taxed the ingenuity of the ancient Greeks. Leonardo took over a solution which had been developed in the thirteenth century by Buridan that a force was infused into the ball which kept it moving for a limited period. In his early notes, e.g., *Codex Trivulziano* and MSS B and A, Leonardo describes impetus but does not use the term. For example, 'A percussed body keeps in itself for a time the nature of the movement of percussion so much the less or more in proportion to the force of percussion' (Triv 43 r). By 1490–1493 he is testing out the concept, 'Test whether the impetus of the revolving wheel acquires force from its mover' (Forster III 48 r). At this same time his experiment creating wave motion by the percussion of a stone thrown into water reveals to him that in this case impetus is carried by wave motion as opposed to the direction of the actual movement of the particles of water. Thus, whilst heavy objects falling in water or air create waves of impetus, light things do not. 'Light bodies falling, like a thick sheet of paper, do not prepare in front of them fleeing waves of air and they do not cause impetus' (Madrid I 190 r). He comes to describe impetus thus: 'Impetus is a power transmitted from the mover to the movable thing maintained by the wave of air within air which this mover produces' (CA 219 va). And he affirms that 'Impetus is none other than a wave, which is generated in air, and water' (CA 173 rb). 'In each degree of time it acquires a degree of diminution because of resistance of the medium however rare' (CA 168 vb).

With the passing of the years Leonardo gives impetus an ever-increasing importance. It plays a dominant part in the movements of fish in water, the flight of birds in air and in the movements of man. 'Impetus', writes Leonardo, 'transports weight beyond its natural place' (CA 223 vb). Impetus increases the velocity of motion of any object (Madrid I 190 v). 'Accidental weight' or force is increased with impetus. Thus, although Leonardo never formulates the relationship clearly, impetus is, like momentum, a product of weight and velocity fitting into his scheme of the conservation by transmutation of all forms of energy. When a man jumps, for instance, his muscular energy is released like that of a spring into upward and forward movement of the weight of his body by impetus. Studying the movements of wind, he observes that waves of impetus may be dissociated from local movement, i.e., they may be found in stationary objects. Such waves may 'flee from the place of creation without water changing its position, in the likeness of waves which in May the course of the wind makes in cornfields, when one sees the waves running over the fields without the ears of corn changing their place' (F 87 v). Im-

petus promotes faster movement 'in proportion as the medium through which it passes is more rare' (CA 97 va). This statement is illustrated by a typical 'pyramidal' figure.

Later he applies impetus extensively to his physiology of the organs of the human body; to the movements of its parts; to the emission of the semen; to the emptying of the stomach and to the beating of the heart and its movement of the blood.

FRICTION

Leonardo early came to recognise that one of the main causes for the impossibility of perpetual motion and the loss of mechanical efficiency in machines was friction, or resistance. He found by experiment that friction was related to three factors, roughness of surface, weight of object and 'obliquity or steepness of the inclined plane'.

Roughness of surface first attracted his attention. This he diminished by inserting lubricants (he was probably the first to include systematic greasing and oiling points into machines); secondly, he sought for surfaces which made as small and smooth a contact as possible. This requirement he found was best met by a smooth sphere – the ball bearing. He demonstrated the difference between 'sliding' and 'rolling' friction in a neat word picture comparing a cart carrying a load on rolling wheels with the same cart carrying the same load on fixed wheels (Madrid I 122 r). He suggests loading a weight onto 'balls on an absolutely smooth surface; you will then see that it moves without any effort [see Figure 4.38]. Moreover, since the weight of the balls as they roll remains at their centre, an equal distance from the centre of the world, the resistance created by this friction for the movements of weights is separate from its weight' (Forster II 85 v).

For sliding friction he notes that weight is relevant as well as roughness of surface (see Figure 4.39). He progressively reports loss of power by friction to be '1/2 the weight

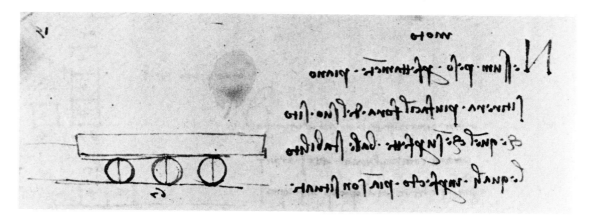

Figure 4.38.   Rolling friction; ball bearings.

Leonardo writes here, 'No perfectly flat weight will be easier to move from its position than one which is rested on perfect balls placed on a perfectly flat plane' (Madrid I 176 v).

By regarding the 'flat weight' as the ball of the foot, and the ball bearings as sesamoid bones, one can see how Leonardo came to describe the sesamoid bones of the foot as 'preventing great damage from friction' (W 19000 r; K/P 135 r).

moved' (Madrid I 176 a). 'On solids it is always 1/3 for a rough body (Madrid I 73 r). Later he states, 'All bodies resist by friction with a power equal to 1/4 of their weight with level motion and the surfaces dense and polished' (CA 72 r). This progressive approximation to a coefficient of friction is convincing evidence that he reached his conclusions on friction by quantitative experiment. Leonardo translated his coefficient of friction into practice, as for example in the axles of wheels, constantly allowing for a loss of one-fourth of applied power in friction.

As regards 'obliquity', Leonardo makes a calculation of frictional loss in relation to angle of obliquity. Loss is maximal when the contact of surfaces is horizontal; it is nil when they are vertical. At angles between, the loss is directly (pyramidally) proportional to the angle, i.e., it is 'half the quarter at half the right angle', etc., whilst at a right angle 'it resists by nil' (E 78 v).

Leonardo is acutely aware of the importance of friction in the human body. For example, he applies it to the movements of tendons over joints and bones, to the production of voice by the friction of air on the vocal cords, and to the creation of the heat of the body in the heart by the friction of blood against its walls. This depends on the fact that 'rapid friction between two dense bodies produces fire' (F 85 v).

HEAT AND COLD

Heat and cold were considered in medieval tradition as opposites or 'contraries' according to Aristotle's views. Leonardo also took this view. Heat, with light, emanates from the sun. To the sun Leonardo attributes all life-giving

Figure 4.39.   The coefficient of friction.

In the top drawing Leonardo draws a body, *a,* drawn by a cord, *b,* over a flat surface, *cd.* He writes, 'From experiment it is found that a polished object dragged along a polished horizontal surface resists the movement of the mover with a power which is one 4th part of its weight' (CA 198 va).

The drawings and text below describe how this coefficient of friction is modified by roughness and a slope upwards or downwards. At the bottom is a diagram of a further experiment to be made on the friction of a 'curved surface', i.e., a round pole in a hollow. By completing the hollow into a full circle he came to study the friction of a fixed axle in a wheel.

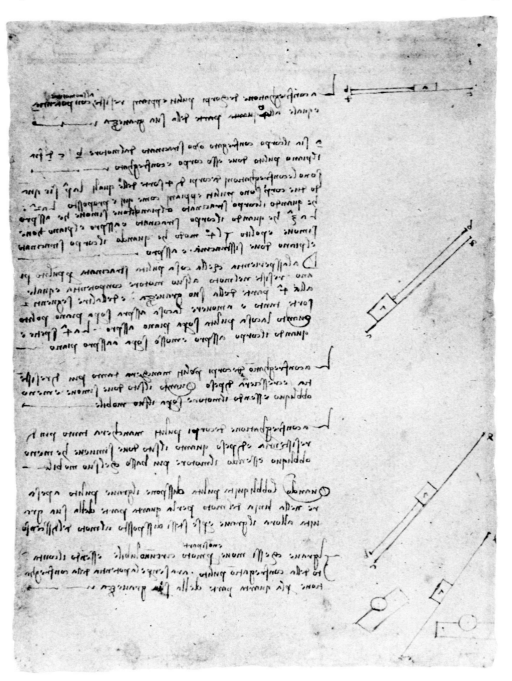

properties through its heat. For this reason more than half his references to heat are contained in reference to the life of the world (macrocosm) or man (microcosm).

'In the whole universe', he writes, 'there is nowhere to be seen a body of greater magnitude and power than the sun. Its light gives light to all celestial bodies which are distributed throughout the universe; all souls [*anime*] descend from it, for the heat that is in living animals comes from their souls and there is no other centre of heat and light in the universe' (F 4 v).

The sun's rays which are made of radiating light, heat and its image, are all 'spiritual' (CA 204 va). These travel as 'swiftly as possible' through the spherical element of air which refracts the rays of both light and heat on to the earth (BM 205 v). This refraction, like that through a curved glass globe or lens, concentrates the light and heat 'pyramidally', as happens also with a 'burning mirror'. This means that 'As many times as the point of the solar pyramid cut across in any part is contained in its base, so many times is it hotter than its base' (A 55 r). Once it has reached the earth this heat 'sucks up moisture' (CA 171 ra), in this way drawing water up to the tops of mountains, as well as in mists to the sky, forming clouds when the moisture cools and condenses (A 55 v). Then it falls again as rain. The analogy between this movement and that of the circulation of the humours in plants and blood in man is repeatedly drawn, for in both, 'heat makes a heavy body lighter by rarefying it, whilst cold makes a body dense and stops its movement' (A 56 r and 57 r). In the body of the living earth, 'The heat that is infused into animated bodies moves the humours which nourish them. The movement made by the humour is its conservation and the vivification of the body which encloses it. Water is that which is dedicated to be the vital humour [blood] of the arid earth; it is poured within it and flowing with unceasing vehemence through the ramifying veins it replenishes all the parts that depend of necessity on this humour. And it flows from the great depths of the great ocean in the hollow broad caverns of the internal viscera of this earth, whence through the ramifying veins against its natural course it flows to the tops of the mountains in continuous ascent, and returns down through the broken veins' (BM 234 r). As mentioned previously, for many years Leonardo considered that the movements of the blood and other humours in man worked on this same principle (see Figure 3.6, p. 86).

Air itself is 'cold and dry'; being rare, the sun's rays heat it but little. When the heat of the sun meets with 'another dense body it gets hotter' (CA 375 vb; Madrid I 180 v). Water vaporises when so heated and ascends. By experiment (Leic 10 r) he finds out how much water expands in vaporising (see Figure 4.40).

Leonardo denies his 'adversary's' suggestion that the heat of the sun's rays comes from the element of fire through which they have to pass, saying, 'Summer and winter teaches us which heats and cools us, whether it is the sun or element of fire' (BN 204 r). He further makes this point by passing sun-rays through cold water. Do they 'put on a mantle of cold'. They do not' (F 93 r).

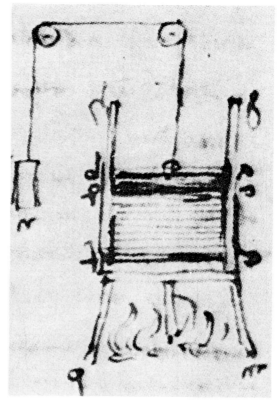

Figure 4.40. The conversion of water into water-vapour (Leic 10 r).

Well aware of the difference between 'air' and 'steam' Leonardo devised this 'Experiment to make a rule as to how much water increases when converted into steam'. He fits a bag of calf-skin into a cylinder; puts a little water inside it, and on top of it a piston, *abcd*. He heats the calorimeter from below and measures the expansion by the rise of the piston and the fall of the counter-weight, *n*. The volume of this calf-skin being insufficient, he repeated the experiment (Leic 15 r). This time he found that '1 ounce of evaporated water fills one of my bags [wine-skin]'. This suggests a ratio of about 1:1500.

Herein lies Leonardo's theoretical explanation of his invention of the steam gun, 'architronito' (see Figure 1.13, p. 18).

Blood heated in the heart, whether by fire (as postulated on W 19045 r; K/P 50 r) or by friction, produces steam, which, in Leonardo's view, condenses on the skin as sweat (as described on W 19119 r; K/P 116 r).

He uses the concave mirror not only to concentrate to a focus the light and heat of the sun but also to point out that one can focus the heat of a fire in the same way (CA 368 rb; see Figure 3.8, p. 88). He extends this pyramidal concentration of forces to focussing steam (BM 57 r) and goes so far as to observe, 'Cold taken on the surface of a concave mirror is thrown back into the point of its pyramid and is multiplied no less effectively than are the rays of the sun' (BM 85 r).

'All powers', writes Leonardo, 'increase the more they approach the centre, as seen in percussion, weight, heat and in all other natural powers' (D 1 r). He is referring to the pyramidal concentration towards the central line as de-

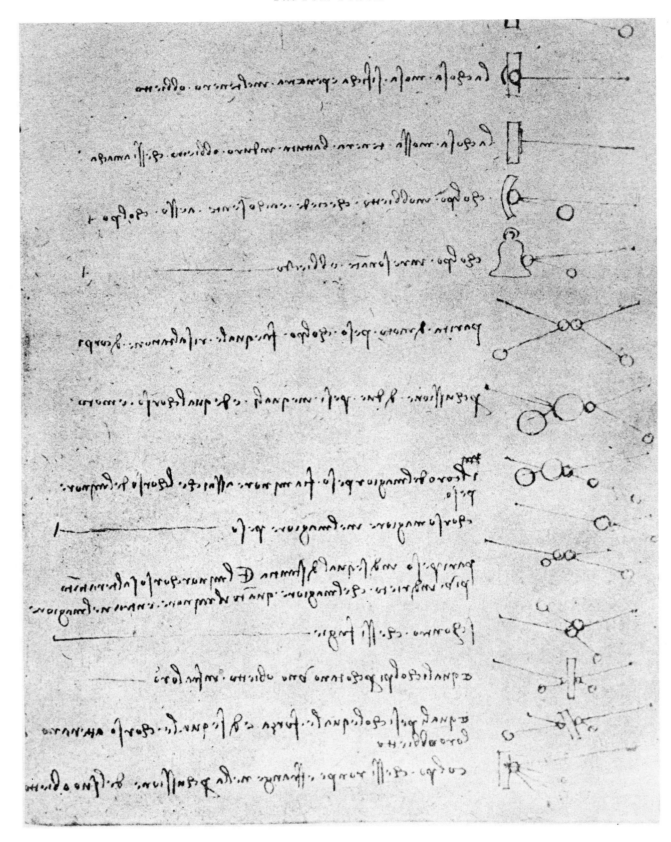

Figure 4.41. Fourteen different effects ot percussion (A 8 r).

From the first figure, where the percussing object rebounds, to the last where it is shattered by the percussed object, the diagrams clearly depict the different effects. Each is described verbally.

scribed first from his investigations on perspective (Figure 2.19, p. 59). Later he says, 'The same power is so much the more powerful as it is more concentrated. This is true of heat, percussion, weight, force and many other things' (G 89 v).

Two sources of heat other than the sun are acknowledged by Leonardo. Dissatisfied with his theory of circulation of the elements in the earth and man by the heat of the sun, he considers the alternative that this heat comes from the fires in the centre of the earth or human body. In a later word picture of the physiology of the macrocosm Leonardo describes the earth as 'sentient . . . its flesh is the soil; its bones the successive strata of rocks which form the mountains; its cartilage the tufa stone; its blood the veins of water. The pool of blood which lies round the heart is the ocean; its breathing and the increase and decrease of blood in its pulses is represented in the earth by the ebb and flow of the sea. And the heat of the soul of the world is the fire which pervades the earth; and the residence of the vegetative soul is in the fires which breathe out through different places of the earth in baths, and in mines of sulphur, in volcanoes and at Mount Aetna in Sicily, and many other places' (Leic 34 r).

Friction and percussion can also be transformed into heat. 'If you beat a thick bar of iron between the anvil and hammer with frequent blows upon the same place you will be able to light a match at the place percussed' (CA 351 vb). Similarly heat is created by the friction and percussion of small particles, as, for example, in churning milk (W 19119 r; K/P 116 r) and in the human heart by the percussion and friction of the blood on its walls. Thus Leonardo reaches his final theory for the creation of body heat (W 19119 r; K/P 116 r).

PERCUSSION

'Percussion is born of the death of motion', writes Leonardo in MS A 34 v, on a page crammed with different definitions and descriptions of relations between the four powers. This force of 'impact' Leonardo considered to be the greatest of the powers because 'Percussion in equal time exceeds any other natural power' (CA 11 vb) – in short its energy–time ratio is maximal. He often demonstrated his investigations of percussion by experiments with balls as on A 8 r (see Figure 4.41). When a moving object such as a ball meets resistance three things may happen: the movement may be continued in the percussed object, the percussing ball may rebound, or it may come to a standstill whilst the percussed one flies off (Leic 8 r). In any case the power of percussion is inversely related to the time it lasts. 'Percussion of twice the power takes half the time' (CA 65 ra). The power of percussion is the product of the weight of an object such as a ball and its velocity, being directly proportional to both. For example, 'a ball dropped twice the distance gives twice the percussion' (M 52 v).

One of the most fundamental rules about percussion which Leonardo repeatedly asserts is that the angle of incidence equals the angle of reflection. This is very clearly

Figure 4.42. Sound is percussion of the air, which percusses the ear.

Leonardo writes, 'I say that the sound of the echo is reflected by percussion to the ear, just as the percussion made in mirrors by the images of objects are to the eye' (C 16 r).

Percussion is here vividly and symbolically represented by three little hammers hitting the bell, the wall and the ear.

illustrated in Figure 2.13 (p. 55), where it is used to illustrate not only the rebound of a ball from a wall but also that of the sound of an echo, both following the same laws of movement, percussion and reflection. This rule he applied to the reflection of light in a plane mirror. In all cases the 'central line of incidence [perpendicular]', is the most powerful.

At this time Leonardo studied the properties of percussion in considerable detail. For example, he compared not only the angles of percussion but also the effects of the hardness and softness ('density' or 'rareness') of the percussing and percussed objects as well as their sizes, shapes and weights. Each of the fourteen variations depicted on A 8 r served Leonardo as a point of departure for his investigations of percussion. In *Codex Atlanticus* 79 rab Leonardo lists sixty headings for his 'Book on the percussion of water'. This serves to show how impossible it is in the space here available to give anything like an adequate description of the extent of his researches into this subject. All aspects of percussion were probed experimentally. For example, the second drawing in Figure 4.41, illustrating that 'the moved object is fixed in, and penetrates, the soft object', is used by him to measure the difference between the impression made by a static weight and a blow from its percussion. The fifth drawing, showing the percussion of a bell, is notable for being the origin of his researches on 'sensitive' motion (*moto sensibile*), that is, oscillatory or wave motion as in the production and reception of sound (see Figure 4.42). The last drawing in Figure 4.41, depicting 'damage' by percussion, sets the trail for research on cannon balls, their destructive potential and the form of fortification best shaped to meet or diminish the effects of their percussion. It also provides the basis for his conception of the production of the sensation of vision, hearing and tissue destruction with bodily pain.

Early in his notes Leonardo records the destructive effects of percussion. For example, the 'thunder of the bombard aimed over water kills all the animals in the water'

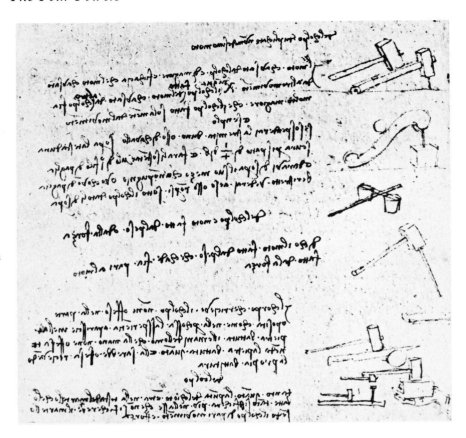

Figure 4.43. Some damaging effects of percussion.

A series of experiments on percussion. The second from the top demonstrates the ease with which 'a bone of a horse' can be broken by a light blow 'at its middle'. This effect is likened in the drawing below to breaking a stick suspended between two pails of water without spilling any water. At the bottom Leonardo demonstrates how 'a stone which is beaten whilst in the hand of a man does not damage the hand holding it as much as the hand would be damaged if it received the blow directly' (A 53 v).

(A 52 v). He has already appreciated that the bombard or mortar produces one of the most powerful examples of percussion by the sudden expansion of the element of earth in the form of solid gunpowder into the element of fire (see Figures 1.6 and 1.11). This sudden force produces recoil behind it and projects a cannon-ball in front of it, into the element of air, as well as the noise (Triv 18 v) (see Figure 1.17c, p. 27, and Figures 10.4 and 10.5, pp. 234, 235). Here, the blast set up by the explosion produces percussion waves of force, which he describes further thus: 'If you discharge a small bombard in a courtyard surrounded by a convenient wall, any vessel that is there, or any windows covered with cloth or linen, will be instantly broken; even the roofs will be lifted up away from their supports, the walls and ground will shake with a great tremor, the spider's webs will fall down and small animals will perish, every air-containing body will suffer instant damage' (CA 363 vd). It is evident that he appreciated the destructive effects of a blast wave as well as other forms of percussion (see Figure 4.43). He describes similar effects from percussing a rock in water, 'The blow given to any dense or heavy body passes naturally beyond this body and damages everything around, whether dense or rare' (A 31 r), and he describes how fish so stunned will float to the surface after such a blow (see also Figure 10.4, p. 234).

The better-known experiment where he drops two stones into still water and notes the spread of circular waves of impetus from the points of percussion intersecting each other, has been previously described in Figures 2.12 and 2.18 (pp. 54, 58). All wave formation, whether presented as the effects of blast, light, sound, heat, magnetism or mechanical spread in water, follows similar laws. Waves spread out in spherical form from their source and lose their power 'pyramidally' in proportion to their distance from the point of their origin. This is the law of linear perspective applied to all the powers of nature. It is the law that Leonardo calls pyramidal because of its geometrical form of steady continuous diminution with every 'degree' of distance. It is modified in different elements. Air, for example, modifies it by being condensible and elastic, whilst water is incompressible, but the geometrical principle of spread in both is 'pyramidal'. Thus the quantitative laws that Leonardo finds for the spread of the power of percussion are those of simple arithmetical proportion, as for linear perspective.

Conversely, the perspectival or pyramidal law explains the concentration of percussive power. This Leonardo illustrates in the example of the concentration of the heat of the sun by a concave mirror (A 20 r). He expresses this in more general terms later thus: 'Nature did not make visual power uniform but has given this sense greater power in proportion as it is nearer its centre. And this she has done in order not to break the law given to all other powers, which become more powerful in proportion as they approach nearer to their centre. This is seen in the act of percussion by anything . . . it is seen with heat and all the other natural powers' (D 1 r). That this idea of pyramidal concentration of percussion and all other 'powers' persisted to the end of his life is shown by another note written about 1513. In

Figure 4.44, under the heading 'Power', he writes, 'The same virtue or power is so much the more powerful as it is more concentrated. This is true of heat, percussion, weight, force and many other things. Let us speak first of the heat of the sun which impresses itself on a concave mirror, and is reflected by it in a pyramidal figure which acquires proportionately so much more power as it narrows . . . Furthermore the percussion of pyramidal iron will penetrate the penetrable thing percussed by its point to a greater extent in proportion as the point is narrower. A heavy substance also when narrowed into less space pyramidally is of greater weight because a less quantity of air offers resistance to it' (G 89 v). Once more in the last years of his life Leonardo is speaking in terms of the laws of perspective which he worked out over twenty years previously (see Figures 4.5, 4.6 and 4.7), in relation to the pyramidal form of the power of vision concentrated around the 'central line' as constructed in Figures 2.32 and 2.33 (pp. 70, 71).

Since force, weight, percussion and motion all generate each other (Madrid I 52 v), Leonardo searched for some quantitative rule whereby this mutual transformation is realised. In particular he concentrated on the conversion of percussion into weight. He tried to measure this in various ways, for example, by letting weights fall down inclined planes and measuring the 'dent' at the bottom. He also measured the depth of the dent made by beating lead with a

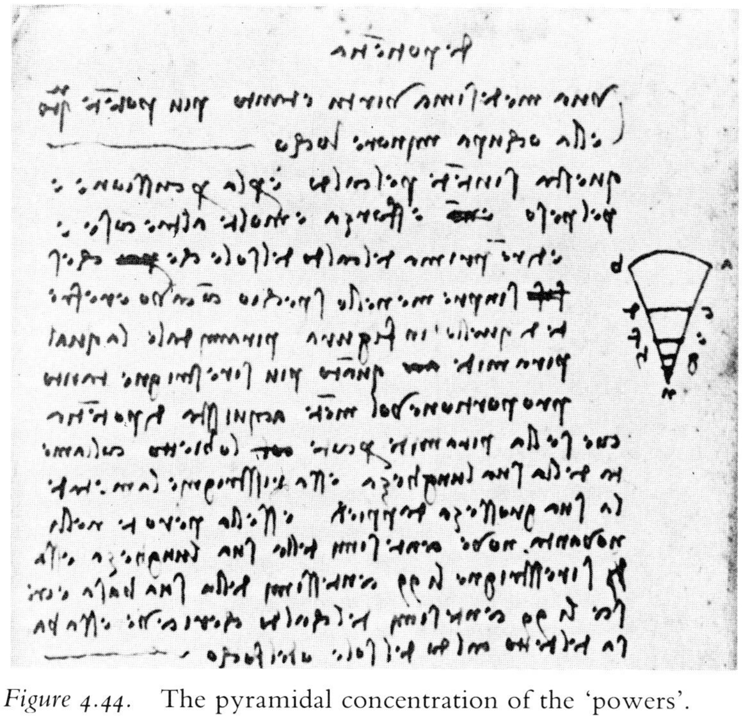

Figure 4.44. The pyramidal concentration of the 'powers'.

Here Leonardo describes the pyramidal concentration of power using a very similar drawing to that in Figure 2.19 (p. 59). As the power nears the point *n*, it nears an infinite concentration of its source at *ab* (G 89 v).

For the script alongside the drawing, see text.

hammer (A 4 v; Madrid I 59 r, 188 v). He then let a known weight fall on a balance-pan from different measured heights and measured how much weight it is necessary to add to the opposite pan to resist such percussions (CA 20 va and Figure 4.1, p. 94). He observed the trajectory of a spherical weight thrown off from the opposite scale-pan (CA 337 rb). These by no means exhaust his different experimental efforts to find the conversion factor between weight and percussion.

He concluded from such experiments that a falling weight will raise as much more weight than its own as is equal to its percussion, and on a number of occasions, as in Figure 4.1 (p. 94), he tried to apply this to the percussive effect of falling water in relation to the power of waterwheels. From these experiments he reached the conclusion that the 'power' of percussion is directly related to the height of fall and the weight which falls (M 52 r). Since Leonardo was aware of gravitational acceleration one can legitimately interpret this finding as an approximate expression of the fact that gravitational potential energy equals the product of the weight of a falling body, its gravitational acceleration and the height of its fall. However, Leonardo was unable to find in his pyramidal geometrical figure an adequate expression of such a relationship.

There is, perhaps, one partial exception to this last statement, for in his typically concrete way Leonardo on at least three occasions refers to the transformation of a cube into a sphere by percussion. First described on Madrid I 159 v, the process is more clearly expressed on W 12664 and BM 188 r. Both these studies were made after he began his book on transformation in 1505 (Forster I). 'On percussion made in dilatable material', he writes, 'The cube of extensible material will be dilated by percussion; and if this percussion is uniformly uniform will remove its edges from square to circular' (BM 88 r).

In relation to the human body it is evident that Leonardo saw innumerable applications of percussion to both the sensory and motor sides of human and animal physiology and psychology. On the sensory side we have already seen that percussion was involved in vision, hearing and the other senses, including pain.

On the motor side Leonardo applied the principle of percussion to human labour in innumerable ways. 'Amongst the natural powers in the same time percussion exceeds by a great excess any other human action or natural device' (CA 370 rb). These words are written on a page devoted to methods of excavating a trench where a mechanical digger is elevated like the hammer of a pile-driver by manpower and then dropped, so that its weight accelerates and successfully excavates a greater depth of earth than could any direct human percussion.

On a number of occasions Leonardo raises the question of the nature of a man jumping and landing on the ground. 'The jump is the end of a blow [its rebound]' (CA 144 ra). He analyses the act of landing on the heels and the toes, in relation to both the noise and the 'spring' obtained by bending the ankle and knees (CA 338 ra; see Chapter 6,

Figure 4.45.   Destructive cataclysm of the Four Powers (W 12380).

This drawing, one of the so-called 'Deluge' series, illustrates Leonardo's idea of the chaos produced by the liberated powers of an explosion within the four 'elements'. An exploding mountain expels rocks in spreading wave-rings amongst violent eddies of gases, wind and rain.
Unleashed, these Four Powers exert their destructive forces on the trees and houses below. It
is the final cataclysm of our world.

pp. 179–181). He analyses the different effects on the body
of 'direct' percussion and 'indirect' or 'compound' percussion through intervening objects such as stones held in the
hand (see Figure 4.43). He analyses the flight of the bird by
the percussion of its wing condensing air so that the resistance of the air becomes greater than the downward
movement of percussion, enabling the bird (and man?) to
rise in the air (see Figure 4.33, p. 118).

Finally, in the field of art one must refer back to the
agony of *St. Jerome* (Figure 1.1, p. 10), in which the Saint
beats (percusses) his chest with a stone; and to the moment
of stillness between impact and rebound of psychological
combined with physical percussion depicted in his *Battle of
Anghiari* (Figure 1.26, p. 34). In *The Last Supper* (Figure
1.16, p. 24) he depicts a wave of percussion of the voice in
silence and the psychological percussion of emotional
shock. The patterns of the effects of percussion are power-

fully portrayed in their elemental beauty in his drawings of
The Deluge (Figures 4.45 and 5.26).

We should not be surprised therefore to find Leonardo
introducing his anatomical researches into human 'powers'
with the phrase 'Percussion is the immense power of things
which is generated within the elements' (W 19060 r; K/P
153 r).

REFERENCES

1. Reti, Ladislao, and Bedini, S. A. 'Horology', in *The
   Unknown Leonardo,* ed. by Ladislao Reti. London,
   Hutchinson, 1974.
2. Clagett, Marshall. 'Leonardo da Vinci: Mechanics', in
   Dictionary of Scientific Biography, ed. by C. C. Gillispie.
   New York, Scribner's, 1973. Vol. 8, p. 215.
3. *Ibid.,* p. 217.

Chapter 5

Leonardo's Scientific Methods and the Mathematics of His Pyramidal Law

Whether he was investigating the macrocosm of the universe or the microcosm of man, Leonardo's methods were the same. 'Nature', he writes, 'cannot give the power of movement to animals without mechanical instruments . . . for this reason I have drawn up the rules of the four powers of nature' (W 19060 r; K/P 153 r). Some pages later in this same notebook give some idea of his methods of pursuing the study of anatomy. They are a little surprising; not those which we would focus upon today. Among the obstacles to be overcome by the would-be investigator Leonardo begins with the natural 'repugnance of living during the night hours in the company of quartered and flayed corpses'. He continues with the problems of illustration and mathematics: 'perhaps you will lack the good draughtmanship which appertains to such representation; and even if you have the skill in drawing it may not be accompanied with a knowledge of perspective; and if it were so accompanied you may lack the methods of geometrical demonstration and methods of calculating the force and strength of the muscles' (W 19070 v: K/P 113 r).

Leonardo is citing here some of the qualities which he considered essential in all scientific studies, as well as that of the science of man. These he evolved from his theory and practice of painting. From the first he was convinced that the painter interprets and rivals nature. Painting achieves this goal through its scientific principles, and 'the scientific and true principles of painting first determining what an opaque body is, what primary and derived shadows are, what illumination is, i.e., darkness, light, colour, body, form, location, distance, nearness, motion and rest, are principles which are comprehended by the mind alone . . . from it is born creative action which is of much more value than the reflection or science just mentioned' (TP I 80). Elsewhere he speaks of painters studying 'such things as pertain to the true understanding of all the forms of nature's works . . . This is the way to understand the Creator . . . and to love such a great Inventor. In truth great love is born of great knowledge of the thing loved' (TP I 80).

Thus, in Leonardo's view painting 'compels the mind of the painter to transform itself into the very mind of Nature, to become an interpreter between Nature and the art. It explains the causes of Nature's manifestations as compelled by her laws' (TP I 55).

'The eye, the window of the soul, is the chief means whereby the *senso comune* can most fully and abundantly appreciate the infinite works of Nature' (A 99 r). By *senso comune* we realise Leonardo refers to the seat of the soul, to which the eye is linked by the optic nerve and the *imprensiva* (see Chapter 2). This concrete, physiological relation between the eye and the soul is combined at this time with a number of vehement appeals to 'Experience', which increase their significance when one appreciates the physiological context within which Leonardo uses the word experience. He writes, 'To me it seems that all sciences are vain and full of errors that are not born of Experience, mother of all certainty, and that are not tested by Experience; that is to say, which do not at their beginning, middle or end, pass through any of the five senses' (TP I 19).

Moreover, after his early anatomical research into the apparatus of vision, tracing it from eye to *senso comune*, Leonardo now believes that 'experience', particularly visual experience, has as rational an origin as have the images which he has drawn on the vertical glass pane in his studies of perspective. Therefore, 'Experience never errs; it is only your judgement that errs in promising results which are not caused by your experiments. Because given a beginning, it is necessary that what follows must be its true consequence unless there is an impediment. And should there be an impediment the result which should follow from the aforesaid beginning will partake of this impediment in proportion as it is more or less powerful than the aforesaid beginning.

'Experience does not err; only your judgement errs by expecting from her what is not in her power. Wrongly do men complain of Experience and with bitter reproaches accuse her of leading them astray. Let Experience alone and turn your complaints against your own ignorance which causes you to be carried away by your vain and foolish desires as to expect from her things that are not within her power, saying that she is fallacious. Wrongly do men complain of innocent Experience, constantly accusing her of deceit and lying demonstrations' (CA 154 rb).

Our next step is to ascertain what Leonardo included in 'Experience'. We need not emphasise the self-evident fact that he was a good observer. His few surviving paintings, added to his verbal descriptions of such subtle and transient events as the movements of a bird's wing, establish it.

However, Leonardo was a creative inventor as well as an artist, and he set out to expand the field of ordinary, naked-eye observation in many ways. His early studies of 'global' lenses, of convex and concave mirrors, led to his grinding parabolic concave mirrors (CA 32 ra), and so to his first appreciation of the pyramidal shape of focused light and heat (CA 292 vb). Such observations initiated in him the idea of experiment as a form of observation, and obviously his tests of machinery of all kinds must have enlarged his extraordinary innate capacity for experimental design and execution.

Leonardo used experiment for two different purposes: to discover the result of some new combination of forces, for example, the rate at which a man can make a wing descend in relation to raising his own weight; and to demonstrate an ascertained 'rule'. Like most brilliant experimenters, Leonardo says little about the art of designing intelligent and practical experiments; he just performed them. That he did actually perform many of the experiments he describes is verified by entries on a number of occasions of the word 'experimented'. Even more significantly perhaps is the entry alongside some drawings, 'not experimented' (see, e.g., Forster II 135 r and 67 v). He summarises his lifelong use of

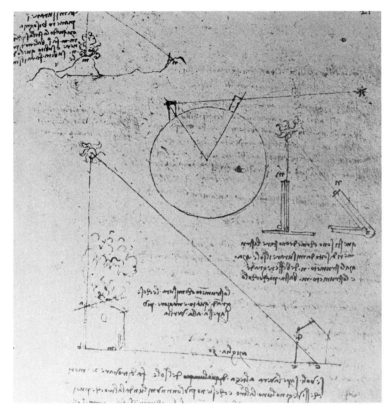

Figure 5.2. Measurement of the height of the sun (A 21 r).

'If you want to measure the true height of the sun,' writes Leonardo, 'find two mountains which are a long way from one another' These are drawn in the top diagram. Below, points he insists on are that the observations of the sun (or planet) should be made through very small holes (*spiracoli*) and that as soon as the sun is seen by one observer to be directly above, i.e., 'through a perpendicular line', he should notify this to the other observer by lighting a fire, the smoke of which he draws.

In the central drawing he measures the height of the pole star from two points on the circumference of the earth. From the different angles of elevation Leonardo calculates the distance to the centre of the earth. This method is developed on W 19148 r.

It is significant that both sets of observation involve the laws of perspective. 'The science of visual lines has given birth to the science as astronomy, which is simple perspective since it is all visual lines and sections of pyramids' (TP I 5).

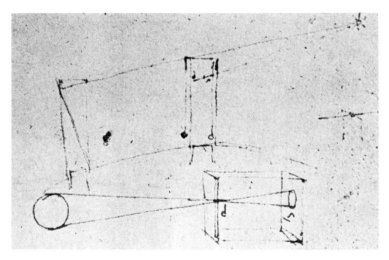

Figure 5.1. Measurement of the size of the sun (A 20 v).

In the lower drawing the sun, represented by the circle on the left, casts its image through a very small hole (*spiracolo*) at *b*; its image within the camera obscura is caught on the wall at *a*. Leonardo writes, 'A method of knowing how large the sun is. Make it so that from *a* to *b* is 100 *braccia* [i.e., about 200 feet or 61.0 metres] . . . and note how much the ray increases in its percussion'. If one knows the distance of the sun one can calculate its size.

In the drawing above Leonardo is taking two bearings on the pole star (top right) from different points on the curved surface of the earth, in order to calculate the distance to its centre. See also Figure 5.2, which is taken from the next page of MS A.

experiment thus: 'First I shall test by experiment before I proceed further, because my intention is to consult experience first and then with reasoning show why such experience is bound to operate in such a way. And this is the true rule by which those who analyse the effects of nature must proceed. And although nature begins with the cause and ends with the experience we must follow the opposite course, namely begin with experience and by means of it investigate the cause' (E 55 r). That he was well aware of experimental error emerges from such remarks as, 'The experiment should be made many times so that no accident may occur to hinder or falsify the test; for the experiment may be false whether it deceives the investigator or not'

(M 57 r). Again, 'But before you make a general rule of this [experiment] test it two or three times and observe whether the tests produce the same effects' (A 47 r). He uses experiment deductively when he summarises the geometry of perspective thus: 'Perspective is a rational demonstration whereby experience confirms how all objects transmit their images to the eye by pyramidal lines' (A 10 r).

Leonardo did not confine his observations to the unaided, naked eye. He assisted his visual acuity by various means. Two of the most fruitful were the use of the *spiracolo,* a very small hole, and the projection of the outlines and colours of objects onto a vertical glass pane interposed between the object and the eye.

The *spiracolo* technique he used early for observing an eclipse of the sun (which has been dated as occurring 26 March 1485). In order to observe this 'without hurting the eye', he advises, 'Take a piece of paper and prick a hole in it with a needle, and look through it at the sun' (Triv 6 v). This technique he used extensively in his investigations on vision and in his analysis of the propagation of light. On a number of occasions he reports how light enters through a small hole, whether this be a window in a wall, a perforated iron plate or the pupil of the eye, in such a way that an image is formed behind the *spiracolo* in the dark place. This technique has since been dignified with the term *camera obscura.* Here, at a proper distance, Leonardo observed the image of the illuminated object, inverted and coloured, and retaining its 'real' proportions. These experiments he saw as demonstrating that the pyramids of light crossed at a point in the region of the *spiracolo.* Such objective real images were suitable for measurement. This he did, expressing his measurements in terms of their proportion to the actual size of the object and working out the relationship by using his pyramidal law of simple proportion.

No object was too large for this procedure, or too distant, for he proposed to measure the size of the sun, moon and stars by it (see Figure 5.1, A 20 r). He also applied his *spiracolo* technique to surveying, improving the accuracy of the observed angles by applying this to measurements of the height of the sun (see Figure 5.2, A 21 r).

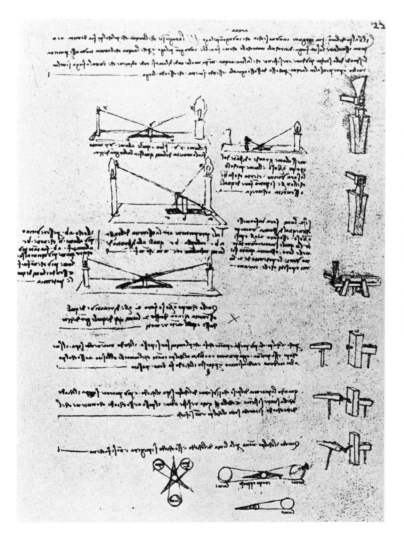

Figure 5.3. The measurement of light; photometry (C 22 r).

The practical experimental brilliance of Leonardo is well shown by the four drawings by which he illustrates these studies of photometry. Unfortunately, he equated luminosity with the size of a candle flame. Under the middle drawing he states his conclusion: 'The darkness of the shadows at the places *ab* and *bc* will be proportional to the distances of the light sources at *n* and *f*'. Thus he equates light intensity perspectivally with his pyramidal law, inversely as the distance and not inversely as the square of the distance (C 22 r).

(The drawings down the right margin are not relevant here, but it is worth noticing that *all* the studies on this page are for Leonardo studies of percussion, whether of light, a sharp axe, or a blunt hammer on a nail.)

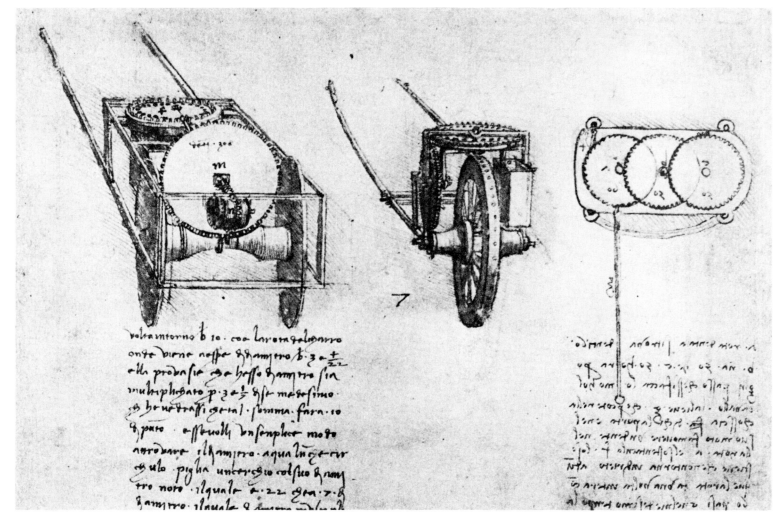

Figure 5.4.   The measurement of distance travelled; the hodometer (CA 1 ra).

Of the three designs here drawn that on the right is a pendulum-type pedometer; 'the lever *g* percusses the thigh of him who carries it, and in its movement moves a tooth of the wheel *a*'.

The other two hodometers are designed to rotate their wheels once in ten *braccia,* and at a distance of one mile it 'makes the ear hear the sound of a little stone falling into a basin made to receive it'.

These instruments were used by Leonardo for map-making.

Leonardo used the optical aid of the vertical glass pane for perspective, performing meticulously detailed experiments to establish the variations of the position of eye, glass pane and object quantitatively, as has been outlined in Chapter 2 (Figure 2.1, p. 44). It remains here only to re-emphasise the way Leonardo utilised this whole body of observations and experiments for quantitative estimations which depended on geometrical figures. In this way he established the 'rule' that the size of an object was inversely proportional to its distance from the eye, and not to the angle it subtended in the eye, as Euclid had declared. It is to be emphasised that he extended his perspectival studies to the production of perspectival distortions or anamorphosis, which he called 'accidental perspective'. Thereby he defined conditions under which pyramidal perspective was valid or not for the purposes of measurement. He used this perspectival method, too, for attempting to measure the size of the sun and moon (CA 243 rb).

The importance which Leonardo himself attached to his perspectival method is reflected in his emphatic assertion that 'Mathematical sciences are those which through the senses have a first degree of certainty. There are only two of them, of which the first is arithmetic, the second geometry. The one deals with discontinuous quantities, the other with continuous ones. From them perspective arises which deals with all the functions and delights of the eye, with varied speculations. Of these 3 mentioned, that is arithmetic, geometry and perspective, is born astronomy which by means of visual rays with number and measure establishes the distance and measurement of celestial as well as terrestrial bodies' (Madrid II 67 r).

The intensity of light and shade form the main content of the whole of MS C. In Figure 5.3 (C 22 r) Leonardo attempts to measure objectively the intensity of light as explained in the legend. This is remarkable for its resemblance to the Rumford photometer, introduced some three centuries later.

Thus Leonardo aided his naked-eye observations with devices which were designed to give quantitative measurements of those observations, using surveying, the vertical glass pane, the camera obscura and the photometer for this purpose. Amongst many other measuring devices may be added hodometers (Figure 5.4), the hygrometer and anemometers (Figure 5.5). To these methods for enhancing the

accuracy of his observations of stationary objects Leonardo added two invaluable methods for measuring the movements of objects, making systematic use of markers and models, combined with many attempts to improve the accuracy of timepieces for the measurement of time.

The simplest and earliest use of markers is described by Leonardo in *Codex Trivulziano* 31 r. It is used for measuring 'the rate of fall of a river per mile'. For this purpose Leonardo describes in detail the use of 'an oak-apple or cork'. 'Observe', he adjures, 'how many beats of time the aforesaid object takes to arrive at the end of a journey of a hundred *braccia*. Then you must measure other reaches of 100 *braccia*, and then you will be able to calculate how much the river falls per mile, checking this with an instrument for observing levels'. Such an instrument for measuring levels is drawn and described on B 65 v (Figure 5.6). A 'beat of

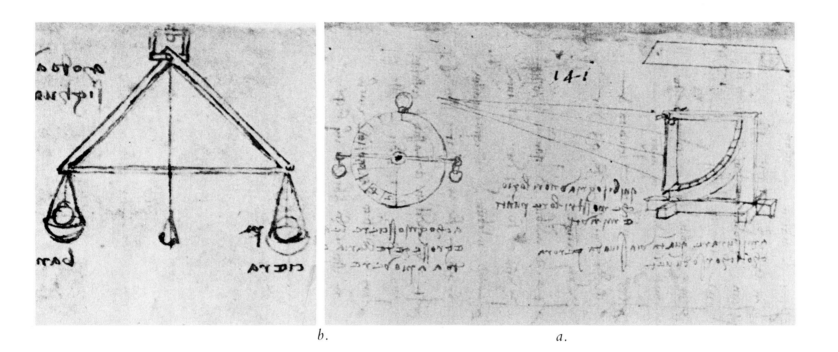

b.   *a.*

Figure 5.5.

a. The measurement of wind speed; anemometer (CA 249 va).

Below the right-hand drawing Leonardo writes, 'To measure how much the course of the wind blows in an hour'. The flat plate, hinged at the top, is blown up along the curved scale in proportion to the speed of the wind.

b. The measurement of humidity; the hygrometer (CA 8 vb).

The two other instruments are hygrometers. In the left-hand pan of the balance cotton wool or a sponge is placed, in the right, wax. In a humid atmosphere the wool soaks up moisture, increasing its weight, whilst the wax does not. Thus, as Leonardo writes, this is 'a method of seeing when the weather is turning bad'.

The left-hand hygrometer, a self-indicating balance, was unknown before Leonardo.

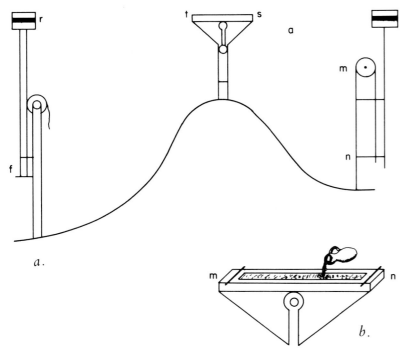

a.

a. On this page Leonardo writes, 'If you wish to know how much one plane is higher than another do thus: Place your level between the two planes of which you wish to know the depression of one below the other; place on the lower position the graduated rod and raise or lower it until your eye at *a* and the two ends *s* and *t* are in the same horizontal plane as the black bands on the upper drum *r*. Mark the rod below at the point *f* with a nail. Then carry the rod to the other side and do the same again; look through the instrument the other way *ts* and put in another nail. If the first nail is at *n* and the second at *m* then this is the difference in levels' (B 65 v).

b. He describes the construction of the level by filling a channel two *braccia* in length with water and placing rods at the same horizontal level, *m* and *n* either end.

Other designs for levelling instruments based on self-indicating balances are to be found in Windsor Drawings 12668 and *Codex Atlanticus* 346 vb. The simplified figure here is from Ivor B. Hart, *The World of Leonardo da Vinci* (London, Macdonald, 1961), p. 180.

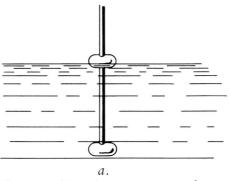

b.

Figure 5.6.   The measurements of levels (B 65 v).

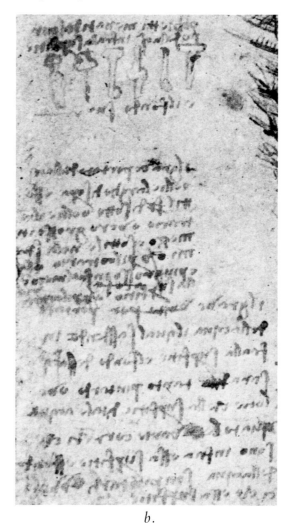

a.

a. Markers; floats marking the movements of water (A 75).

Leonardo writes, 'Test to know if water runs more above than below'. On the surface the float contains a bag of air; below, at the bottom, it is weighted with a stone. If the float inclines towards the oncoming current the water flows faster below, and inversely.

This sheet is taken from an apograph made by Venturi in 1796; the original is since lost. J. B. Venturi, *Essaisur les ouvrages Phỳsico-Mathematiques de Leonardo de Vinci* (Paris, Duprat, 1797).

b. Markers; variously shaped floats for use in water (W 19108 v; K/P 185 v).

Leonardo writes, 'Objects carried by a current of water between the surface and the bottom'. He describes in detail how each float is weighted and shaped in order to float at different depths of water. Thus, 'The weight carried by the water which is situated between the surface and the bottom will be as much slower or faster than the surface of the water as the different currents between the surface and the bottom are slower or faster'.

b.

Figure 5.7.

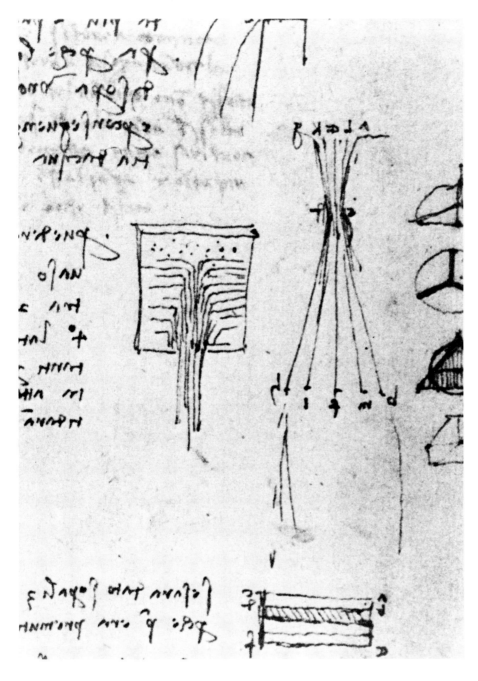

Figure 5.8. Markers using millet seeds to show the movement of water (CA 81 ra).

The object is to tell which water empties out of the hole in the bottom of the glass vessel on the left. Leonardo notes that his millet seeds drift in from the sides, as his drawing indicates. He therefore in another experiment (CA 170 vb) uses black seeds in the middle and white seeds peripherally, in order to see which pass out of the hole first.

On the right he observes that the outflowing stream expands pyramidally at first before tapering off. Only the central line of water, *ef,* runs undeviated through the hole.

Leonardo used this same marker, grass seed, to study the passage of blood through the aortic valve of the heart.

time' he describes as 1,080 per hour (W 19081 r; K/P 164 r); a *braccio* he defines as 'the measurement from the wrist to the shoulder', approximately two feet (W 19542 v; K/P 100 v).

MARKERS

Leonardo used markers of all shapes and sizes for tracing the movements of water and wind in order to detect the geometrical shapes of their lines of movement. His first analysis of water currents was clearly facilitated by the use of a rod dropped perpendicularly into the water in order to determine by its inclination whether the flow is 'more above than below' (Figure 5.7a). Variations of such floats are shown in Figure 5.7b.

When dealing with small quantities of water, particularly in his models, Leonardo commonly used millet seeds or coloured inks to detect movements (see Figure 5.8). When dealing with rivers he observed the movements of leaves, stones, pieces of wood, etc. He was particularly anxious to compare the movements of water at different levels from the surface to the bottom of rivers and at different distances from the banks. For this purpose he used oil for surface movement and designed floats constructed to be suspended at different depths. He draws such a set of floats in Figure 5.7b. Such subtly designed markers show his grasp of the significance of specific gravity.

As markers for the movements of wind Leonardo used clouds and smoke, as illustrated in Figure 3.9 (p. 90), where he has used the smoke from a burning town; he also utilised

a.   *b.*

Figure 5.9.   Markers; the use of sand and birds as markers.

a. Markers; the use of sand.

Two currents of wind are shown at the top, meeting at right angles. This gives rise to the rotation drawn below and described by Leonardo thus, 'How eddies of wind at the mouths of certain valleys percuss the waters hollowing them out into a great hollow, and carrying the water into the air in the form of a column the colour of a cloud. And I saw this very thing happen on a sand-bank in the Arno where the sand was hollowed out to a depth greater than the height of a man, and the gravel was removed and thrown a great distance, and it appeared in the air in the form of an enormous bell-tower, and the top grew out like the branches of a great pine-tree; and then it bent on meeting the direct wind which was passing over the mountains' (Leic 22 v).

Using the sand, gravel and water as his markers, Leonardo did not fail to describe and draw the shape of this waterspout (Leic 30 v).

b. In his studies of bird flight, Leonardo uses even birds as markers.

Around the two birds drawn above he writes, 'Definition of waves and the impetus of the wind against flying objects. A bird is supported with imperceptible balance in the air near mountains or high rocks in the sea. And this happens because of bending the winds which percuss these globular bodies. Being forced to preserve their first impetus they bend their straight course towards the sky with different revolutions, on the fronts of which the birds rest with their wings open, receiving beneath themselves the continued percussions of the reflected currents of wind; and by the obliquity of their bodies they make as much weight against the wind as the wind exerts force against their weight' (E 42 v).

Aware of the dynamic form of ascending winds Leonardo makes the resting birds markers of the equilibrium or rising force and descending weight. He was not aware of the thermals so valuable to the glider.

smoke from guns and devised an experimental tube to control the smoke output (CA 79 rc). He cleverly put to this use leaves, dust, sand, the 'cat's-paws' on the surface of fields of corn or water, even the bodies of birds.

It was by such observations that he came to appreciate the nature of turbulence in both water and air. In both he found vortices of 'pyramidal' or conical shape. In water such vortices usually bored downwards to a point as shown in Figure 4.11 (p. 100); they may, when taken up by air, rise as whirlwinds or waterspouts in spiral or cylindrical columns (see Figure 5.9a). When reflected upwards by hills, wind may balance the weight of birds (see Figure 5.9b).

By observing markers Leonardo came to classify eddies into three types: those level with the surface, those with raised centres, and those with concave centres (BM 30 v). See Figure 5.24. Eddies, by their curving reflected movements, serve to stabilise the main current of water, and to preserve its continuity of movement. They 'balance excessive velocity' (BM 30 r). They do not increase its movement (G 85 v).

This remarkable insight into eddy formation Leonardo used in studies of the movement of blood in the human body.

MODELS

Of Leonardo's very numerous models a number remained as ideas on paper, but many were constructed, such as those he used for clarification of his sciences. Figure 5.10

a.   *Figure 5.10.*   *b.*

a. The measurement of water in relation to its velocity and size of aperture.

This apparatus Leonardo used to make experiments for indicating the rules of movement of air and water. It will be noticed that the velocity of air and water is measured by the same anemometer shown in Figure 5.5 (p. 135). He performs the experiment by letting the fluid air or water flow first through the hole *a* with the hole *b* stopped up, then through the larger hole *b* with *a* stopped up. The flow is measured by the force exerted on the little water-wheel, with weights attached to its axle. With constant velocity and equal apertures at the ends of his 'pyramidal holes' he finds equal quantities of water. An imaginative twist to the experiment is introduced by reversing the direction of flow in the lowermost drawing at the aperture *c* (BM 241 r).

b. By replacing the wheel and weight with a vessel full of water Leonardo proceeds to make a miniature wind-tunnel in which he tests the formation of waves and their depth of action.

This experiment is shown in the drawings in the right-hand margin of Leonardo's page. Above the upper drawing, he writes, 'Experiment'. Below it, he writes, 'Make a test with your mouth whether the wind going from *a* to *b* pushes the object *n* on the bottom. I judge that the object will return to *m*, i.e., it will make a movement contrary to the movement of the wind'. The experiment is performed with grass seed (Leic 9 v).

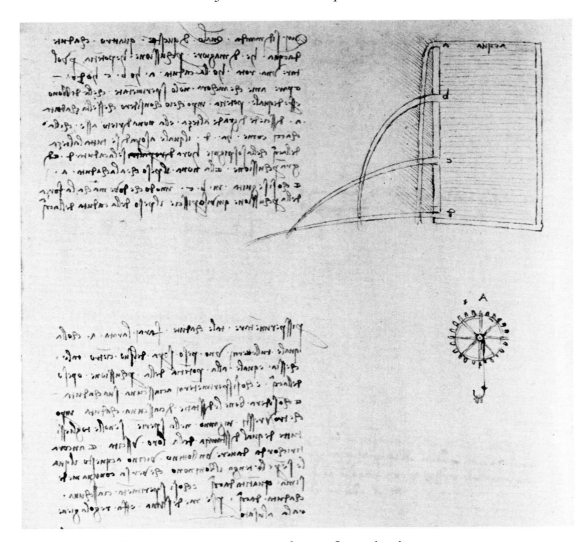

Figure 5.11.   Measurement of water flow related to pressure.

In this experiment Leonardo depicts the trajectories of spouts of water produced under different pressures. He compares the total 'power' of each jet; this power consisting of two components: the weight of water falling to the ground and the percussion of the water. 'Which of these four waterfalls has more percussion and power?' he asks, and answers, 'It seems to me that they must have the same power'. The fall *a* has great weight but no percussion; in contrast, the fall *d* has great percussion but no weight.

Below this is drawn a paddle-wheel with a weight attached to its axle (compare Figure 5.10a). Leonardo explains this: 'To experiment with these falls you will construct the wheel *A* to the centre of which you will attach a weight equal to the power of percussion or to the weight of water' (Madrid I 134 v).

Thus the total 'power' or, more precisely, 'energy' involved in all cases is, in Leonardo's view, equal. The experiment is described again later in MS F 53 r.

(For 'weighing percussion', see Figure 4.1, p. 94.)

shows a remarkable device designed to relate the flow of water or air to the size of aperture and velocity of flow using the anemometer shown in Figure 5.5 (p. 135). This is developed into a model wind tunnel to test the effect of waves on submerged objects (Figure 5.10b). A method of relating water flow to pressure is illustrated on Madrid I 134 v (Figure 5.11).

About 1498 he constructed glass chambers for the purpose of observing the movements of water entering or leaving them. On I 89 v (Figure 5.12) he constructs such a reservoir, 'in order to experiment on the issue of water from its base', and he uses markers consisting of small bits of paper. 'Note the movement of the particles', he admonishes, 'in their flow'.

On I 115 r (Figure 5.13) he once more uses small pieces of paper as markers, this time to follow the course of water entering a glass chamber from a pipe. In his drawing he notes that the angle of incidence of the stream of water approximately equals the angle of reflection from the glass walls, but at each percussion some of the water meets resistance and peels off as an eddy.

On CA 81 ra (Figure 5.8, p. 137) an experiment using millet seeds as markers is illustrated. As Leonardo puts it, they 'tell which water flows out of the glass vessel with a hole at the bottom'.

These experiments with glass chambers and markers can be seen to be logical precursors of the final refinements of Leonardo's construction of a glass model of the base of the human aorta, fitted with cusps to represent the aortic valves, and his injunction to 'Make this test in the glass model and move inside it water and panic grass seed' (W 19082 r; K/P 171 r).

One cannot leave Leonardo's use of his skill in constructing models without mentioning his application of this art to his research into the shape of the cerebral ventricles, whereby he revealed for the first time their approximately true

Figure 5.12. Experiment on the flow of water from a tank (I 89 v).

Within the glass tank Leonardo writes, 'This part is of glass, behind, it is of wood'. Beside the tank he writes, 'In order to experiment on the outflow of water from the bottom of a reservoir . . . make this vessel of straight sheets of glass, as you see, and stir up the water with mashed up bits of paper, and note the movement of its particles in their flow'.

In the drawing above Leonardo illustrates how the central part of water issuing from a pipe flows further than the side portions. The latter, he points out, are impeded by friction on the walls of the pipe. Leonardo draws a very similar figure in his study of the outflow of blood from the aorta. See Figure 15.19.

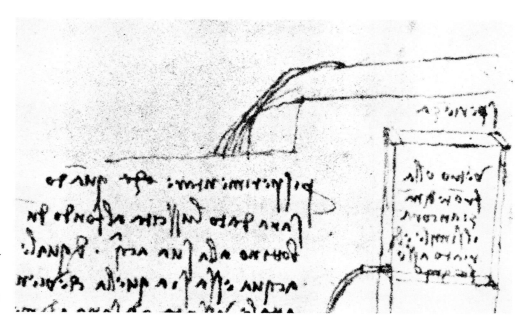

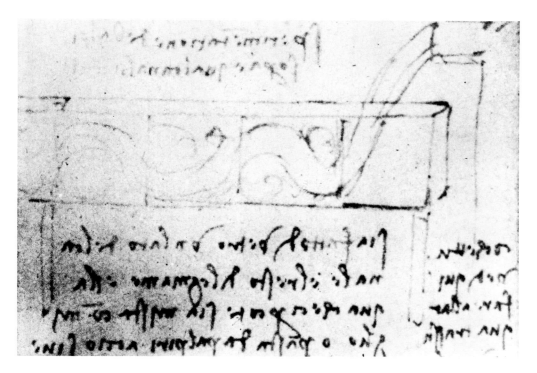

Figure 5.13. Experiment on the flow of water into a tank (I 115 r).

This glass model has wood behind, and one is instructed, 'Let the water that percusses have millet seed or papyrus mixed in it to see its course'. This course is notable for the angle of incidence being approximately equal to the angle of reflection; as well as for the resistance which is manifested by the formation of eddies after each rebound.

A similar model was made for the observation of the movements of blood in the aorta. See Figure 15.20.

conformation in wax, a pioneer model, demonstrating how he used models as bridges between subjective idea and objective reality (see Figure 2.27, p. 65).

MATHEMATICS: NATURAL GEOMETRY

All these methods show Leonardo seeking to measure the shapes of nature, whether in static or dynamic mood. With Leonardo, physics did not divide up neatly into sub-divisions of statics and dynamics which have since been imposed, for, as we have seen, his whole natural world consisted essentially of the geometry of movement or trans-formation. See Chapters 2 and 4.

Leonardo himself drives home this conviction in the very first paragraph of his *Treatise on Painting,* which begins thus: 'Science is that mental analysis which has its origin in ulti-mate principles, beyond which nothing in nature can be found which is part of that science. For example, continu-ous quantity, that is, the science of geometry, beginning with the surface of bodies, is found to have an origin in line, which is the boundary of surface. We are not satisfied with this because we know that the line ends in the point, and the point is that than which nothing can be smaller. Therefore, the point is the first principle of geometry, and nothing in nature or the human mind can be the origin of the point'. He then goes on to deny that the creation of the point is due to 'contact of the very fine point of a stylus with a surface'. This contact is a surface, and within it is the residence (*residentia*) of the point. 'The point itself is not part of the sub-stance of the surface. Neither it, nor all the points in the universe together, could they be united, would be capable of composing any part of a surface'.

'No human investigation can be termed true science if it is not capable of mathematical demonstration. If you say that the sciences that begin and end in the mind are true, this is not conceded but is denied for many reasons; foremost among these is the fact that the test of experience is absent from these mental discourses, and without this experience there is no assurance of certainty' (TP I).

Figure 5.14. 'The point being indivisible occupies no space'.

Leonardo illustrates this statement with regard to the centre of a circle, asserting, 'If one single point placed in a circle can be the beginning of an infinite number of lines and the termination of an infinite number of lines, from such a point there are an infinite number of separated points, which when reunited together return to one again. Whence it follows that the part may be equal to the whole' (BM 131 v).

When one recalls that for Leonardo 'a line is generated by the movement of a point', one realises how movement itself becomes the basis of his natural geometry.

Here once more, as we noted previously in Leonardo's definition of the centre of gravity of a body or the world, or even of the soul, the 'first principle' of everything, like the mathematical point, is spiritual; it does *not* occupy space but *resides* in a space such as that formed by the 'natural point', as opposed to the mathematical point. His 'natural point' is equivalent to the smallest conceivable unit occupying space—an atom. Leonardo repeatedly stresses the 'nothing-ness' of a point. A simple example is shown on BM 131 v (Figure 5.14), where he illustrates how the centre of a circle may be the beginning of an infinite number of lines created by movement to or from this central point, which occupy its whole surface. 'Whence it follows that the part may be equal to the whole.'

However, this geometrical point by its movement cre-ates lines, surfaces and bodies which constitute visible ob-jects and their movements. On the observation of these, therefore, science is based. Science consists of knowledge of the complex patterns of geometrical figures hidden in the changing bodies of natural things. Leonardo saw his scien-tific investigations as a search for those patterns and the laws that organised them.

Leonardo's studies in perspective provided his gateway into a scientific hypothesis. This shaped that gateway into the geometry of the 'pyramid' or cone. His experiments with stones percussing pools of water added the dimension of circles to his natural geometry. The spread of waves from the point of percussion, diminishing perspectively in power along the radius of its circular spread, combined his pyramids and circles into one basic pattern. This pyramidal or perspectival pattern was amenable to measurement, which Leonardo undoubtedly used successfully for small-scale observations such as the analysis and invention of machines. His perspectival drawings of mechanisms dem-onstrate this. But measurement was not for him the success one might have at first sight expected, by reason of the lack of a language of agreed units in which to express lengths, weights, etc., let alone force or work. Both the *braccio,* which he took as his standard of length, and the pound, which he took as his standard of weight, varied in different Italian cities such as Florence, Milan and Rome as well as the different countries of Europe. This in addition to the cru-dity of his apparatus (most of which he must have made himself) and his pioneering methods often rendered Leonardo's measurements approximate only. Often he abandoned absolute measurements in favour of propor-tional representation. Geometry, after all, is primarily con-cerned with proportions rather than absolute measurement.

Leonardo's choice of triangles and circles upon which to base his natural geometry was in some ways fortunate in that these two figures possess the same fundamental posi-tion with regard to 'similarity'. Given the magnitude of angles of a triangle, the relative magnitude of the sides logi-cally follows. Among curvilinear figures the circle holds the same fundamental position. Any two circles are similar fig-ures; the lengths of their circumferences are proportional to the lengths of their radii or diameters. Further, if two circles

Figure 5.15. Acceleration due to gravity is of pyramidal and arithmetical proportion (M 59 v).

In his studies of the fall of heavy bodies in air Leonardo expresses their acceleration in the form of a pyramidal section or triangle. In this drawing he adds figures beside each of eight 'degrees' of fall and appreciates that the increases of velocity are in both pyramidal and arithmetical proportion.

have the same centre, the arcs cut off by two radii forming equal angles at their common centre are also proportional. Leonardo was aware of these facts. See Figures 2.12 and 2.18.

Leonardo early became aware, too, that his perspectival or pyramidal proportion could also be expressed arithmetically as 'the rule of three'. For example, he exercises his skill in this rule in *Codex Arundel* 11 r, demonstrating that 'just as 3 is 3/8 of 8, so 12 is 3/8 of 32'. Other examples are found in a number of places. That on MS I 135 v carries with it a wording enthusiastic enough to suggest his recent mastery of the procedure, 'If you want a method approaching truth, which escapes you, giving the truth of a fraction, in order to know what part of the whole it is, do this', and he applies the rule of three procedure. It is to be noted that this involved his capacity for multiplication and division.

Leonardo's investigation of the acceleration due to gravity led him to express the equivalence of 'pyramidal' and 'arithmetical' proportion, though, as we have seen in Figure 4.1 (p. 94), he does not like using the term *arithmetical* for what is to him primarily a geometrical, pyramidal figure. However, he overcomes this ambivalence in the text in Figure 5.15 alongside a 'pyramidal' drawing marked off in horizontal lines into eight equal 'degrees' with the velocity of each degree recorded to one side. He describes the draw-

ing thus: 'Such movement, as it acquires power, shows itself of pyramidal shape because the pyramid similarly acquires in each degree of its length a degree of breadth, and such acquisition of proportion is found in arithmetical proportion because the parts that exceed are always equal' (M 59 v). This analysis of degrees of pyramidal proportion is carried further in Figure 4.27a (p. 113). Here for the first time he seems to suspect that his 'pyramidal law' might also be interpreted as geometrical proportion; for to the left of the middle of this triangular figure he writes, '1.2.4.8.16.32.64.128.256' (Madrid I 88 v).

In describing pulleys Leonardo speaks of pyramidal movement as being 'uniformly difform' (G 87 v). He uses the rule of three also to calculate the height of pyramids with unequal bases, e.g., if the base of a pyramid of one *braccio* has a height of ten *braccia,* what will be the height of a pyramid with a base of sixty-five *braccia?* 'Proceed by the rule of three', Leonardo writes (M 53 r). Thus these three terms, *pyramidal, uniformly difform* and *arithmetical proportion,* are equivalents in Leonardo's terminology. His applications of these mathematical methods to particular problems are innumerable. Two of the strangest are his use of the rule of three to time and motion studies of men digging trenches (CA 189 vb) and to an estimation of the rate of growth of the gravid uterus (W 19101 v; K/P 197 v).

These examples emphasise what a long way Leonardo travelled on his geometrical path from his early studies of perspective.

TRANSFORMATION

We have previously mentioned how important to Leonardo was his intimate friendship with Luca Pacioli, the mathematician. Their collaboration in matters mathematical lasted for about ten years, probably not ceasing until Leonardo left Florence for the second time about 1506. Leonardo illustrated Plato's five regular solids for Luca Pacioli's book *De divina proportione.*

During this period Leonardo embarked on an enthusiastic trail of mathematical exploration. This we learn from *Forster* I 3 r, which bears the heading 'A book entitled on transformation of one body into another without diminution or increase of matter'. The first forty folios (eighty pages) of this notebook are filled with examples of such transformations of solid bodies. 'All bodies have three dimensions', says Leonardo and asks, for example, 'How much can an oblong board be lengthened and narrowed with a given space without altering its depth'. His answer is expressed by illustration (Forster I 13 v). This transformation of bodies bears obvious relevance to the changing shapes of movement in living bodies, but of immediate interest to Leonardo is the rich field provided by transformations of different regular solid bodies into what he calls 'equal' pyramids, and of pyramids into other shapes.

He transforms a board-shaped piece of wax into an 'equal' cube, and then into an 'equal' pyramid – a typically practical Leonardian test demonstrating the possibility of transformation of shape without loss or increase of matter (CA 299 vb). We have previously mentioned how he hammered a cube of wax into a sphere – another transformation.

It is quite in character to find Leonardo making geometrical play with the five regular Platonic solid bodies described in Plato's *Timaeus*. His procedure is to cut through the angles in such a way that each section forms the base of a solid pyramid. He puts it like this: 'Regular bodies are 5, each angle when cut through reveals the base of a pyramid with as many sides as there are faces of such a pyramid. And as many angled bodies remain as sides. These angles can again be sectioned, and so on. One can proceed to infinity because the continuous quantity can be infinitely divided' (Forster I 15 r). Thus Leonardo systematically constructed dual configurations of the Platonic regular solid bodies, analysing them into pyramidal figures, i.e., tetrahedrons. Since Leonardo found 'infinite' pyramids locked within the five regular solids, it is interesting to see him attacking Plato's choice of the cube to represent earth as the primary element. 'To Plato I would reply', writes Leonardo, 'that the surfaces of the figures which the elements would have according to him could not exist . . . They say that the earth is cubical, that is to say a body with six bases, and they prove this by saying that there is not among regular bodies a body of less movement or more stability than the cube. And they attribute the tetrahedron to fire, that is a pyramidal body, as being more mobile, according to these philosophers, than the earth. For this reason they attribute the pyramid to fire and the cube to earth. Now if one considers the stability of a pyramidal body and compares it with that of a cube, the cube is incomparably more capable of movement than a pyramid; and this is proved as follows'. Leonardo's proof here meticulously compares figures showing the four sides of the pyramids and the six sides of a cube. 'Therefore', he concludes, 'the cube will turn over completely with the change of four sides on the same plane while the pyramid will turn completely with three of its sides on the same plane. The pentagon turns completely with five sides; so the more sides there are the easier is the movement, because the shape approaches more nearly to a sphere. I wish to infer therefore that the triangle [pyramid], being of slower movement than the cube, one should take the pyramid and not the cube for the earth' (F 27 v; see Figure 5.16).

Thus Leonardo sees the earth composed of infinitely small pyramidal corpuscles or 'atoms' combined into the forms of Plato's five regular solids, in magnitude varying from the smallest corpuscular to the largest terrestrial masses. He was, of course, aware of crystals (he used rubies and garnets for making red pigments), and he describes a crystal in his unique way thus: 'This transparent [body] is that which shows the whole of itself along the whole of its side, and nothing is hidden behind it' (CA 132 rb).

It is curious that Leonardo's visualisation of the structure of the earth should include so many of the crystalline forms of which we now know the earth to be composed – all inscribed within a sphere of water.

TRANSFORMATION OF STRAIGHT PYRAMIDS TO CURVED PYRAMIDS AND SPIRALS

Rectilinear pyramidal transformations might suffice as a basis for Leonardo's permutations of the forms of straight light rays and the relatively static forms of earth, but they did not meet his requirements for the movements of its flowing fluid elements: water, air and fire.

The geometrical transformation necessary to meet these fluid requirements involved curves. About 1500–1504, when his interest in the movements of water became intensified, we find him constructing curved pyramids 'equal' to straight or rectilinear pyramids. His early method of performing this is easy to understand, though later he carries his procedures to extremely refined complexity.

It will have been noticed that one of the first examples of the pyramid which Leonardo observed was the focussing of light and heat from a concave mirror (see Figure 2.19, p. 59). It had not escaped him that the surface of this mirror was curved. Thus even at this early date he draws a pyramidal figure with its base curved. Pyramids with curved sides he calls 'falcates', a term relating the curve to that of a falx or scythe. This shape is in contrast with his perspectival pyramid, which is purely rectilinear, consisting only of straight lines. Since light travels in straight lines, such pyramids served to represent transmission of linear perspective of light and shade and even the greater part of the physiology of vision. But with water, although some of the movements follow straight lines, a great many do not. Waves themselves are curved, and when they form eddies of infinite variety it is clear that rectilinear pyramids no longer apply. On A 60 r (Figure 5.17) Leonardo describes by word

4	tetracedon ——
6	evsacedron
12	duodecedron
8	ottocedron
20	icocedron

4	Tetrahedron
6	Hexahedron
12	Dodecahedron
8	Octahedron
20	Icosahedron

Figure 5.16. The five regular solid bodies of Plato (F 27 v).

These figures illustrate the passage quoted in the text.

The five regular solids appear much elaborated in the woodcuts of Luca Pacioli's *De divinia proportione*, printed in 1509. See Carlo Pedretti, *The Literary Works of Leonardo da Vinci: A Commentary to Jean Paul Richter's Edition* (London, Phaidon, 1977), vol. 2, p. 335.

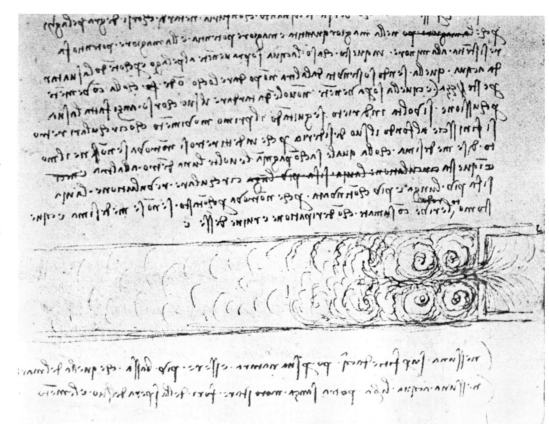

Figure 5.17. The reflection of water along curved lines or eddies (A 60 r).

Water issuing swiftly from a narrow opening percusses the slow-moving water in front and is reflected, not in straight lines as are light and solid bodies, but in equivalent curved lines.

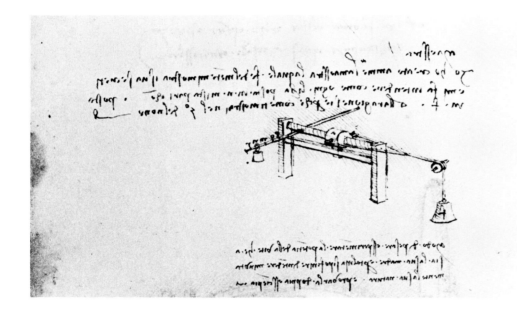

Figure 5.18. Testing the power of a screw with a lever and weight.

The test is performed by varying the weights on the lever passing through the screw and observing the effects on raising the weight on *f*. The screw passes through a nut at *a*. Below the drawing Leonardo writes, 'Experimental method of weighing and testing the power of a screw. The nut is *a*. And in this way one will be able to understand perfectly every detail of its nature' (Madrid I 4 v).

and illustration the formation of eddies when water passes swiftly through a narrow opening into a wider space. He accounts for this eddy formation by the flowing water percussing the slow-moving body of water in front and being reflected in a curve rather than in a straight line until its force has 'consumed itself'. The length of a curved or straight line of reflection is the same in both cases. The resemblance of these eddies to the shapes of screws must have struck him

forcibly. Screw mechanisms, whether in the form of the Archimedes screw for raising water, or as mechanical contrivances, had already received much practical study. On Madrid I 4 v (Figure 5.18), for example, one finds an experimental set-up of lever, screw and weight designed to 'make tests on the power of screws'. The pitch, shape and inclination of the screw-thread are all studied in relation to its power and rate of lifting an attached weight. Action always

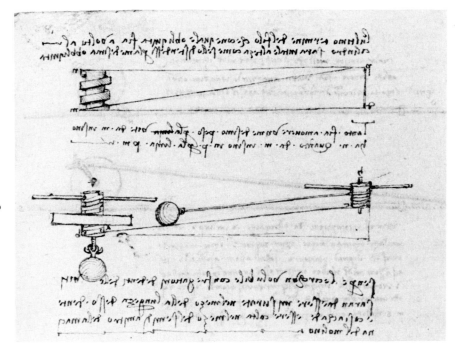

Figure 5.19. The screw; a curved spiral equivalent to a straight inclined plane.

At the top of the page Leonardo writes, 'The end of a cord wound with equal obliquity around a cylinder will reach the same height as if it were stretched out along the same obliquity'.

This is illustrated in the drawing of the screw, below which he notes, 'It comes to as much to move the same weight along the screw from *m* to *n* as it does to move it from *m* to *p* along the line *pm*'.

In the drawing below he emphasises that this relation depends upon the 'obliqities of *cb* and *ab* being the same' (Madrid I 86 v).

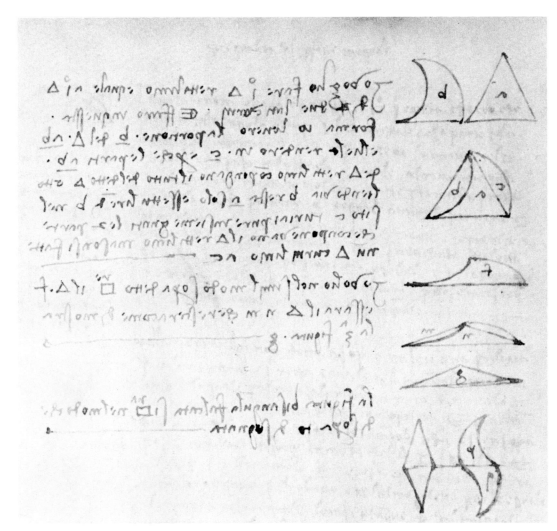

Figure 5.20. Transformation of a rectilinear triangle into a falcate.

Leonardo writes, 'I wish to make a rectilinear triangle equal to a triangle with two curved sides. And I shall proceed as follows: I shall take away portion *b* from triangle *ab* and I will return it at *c* [second drawing]. And because parts *ab* of the rectilinear triangle compose the whole triangle, and taking away *b* leaves only *a* if you return *b* at site *c* you will have joined together the two parts of which the rectilinear triangle was composed, but they are now transformed into a curvilinear triangle *ac*' (Madrid II 111 v).

It will be noticed that Leonardo has doubled this procedure in the two falcates *p* and *q* (lowermost drawings on the margin), an example of his method of symmetrical duplication.

being equal to reaction, Leonardo also describes the rotation of a screw with such velocity that the reaction of its force on the air raises it from the ground – the helicopter (see Figure 1.25, p. 33). 'I find', he writes, 'that if this instrument made with a screw be well made . . . and turned swiftly the said screw will make its spiral in the air and rise high' (B 83 v). That a screw with its thread 'unwound' becomes a triangle is illustrated on Madrid I 86 v (Figure 5.19). On Madrid I 176 v he equates wedge and screw power with the pulley. Such practical considerations, particularly the last, led him to search for a mathematical principle whereby he could transform straight pyramids into curved ones. The principle Leonardo applied was simple: 'If I give back to a surface what I have taken away, the surface returns to its former dimensions' (Madrid II 107 v).

He applies this rule to a plane triangle on Madrid II 111 v (Figure 5.20). Here beside his illustration he writes, 'I wish to make a rectilinear triangle equal to a triangle with two curved sides [top drawing]. I shall proceed as follows: I shall take away the portion *b* from the triangle *ab* and I will return it at *c*. And because parts *ab* of the rectilinear triangle compose the whole triangle, and taking away *b* leaves only *a*, if you return *b* at site *c* you will have joined together the two parts of which the rectilinear triangle was composed; but they are now transformed into a curvilinear triangle *ac* [second drawing].' Similar procedures are applied to the other drawings below on this page and W 19145, K/P 128 v. See Figure 5.27.

The same procedure is applied to the three-dimensional pyramid shown on Madrid II 107 r (Figure 5.21). This time the process is preceded by three emphatic underlined statements of general 'law'. 'If what has been removed is replaced, the thing taken away will not be missing. That quantity lacks nothing to which what has been taken away is restored. That part is reconstituted in its entirety to which its remainder has been restored'. This rephrasing provides a good example of the efforts Leonardo constantly makes to frame a general law clearly. (We shall see him apply this law to the movements of the diaphragm and abdominal wall in Chapter 15.) In Figure 5.21 the two drawings in the top right margin illustrate the law applied to a solid pyramid. 'Therefore', he continues, 'I will take away *acfge* from pyramid *acbde* and I will put it back on the opposite side, that is at *bde*. And from a rectilinear pyramidal body I will have made a pyramid of curved lines and sides, which is shown at *acbde* [lower drawing]. And according to the concept proposed above, this curved pyramid is exactly equal to the rectilinear pyramid proposed above' (Madrid II 107 r).

Below these two drawings in Figure 5.21 are two corresponding drawings of a curvilinear and a 'straight' cone; each is then inscribed in straight and curved cylinders. The 'central line', to which Leonardo attached so much importance, is not omitted from his curved pyramid. On Madrid II 136 r he describes its construction.

To return to Figure 5.21, the three central drawings show curvilinear pyramids arising from similar bases of rectilinear figures between two circles, i.e., circular, equi-

distant, parallel lines. Such curved 'pyramids' or falcates, strange as they may look, are now 'transformed' into figures 'equal' to the rectilinear triangle or parallelogram.

The lettered drawing below, in Figure 5.21, is of particular interest in reflecting Leonardo's thought. He sets out to prove that the rectangle *abcd* is of the same magnitude as the curved space *cdef*. He proves it on the thesis that 'a moving object acquires as much space as it leaves behind'. 'Hence having removed that part of the surface *bdf* from its remainder *dfg* the whole space *abcd*, by necessity the space *cdef* equals the space *abcd*'.* 'This method of proceeding', comments Leonardo, 'is as much subordinate to and part of philosophy as geometry, because the proof is obtained by means of motion, though in the end all mathematics consist of philosophical speculations'. He therefore sets out another 'proof' of this proposition, saying, 'Moreover we will prove that the surface *cdef* is equal to the quantity *abcd* without using the method of motion but based only upon the concept that says if from 2 equal things equal quantities are taken away, the remainders are equal. Therefore I have here two equal surfaces, that is *ace* and *bdf* which I propose to be equal and similar, and I take away from the one and from the other part *bde* which I imagine to be duplicated because they are superimposed upon each other. Therefore taking away from both surfaces the aforesaid parts the remainder, *efcd* is left equal to the remainder *abcd*' (Madrid II 106 v).

From the drawings in Figure 5.21 one can perhaps see that with further extension of the curvilinear pyramid one will reach the figure of a spiral. Leonardo saw this. In Figure 5.22 he shows how this was done. He declares his principle at the top of the page as a heading: 'All uniformly difform triangular figures with curved sides (as with straight lines) made between parallel lines are always equal'. Below, in the centre of the page, one sees a typical falcate, scythe-like figure reminiscent of that in Figure 5.21. Above this is a spiral with its base formed by one side of its enclosing square. To the left is a drawing, the importance of which is stressed (by a line) by Leonardo himself. The rough sketch shows four triangles arising from the same base *bc*. The simplest is a right-angled triangle, *abc*. From *a* a line *ade* runs parallel to *bc*. The second drawing is another triangle, *bcd* equal to *abc* because it has the same base and is between the same parallel lines. The same is true of the even more oblique triangle *bce*. The fourth drawing, unlike the curved pyramids we have seen, is extended to a spiral, arising also from the base *bc*, curling round as it narrows to its apex at *a'*, within the circle with radius *ab*, the distance between the parallel lines.

Leonardo writes, 'It is proved because the helix [spiral] is equal to the triangle *abc*; and the triangle *bce* and the triangle *abc* are equal. This is because when things are equal to a third

---

*Leonardo's brevity, due to this being but one of a number of similar propositions on these pages, may be clarified by noting that 'moving' the curved figure *cea* to *dfb* empties the space *abcd*, except for a small area now common to both *abcd* and *cdef*. These figures are therefore equal. This same procedure is drawn in three dimensions on the right lower margin of this page (Madrid II 107 r, Figure 5.21).

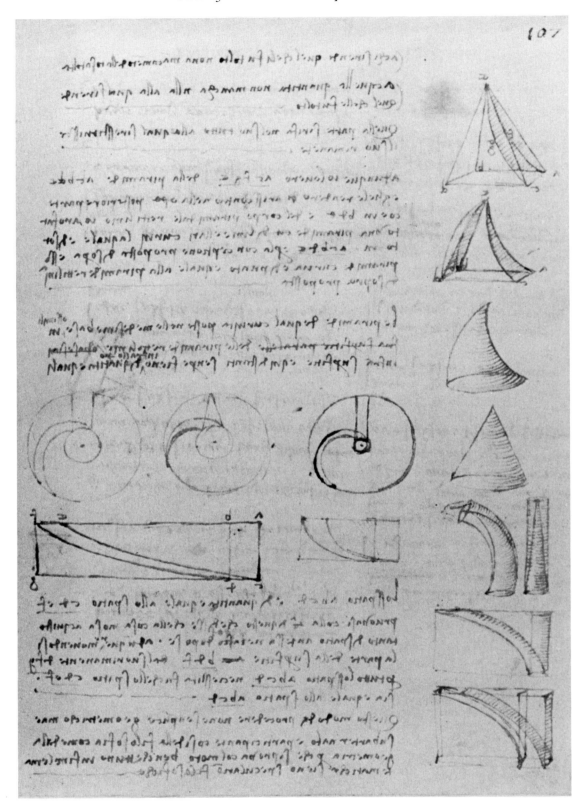

Figure 5.21. Transformation of solid rectilinear pyramids into curved pyramids and other forms (Madrid II 107 r).

The central drawings use the method of transformation to confirm Leonardo's thesis that everything that moves in space acquires as much as it loses. Compare Figure 4.10 (p. 100). The text quoted from *Madrid Codex* 106 v applies to the left-central lettered drawing on this page.

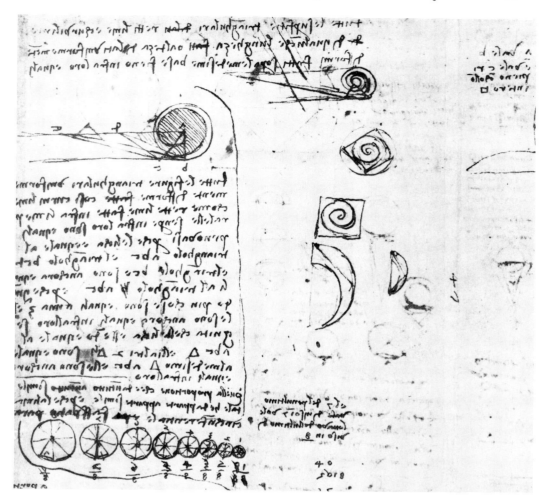

Figure 5.22. Transformation of rectilinear triangles into a spiral.

Advancing from the geometry in Figure 5.21, Leonardo develops the curvilinear falcate into a spiral equal to triangles with equal bases and between parallel lines. This he does by making 'the circumferential line at a parallel distance from the centre of a circle of which the radius is equal to the rectilinear distance between the parallel lines. And what is made between rectilinear parallels holds for curvilinear parallel distance' (CA 235 vb).

they are also equal to one another. It follows that since the helix is equal to *abc* and the other triangles are equal to the same triangle *abc* they are equal to one another' (CA 235 vb).

Leonardo's 'rules' for the construction of 'equal' curved and rectilinear pyramids are often repeated. He expressed it thus, 'All rectilinear pyramids and those of curved sides formed upon the same bases and varying uniformly as to breadth and length between parallel lines are equal' (K 80 r).

The spiral Leonardo described as follows: 'Spiral movement is formed out of slanting and curved lines, in which lines drawn from the centre to the circumference are all found to be of varying length; and it is of four kinds, namely convex spiral, level spiral, concave spiral, and the fourth kind is a spiral of cylindrical form' (E 42 r).

It will be noticed that this classification of spirals is the same as Leonardo's classification of eddies in water as far as convex, level and concave eddies are concerned. The fourth type, 'the cylindrical spiral', constitutes a screw. He describes it in water as a 'longitudinal, columnar or cylindrical wave'. These are well illustrated in Figure 5.23 (W 12660 r) and MS F 93 r and 91 r. The three types of eddy correspond to the three types of helical spiral illustrated diagrammatically in Figure 5.24. Increasing complexity of geo-

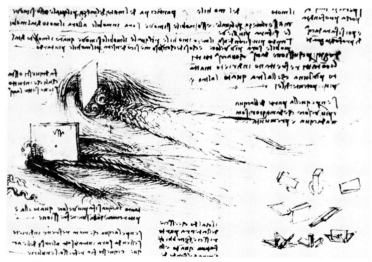

Figure 5.23. Types of eddies in water; composite and longitudinal.

By inserting planks into currents of water at different angles (see small drawings to the right), Leonardo created eddies of all types. In the top experiment the eddies are 'composite'. In the lower, most are longitudinal cylindrical spirals thus resembling screws. Below in his text Leonardo describes this type of wave as 'stretched longitudinally in the shape of a long spiral which goes rotating around its central line' (W 12660 r).

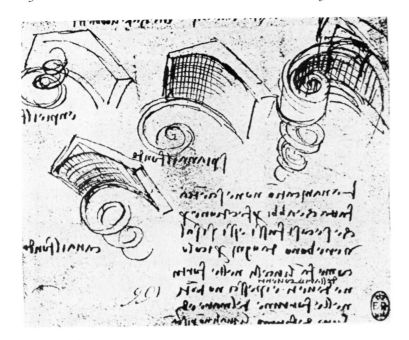

Figure 5.24.   The three types of spirals or eddies (W 12666 r).

The three sketches on the left show convex, level and concave spirals or eddies. Leonardo describes them as 'filling from the base', 'level with the base' and 'hollowing the base'. The fourth sketch shows the mode of formation of a concave eddy more clearly.

Figure 5.25.   The formation of a complex eddy in water (W 12660 v).

In this drawing, renowned for its beauty, Leonardo greatly elaborates the experiment illustrated in Figure 5.13 (p. 141). This time the formation of the eddy is traced by air bubbles used as markers. Their tracks in the depths of the eddy show the angles of reflection, their bursts when they reach the surface make patterns like flowers. From the drawings above this it is clear that Leonardo has also used the experiments drawn in Figure 5.23 in the construction of this pattern.

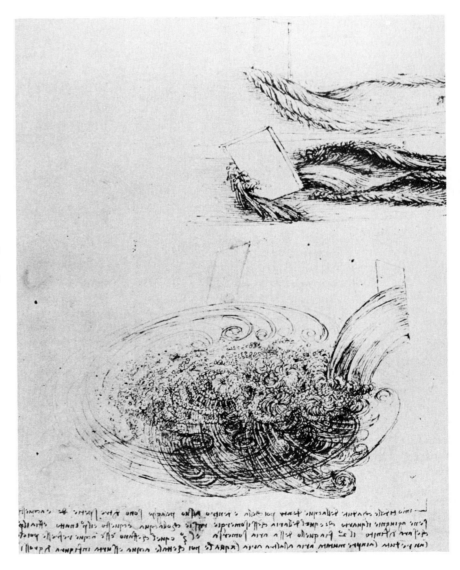

Figure 5.26.   The Deluge (W 12382).

In this drawing all four types of eddy construction are shown blasting a landscape, so creating a very powerful emotional effect of 'percussion' or impact.

metrical structure with beautiful rebounds and spirals made by eddies of water containing air bubbles as markers is illustrated in Figure 5.25.

Transformation of the forces arising from percussion waves and all forms of spiral eddies reach the realm of almost complete abstraction in the series of drawings entitled *The Deluge* (Figure 5.26). Leonardo produces two word pictures of these drawings. One describes the human side of the picture: terrified animals, weeping women, men's frantic gestures of despair, etc. (W 12665 v). The other describes the macrocosmic actions of 'the four powers': the darkness, wind, tempests, floods of water, forests on fire, bolts from heaven, earthquakes, ruins of mountains, levelling of the cities, and the like. Here he

composes this description of these events in nothing but physical terms describing turbulence, eddies, percussion, with angles of incidence equalling angles of reflection, etc. (W 12665 r). This second verbal version provides the subject matter of nearly all the series of drawings. The patterns of percussion waves spreading in circles of destruction, spiral eddies of water and air combined, mountains exploding – all these are woven into one horrific vision of destructive forces created out of Leonardo's natural geometry, now transformed into a figure of monstrous power. See also Figure 4.45.

Leonardo's transformation geometry of nature not only underlies his important discoveries of the pattern of the movement of the blood in the aorta but also provides

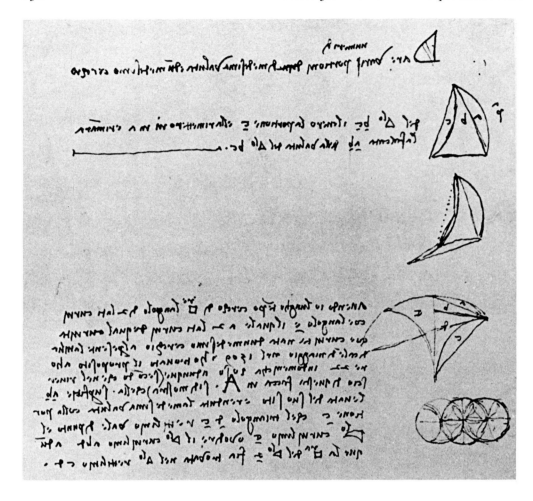

Figure 5.27. Transformation of a rectilinear triangle into a falcate and the quadrature of a falcate (W 19145; K/P 128 v).

The upper drawing illustrates how 'I will lift off the portion *c* of the triangle *bc* and put it back at *a;* and the falcate *ab* will remain equal to the value of the triangle *bc*'.

Below, in a passage containing the date 'in the year 1509 on the eve of the calends of May', Leonardo demonstrates that the falcate *e* equals the rectilinear triangle *cd*. To do this he refers to a passage marked *A* on another sheet.

the basic principle upon which he constructs his illustrations and descriptions of the movements of the human body.

TRANSFORMATION OF LUNULES

The replacement technique was one of Leonardo's favourite methods of geometrical transformation. Others involved rotation of forms round an angle, or symmetrical duplication of figures. We will confine ourselves to showing the beginning of one important path, that to lunules.

If we retrace our steps to W 19145, K/P 128 v (Figure 5.27), we see here in the top drawing an illustration of the now familiar process of converting a rectilinear triangle into a curvilinear triangle by taking the part *c* from the one side and adding it to the other side at *a* so that 'the falcate *ab* remains of the same value as the triangle *bc*'. The text below refers to Leonardo's attempt to square 'an angle of two curved sides marked *E*'. It is interesting that Leonardo himself attached so much importance to this exercise as to write in the text, 'On the calends of May 1509 at the 22nd hour on Sunday I have found the proposition'. It also serves to illustrate the complexity of some of Leonardo's transformations.

The large letter A in the middle of this passage refers to studies in Figure 5.28b upon which the 'solution' depends. This was originally attached to Figure 5.27 but became

separated. In Figure 5.28b, he deals with the quadrature of falcates with unequal curved sides. This in its turn depends upon duplicating octagons. The complexities of these problems have been admirably clarified by J. McCabe.[1]

At the top of Figure 5.28b Leonardo draws a small rough diagram of the first quadrature of a lune by Hippocrates of Chios. In this the segment on the diameter of a semicircle (Figure 5.28a) is proved to be equal to the sum of the two smaller segments. By adding to both the portion of the isosceles triangle above the arc of the segment on the diameter the lune is shown to be equal to the right-angled isosceles triangle inscribed within the semicircle.

From this beginning Leonardo constructed a very large number of valid variants of the lunule. The most significant figure has been called by McCabe the 'key figure'. In this Leonardo duplicates circles by inscribing a circle in a square, doubling the square and inscribing a second circle within this, as in Figure 5.29. Thus he has duplicated the square and the circle; the circles are proportional to the squares within which they are inscribed. As Leonardo puts it, 'The smaller circle drawn here enters 4 times into the greater circle which surrounds it'. The diagram proves that doubling the square also doubles the circle's diameter. From this figure Leonardo not only derived many further variants of lunules but also showed that circles are proportionate to the squares of their diameters. Thus, if a section of a cone or 'pyramid' is

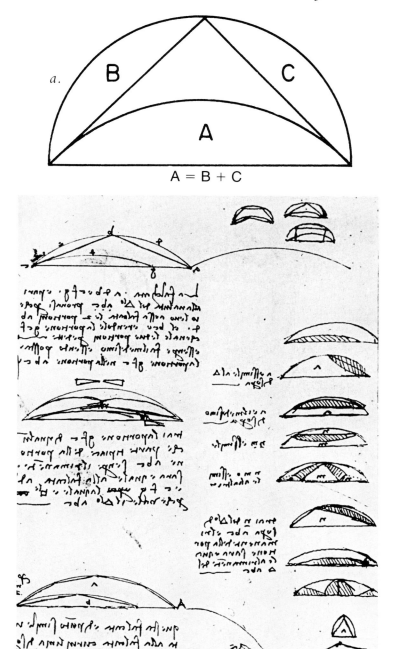

$$A = B + C$$

Figure 5.28.

a. The first quadrature of the lune by Hippocrates of Chios.

b. This figure is drawn by Leonardo in the top right-hand corner of the page, from which he derives a number of transformations of the lune (W 12658 v; K/P 129 r).

taken so that its diameter is one-half of another section, its area is one-quarter, i.e., the area of the cross section of a cone is proportional to the square of its diameter.

Reference to one of Leonardo's figures of a cone (e.g., that in Figure 5.21, p. 148) immediately shows how relevant this is to Leonardo's perspectival 'pyramidal' law governing the four powers. By observation he has ascertained that the perspectival alteration in the 'size' of an object is

inversely proportional to distance. But his 'key figure' shows that the area of an object is inversely proportional to the square of the distance. Can both laws be true? There is good evidence that Leonardo came to equate his pyramidal law with geometrical progression, i.e., an exponential rather than simple arithmetical progression (see Figures 5.30 and 4.27a, p. 113).

A point needing explanation is Leonardo's habitual use of the term *pyramid* when he appears to be referring to a triangle. Ptolemy's great works on astronomy and optics came down through translations into Arabic. In the twelfth century these were translated into Latin by Eugene of Sicily.[2] The Greek word *konos* was now rendered *piramis*. Hence the visual 'cone' of Ptolemy became the medieval visual 'pyramid'. Alhazen (in translation), Roger Bacon, Vitello, Pecham, Biagio Pelcani and Alberti – all predecessors whose works were known to Leonardo – used the term *pyramid* for cone. Moreover, they drew it as a triangular figure representing a section from apex to base. Thus Leonardo's use of the term *pyramidal* to describe the phenomena of perspective, both verbally and geometrically, was indeed traditional. That he used the word broadly to include the cone is illustrated in the series of drawings in Figure 5.21 (p. 148) and elsewhere. And in *Codex Arundel* 176 v he writes, 'a pyramidal body is that in which all the lines parting from the angles of its base converge to a point. And such a body can be invested with infinite angles and sides'.

Insight into the fact that his pyramidal law included arithmetical and geometrical progression with distance came to him later. The development of this realisation was reached gradually by stages.

In Figure 5.29 he demonstrates that a circle inscribed in a square is half that of one circumscribing a square. On many occasions between 1492 and 1513, for example in *Codex Atlanticus* 318 ra, 44 va and 89 va, he draws and explains such figures. With regard to the pyramidal law, the most significant late applications of geometrical progression are made in the form of tables, in Figures 5.30b and 531. Under the title 'Perspective', in Figure 5.31, three columns are headed: Pyramid; Base; Power (*potentia*).

Pyramid	Base	Power	
8	4	1	
4	2	4	[4^1]
2	1	16	[4^2]
1	1/2	64	[4^3]
1/2	1/4	256	[4^4]
1/4	1/8	1024	[4^5]
1/8	1/16	4096	[4^6]
1/16	1/32	16384	[4^7]

Note: It will be noticed that this table of "powers" takes the form of an exponential series to the base 4. Though acquainted with algebra, Leonardo never once used an algebraic expression in his mathematics; hence, the use of a convenient number, 4, in this context. This table is commensurate with the inverse square law.

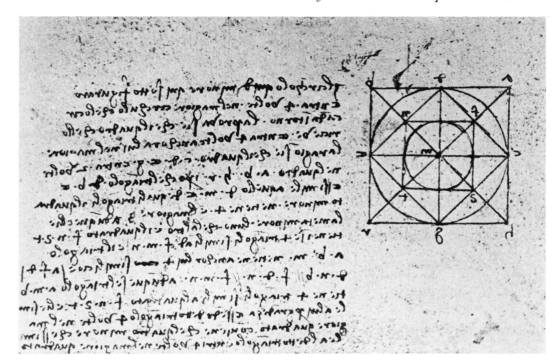

Figure 5.29. The duplication of a circle and a square (A 14 r).

From this drawing one can see how by drawing segments on each side of a square Leonardo derived many transformations of lunes on the principle of Hippocrates of Chios, as illustrated in Figure 5.28b. In this figure he also demonstrates that the area of a circle is proportional to the square of its diameter.

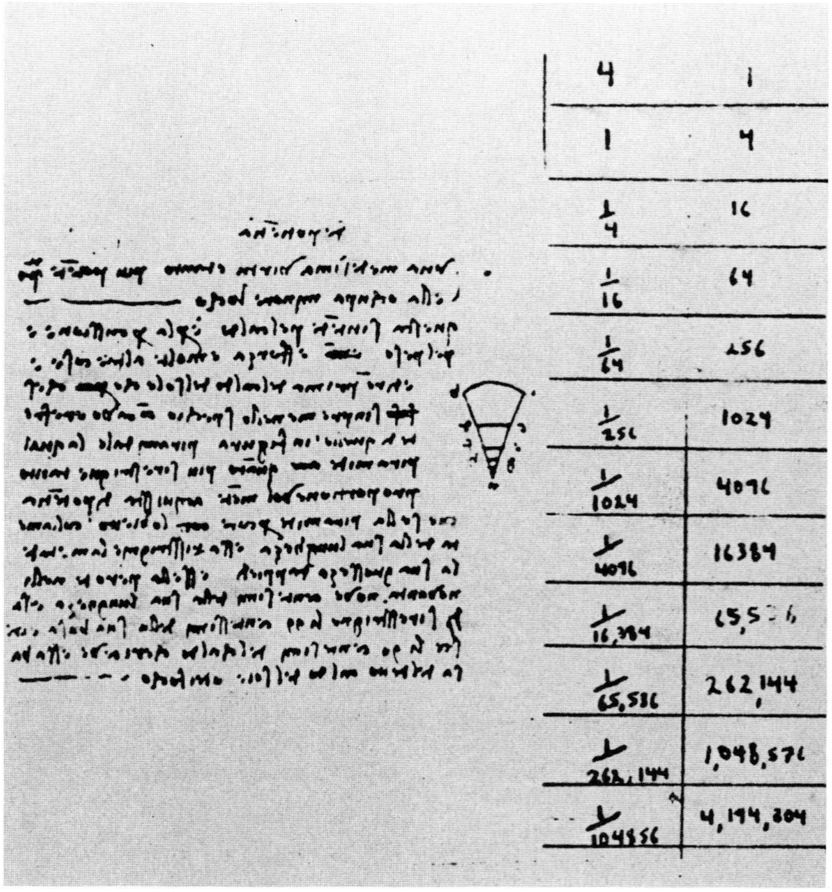

Figure 5.30. The pyramid as expressing exponential proportion.

a. Alongside the pyramidal drawing Leonardo writes, 'On Power'.

'The same virtue of power is so much the more powerful as it is the more concentrated. This is true of heat, percussion, weight, force, and many other things.
Let us speak first of the heat of the sun which is imprinted on a concave mirror, and is reflected by it in a pyramidal figure, which pyramid acquires proportionately as much more power as it is narrowed' (G 89 v).

b. Here the proportion of this concentration of power is laid out as a table.

The left-hand column denotes the 'base' or 'diameter' of the pyramid or cone; the right-hand column, the 'power'. Leonardo writes under the heading, 'On the nature of heat'. 'If the base of 4 *braccia* emits 1 power in the space of 1 *braccio* the heat from the base increases to 4; and 16 times if this base is reduced to ¼ *braccio*. And if this base (in turn) is reduced by a quarter it acquires a power of 64 degrees over its base. Such diminutions and increases of power are laid out here below'. Then follows the table shown in Figure 5.30b.

It will be seen that the 'power' increases by geometrical progression as the base of the pyramid narrows. Leonardo calculates this increase of the power of heat to 4,194,304 times, in terms of an exponential series of powers of 4 (G 85 r).

This figure is composed from G 89 v (Figure 4.44, p. 129) and the table on G 85 r, which is contemporary. See also Figure 2.19.

4	1
1	4
$\frac{1}{4}$	16
$\frac{1}{16}$	64
$\frac{1}{64}$	256
$\frac{1}{256}$	1024
$\frac{1}{1024}$	4096
$\frac{1}{4096}$	16384
$\frac{1}{16,384}$	65,536
$\frac{1}{65,536}$	262,144
$\frac{1}{262,144}$	1,048,576
$\frac{1}{1048576}$	4,194,304

a.   *b.*

Figure 5.31. The relation of the height of a pyramid, its width and power.

Beneath the heading 'Perspective' Leonardo sets out a table relating height and width of the base of a pyramid with power. In it he calculates that power increases in geometrical progression as the pyramidal section grows shorter and its base grows narrower (BM 280 v).

By citing the first eight enumerations it will be seen that Leonardo has assimilated geometrical progression as opposed to arithmetical progression into his pyramidal figure. This is further demonstrated in Figure 5.32, where the dimensions of the bases and heights of the suggested pyramids are placed alongside the familiar triangular figures. This page contains the explanation, 'Let a given section of a pyramid be given all the force of its base. Let the base be 4 *braccia* and the pyramid 8 miles in length and a section of this pyramid be made at a distance of 7 1/2 miles of the pyramid from its base. This section will be 1/4 *braccio*, that is 1/16 of the diameter of the base and it will be 256 times more powerful than the base, and the remainder of the pyramid will be 1/2 mile. And if this section too diminishes its diameter by 1/2, its base will be 1/8 *braccio*, and its whole length 1/4 of the antecedent section which was 1/2 *braccio* and 4 times more powerful. Therefore if the previous power was 256 times more powerful than the base, this 1/8

braccio will be 1,024 times more powerful than the base' – and Leonardo diligently calculates correctly the power of each section as laid out in the table in Figure 5.31 until he has concentrated the original power 16,777,516 times.

In *Codex Arundel* 279 r Leonardo permits us to glimpse what particular power he has in mind. In a fragmentary passage one finds the words 'the pyramid given the nature of its base, that is of cold or heat'. On the same page he vastly prolongs the suggested length of his pyramids into a new project: 'For seeing the nature of the planets, open the roof and show at the base [of the pyramid] one single planet. And the reflected movement on this base will show the nature of the said planet. But make it so that the base does not observe more than one at a time'.

In *Codex Arundel* 78 v the sun is revealed as the object of this pyramidal investigation. On this page of muddled notes there occur three circles of diminishing size. Near them Leonardo asserts, 'Such is the proportion of circle

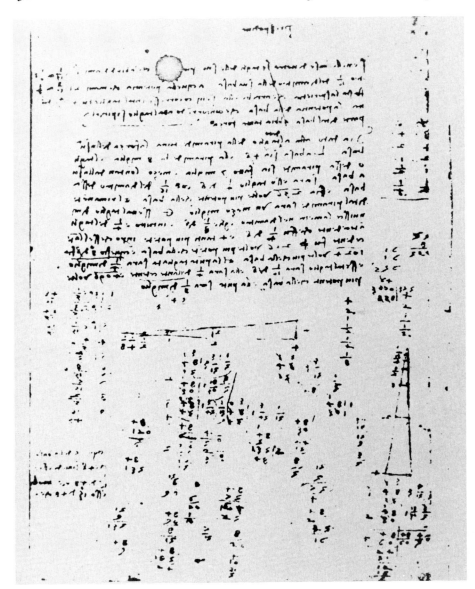

Figure 5.32. Perspective and the concentration of power.

On this page pyramids (drawn as triangles) show the proportions enumerated in the table in Figure 5.31. In this case the concentration of power as demonstrated by the diminution of width and increase of height occurs by geometrical progression (BM 279 r).

For translation of the script, see text.

Figure 5.33. The pyramidal concentration of heat and light (CA 277 ra).

Parallel rays from the sun are here concentrated pyramidally by a huge concave mirror, so huge that the pyramid *cde* equals four miles. Such rays focused at *e* would produce enormous heat and light. The tube leading from *enm* bears some slight resemblance to Newton's reflecting telescope. But there is no suggestion of a reflecting mirror at *e*, which is described as a support.

See Carlo Pedretti, *The Literary Works of Leonardo da Vinci: A Commentary to Jean Paul Richter's Edition* (London, Phaidon, 1977), vol. 2, p. 20.

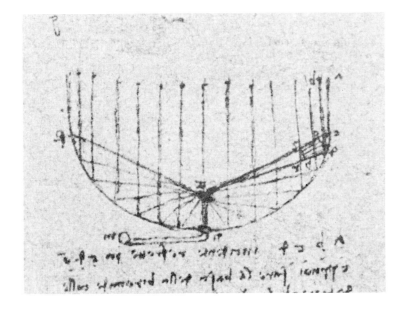

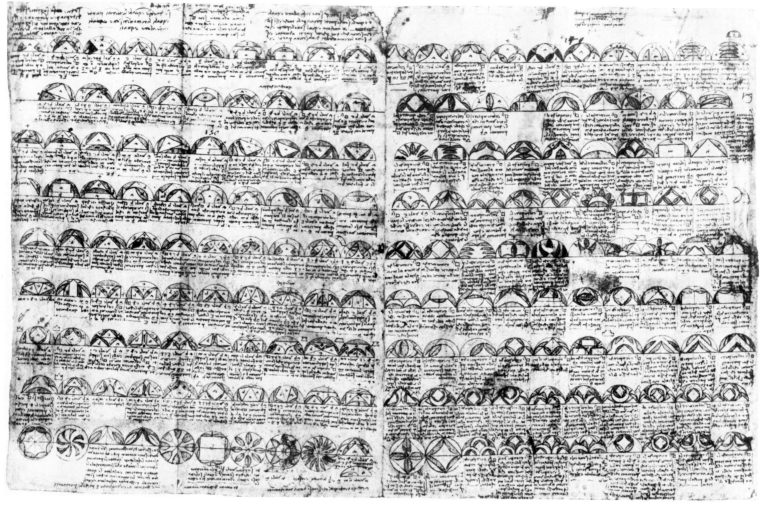

Figure 5.34. An exercise in geometrical transformations (CA 167 rab).

to circle as is that of the square to the square of their dia-meters'. This is a translation of Euclid, Book XII, Proposi-tion 2. It also reflects Leonardo's many figures of circles inscribed in squares (see Figure 5.29, p. 154).

Leonardo goes on to say that if the large circle has a diameter of 12 inches and the small one 1/12 of this, then this is $1/(12 \times 12) = 1/144$ of the larger, and the smallest $1/144 \times 144 = 1/20,736$, etc..

It is significant that on this same page Leonardo writes, 'Take the measure of the sun at the solstice in mid-June and make its pyramid to the centre of the world. And cut its pyramid in the diameter of a mirror and observe how much the parallels *ab df* are narrowed in the space of one *braccio ab*. But see that you have the diameter of the sun and its dis-tance, and make your calculation'.

These inverse square ratios apparently led him to attempt to construct a concave mirror of enormous size by which he hoped to concentrate solar energy in the form of heat and possibly light as well in the case of planets.

In Figure 5.33 he writes of making a mould (*sagoma*) 'for generating pyramids which come to a point, that is, a con-cave mirror'. This is accompanied by a drawing showing

parallel rays converging to a focus at *e*. From this point runs a right-angled tube, *nm,* designed either to hold an object or lead off the heat energy focussed at *e*.

Leonardo wonders 'if the pyramid can be condensed and compelled to so much power at one single point'. If it can be done, then 'with this one can make any cauldron in a dye-works boil – and with this . . . there will always be boiling water' (CA 371 va).

Thus Leonardo's late realisation that his pyramidal law includes the inverse square law as a form of his perspectival law leads him to contemplate the possibility of industrial use of solar energy for boiling water.

These rewarding insights, about the range of his pyrami-dal law were clearly gained with great difficulty. That his pyramidal law of power could include not only arithmetical but geometrical progression, even the inverse square law, was realised by Leonardo relatively late in life. He expressed it clumsily and fragmentarily, and only in terms of the concentration of heat and light. Undoubtedly with time he would have clarified and perhaps extended these forms to their place in application to the vast number of observations and experiments performed decades earlier. But it was too late, and his work in science shared the fate of so many of his works of art in remaining unfinished.

We are fortunate in possessing a declaration from Leonardo himself on the value of mathematics to science entitled 'Of mathematics and of those that are primary and those that are derivative'.

'There are 2 mathematical sciences of which the first is arithmetic, the second is geometry. These deal with the 2 quantities, that is, the continuous and the discrete. There is no other kind in nature. Consequently these embrace everything in the universe. But arithmetic never needs geometry, whilst geometry in many cases makes use of arithmetic.

'There follows perspective, first child of geometry, which is descended from geometry inasmuch as its function extends into visual lines that extend between the object and the eye. These lines must be numbered amongst the continuous quantities. Perspective gives birth to astronomy because by means of visual lines the height and magnitude of celestial bodies are measured with the astrolabe. There are also the lines of natural motions by which the world is measured. This primary perspective which serves and deals with the function of the eye must be placed far before music which serves only the function of the ear, a sense of less dignity than the eye with which geometrical bodies are represented' (Madrid II 62 bis r).

We have glimpsed some of the simpler geometrical transformations built up by Leonardo into visual natural geometry. Towards the end of his life these geometrical transformations became ever more complex – so complex as to be incomprehensible in terms of Nature. Apparently Leonardo proposed to compose such transformations into a unified work called *De Ludo Geometrico*. Examining such pages as the *Codex Atlanticus* 167 rab (Figure 5.34) one can-not help feeling that Leonardo had some deeper objective than that of a merely visual geometrical game. These pages suggest geometrical concepts which remind one vividly of the patterns of growth in flowers, of the diffraction patterns in crystallography and of embryological transformations.

These impressions resonate so forcefully in the mind of the author that he cannot resist the feeling that Leonardo was visualising his science topologically from the mathematical end and, like some mathematicians of today, suggesting mathematical possibilities which so far he had not found in Nature. His many studies of transformations and lunulae, particularly the 176 geometrical transformations presented in an organised sequence on this page, strongly remind one of three of Leonardo's descriptions of science: 'Science is knowledge of things possible, present or past; prescience is the knowledge of things that can come to pass even though slowly' (Triv 17 v). 'Nature is full of infinite causes which have never occurred in experience' (I 18 r). And 'O observer of things do not praise yourself for knowing the things that Nature ordinarily brings about. But rejoice in knowing the end of those things which are designed by your own mind' (G 47 r).

REFERENCES

1. McCabe, James E. *Leonardo da Vinci's De Ludo Geometrico*. Los Angeles, University Microfilms Library Services, Xerox Corporation, Ann Arbor, Mich., 1972, p. 157.
2. Lejeune, Albert. *L'optique de Claude Ptolémée dans la version Latine d'apres l'arabe de l'emir Eugene de Sicile*. Edition critique et exégétique. Universitè de Louvain, Louvain, 1956.

Chapter 6

The Movements of Man and Animals

Impossible as it is to say that Leonardo specialised in any particular subject, one is nevertheless justified in asserting that his interest in the human body was lifelong. There is not a painting which does not contain the human figure as its central theme. Moreover, Leonardo saw his own body as the instrument of his soul with his eye as its window. For this reason his studies in perspective were followed through to studies of the physiology of vision and other sensations.

Even in his early anatomical work, about 1489, there are studies of the muscles, some surprisingly and beautifully accurate, although he was then concentrating mostly on the nervous system; in his painting *St. Jerome* (Figure 1.1, p. 10) his knowledge of surface anatomy is evident. Nevertheless, Leonardo gave up his directly anatomical research for many years before returning to his final attack about 1506. Why?

During the long interval he had, as we have seen, been developing his physical 'rules' of the macrocosm and his concept of the 'four powers' and their patterns of action on the elements of earth, water, air and fire. These patterns he had woven into the nature of the outside world. He had built them into innumerable machines; he had established them by his own inventions and found some truth in his rules. In this way he had answered some of his own questions. 'The result or effect of my rules; what do they produce, what good are they? They are the bridle of engineers and investigators, not letting them promise themselves impossible things' (CA 377 rb). And as he discovered these rules of movement, weight, force and percussion, this pioneer bio-engineer applied them methodically to the actions of that most complex of machines, the human body.

THE PHYSIOLOGY OF HUMAN MOVEMENTS

All the movements of man and animals, whether expressing emotion, work of the limbs, or 'local' movement from one place to another, have a common source of origin – life. And 'Our life,' says Leonardo, 'is made by the death of others. In dead matter life deprived of sense [*dissensata*] remains, which, reunited to the stomachs of living beings, resumes life, both sensitive and intellectual' (H 89 v).

Leonardo had previously described how 'The five senses are the servants of the soul' which resides in the middle ventricle of the brain with the *senso comune* (W 19019 r; K/P 39 r). Here also originates 'voluntary motion' and any kind of 'local movement' (W 19048 r; K/P 49 r).

All human force 'has its origin in spiritual motion flowing through the limbs of sentient animals, and broadens their muscles. Thus broadened, these muscles become shortened and draw back the tendons which are connected with them. And this is the cause of the force of human limbs' (BM 151 r). 'Consequently material movement springs from spiritual' (BM 151 v).

From the painter's point of view, the most important motions are those which express the emotions of the mind or soul. 'The good painter has two principal things to paint, man and the intention of his mind' (TP I 248) writes Leonardo, and 'let the attitudes of men with their limbs be disposed in such a way that these display the intention of their souls' (TP I 299). He adds that a painting of man must 'express in gestures the passion of the soul' (TP I 400). We are reminded that 'gestures' exert 'force,' and Leonardo's 'Definition of force and movement in animals' is as follows: 'I affirm that the said force of movement is based upon different points of support [*poli*]. Force is produced by the lessening and contraction of the muscles which draw back and by the nerves which reach as far as the stimulus [*sentimento*] communicated by the hollow nerve dictates' (B 3 v).

In this description of the origin of movement in man and animals Leonardo, it can be seen, has built up what today we would call a kind of reflex arc. The afferent sensory part brings the sensations to the soul in the third ventricle of the brain, the efferent motor part flows thence down the nerves to the muscles which then contract (see Chapter 2). That he considered this the normal way in which sensation and movement are connected is made all the clearer by Leonardo's recognition of another, pathological path. If the soul in the third ventricle is cut off from the motor nerves, they may 'act on their own.' Leonardo describes this unhappy situation thus: 'How nerves sometimes operate by themselves without any command from other functioning parts or the soul. This is clearly apparent for you will see paralytics and those who are shivering and benumbed by cold move their trembling parts, such as their heads or hands, without permission of the soul; which soul with all its forces cannot prevent these parts from trembling. This same thing happens with epilepsy and with severed parts such as the tails of lizards' (W 19019 v; K/P 39 v). To these examples he adds his investigation of the 'cause of movement' in the frog. He writes, 'The frog retains life for some hours deprived of its head and heart and all its bowels. And if you puncture the said nerve [spinal medulla or medulla oblongata] it immediately twitches and dies. All the nerves of animals derive from here. When this is pricked the animal dies at once' (W 12613 v; K/P 1 r). Over the page he adds, 'Here, therefore it appears, lies the foundation of movement and life' (W 12613 r; K/P 1 v).

Thus Leonardo established to his own satisfaction that life and all its movement, emotional or otherwise, is rooted at the base of the brain whence all the nerves arise. He also discussed the effects of deprivation of the senses of sight and

hearing on the movements controlled by the mind or soul. He is, of course, emphatic about the terrible results of the loss of sight. 'The eye is the window of the human body through which the soul views and enjoys the beauties of the world. Because of it the soul in its human prison is content, and without it this human prison is its torment . . . The eye carries men from the east to the west, it has discovered navigation, it surpasses nature in dealing with things that are simple, natural and finite, for the works which our hands do at the command of our eyes are infinite, as the painter shows in his representation of the infinite forms of animals, plants and places' (TP I 34). Thus with the loss of sight we lose a whole world of human movement, one far larger, says Leonardo, than that lost by deafness.

The deaf-mute, on the other hand, presents the observer with unusual opportunities for learning about human movement. Repeating his admonition to learn what gestures of men are appropriate to their thoughts, Leonardo adds, 'This can be done by copying the motions of the

Figure 6.1.   The emotion of anger expressed in the faces of man, lion and horse (W 12326 r).

Studies of the anatomy of the expression of fury made in preparation for the depiction of *The Battle of Anghiari* (Figure 1.26, p. 34).

Figure 6.2. Anger expressed by the body and face of the horse (W 12326 r and 12327 r).

The rearing horse with curved neck and striking hoofs represents the whole bodily expression of fury.

The horse's face in frontal view makes very clear the symmetrical musculature which curves back the upper lip so baring the teeth in bestial anger.

dumb who speak with the movements of their hands and eyes and eyebrows and their whole person in the desire to express the idea that is in their souls . . . for these men are masters of movements and understand from afar what one says when one fits the motions of the hands to one's words' (TP I 250).

Leonardo divided his investigations of the movements of the body into three main categories: (1) those movements which express emotion; (2) movements or actions of the body contained in itself, e.g., breathing, pulling, lifting, etc. (*moto actionale*); and (3) movements from one place to another, i.e., local movements (*moto lochale*), such as walking, jumping, swimming, etc. (see TP I 355).

EXPRESSION OF THE EMOTIONS

This class of movements was of intense interest to Leonardo as an artist from the days of his youth. Since emotion may involve both work and 'local movement', this category overlaps the other two, yet it constitutes a separate problem of motivation. Leonardo's *St. Jerome* (Figure 1.1, p. 10) movingly illustrates his early concern with

the anatomical analysis of emotional expression in the face and body of man. The muscles of the face particularly poignantly express both physical and mental pain. The outline of the roaring lion in this picture appears to depict sympathetic distress, perhaps the result of Leonardo's close observation of the lions kept behind the Palazzo Vecchio in Florence. Elsewhere he depicts roaring lions associated with men shouting in anger rather than pain (W 12276 r and v). This comparative anatomy of expression reaches poignant fulfilment in Figure 6.1, a comparison of ferocity in the faces of horses, man and lion. There is no more than a slight suggestion of anatomy in the faces of the man and lion, whereas that of the horse, in marked contrast, displays a beautiful dissection of the facial anatomy of fury. Here the upper lip is curled right back by the action of levator superioris proprius, and the dilator naris lateralis is clearly seen drawing back the nostril and central upper lip; this compares with the lateral lip movement of the snarl produced by the caninus muscle in man. The buccinator and zygomaticus muscles drawing back the corner of the mouth are clearly delineated. In the view shown more from the front in Figure 6.2, the paired levator superioris proprius muscles which evert the upper lip can be seen arising from each side of the nose, converging to meet in the central line. These fierce horses' heads appear again, their anatomical reality diluted, in the far less expressive copies of *The Battle of Anghiari* by Rubens (Figure 1.26, p. 34). Here, however, one can glimpse the power of the rearing horse (also sketched in Figure 6.2). One almost feels the power of the head-on percussion as the massive gluteal muscles of their hindquarters force the horses into violent impact. Their bodily shapes are transformed by Leonardo's 'rules' into the expression of that human and equine 'bestial madness', which for him constitutes war.

One cannot help comparing this muscular violence with the muscular quietude of the smooth, smiling faces in the *Mona Lisa, St. Anne* and *St. John the Baptist,* so subtly depicted as to defy any accepted description of the emotion depicted. In these figures Leonardo illustrates contemplation. He describes how purely mental actions 'move a person in the first degree of facility and ease. Mental movement moves the body with simple and easy actions . . . because its object is in the mind when it concentrates upon itself'

Figure 6.3. The Five Heads (W 12495).

These five grotesque heads have given rise to many different interpretations. The central head, crowned with oak leaves, suggests a noble, aged figure; the proportions of the other faces express various grades of changing expression suggesting madness. In fact, like many of Leonardo's so-called caricatures, they may well represent how proportional anatomical change is reflected in mental change.

(TP I 409). In such cases, he advises, 'Do not paint muscles with harsh outlines but let soft lights fade imperceptibly into pleasant delightful shadows' (TP I 413).

The muscles of the face and lips received a great deal of Leonardo's attention between 1508 and 1512 with regard to their detailed anatomical description. He makes a special note to delineate the muscles of anger and despair and the part facial muscles play in 'whistling, laughing and weeping', adding his intention 'to describe and illustrate in full these movements by means of my mathematical principles' (W 19046 r; K/P 51 r). To the last Leonardo remained convinced that no movement, even the subtlest of human expressions, was beyond the reach of his geometrical transformations. He keeps his anatomical promise in the later detailed dissections, drawn in Figure 10.2 (p. 232), where once more he compares the muscles of the face of the horse with those of a man, particularly those which 'raise the nostrils'. We shall mention these dissections later, for they contain Leonardo's final achievements in the detailed anatomy of expression by speech and facial movements. Before he reached this stage Leonardo carried out extensive studies of the movement of the body-machine as a whole.

Leonardo is, of course, keenly aware of the infinite variety of shapes assumed by the body when moved by different emotions. 'Some emotions', he writes, 'are without bodily gestures, and some with bodily gestures. Emotions without bodily action allow the arms and hands to fall, and every other part which shows action. But emotions that have corresponding bodily actions cause the body and its limbs to be moved appropriately with the motion of the mind . . . there are, too, those movements of the senseless, or rather those deprived of their senses, and this is to be put into the chapter on madness, buffoons and their morrisdances' (TP I 408). By 'senseless' Leonardo is referring here to the blind and deaf, mentioned previously. No chapter on madness has been found in his notes. However, in relation to the effects of wine, Leonardo says, 'As soon as wine has entered the stomach it begins to ferment and swell; then the soul of the man begins to abandon his body; then it turns towards the sky, and the brain is the cause of its parting from the body. Then he begins to be degraded and is made to rave like a madman; and then he does irreparable evil, killing his friends' (CA 67 rb). In this instance Leonardo presents 'madness' as the result of the departure of the coordinating soul from the body. The subsequent incoherence and discord between the senses and movements leads to what Leonardo has already stigmatised as 'bestial madness', the killing of other men.

The five heads in Figure 6.3 seem designed to illustrate a series of emotional expressions ranging from dignified sanity, through libidinous fatuity to howling depravity. They have been variously diagnosed. The drawing was made about the same time as the effects of wine were described in the above-quoted passage. Might not the figures depict degrees of degradation or madness produced by the soul's abandonment of the body? This interpretation is supported by Leonardo's remark on the verso of the page, 'This spirit returns to the brain whence it had departed'.

Figure 6.4. Men at work (B 51 v).

Below this rough sketch Leonardo writes, 'A method of doing work quickly'. His interest in time and motion studies of working methods was life-long. Such studies were performed in planning the digging of trenchworks and diverting the river Arno.

THE MOVEMENTS OF MAN, THE MACHINE

Just as Leonardo the artist focused his attention on the emotional movements of man, so Leonardo the engineer focused on man the machine. When the work done by this machine did not involve change of place he described it as 'action' (*moto actionale*). Such actions included not only external bodily movements but also physiological visceral movements such as breathing, urinating, etc. (W 19018 v; K/P 41 v).

During his first Florentine period until about 1482, there are many drawings of men at work (see Figure 6.4). Illustrations of the mechanical principles of man the machine appear after his move to Milan. Those inventions which figure so prominently in his letter to Ludovico Sforza incorporated the bodies of men into many of his war machines, and Leonardo must have felt impelled to take into consideration the mechanics of this essential human component of his machines. For example, men's actions in shooting arrows, charging with multiple spears and providing 'tank' power become an obviously essential part of his invention.

Leonardo's interest in the mechanics of the movements of man in water developed into the invention of flippers, life-belts and studies of diving apparatus (see Figures 1.20 and 6.28, pp. 29, 184). Here they reach surprisingly mature concepts, providing means of staying under water for hours with air tanks, graduated weight attachments and goggles. To these refinements he adds an apparatus for boring holes into the hulls of ships. All this appears to have come up to the surface of possible practical application on Leonardo's visit to Venice in 1500. Models of the astonishing diving apparatus are to be seen in the Science Museum in Milan.

One of the most dramatic studies of the bodies of men being used as instruments of industrial power is found in

Figure 6.5.   Men working in a cannon foundry (W 12647).

The mechanical forces exerted by men in the process of manufacturing a large cannon are here illustrated in all their variety. For men doing agricultural work, see Figure 13.5.

Leonardo's drawing of the cannon foundry, Figure 6.5. Here he illustrates not only the stages of the manufacturing process with its array of apparatus but also the working forces of men pulling, pushing, lifting and pressing down. All are depicted in vigorously naked objectivity, revealing perhaps for the first time the part men have had to play in industrialisation.

As if not content with studies of the movements of man on the ground and in water, Leonardo also launched into attempts to enable man to move in the air. His early drawings of such a possibility are, like his diving apparatus, surprisingly developed. For example, Figure 4.33c (p. 118) opens with the remark, 'As much force is exerted by an object against the air as by the air against the object'. This, one of Leonardo's most important generalisations, governs the contents of the rest of the page on which he not only describes a man standing on the pan of a balance forcing down a wing against the air to see if he can exert enough power to raise his own weight but also depicts the parachute in dimensions which work. Provisionally dated 1485, the whole nature of drawing and text suggests a scientific approach to his problem. This seems to differ from his spec-

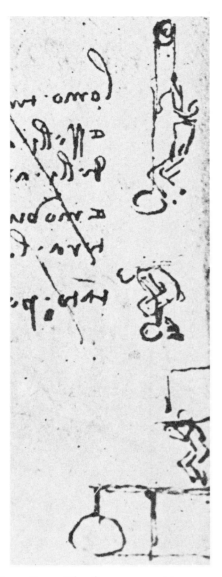

Figure 6.7. The force exerted by a man.

Besides the two upper sketches Leonardo writes, 'A man pulling a weight balanced against himself cannot pull more than his own weight'.

The lowermost drawing he explains thus: 'The greatest force a man can apply with equal quickness and movement will be when he sets his feet on one end of a balance and presses his shoulders against some stable object. This will raise at the other end of the balance a weight equal to his own weight which is as much weight as the force of his shoulders carries' (A 30 v).

Thus, if a man weighs 200 pounds, he has the capacity of carrying another 200 pounds, 400 pounds in all.

Figure 6.6. The force exerted by a man.

Under this sketch of a proposed flying machine Leonardo writes, 'You will exert a force equal to 400 pounds as quickly as the movement of the heels' (B 79 v).

ulation in *Codex Atlanticus* 377 vb, where he describes the flight of the moth with two front wings moving alternately with its two hind wings. He then draws a similarly winged man supplied with rollers and cranks to perform similar movements in Figure 6.6. Such a winged man is more clearly seen in Figure 6.23c (p. 180). These latter drawings suggest the empirical inventor whilst *Codex Atlanticus* 381 va

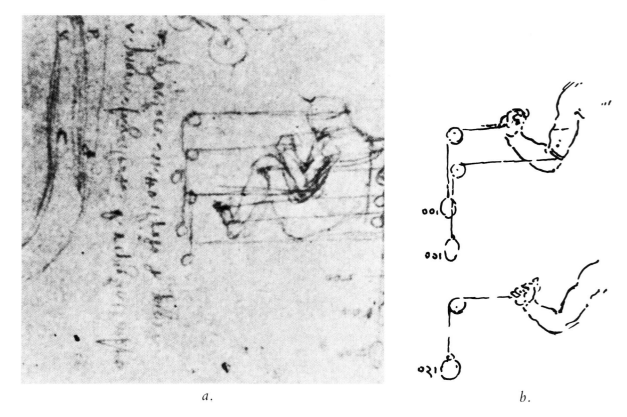

a.   *b.*

Figure 6.8.

a. Measurement of the forces of muscular movements (H 43 v).

Cords attached to the head, trunk, arms and legs are run over pulleys. Attached weights give a measure of the power exerted.

b. Measurement of the power of the muscles of the arm (H 44 r).

The cords just mentioned are here more clearly seen. The attached weights measured in pounds represent the power of the movement being tested, e.g., flexion of the elbow.

These experiments represent the first known systematic use of the dynamometer (or ergometer) in human physiology.

(Figure 4.33c) suggests the emergence of the scientist. They confirm the view that sometime between 1483 and 1490 – after he went to live in Milan – Leonardo gradually matured from the empirical inventor into a scientist with a set of provisional hypotheses upon which to base his further investigations. The fact that in the *Codex Trivulziano* and MS B one finds scattered among many empirical observations some quantitative measurements of forces, including those of man, brings additional evidence of a gradually changing outlook on Leonardo's part.

For instance, in Figure 6.6 Leonardo states that 'a force of 400 lbs can be made with feet and hands'. In Figure 6.7 he shows how this measurement was obtained. First, demonstrating that a man 'cannot pull more than his own weight' over a pulley, he proceeds, 'The greatest force a man can apply with equal quickness and movement will be when he sets his feet on one end of a balance and presses his shoulders against some stable object. This will raise at the other end of

the balance a weight equal to his own weight which is as much weight as the force in his shoulders carries', which is at least that of another 200 pounds (it should be noted that Leonardo's Florentine pound equals four-fifths of an English pound). On MS B 88 r Leonardo shows how a man weighing 200 pounds can 'very rapidly exert a force of 1200 lbs for 1/2 *braccio*' with the aid of favourable leverage. This is the brief peak of 'work' available to the body of man. In Figure 4.33b (p. 118) he depicts the experiment sketched also in *Codex Atlanticus* 381 va, this time with given measurements, the wing being twenty *braccia* long and twenty *braccia* wide, and the beam on which the weight of the man (200 pounds) exerts leverage. Here, too, Leonardo points out that action equals reaction, and if the wing descends with sufficient velocity, the man will be raised up. This observation gives him grounds for believing that energy equivalent to a weight of 200 pounds is available to man for the possibility of flight in the air.

In his endeavour to quantitate the ratio between wing-span and load Leonardo measures the body of a bat (two ounces) and its wingspan (1/2 *braccio*), comparing these measurements with those of an eagle (B 89 v). Later he repeats these measurements on a pelican, finding its wing-span 5 *braccia* and its weight 25 pounds, noting that 'its measurement thus expanded is the square root of the weight' (CA 302 rb). Therefore, he concludes that a man exerting a total force of 400 pounds should have a wingspan of 20 *braccia*, i.e., about 40 feet. Such were the early calculations with which Leonardo set out on his unsuccessful attempt to achieve the flight of man. The details of this fascinating project cannot be pursued further here. They have been presented for two reasons; not only did they initiate Leonardo's measurements of the forces of man, but they also gave him a measure of the maximal summation of the 'powers' of the human body. Regarding the human body as a possible flying machine, Leonardo had ascertained that it could raise itself from the ground by a force equal to its own weight, i.e., 200 pounds, using both arms, legs and neck.

The next logical step was to ascertain how this total 'work' was distributed between the limbs of the body. As the little sketches in Figure 6.7 make clear, a man can lift a weight off the gound equal to his own body weight, and no more, whether directly or over a pulley.

Leonardo then turned to an assessment of the muscular powers of different parts of the body separately. In Figure 6.8a and 6.8b he illustrates his method of measuring the force exerted by the neck, arms and legs. It will be seen that a cord attached to the head runs forward over a pulley carrying a weight. This measures the power of extension of the neck. Other cords are similarly attached to the wrists, elbow, shoulder, hip, knee and foot. The cords running forward over pulleys shown in front of the figure measure the power of flexion of these joints; those running backwards over pulleys behind the figure measure their powers of extension. These constitute the earliest investigations in ergometry. Later he applied these experiments to the problem of flight as shown in *Codex Atlanticus* 369 ra.

The ergometer experiments on wrist and elbow are separately illustrated in Figure 6.8b. Here the arm, like the rest of the body, is being treated as a compound lever system. Lever systems depend very closely on the relative proportion of the lengths of their parts. No one was more conscientious and detailed in his investigations of the proportions of the parts of the human body than Leonardo. His careful measurements have therefore mechanical as well as artistic significance. Here, as usual, Leonardo fuses his art with his science.

Leonardo's interest in the powers of the human body was primarily directed towards its efficient use as a machine. But the efficiency of the human machine depends also on its mode of application. For example, on Forster II 73 v (Figure 6.9a) Leonardo points out how the angle at which a man pulls on a weight over a simple pulley affects the power of the pull. 'The man standing at *a* will pull as much weight by himself as will two men at *b*'. Again, on H 80 v (Figure 6.9b) Leonardo shows how the power on the rope over a pulley can be increased by dividing its end into four strands, each pulled by one man. This is applied to pile-driving on B 70 r (Figure 6.9c).

This simple illustration shows how Leonardo integrates the internal powers of the body-machine (the microcosm) with the external mechanical principles of the macrocasm. Its further development into the field of power in pulley systems is touched on in Chapter 4.

How extensive were his studies of the work performed by the body of man? This question is answered in part by Leonardo himself. He makes a list of 'the 18 operations of man'. This list consists of 'Rest, movement, running, standing, supporting, sitting, leaning, kneeling, lying down, suspended, carrying or being carried, pushing, pulling, beating, being beaten, pressing down or lifting up' (BN 2038 29 r). Every one of these 'operations' and many more added later, such as breathing and speaking, Leonardo analysed into their mechanical components according to the principles of the lever or balance and his tests of muscular power. Once more one is confronted with the impossibility of doing justice to the extent of Leonardo's investigations in a particular field. A number of pages summarise in visual form these '18 operations'. Some of them are to be found in the third part of his *Treatise on Painting*. A few examples from Leonardo's list and the *Treatise on Painting* must suffice to represent the remainder.

A balanced, standing human body will automatically place equal weights on opposite sides of its centre of gravity. If this figure moves an arm, the centre of gravity will change to compensate, e.g., if an arm is thrust forward, the centre of gravity shifts backwards. When a man stands on his feet, the centre of gravity is always between his feet. 'The balance or equilibrium of man is divided into two parts', writes Leonardo, 'that is simple and compound. Simple balance is that made by a man on his two motionless feet, above which that man opens his arms at different distances from the centre of gravity, or bends while standing on one or both feet. But the centre of gravity is always in a perpendicular line through the centre of that foot on which the weight rests; and if it rests equally on both feet then the weight of the man will have its centre of gravity on a perpendicular line in the middle of a line which divides the space between the centre of the feet (Figures 6.10a and 4.9a).

'Compound balance is understood to be that of a man who sustains a weight above him by various motions as in the figure representing Hercules who crushes Antaeus suspended above the earth between Hercules' chest and arms. Place his figure as far behind the line through the centre of his feet as Anataeus' centre of gravity lies in front of those feet' (see Figure 6.10b).

On folio 128 r of the *Treatise on Painting* (*Codex Urbinas*) Leonardo describes 'The motion of figures pushing and pulling. Pushing and pulling are one and the same action, for pushing is only an extension of the limb and pulling is a

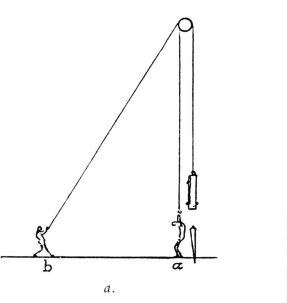

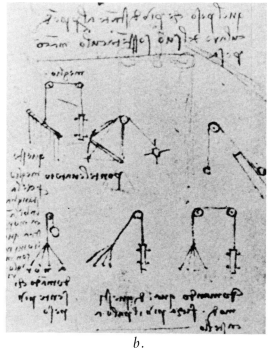

Figure 6.9.

a. The force exerted in raising weights.

A man at *a* straight under the pulley can raise his own weight. But the further the man *b* is away from this line, the less weight he can raise. Leonardo puts it thus: 'By the 5th on Theory *a* pulls as much weight as his own as would two men at *b* . . . and the more obtuse the angle at *b* the less weight will he raise' (Forster II 73 v).

This drawing develops the idea expressed on A 30 v (Figure 6.7, p. 165).

b. Improved methods of raising weights. (H 80 v).

Heavy weights, as in pile-driving, cannot be raised by one man. A method labelled 'better' by Leonardo is shown top left, where a bar with ten strands of rope attached is made for ten hands. The provision of two pulleys keeps the force perpendicular. Further variations are drawn below, beneath which Leonardo notes that these methods are useful for pile-driving.

c. Pile-drivers (B 70 r).

A comparison is made here between the power exerted by two men on the pile-driver and that exerted by a man using a pulley or a lever to raise it to a height in order to increase the impact of its percussion.

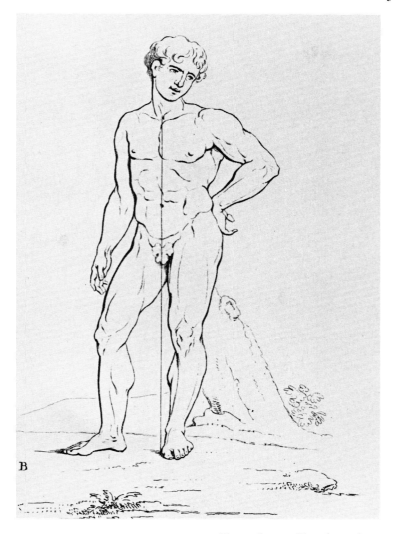

Figure 6.10. Simple and compound balance or equilibrium.

The lines of the centre of gravity are drawn through a man standing motionless on his feet, showing simple balance. A line drawn through the combined centres of gravity of Hercules crushing Antaeus shows compound balance.

These drawings illustrate the text in *Codex Urbinas* 128 v. They are from Leonardo da Vinci, *A Treatise on Painting,* trans. by John Francis Rigaud (London, J. B. Nichols & Son, 1835), plates 11, 5.

contraction [flexion] of this limb, and to both powers is added the weight of the mover against the thing pushed or pulled . . . Pushing and pulling can be done along different lines in relation to the centre of the power of the mover, a centre which with respect to the arms, is in the place where the sinew of the humerus of the shoulder, that of the breast and that of the shoulderblade [scapula] opposite the breast, are joined to the bone of the upper shoulder'. Leonardo's original note which Melzi copied into the *Treatise on Painting* is to be found on Madrid II 78 v, where it is accompanied by two rough figures. That reproduced in the *Treatise* is on *Codex Urbinas* 120 v (see Figure 6.11).

Obviously, lifting, pulling, pushing with the whole body constitute 'work'. Leonardo was keenly aware of the time element in the performance of work. One notices this

in the calculation that a man can exert maximal pressure of some 1,200 pounds over 1/2 a *braccio* 'very rapidly', i.e., for a very short time. It is one of Leonardo's remarkable insights into the problems of manual workers that he should study and compare various labour-saving methods of performing the same job. He gave particular attention to excavation, since digging trenches for canals, moats, etc., was one of his ever-recurring problems. In short, just as he was concerned to increase the working potential of the body of man, so he sought to obtain the utmost mechanical efficiency from the powers of that body.

Not only did he devise means of more efficient excavation as, for example, an improved digger using the power of percussion generated by acceleration due to gravity, mentioned in Chapter 4 (Figure 6.9c, p. 168), but he also studied the arrangement whereby men are best able to excavate a given area of trench. He measures the time taken for a man to take a shovelful of earth from *a* to *b* and calculates by 'using the rule of 3' how much earth is carried in one hour (1,080 *tempi*) (BM 207 r).

In *Codex Atlanticus* 189 vb he makes the remarkable statement that 'The number of men digging out a trench must be pyramidal, putting the base of the pyramid towards the place where the earth is deposited; otherwise the bottom of the trench will not descend at the same level'. A self-explanatory pyramidal figure is accompanied by a detailed description of each man's performance in digging and transporting the earth. Leonardo appears to like this 'pyramidal' formation, for he repeats the statement on the cover of MS L and in MS L 77 r. He discusses the problem of excavating a trench at length on Madrid II 9 v, 10 v and 10 r in connection with his work at Piombino. On folio 10 v he again illustrates his pyramidal arrangement of the excavators.

But the manual excavations of trenches, moats or canals was clearly only to be used when dealing with unfavourable circumstances such as hard, rocky soil. In more favourable circumstances Leonardo used revolving derricks for transporting the soil, as shown on CA 363 vb (Figure 6.12a).

Here he replaces man by the heavier ox. This mounts the steps onto a platform which as it descends raises the load of excavated soil; the crane then swings round (as shown on the left) to deliver the soil onto the bank. In the little sketch below and to the right he replaces the straight steps by a spiral stairway which, he points out, for every thousand journeys saves one and a third miles. So pleased is Leonardo with this saving of man-power that he repeats the drawing on CA 164 ra (Figure 6.12b), allowing himself praise for his scheme in such phrases as, 'Here it would be impossible to load more quickly'. 'Here the traction could not be done with greater speed'. And conscientious to the last about expense, he adds, 'There will be expense for the woodwork but all of it can be used again in building'.

This page is made all the more relevant to the subject of man the machine by virtue of the rough sketches of a man pulling a load up an incline with the axes of the spine and legs marked out in relation to his centre of gravity. Leonardo's sketch of flexion of the knee-joint is so crude that it must have suggested to him his need for a better knowledge of the anatomy of joints.

In the bottom left-hand corner of Figure 6.12b is a faint sketch of a water-wheel, below which Leonardo empha-

Figure 6.11. 'The compound force of a man; and first that of his arms'.

Leonardo writes, 'Whether a man is more powerful in pulling or in pushing is proved by the ninth proposition on weights, where it is stated; Of weights of equal power that one appears more powerful which is further away from the fulcrum of the balance'. He then shows that the muscle *HC* has more power because it is attached to the arm at *C*, 'a position more remote from the fulcrum *A* at the elbow than is *B*'. He describes this as a simple not a compound force. Compare with Figure 12.1.

'A compound force', he adds, 'is that which, when the arms are at work has a second power added, that of the weight of the person, and of the legs, as in pulling and pushing . . . as would be the case of two men at a column, one of whom pushes the column while the other pulls it' (Facsimile of CU 120 v).

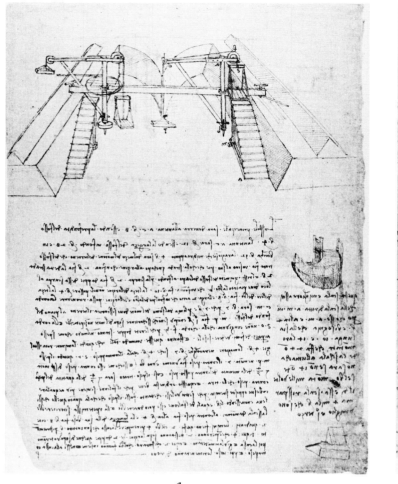

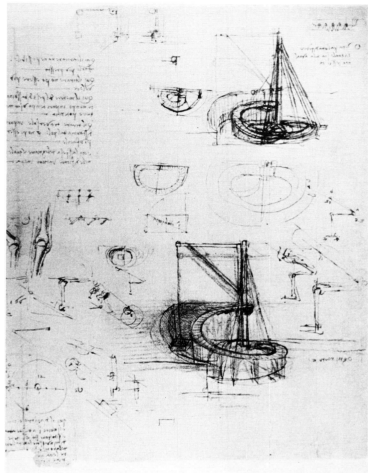

a. *b.*

Figure 6.12.

a. Excavation of canals (CA 363 vb).

On this page Leonardo proposes using the weight of oxen for raising the soil which is deposited by revolving derricks onto the banks. The spiral ramp drawn on the right margin saves time and is developed in the accompanying drawing.

b. Excavation of canals; the actions of men pulling loads up an incline (CA 164 ra).

Here the platform bearing the animal is designed to descend exactly to the place where it begins to ascend the spiral ramp. Rough sketches of the leverage of a man's actions in pulling a load up an incline concentrate on the knee-joint (both margins). An external treadmill is sketched in the lower left corner.

sises a theme repeatedly stressed elsewhere, that a man standing on the outside of a wheel will exert much more efficient leverage than one standing inside, which was the traditional position. In this case he marks out the advantage in leverage length. His principle is dramatically illustrated on CA 387 r (Figure 6.13), where the wheel is being rotated by a number of men outside for the purpose of rapid loading and firing of four crossbows by an archer seated at the centre.

As a result of his many quantitative studies of work Leonardo reached the conclusion that by whatever instruments work was performed 'equal force applied at equal velocity will perform the same amount of work' (Madrid I 35 r; see Figure 4.37, p. 122). He expresses the same rule as 'he who lightens the work, prolongs the time' (W 19139 v; K/P 31 v).

'LOCAL' MOVEMENT OF MAN AND ANIMALS

As we saw in Chapter 4, Leonardo considered movement as the basic form of expression of force or energy. He asserts that 'All action must be exerted through movement' (Triv 36 v) and that 'Movement is the cause of all life' (H 141 r). It is interesting to see him follow this with the asser-

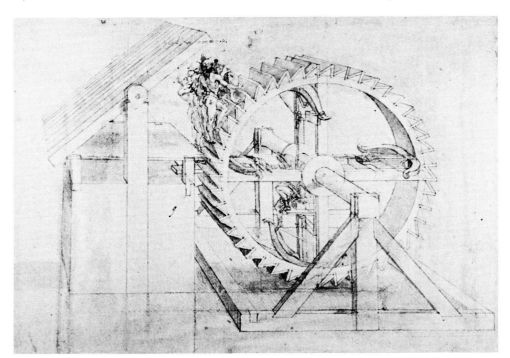

Figure 6.13.   The external treadmill (CA 387 r).

From his earliest studies of men at work Leonardo favoured the advantage in leverage obtained from an external as opposed to an internal treadmill. Here he makes a sketch (right) and a more finished drawing of men working on an external treadmill in order for a man seated at the axle to fire four crossbows in rapid succession.

tion 'Nothing without life will have durability of accidental movement' (W 12349 r), and 'although with impetus every body retains in itself for some distance the nature of its movement' (Triv 43 r). 'Movement is not durable in dead things' (W 12349 r). Though these are early assertions, Leonardo modifies but never deviates from them from about 1492 to the end of his life.

About the same time he contradicts the possibility of perpetual motion thus: 'Against perpetual motion. No insensible [*insensibile*] object will move on its own; consequently when it is moved it is moved by unequal power, unequal that is in time or velocity or of unequal weight; and when the desire of the prime mover ceases, the second will cease at once' (A 22 v). In his attempt to classify movements about this time his first three categories are the movement of time, the movement of life and mental movement 'which exists in bodies with souls [*corpi animati*]' (CA 203 va).

Thus all 'local' movement, whether in animate or inanimate bodies, arises from unequal power, velocity or weight distribution. In the case of man and animals the movement of the special spiritual powers flowing along the nerves from the soul may affect the mechanical situation; it may in fact convert 'spiritual movement into material movement' (BM 151 v). Such a conversion takes place, for example, when the emotion of fear is manifested in flight or fight, and similarly with any simple voluntary movement such as walking or running.

The body of a man or animal is stationary as long as its centre of gravity falls in the space between its feet. We have seen that many vigorous movements of limbs and trunk can be made with compensatory adjustments of the centre of gravity which do not involve movement from one place to another. Even when Hercules crushes Antaeus, the compensatory shift of his centre of gravity is achieved by a lordotic, backwardly bent spine, without the necessity of his stepping backwards in 'local' movement. Thus even in a stationary man performing work with his limbs, the centre of gravity is a mathematical point shifting from place to place, just as the centre of gravity of the earth shifts with the movements of the four elements on its surface.

When this shift of the centre of gravity goes beyond a line falling between the feet, 'local' movement occurs and the man moves his feet and body forwards, backwards or sideways. The situation in four-footed animals is obviously modified, though in principle it is the same. Leonardo's particular interest in the horse leads him to study this. In birds the situation is grossly different in flight. This particular problem alone occupied some 160 pages of Leonardo's notes. To fish and swimming, however, he gives much less attention. These movements of animals are all observed by Leonardo with a view to illuminating the actual or possible movements of man in the elements of air or water as well as on earth; they are therefore relevant to his science of man.

Moreover, Leonardo sensed the homologous structure of many animals, saying, 'It is an easy matter for anyone who knows how to represent man afterwards to acquire universality, for all animals which live upon earth resemble each other in their limbs, that is in muscles, sinews and bones, and these do not vary except in length and thickness, as will be shown in the Anatomy. There are also aquatic animals of many different kinds; but . . . these are of almost infinite variety; and the same is true of the insect world' (G 5 v). Even Leonardo, it appears, was daunted by the prospect of analysing the movements of 'aquatic animals' such as the octopus, and insects, though he by no means entirely neglected the movements of fish, moths and flies (see, e.g., Figure 6.14).

On the other hand, he perceived that 'Walking in man is always after the manner of the universal gait of four-footed animals; because just as they move their feet cross-wise as a horse does when it trots, so a man moves his four limbs cross-wise; that is, if he thrusts the right foot forward in walking he thrusts the left arm forward with it, and so it always continues' (CA 297 rb). And he adds, 'Make a separate treatise on the description of the movements of animals with four feet, among which is man who also in infancy goes on four feet' (E 16 r). Thus, 'Movement is created by the destruction of balance, that is, of equality of weight, for nothing moves of itself which does not go out of balance, and that moves more rapidly which is further from its balance' (TPI 346; see Figure 6.15). And 'When a man is moving he will have his centre of gravity above the centre of the leg which is placed on the ground' (TP I 384; CU 112 v). Moreover, 'When a man or other animal moves either rapidly or slowly that side which is above the leg that sustains the body will always be lower than the opposite side' (TP I 347; CU 111 v).

Leonardo now analyses what takes place as the velocity of movement increases. 'That animal will have the centre of its supporting legs so much the nearer to the perpendicular of the centre of gravity, which moves more slowly. Con-

versely that will have the centre of its supports further from the perpendicular of the centre of gravity which moves faster' (TP I 349; CU 112 r; see Figure 6.16). 'Of more or less rapid local motion', he writes, 'The local motion made by man or any other animal will be of so much the greater or less velocity in proportion as the centre of gravity is further away from or nearer to the centre of the sustaining foot' (TP I 350; CU 129 v). 'That body which moves by itself will be so much the faster as its centre of gravity is more distant from the centre of support' (TP I 353; CU 138 v). And, 'That figure will appear swiftest in its course which is about to fall forward' (TP I 352; CU 138 v).

In relation to 'a figure that moves against the wind', he writes, 'The figure which moves against the wind along any line of direction never preserves its centre of gravity in proper relation above the centre of its support' (TP I 354; CU 137 r).

Of interest in this series of observations is Leonardo's apparent failure to distinguish between the posture in acceleration, so evident at the beginning of a sprint, and the relatively upright posture of the body after a constant velocity has been reached. When such a posture is adopted Leonardo appears to attribute it entirely to wind resistance (see Figure 6.17b).

Leonardo was aware of the undulatory movement of the spine and body brought about as the centre of gravity shifted from a position above one foot to the other. This is reflected in the remark, quoted earlier, that 'that side which is above the leg that sustains the body will always be lower than the opposite side' (TP I 347; CU 111 v). To this he adds, 'There will be the greatest difference in the height between the shoulders or sides of man or other animal when the whole is in slower motion. The converse follows that there will be least difference in height in those parts of an animal whose whole is in more rapid motion. This is

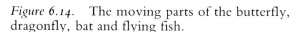

Figure 6.14. The moving parts of the butterfly, dragonfly, bat and flying fish.

To the left of the flying fish (gurnard) Leonardo writes, 'Animal as it flies from one Element into another' (B 100 v).

All these animals are drawn with their wings outspread in flight.

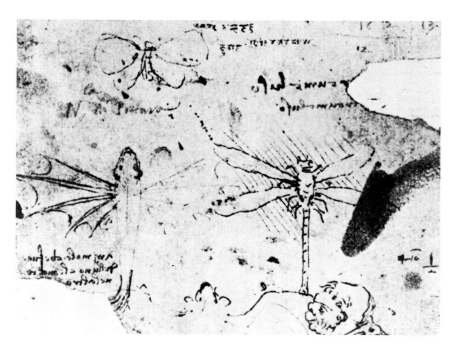

proved by the ninth proposition on local motion where it is stated, every weight weighs through the line of its motion. Therefore when the whole moves towards some place the part united to it follows the shortest line of motion of the whole, without itself giving its own weight to the lateral parts of that whole' (TP I 351; CU 11 v).

Here Leonardo is referring to the creation of 'accidental weight' by momentum or impetus, as described in Chapter 4. As the 'accidental weight' increases, so 'natural weight' decreases. Therefore the faster the animal runs, the less it weighs on the ground (Forster II 50 v). Leonardo puts it this way, 'The horse in running feels less of the weight of the man he carries. Whence many wonder that in running the horse can support itself on one foot only. Therefore one can say of weight in transverse motion, the faster its velocity the less it weighs perpendicularly towards the centre [of the world]'. And the less weight is put on its feet, the less is the rise and fall of its shoulders. This applies to man as well as to the horse.

The alternate rise and fall of the shoulders reflects an undulatory movement of the spine during walking or running. Such a wave movement along the spine, Leonardo appreciates, depends upon the spine being composed of many small bones strung together – the vertebrae. He refers to 'The serpiginous movements and balance in human and animal figures' and adjures the artist to make a 'graceful counterbalancing and balancing in such a way as the figure shall not appear like a piece of wood' (BN 2038 22 v). In his early years his most vivid illustrations of these serpiginous movements seem to be reserved for dragons, such as those fighting a lion in a drawing in the Uffizi Gallery. Later however, his vivid understanding is depicted in the series of horses, cats, dragons and a man in Figure 6.18 under the title 'The serpentine winding of the attitudes and principal action . . . is double in the movements of animals, the first of which is in their length [longitudinally], the second in their breadth [laterally]'. A companion drawing shows a series of cats with 'serpiginous' spines freely curved longitudinally and laterally. Amongst them is another very curly spined dragon providing the symbolic signature of Leonardo's intention. Beneath the whole he writes, 'Of bending [the spine] forwards and backwards. The species of these animals of which the lion is the chief for having the joints of their backbones suitable for bending'. Both drawings are dated about 1513, when Leonardo had completed his dissection and anatomical demonstration of the curvatures of the spine in man and animals (see Chapter 13).

Leonardo's sensitivity to any production of wave formation led him to detect both lateral (serpentine) and longitudinal (up and down) wave forms in the body movements of vertebrate animals. These are, of course, more evident in quadrupeds than in man and therefore are more easily

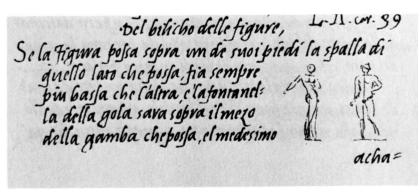

a.

Figure 6.15.

a. Sketch from the *Treatise on Painting* of a man standing on one leg, as one does to take a step forward (CU 113 v).

The line of the centre of gravity is drawn through the pit of the throat, the umbilicus, the genitalia, and down the leg to the foot. In this sketch (not made by Leonardo) the shoulder above the weight-bearing foot is not significantly lower than the other.

b. A similar figure as drawn in the first English edition of the *Treatise on Painting*.

Here it will be noticed that whilst the difference in the level of the shoulders is represented, the line of the centre of gravity deviates to one side of the genitalia and to the inside of the weight-bearing foot.

Figures from Leonardo da Vinci, *The Treatise on Painting,* trans. by J. Senex (London, 1721), figure 13, p. 114.

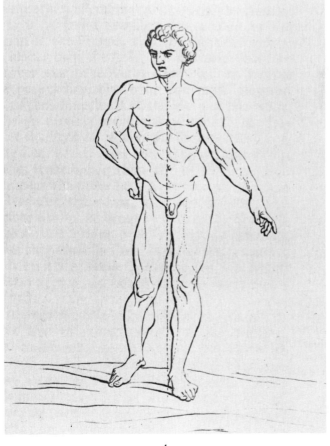

b.

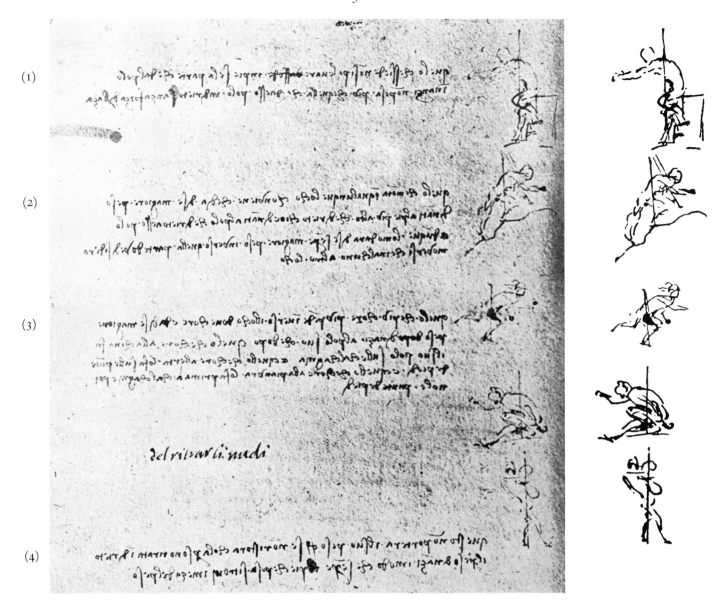

Figure 6.16. Variations in the distribution of the weight of the body during the movements of man.

Note how the weight of the body always passes in front of the line of the supporting foot when the body moves forwards.

Translation of the text to the illustrations in the margin:

(1) 'He who sits cannot raise himself to his feet if that part of his body which is in front of his axis [fulcrum] does not weigh more than that which is behind that axis, without using the force of his arms'.

(2) 'He who ascends any slope finds it necessary to give most of his weight forwards in front of the higher foot rather than behind – that is, in front of his axis of balance rather than behind this axis. Therefore a man will always present more of his weight towards that point to which he desires to move than to any other place'.

(3) 'The faster a man runs the more he leans towards the place to which he runs, and gives more of his weight in front of his axis of balance than behind. He who runs down a slope has his axis on his heels; and he who runs uphill has it on the toes of his feet; and a man running on level ground has it first on his heels and then on the toes of his feet'.

(4) 'Fifth drawing in the margin: 'This man will not carry his weight unless by drawing his body back he makes up for the weight in front in such a way that the foot is always placed so that it is in the centre of the total weight [i.e., at the centre of gravity of the combined weight]' (A 28 v).

a. A man moving against the wind (CU 137 r).

Movement against the wind disturbs the normal relation between the centre of support (the axis or fulcrum) and the resting centre of gravity of the body. Note the position of the arms.

a.

b.

b. A series of positions of the body of a runner at one-third second intervals at the start of a race.

At the end of two seconds, the body, having completed acceleration, has become almost vertical. Leonardo appears not to have illustrated this as a general rule. Figure from R. A. R. Tricker and B. J. K. Tricker, *The Science of Movement* (London, Mills and Boon, 1966), Figure 18.5.

Figure 6.17.

demonstrated pictorially in horses and cats than in man, but he saw them in man also.

This oscillation of the body in man is related in particular to the position of the pit of the throat. 'The pit of the throat lies above the foot, and if one arm is thrown forward the hollow moves out of the line of the foot; and if the leg is thrown backwards the pit of the throat moves forwards; and so it changes with each attitude' (BN 2038 20 v).

The movements of a man walking uphill are studied, particularly in relation to the shift of centre of gravity with the positions of the limbs. For example, in Figure 6.19 the contrast is drawn between the long steps of the man climbing the hill with his weight thrown forward and the short steps of the man descending the hill with his body weight thrown backwards. These sketches may be compared with that in Figure 6.12b (p. 171). They are again clearly shown in Figure 6.16 (p. 175), in a series devoted to the different postures and movements of man, in which it will be noted that the running man has his arms drawn across his chest, not stretched forward as in *Codex Urbinas* 137 r.

Mounting steps or stairs is the next logical pattern of movement to be studied. Examples appear on W 19038 v, Forster II 45 v, L 27 v and H 75 r. In Figure 6.20 Leonardo writes at length in a typical passage around his accompanying sketches. 'A man who goes up steps', he writes, 'puts as much of his weight in front and to the side of the upper foot as makes a counterweight to the lower leg, whence the work of the lower leg only extends to moving itself. The first thing a man does in mounting steps is to unload from the leg he wants to lift the weight of the body which was placed on this leg; and in addition to this he loads on the opposite leg all the remaining mass of the man together with that of the other leg. Then he raises this leg and puts the foot on the step on which he wishes to raise himself. Having done this he gives back to the upper foot all the weight of the body and of the other leg, supports his hand on his thigh, thrusts his head forwards, and makes a movement toward the point [toes] of the upper foot, quickly lifting the heel of the lower foot; and with this impetus he lifts himself up, and at the same time extends the arm which

Figure 6.18. Flexion, extension and lateral flexion of the spinal vertebrae (W 12331).

At the top is a cat showing longitudinal flexion of all the spinal curves including the tail. Below this a horse with its cervical vertebrae so bent and twisted as to have broken its neck. Five sketches of St. George killing the dragon give Leonardo ample scope for illustrating all varieties of active curves of the spines of man, horse and dragon. The remaining drawings illustrate these curves in the horse. Compare with Figure 13.16.

was supported by the knee; and this extension of the arm pushes the body and the head upwards, and thus straightens the curve of the back. In proportion as the step by which a man raises himself is of greater height by so much the more will his head be placed in front of his upper foot. Since *a* weighs more than *b,* this man will not get on to the step *m,* as the line *gf* shows'. The line *gf,* it will be noticed, passes through the foot of the man in such a way that the greater weight of his body is clearly behind it, not in front. Therefore he will not succeed in mounting the step.

This lengthy detailed description is typical of Leonardo's meticulous descriptions of so many human actions. It is given here in full, both because of the beautiful analysis of the movement which can be tested on oneself, and also in order to re-emphasise the necessity for the much abbreviated accounts of Leonardo's observations throughout this book. Variations, some very interesting, of this description are given on H 75 r (Figure 6.21) and MS L 27 v in which the muscles employed are further described.

On Forster II 45 v (Figure 6.22) Leonardo considers a mechanical aspect of man's movements ascending a step-ladder. Here a man is shown at the top of a ladder. Alongside this drawing Leonardo writes, 'I ask this; the weight of a man, at every step of movement on this step-ladder, what weight does he give to *b* and *c?* Observe his perpendicular line below the centre of gravity of the man'.

Typically, one finds Leonardo dealing with his own question in another notebook, MS I 14 v. He heads this passage, 'On movement and percussion'. 'If someone', he

Figure 6.19. The steps of a man ascending and descending a hill (W 12639 r).

Alongside these two sketches Leonardo writes, 'He who descends takes small steps because his weight rests upon the back foot. And he who ascends takes big steps because his weight rests upon his front foot'.

The problem of leverage is compounded in these sketches by making each man carry a load on a stick, the fulcrum of which is made by the shoulder.

Figure 6.20. How a man mounts a step.

The two sketches on the right show this action from the front and side with the lines of their axes or fulcra about which the movements takes place. This is described in detail (see text). The whole movement is governed by the rule that 'The centre of gravity of a man who raises one foot from the ground always rests on the centre of the sole of the other foot'.

How a running man stops himself.

This is illustrated in the lower left drawing, underneath which Leonardo writes, 'When a man wants to stop running and use up his impetus necessity makes him lean back and make small quick steps' (W 19038 v; K/P 80 v).

Figure 6.21. Going upstairs.

Leonardo explains this drawing thus: 'When going upstairs if you place your hands on your knees all the work acquired by the arms is taken away from the tendons under the knees' (H 75 r).

Figure 6.22. The weight of a man climbing steps.

Besides this drawing Leonardo asks himself a question, 'I ask this, the weight of a man in every step of movement on this stepladder, what weight does he give to *b* and to *c*? Observe his perpendicular line below the centre of gravity of the man' (Forster II 45 v).

This question is answered on Ms I 14 v, for which see text.

writes, 'descends from one step to another by jumping from one to the other, and then you add together all the forces of the percussions and the weights of these jumps, you will find that they will equal the whole percussion and weight that this man would produce if he jumped down by the perpendicular line from the top to the bottom of the height of the steps.

'Furthermore if this man should come down from the height, percussing step by step, objects which would bend, such as a spring, so that the percussion from one step to the other was slight, you will find that at the end of his descent this man's percussion will have diminished the spring by as much as it would have done [if he had jumped] in a free perpendicular line. That is if the diminution of the spring at each descending step with all the percussions were added together'. This example provides a good instance of Leonardo's intensive analysis of the act of jumping.

JUMPING

One human movement which particularly attracted Leonardo's attention was jumping. This being the movement by which man got off the ground into the air and landed again, represented for Leonardo a miniature flight. To fly, all one had to achieve was to prolong the jump indefinitely. The analogy he uses for both take-off and landing is the action of a spring.

The fourth figure down the series of human actions drawn in Figure 6.16 (p. 175) shows a man rising from sitting on the ground. Leonardo adds no explanatory text, but he does describe a similar drawing, Figure 6.23a, in the *Treatise on Painting*. Here he says, 'When a man is seated on the floor the first thing he does in rising is to pull in the leg and place the hand on the ground on the side on which he wishes to rise, and then throwing the weight of his body on that arm, he puts his knee to the ground on the side on which he wishes to rise' (TP I 378; CU 127 v). The man is now in a crouching position, like a coiled spring, ready to 'rise up'.

The description of 'rising up' is the next note on the same page of the *Treatise on Painting*. It is headed, 'Of jumping and what increases a jump. Nature teaches and works', he writes, 'so that without any thought on the part of the jumper, when he wishes to jump he raises his arms and shoulders with impetus, which following the impetus moves him together with a large part of his body, lifting him up high until the impetus in him is exhausted. And this impetus is accompanied by rapid extension [straightening] of the bent curve of the spine of the body and the joints of the hips, knees and feet. This extension is made obliquely, that is, forwards and upwards, so the movement used for going forward carries forward the body which jumps, and the movement of going upwards raises the body and makes it describe a big arc, so increasing the jump' (TP I 377; CU 127 v).

It will be remembered that impetus for Leonardo is imparted by weight and velocity (see Chapter 4), and it will be noticed that the sudden extension of all the flexed joints, ankles, knees, hips and spine acts like the straightening of the coils of a compressed spring. The upward impetus and movement so produced is enhanced by throwing up the arms. Jumping is thus an example of Leonardo's general statement that 'Impetus may be acquired and augmented by accidental as well as natural movement through space, but much less upwards than horizontally or downwards' (CA 202 vb). He analyses the movements of the jumping body even more finely thus, 'Impetus of a part that is moved moves with itself a greater part which was at first immobile. Just as a person who jumps up with impetus, at the time that his knees move up his legs are immobile, then they move' (CA 373 vb).

Of course a man cannot jump off a surface which slides or gives way under him. The force exerted by the man's feet in jumping Leonardo looks upon as percussion and the leap as two components of rebound, one forward and one upward. 'Impetus', he writes, 'moves a movable thing after percussion has stopped movement and been reflected . . . But in the case of weights of equal size, weight and substance, the movable thing, after having percussed the object, remains in the position where it was when it made the percussion . . . here reflected movement is not produced because the object percussed immediately flies away bearing with it the power and impetus of its percussor. This percussor does not spring back because it has nothing to serve as a foundation for its spring; like a man who wishes to jump from a board placed on the ground on top of several sawn-up logs. As he gathers impetus for his leap the impetus communicates itself to, and unites with, the board which flies away as if it were on wheels. And he who would fain leap, deprived of the impetus of the leap, is left in the same position' (Leic 29 r; Figures 6.23b and 6.24).

A similar result of percussion in relation to the jump is described in Figure 6.24b, where he discusses 'a ball percussed by another ball which transfers its own motion to the ball which it hits and remains motionless at the position of the ball which flees from it. It is certain that if the ball which flew away had been resistant and immovable the hitting ball would have jumped back. But by moving the struck ball, the ball which strikes has not the firmness nor support to make its leap; like a man standing on a board who wants to jump forward. The board flies back without waiting for the completion of the leap. Therefore the jumper falls back on to the place from which he rose up' (Madrid I 113 r).

So much for the forward component of the jump. Leonardo describes the downward component of the force by an ingenious experiment involving two men at each end of a board which is balanced on a round log that acts as the fulcrum of a balance. 'When 2 equally heavy men are placed in balance at the opposite ends of a plank and one of them wants to jump upwards then the jump will be made downwards at his end of the plank. This man will never raise himself upwards but will remain in the same place until the man opposite in his turn beats the plank with his feet' (Leic 8r; see Figure 6.23b). In this way Leonardo expresses by

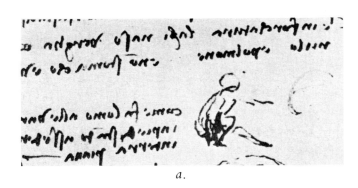

a.

a. Rising up from the ground (K/P 113 v).

Leonardo writes of this drawing, 'How a man lifts himself up from a sitting position on level ground'.

This drawing should be compared with the first and fourth the margin of A 28 v (Figure 6.16, p. 175).

b. A man jumping off a plank (Leic 8 r).

The man on the left is preparing to jump from a plank balanced on a log. Another man at the other end keeps the plank level.

The conditions upon which the jumping man will be able to take off are described in the text.

b.

c.

c. The leap into the air with wings (CA 276 rb).

The upper drawing with wings is shown in the same position as the man on the plank in Figure 6.23b. The wings are intended to flap down as the legs are extended, as in the lower drawing.

Figure 6.23.   The jump into the air, three stages.

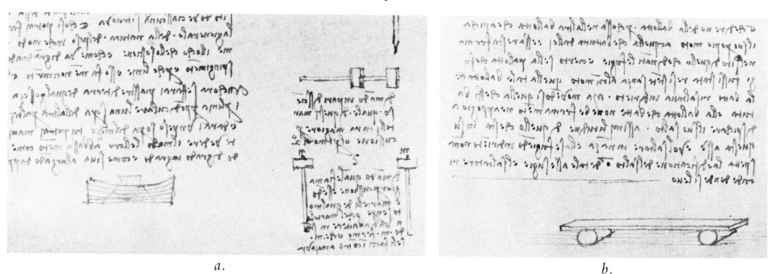

a.   *b.*

Figure 6.24. Two conditions which impede the percussion of a jump-off.

a. On the left margin the lines of weight of an object resting on 'earthen layers of equal thickness' are drawn. The subsidence of such a surface will impede the rebound of percussion on which a jump depends (Madrid I 112 v).

b. Leonardo describes above the board on rollers what happens to a man standing on the board who wants to jump forward. 'The board flies back without waiting for the completion of the leap' (Madrid I 113 r).

experiment the previous assertion that 'When a man jumps he doubles his weight on the place from which he takes off' (Madrid I 90 r).

To increase the length of the jump Leonardo has already mentioned using the arms. This arm movement he analyses in his typically detailed way, 'In order to make a bigger jump first throw the fists straight behind and with fury bring them to the front in taking the leap, and it will be found that this movement makes the leap greater (I 104 v).

So much for the power of jumping or springing upwards. It is fascinating to find, in Figure 6.25, at the base of one of Leonardo's designs for a flying machine a large steel spring. Inside a plan of the mechanism of the spring and pulley to the left of the main drawing are written the words *fondamento del moto,* 'beginning of the movement'. Here a mechanical spring replaces a man's jump.

Once launched into the air by springing up from a suitable place, the question of prolonging and maintaining flight was studied mostly through some 160 pages on the flight of the bird, which do not form part of our present subject of jumping. The problem of alighting on the ground again does, however, constitute an essential part of the leap. It too received meticulous study in terms of mechanics and human physiology.

Several times Leonardo asks why a leap onto the toes is more silent than that onto the heels, finally deciding that there must be less percussion when one lands on the toes. On CA 338 ra (Figure 6.26a) he answers the question, comparing the landing on the toes with a spring being compressed. The explanation is accompanied by a diagram:

'The body *dab* falling down with the part *b* on *c* does not give its weight at the time of the blow on to the percussed

place. If *b* weighs 10 pounds and *a* 10 pounds, and the whole body 20 pounds, it [*b*] will make a blow of 10 pounds. The reason is that when *b* has percussed *a* still falls and as it falls it does not weigh at the moment of the blow on the place percussed, *b*. The same happens to a man who jumps on to the point of his toes, who, at the time of percussion bends his knees and all the joints above and behind the blow. As much percussion is made above the first place of percussion as occurs up to the part in which it is stopped, that is from the knee to the foot. Some would say from the heel to the point of the toe since in the blow on the point of the toe the heel falls somewhat, and therefore does not make a blow' (CA 338 ra; see Figure 6.26b).

It will be remembered that Leonardo described percussion as 'the end of movement' (see Chapter 4); therefore it may continue from one body to another until it stops, in this case from the bones of the toe, the foot and the leg to the knee, through all the intervening joints until it stops. This he calls 'compound percussion'. With this 'the blow grows less and less in proportion to the number of obstacles interposed between it and the final resistance [which stops it]. It is just as if someone were to strike a book on its front page when all the pages were touching; the last page would feel the damage very slightly' (BM 82 r).

It is the same when a man jumps onto a coiled spring. 'That weight', he writes, 'will give less percussion as it percusses with parts in the object which consent, like jumping on a spring. Or with parts which consent as with a spring in the percussing object, like someone who jumps on the points of the feet' (M 59 r).

From these few samples of Leonardo's studies of the movements of man on the ground some idea of the sub-

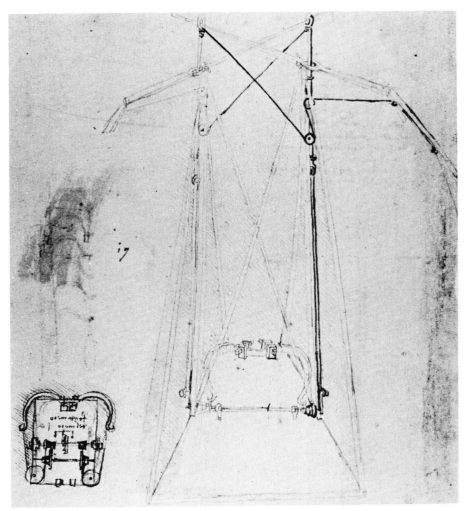

Figure 6.25. Assisted take-off (CA 314 rb).

This scheme presents Leonardo's attempt to substitute mechanical power for that of the legs in take-off. The power used is that of the crossbow, the spring of which when compressed is very powerful but short-lived. Here Leonardo attaches a compressed crossbow to pulleys, cords and two wings of a flying machine. In the left-hand drawing the bow and its mechanism are shown ready for action. Within this drawing Leonardo makes his idea clear by writing the words 'the beginning of the movement'. How this movement of the wings would be continued is another problem.

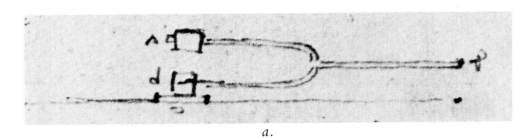

a.

Figure 6.26.

a. A weight landing on a spring (CA 338 ra).

The object *abd* shaped like a tuning fork is envisaged falling on the ground, *c*, making contact at *b*. The ensuing sequence of events is described in the text.

b. Jumping on to the toes (W 19094 r; K/P 190 r).

This action illustrated by the fulcrum at *n* and the lowering of the heel is described in the text describing Figure 6.26a.

b.

stance and intricate texture of his science of the powers and
movements of the body of man on land can be obtained.

MOVEMENT IN WATER

It is not surprising to find that Leonardo first approached
the problems of movement in or on water as an inventor.
CA 7 ra (Figure 6.27), one of his earliest pages of drawing
made when he was about twenty-eight years old, is devoted
to problems presented by water. It contains a 'screw for
pumping water to any height you want', a novel method of
raising water by fire at the top of a well (left margin), a
primitive diving apparatus and a method of 'walking on
water'. One must place in a similar category the ingenious
life jacket described on B 81 v (Figure 6.28) under the heading, 'A way of saving oneself in a tempest and shipwreck'.
This jacket is made of two layers of 'Leather impermeable
to air' and is blown up before jumping into the sea, 'keeping
the tube of air which goes into your coat in your mouth'.
The accompanying illustration, however, shows only a
man in a life belt and webbed flippers, swimming
purposefully.

By this time Leonardo is taking a more scientific interest
in swimming in water, seeing in it a close resemblance to
flying in the air. Discussing the required rate of descent of
an artificial wing (as illustrated in Figure 4.33, p. 118),
Leonardo interrupts himself to say, 'You know that if you
stand in deep water holding your arms stretched out and let
them fall naturally, the arms will fall as far as the thighs and
a man will remain in the same position. But if you make the
arms, which would naturally fall in four lengths of time, fall
in two, the man will quit his position, move violently and
take a position on the surface of the water' (B 88 v). By
'natural' movement Leonardo means their 'natural weight',
and 'violent' movement is that made by force.

This comparison between the mechanics of flight and
swimming emerges in another form on B 100 v (Figure
6.14, p. 173), a tattered sheet on which Leonardo draws a
butterfly, a dragonfly, a bat, and on the left, a flying gurnard with outspread wings launching itself by its beating
tail from water to air. Beside the drawing Leonardo notes,
'Animal which flits from one element into another'. This
comparison is developed in a note about the goose which
can both swim and fly. 'Swimming', he comments, 'illustrates the method of flying, and shows that the widest
weight finds most resistance in the air. Observe the goose's
foot; if it were always open or closed the creature would not
be able to make any kind of movement'. He then compares
the relative narrowness of the goose's foot as it goes
forwards, with its width and propelling power as it moves
backwards. 'As it pushes it back it spreads out and so makes
its own [movement] slower; then that part of the body [of
the goose] that has contact with the air becomes swifter' (M
83 r). Pursuing the analogy between the hand of a swimmer
and the wing of a bird, Leonardo starts from a general
'rule', deducing the particular example. He writes, 'When
two forces percuss each other it is always the swiftest which

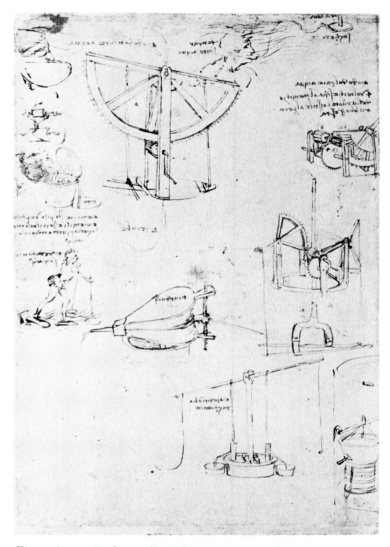

Figure 6.27. Early studies of water pumps, diving apparatus
and water skis (CA 7 ra).

At the centre of the top margin a man's head under the water is
drawn. He breathes through a pipe connected to a float. At the
left top corner is another primitive masked and goggled figure.
At the left margin, centre, is a man walking on water with the
aid of buoyant shoes and sticks, an anticipation of the water
skis of today.

leaps back. So it is with the hand of the swimmer when it
strikes and presses on the water and makes his body glide
forward in a contrary movement. So it is also with the wing
of a bird' (F 41 v).

The *Codex Leicester* is remarkable for being a depository
of innumerable questions, some of which are answered, as
is that on swimming on folio 22 v. This particular mixture
of penetrating questions and illuminating insights so reveals
the intricacies of Leonardo's style in the *Codex Leicester* that
a substantial part of it will be given without breaking the
continuity of his thought.

'How with air-bladders an army ought to cross rivers by
swimming. On the movement of swimming in fish; of the

Figure 6.28. Movement in water.

Above is a webbed glove very reminiscent of a frogman's flipper. Below this is a man wearing a life belt. The buoyancy is now placed below the shoulders. On his right hand the frogman's flipper is drawn; the feet, too, have a suggestive shape attached to them, but the faint lines are indefinite. Beneath the drawing Leonardo writes, 'A way of saving oneself in a tempest and shipwreck at sea' (B 81 v).

way in which they jump out of the water as may be seen with dolphins; and it seems a marvellous thing to make a leap from something that does not stand still before it flies away. Of the swimming of animals of long shapes such as eels and the like. Of the way of swimming against currents and great falls of rivers. Of the way in which fishes of round shape swim. How all the other animals which have feet with toes know naturally how to swim, except man. In what way a man ought to learn to swim. In what way a man should rest on the water. How a man ought to protect himself from whirlpools or eddies of water which suck him down to the bottom. How a man drawn down to the bottom has to seek for the reflected movement which will throw him up from the depths. How he ought to propel himself with his arms. How to swim on his back. How he can stay under water only as long as he can hold his breath. How many with a certain instrument are able to remain for some time under water. How and why I do not describe my methods of remaining under water for as long as I can remain without food; and this I do not publish or divulge by reason of the evil nature of men who would practice assassinations at the bottom of the sea by piercing holes in the bottoms of ships and sinking them with the men in them; and although I will impart others there is no danger in them because the mouth of the tube by which one breathes is above the water supported on bladders or corks'.

Leonardo continues to relate swimming and flying during the time he is collecting together his notes on flight into the small special codex, *On the Flight of Birds*. At the end of a long analysis of the movements of the wings, head and tail of a bird in descent he adds tersely, 'Swimming on water teaches men how birds do up in the air' (CA 66 rb).

Throughout his inquiry into the mechanism of swimming Leonardo is guided by the reiterated principle, 'There is as much power of movement in the water or the air against an object as there is in the object against the air or water' (CA 214 rd). However, he is put into difficulties by the great difference between water and air in their capacities of condensation. Whereas air can be condensed or rarefied, for it is indeed 'infinitely compressible', water is very much incompressible. Yet in both, action equals reaction constantly.

Leonardo, like many others since his day, was acutely aware of the rapid acceleration of which fish are capable. Here he made one of his rare errors of observation, mistaking rapid acceleration for rapid steady velocity. It is to be noted that this error defines the limit of a great field of his mechanical achievements.

'Why', he asks, 'is the fish in the water swifter then the bird in the air when it ought to be contrary since water is heavier and denser than air, and the fish is heavier and has smaller wings than the bird? For this reason a fish is not moved from its place by the swift currents of water as is the bird by the fury of the winds amid the air; also we may see the fish speeding upwards on the very course down which water has fallen abruptly with very rapid movement like lightning amid thick clouds, which seems a marvellous thing. And this happens through the immense velocity with which it moves which so exceeds the movement of the water as to cause it to seem motionless in comparison with the movement of the fish . . . This happens because water itself is denser than air and heavier and therefore fills the vacuum which the fish leaves behind it. Also the water which it percusses in front is not condensed like the air in front of the bird but rather it makes a wave which by its movement prepares for and augments the movements of the fish; and therefore it is swifter than the bird which has to meet the air condensed in front of it' (CA 168 vb).

Because air condenses in front of bodies like birds that move in it, whereas water does not condense but merely 'opens up and closes round the fish, the impetus of the fish is of shorter life than that of the bird in the air, even though the muscles of the fish are most powerful with respect to their quantity. For the fish is all muscle, and this is very necessary in order that it should exist in a body so much denser than air' (E 71 v). Leonardo here attributes the rapid destruction of the impetus of the fish to the same cause which destroys the impetus of a projectile in air – to 'resistance'.

In the end, from the note contained on G 50 v (Figure 6.29), it appears that Leonardo is reduced to attributing the

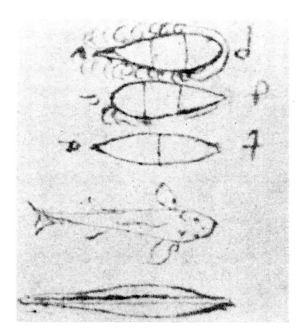

Figure 6.29. The movement of ships and fish.

Comparing the three upper drawings, Leonardo writes, 'These three ships of uniform length and breadth and depth when propelled by equal powers will have different speeds of movement; for the ship that presents its widest part in front is swifter, and it resembles the shape of birds and fishes such as the mullet' (G 50 v).

movement of both ships and fish to the 'Eddies that press against the back two thirds of the ship'. He illustrates this statement with a drawing showing the streamline shape of a schematic fish from the side which 'resembles the shape of birds and fishes such as the mullet'.

MOVEMENT IN THE AIR

No more poignant example of the tragedy of Leonardo's creative science can be cited than his failure to achieve human flight. In this field of endevour Leonardo the inventor outstripped Leonardo the scientist. For Leonardo, however, the endurance of such creative failures was almost a normal part of his life-struggle. It happened to him in painting and sculpture as well as flight. And in all these fields of endeavour the root cause was the same; his passion for creation outstretched his passion for science. Thus we are left with so few examples of his works of art and science, the former extensively recognised, the latter still largely submerged.

This verdict applies to his work on flight, for although he failed in his main enterprise, Leonardo observed and understood many of the factors involved in its achievement. To do justice to those insights would take us too far from our present theme. Here only a few examples can be cited. These will be divided into three groups, each a vital part of his problem: (1) take-off; (2) flight in the air; and (3) landing.

Take-off being the obvious problem of first priority provides the subject for dozens of drawings of flying machines which have come down to us in Leonardo's notebooks. This is not the place to attempt a detailed analysis of these various ingenious devices. Suffice it to point out that they all depended upon Leonardo's conviction that a man had 200 pounds of 'force' available to lift him off the ground and that if this force could be concentrated suitably in velocity, it could keep him in the air. We have already cited experiments demonstrating the act of jumping. Leonardo clinches his conviction thus: 'How a man has a greater amount of power in his legs than is needed for his weight. To show that this is true have a man stand on the shore and consider how deep his footprints are. Then put another man on his back and you will see how much deeper his feet sink in. Then take the second man off his back and have him leap straight up into the air as high as he can. You will find that the impressions of his feet have sunk in deeper by the jump than with the man on his back. Hence we have proved in two ways that man has more than twice the power that he needs to support himself'. See Figure 6.30 for the basis of this experiment. He goes on to argue, 'If you should say that the sinews and muscles of a bird are incomparably more powerful than those of man because all the flesh of the big muscles and fleshy parts of the breast goes to increase the power of the motion of wings, and the breast-bone is in one piece which affords the bird very great power . . . the reply to this is that all this great strength is for the purpose of enabling it, over and above the ordinary action of wings in

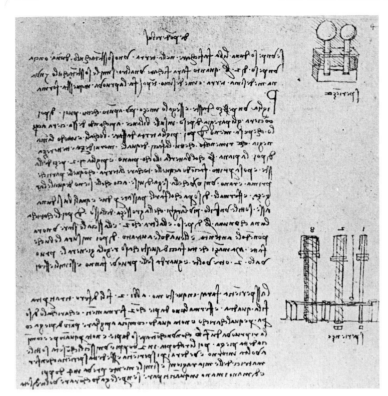

Figure 6.30.   Experiments on weights sinking into the earth.

Beside the top experiment Leonardo writes, 'If a weight of
1 lb makes a sustentaculum fix itself one inch into the earth
how much will a weight of 2 lbs make another similar
sustentaculum fix into the same earth in the same time. Make
the test in this form' (A 47 r).

Under each of these drawings Leonardo writes, '*Sperienze*',
i.e., Experiment.

Systematic variations of this experiment are carried out in the
neighbouring pages. They provide the basis for the experiment
on the imprints in the sand made by a man carrying another on
his shoulders and that made by him leaping upwards on
Sul Vol 16 r.

keeping it up, to double and triple its motion in order to
escape from attackers or pursue its prey . . . They need
little power to keep themselves up in the air, balance their
wings and flap them on the air-currents and steer along their
paths' (Sul Vol 16 r).

Leonardo was well aware of more powerful forces than
jumping for taking off. He had given much attention, for
example, to the effects of gunpowder in creating fire, which
by its sudden expansion gave rise to the movement of projectiles through the air, and he attempted to equate this with
the power of the crossbow. He had even considered 'A
method of sending fire one mile or more up into the air'
(Figure 6.31), using a rocket. His interpretation of the
power of the rocket was 'The flame of a rocket which penetrates the air is not that which drives it in contrary motion
but only that which percusses the air. This is implemented
as it first rushes out of the rocket. The flame does not escape

from the rocket, but the rocket escapes from the flame' (CA
227 rc). In these terms Leonardo recognises jet propulsion.

These methods, however, were too hazardous for
Leonardo's serious consideration. He considered that man
had, like the bird, sufficient power to achieve flight without
further mechanical aid, except perhaps from springs. He

Figure 6.31.   Rocket propulsion into the air (Madrid I 59 v).

The rocket is shown attached to its stick which is loaded with
gunpowder at the point labelled 'polvere'. This in its turn is
attached to the long 'bullet' which is loaded into an arquebus.
Leonardo writes above the drawing, 'A method of sending fire
one mile or more up into the air'. Details of the relative
weights of the parts and their ignition are described in the
passage alongside.

applied his studies of jumping to a bird's take-off from the ground thus: 'When birds wish to commence their flight it is necessary for them to do so in one of two ways, one of which commences by lowering themselves with their body to the ground and then making a leap into the air by extending very rapidly their flexed legs. At the end of this leap the wings have completed their extension and the bird immediately lowers them swiftly towards the ground and ascends in the second degree which is slanting like the first . . . Others first raise their wings slanting forwards, lower themselves as far as they can with their breasts to the ground and in this position extend their legs very rapidly leaping up slanting forward . . . Other birds lower the wings and extend the legs at the same time, and thus the power produced by the first beating of the wings added to the power acquired by extending the legs becomes very great, and this united power is the greatest possible for the beginnings of the flight of these birds.

'The second method employed by birds at the commencement of flight is when they descend from a height, they merely throw themselves forward and at the same time spread out their wings high and forwards and then in the course of their leap lower their wings downwards and backwards, using them as oars' (G 64 r).

This analysis of the modes of a bird's take-off from the ground was, of course, made with a view to applying it to man. The 'second method' is that of the hang glider of today.

'When a bird wishes to rise by beating its wings it raises its shoulders and beats the tips of its wings toward itself and condenses the air between the points of its wings and the bird's breast, and being interposed there again and again it raises up the bird' (Sul Vol 5 v). Why should man not do the same? He points out that 'a man in a flying machine must be free from the waist upwards in order to balance himself as one does in a boat, so that the centre of gravity of himself and of the instrument can be changed and balanced when required to do so by the change of the centre of resistance' (Sul Vol 5 r). Leonardo illustrated a man so placed in a little figure beside this passage in Figure 6.32. A similar figure, equipped with wings to the arms and a waist-girdle, is drawn on CA 59 rb. Here he asks himself, 'Of what kind are the motors, and in what way should they acquire their force?'

Clearly take-off will be facilitated if made from a height. Two such launching pads are designed. Both were conceived years before. One on B 89 r (Figure 6.40a, p. 194) includes a retractable ladder, folded up after take-off. (There is also on the right a primitive shock-absorber to be used on landing, of a pattern elaborated on Madrid I 62 v ([Figure 6.39, p. 193].) One, on B 80 r (Figure 6.33), shows the 'pilot' in a coracle-shaped vessel into which he has climbed using a ladder and hatch. This pilot is certainly not free from the waist up. In fact he is so involved in the ropes and pulley-gear that it is not easy to see him. Clearly this is an early 'invention' and does not meet the necessary conditions demanded by Leonardo in 1505 when he returned to simpler designs. What Leonardo's final launching design

Figure 6.32. The centre of gravity and the centre of resistance.

Beside the little drawing of man in a flying machine bound to it at the waist Leonardo writes, 'A man in a flying machine must be free from the waist upwards to be able to balance himself as one does in a boat, in order that the centre of gravity of himself and of the instrument can be changed and balanced when required to do so by the change of the centre of its resistance' (Sul Vol 5 r). See also Figure 4.19b.

The drawing of the bird below and the two diagrams illustrate the effects of wind on the course of its flight by virtue of changes between the centre of gravity and centre of resistance (Sul Vol 5 r).

was like remains uncertain. We do know, however, that he launched a model. Rightly asserting that 'When a bird's centre of gravity is below its wings it has so much the less risk of being turned upside down', he adds, 'Make a small one [model] to go over the water, and try it in the wind over some part of the Arno . . . There is a suitable place there where the mills discharge into the Arno by the falls of Ponte Rubaconte' (CA 214 rd). Leonardo wrote this note about the time that he had great hopes of success. 'The great bird', he exclaims, 'will take its first flight on the back of the great swan, filling the whole world with amazement and filling all records with its fame; and it will bring eternal glory to the nest where it was born' (Sul Vol inside back cover). And then he repeated, 'From the mountain which takes its name from the great bird the famous bird will take its flight, which will fill the world with its great renown' (Sul Vol 18 v).

The mountain is identifiable as Monte Cecero (the swan) to the south of Fiesole. Clearly Monte Cecero provided the

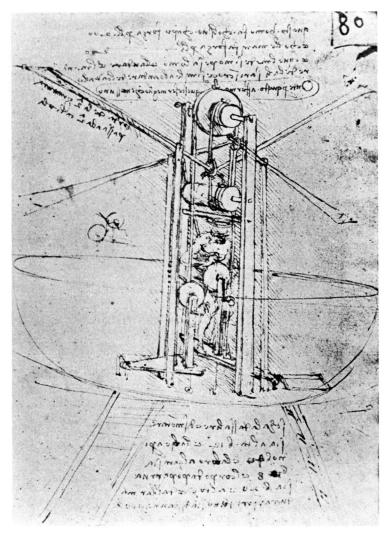

Figure 6.33. Design for flying machine with ladders.

Here Leonardo applies the principle illustrated in Figure 6.7 (p. 165) to flying. He writes, 'This man exerts with his head and hands a force equal to 200 lbs. And this same man weighs 200 pounds. And the movement of the wings will be crosswise like that of the horse. So for this reason I assert that this method is better than any' (B 80 r). He describes the ladder as 12 *braccia* (about 24 feet) high and the wingspan as 40 *braccia* (about 80 feet).

launching pad which he finally chose. This was the 'second method' of a bird's take-off, now suggested for his own bird machine – the hang glider method.

FLIGHT IN THE AIR

Leonardo's interest in birds and bird flight waxed to its height about 1500. It may well be that he realised about then that Leonardo the inventor had not succeeded and that now it was time for Leonardo the scientist to find out more about the theory of flight. Whatever was responsible for this concentration on the flight of birds, the fact remains that as the years passed it never slackened. Notebooks written after the

small codex *On the Flight of Birds* contain many notes on the subject. These include studies of the anatomy of the bird, its wings in particular, as well as an almost obsessional analysis of the variations of the movements of flight in MSS E and G which date from his last years of active work. In all these late notes, however, new inventions of flying machines are conspicuous by their absence.

The study of flight in Leonardo's view involved the correlation of a set of variables consisting of the bird's centre of gravity and its centre of resistance in air, with mechanisms for rising, falling and turning left or right. Since the principle that action equals reaction was fundamental, such correlated adjustments made by the bird could either arise in the live bird itself and act on the air, or they could constitute a response to changes in air flow to winds blowing from all directions. See Figure 4.19b.

'An object makes as much force against the air as the air does against the object', he writes. 'See how the percussions of its wings against the air give support to the heavy eagle in the highly rarefied air close to the element of fire. Observe also how the air in motion over the sea percusses the swelling sails and drives heavily laden ships. From these instances and the reasons given, a man with wings large enough and duly attached, by exerting force against the resistance of the air may be able to overcome it, subjugate it and raise himself upon it' (CA 381 va).

We have already seen Leonardo's early experimental answer to the problem of a man raising himself on air in Figure 4.33b (p. 118), where it is interesting to note that the man is shown being lifted up by the descending wing. The experiment apparently confirmed his expectation of lifting power. If so, it was an unhappy success for Leonardo, for it set him off on the track of studying flapping bird flight rather than gliding flight. The lift for flight is derived from the velocity of the flow of air over the convex wing as well as that below the wing. Flapping flight accelerates both velocities and is therefore a complex extension of gliding flight. Failure to appreciate this, with the tempting example of birds so evident before their eyes, led men to attempt human bird–flight for many centuries. Leonardo was one of those who seems to have fallen into the trap.

Throughout medieval Christendom the acceptance of angels encouraged the belief in 'flapping' flight. Even Leonardo had painted a winged angel in his *Annunciation,* and he certainly had seen Giotto's relief of Daedalus and Icarus on the Campanile of the Duomo in Florence. These influences were undoubtedly significant in delaying the exploration of the potentialities of the simple glider. Yet there are strong signs that Leonardo was on the way to extricating himself. For example, he writes in his treatise on birds, 'I have divided my treatise on birds into four books of which the first treats of their flight by beating their wings; the second of flight without beating the wings and with the help of the wind; the third of flight in general such as that of birds, bats, fishes, animals and insects; the last on the mechanism of this movement' (K 3 r). This passage, written about the same time as his small codex *On the Flight of Birds,* suggests the form of its content. Turning to the codex one finds that a large part of it is in fact concerned with 'flight without beating the wings', i.e., gliding flight. The first three folios (six pages) have no apparent bearing on flight at all. This is deceptive, for they are all concerned with gravity, weight on an inclined plane, and the mechanism of balance – highly relevant aspects of flight to Leonardo, who on folio 3 r announces, 'Instrumental or mechanical science is most noble and useful beyond all others since by means of it all animate bodies that have motion perform all their operations; and these motions have their origin at their centre of gravity which is placed in the middle of unequal weights. And this has scarcity or abundance of muscles, and also lever and counterlever'. To this he adds, 'A bird is an instrument working according to mathematical law, which instrument it is in the power of man to reproduce with all its movements but not with so much power, though it is deficient only in the power of balancing. We may therefore say that such an instrument constructed by man is lacking in nothing except the soul of the bird; and this soul must needs be imitated by the soul of man' (CA 161 ra). And, as we have seen (Chapter 4), Leonardo believed that sufficient expanse of wing should suspend a man in the air.

On the following page of this small codex Leonardo analyses the flexibility of birds' feathers, treating them as flexible beams under mechanical stress. He treats the problems of the centre of gravity and resistance of the bird's wing and body in the same way. He then deals with the parallelogram of forces produced when gliding obliquely down, the wind acting on it from different angles, describing the movement of its wings only in relation to its turning left or right (see Figure 4.19b, p. 107). This 'incident' movement of a glide down leads to a 'reflected' rise in straight or circular movement. On folio 5 v he describes the use of the wings for rising, pointing out that this beating is a fatiguing process usually followed by a restful glide. He then describes 'the four incident and reflected motions of birds under different wind conditions'. They obey the laws of percussion. Such considerations occupy the next few folios, after which he considers the risk of the bird being 'turned upside down' by the winds and its ways of avoiding that fate by movements of the wings and tail.

The motive force for all these movements consists of the condensation of eddies of air beneath and behind the moving bird's wing. When it beats it 'rows' through the air. Turning is produced, as in rowing, by drawing in one wing and beating more powerfully and frequently with the other.

Leonardo attached great importance to the birds' *dito grosso* or 'thumb' projecting up from the front edge of the wing (see Figure 6.34). In his view this 'thumb' 'sets the shoulder of the wing below the wind', so guiding and maintaining direction and balance. 'Nature has made a very powerful bone in that finger to which are joined very strong sinews and short feathers . . . because by means of it the

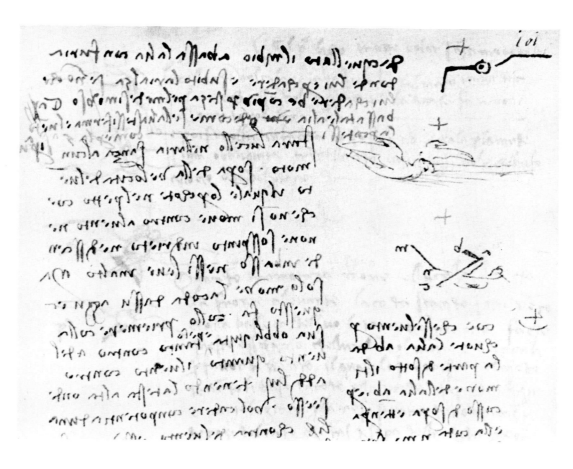

Figure 6.34. The bird's rudder (*dito grosso*) (Madrid II 101 r).

Leonardo attached much importance to the bird's rudder in maintaining equilibrium. This is prominent in the second drawing on this page, beneath the cross marked +. It is again sketched in the diagram beneath, marked *ab*. In the script alongside he describes the interaction between the tail and the rudder in maintaining the balance of the bird.

bird supports itself on the already condensed air with all the power of the wing and force, since it is by means of this that the bird moves forward. The finger serves the same purpose for the wing as claws do for cats when they climb trees' (Sul Vol 13 v). He then calls the 'thumb' a 'rudder, dividing the air high or low for whatever movement the bird wishes to perform. A second rudder is placed at the opposite end, the other side of the bird's centre of gravity. This is the bird's tail' (Sul Vol 13 v). These statements are accompanied by drawings exaggerating the 'thumb' of the wing and incidentally depicting the wings convexly shaped like the aerofoil of an aeroplane.

Leonardo next discusses the general anatomy of the wing. He recognises that the shoulder is a *polo,* a fulcrum, pivot, or ball and socket joint. This is 'turned by the muscles of the breast and back which makes it possible to lower or raise the elbow in conformity with the will and needs of the moving animal'. Here he concludes that 'the rising of

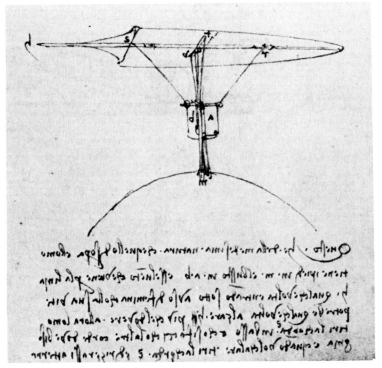

Figure 6.36.   Leonardo's hang glider.

Beneath this drawing of a broad plane from which a man hangs Leonardo writes, 'The man places his feet at *m* and his chest at *ab*. And if the wind comes along the line *h* sometimes entering underneath lifting *h* like the nut of a screw it would raise *h* more than it should. In this case the man should pull the cord *s* down. And he should do the same with the other cords when necessary; and when he wants to descend he should pull the cord *s* and direct himself towards the ground' (Madrid I 64 r).

It will be noticed that this hang glider takes off from the convexity of a hill.

Figure 6.35.   The course of birds in rising flight.

Here Leonardo draws the course of birds rising 'without beating their wings'. This they do in circles which he describes as simple and complex, 'The simple circular movement is a movement forwards with the wind, and a backward movement against the direction of the wind' (CA 308 rb).

birds without beating their wings is due solely to their circular motion within the motion of the wind' (Sul Vol 14 r). This spiral form of ascent is beautifully illustrated on CA 308 rb (Figure 6.35) in 'simple' and 'complex' forms.

Having compared the complexity of the feathered wing in birds with the simple skin of the wing of the bat and pointed out once again that 'the heaviest part of the body is always the guide of movement' (Sul Vol 15 r), Leonardo goes on to discuss the importance of finding the centre of gravity of the bird. 'When the bird goes downwards its centre of gravity is in front of its centre of resistance . . . if the bird wishes to rise, its centre of gravity is behind its centre of resistance' (Sul Vol 15 v). These circumstances are very clearly illustrated (see Figure 4.19b, p. 107).

Leonardo discusses the danger of destruction, recognising two main causes, 'the first being the breaking-up of the machine, the second the turning of the machine edgewise . . . either to the side or tailwise'. Thus he describes stalling. The first danger can be met by strong construction of the 'bird', the second 'must be prevented by constructing the machine in such a way that in descending in any manner the remedy is inherent; this will occur if the centre of gravity is always in a straight line above the centre of the heavy body carried by it, and one centre is always sufficiently distant from the other, namely for a machine 30 *braccia* (60 feet) wide let these centres be 4 *braccia* (8 feet) apart, one beneath the other, the heavier being underneath since in descent the heavier part is always the guide of the movement' (Sul Vol 12 v). This distribution of weight is accurately illustrated in Figure 6.36 – a very modern-looking hang-glider (Madrid I 64 r).

In addition to these precautions he produced a working form of the parachute (see Figure 4.33c, p. 118) and recommended the aeronaut to carry 'bags with which a man falling from a height of 12 feet will not be hurt falling either into water or on to earth; and these bags tied together like the beads of a rosary surround each other'. 'If you fall see that you strike the earth with the double bag you have under the rump', he exhorts (Sul Vol 16 r).

Later, in 1508, Leonardo proposed another treatise on the movement of birds. This time his first book is to be devoted to the subject of 'air resistance (see Figure 6.37); the second to the anatomy of the bird and its feathers; the third to the way the feathers work; and the fourth to the strength of the wings and tail' (F 41 v). These matters are extensively dealt with in MSS E and G. However, quite typically, one of Leonardo's last notes on flight runs, 'In order to give the true science of the movement of birds in the air it is necessary first to give the science of winds, and this we shall prove by means of the movements of water. This science in itself is capable of being received by the senses. It will serve as a ladder to arrive at the perception of things flying in the air and wind' (E 54 r). Obviously, he still felt as if he were just beginning his researches into flight, not coming to the end of his life.

LANDING

Leonardo's chief concern with landing was the prevention of injury or destruction, as we have seen. He does not seem to have considered the possibility of landing at speed. He does, however, consider the effects of rising on 'condensed' air on the descent to the ground. He detects and illustrates the occurrence of up-currents from a wind blowing against a hillside (see Figure 5.9b, p. 138). He also depicts 'wave lift', which occurs when wind blows over hills and mountains. He is all the more ready to detect this since he has noticed a similar wave formation formed at the surface of water by irregular contours in a riverbed. In MS A 59 v he explains, 'Why the surface of flowing rivers always presents protuberances and hollows. The reason for this is that just as a pair of stockings which clothe the legs reveal what is hidden beneath them, so that part of the water that lies on the surface reveals the nature of the riverbed, inasmuch as that part of the water which lines its base, finding there certain protrusions caused by the stones, percusses them and leaps up raising with it all the other water which

a. *b.*

Figure 6.37. Eddies and bending of the wing-tips in a flying bird.

a. The eddies under a bird's wing are drawn; they swirl around the wing, over the bent wing-tips onto their backs (E 47 v).

b. A wind-tunnel picture of a model of Concorde. The patterns of the two air-streams are remarkably similar. However, Leonardo failed to recognize the importance of 'lift' from eddies above the wing. This illustration is from Ritchie-Calder, *Leonardo and the Age of the Eye* (London, Heinemann, 1970), p. 217.

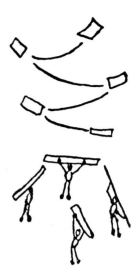

Figure 6.38.   Gliding flight controlled by man.

Under the heading, 'Percussion' Leonardo writes, 'Of things falling through the air from the same height that will give less percussion which descends by the longest route; it follows that that which descends by the shorter route will produce more percussion'. He then describes with his usual care for detail the descent of a 'slightly curved piece of paper' from one position to the next, as drawn in the top part of the figure. He explains this reflected pattern of movement by appealing to lost books on percussion, specifying, 'the ninth of this which says . . . the tenth . . . the eleventh . . . the twelfth . . . and the thirteenth which says; the air that has the greatest velocity of movement moves most – it follows that a man can descend as is shown below'. Here he refers to the lower set of drawings in which a man is attached to a broad sheet of paper or cardboard. Beneath these he writes, 'This man will move to the right side if he bends the right arm and extends the left arm; then he will move from right to left if he changes the extension of the arms' (G 74 r).

Thus Leonardo laboriously reached a method of diminishing the impact of landing and an adequate concept of a man-controlled glider at the same time.

flows over it'. Leonardo makes it clear that air flowing over the protuberances of the earth is similarly affected (CA 180 va). Such winds 'follow the form of the summits of the mountains' (BM 276 r). Just as a man may use the reflected up-currents of water to save himself from drowning, so a bird may use similar up-currents to rise or moderate its descent. But the bird usually lands flying into the wind thus: 'When it has descended from a height a bird beats its wings frequently as it settles in order to break the impetus of its descent, to settle itself on the ground, and diminish the force of percussion' (K 58 r). 'When it is near the end of its flight the bird makes itself slant only a little [from the vertical] in its length and opens its wings and tail very widely; but the wings reach towards the end with frequent small beats in the course of which the impetus is consumed; and as they shorten it remains for a very brief space above the place where it finally alights with a very slight percussion of its feet' (G 63 v).

Another factor using the wind for making a gentle descent arose from Leonardo's observation of the separate movement of the feathers of the wing-tips (see Figure 6.37). 'What difference is there between the tips of the wings of birds which bend and those which do not . . . since one sees that however slightly these tips are cut the bird's power of flight is almost stopped' (K 10 r). He answers his own question by going back to an early experiment on the forces exerted by an arch. This was 'An experiment to show that a weight placed on an arch does not load itself entirely on to its columns; on the contrary, the greater the weight placed on the arches the less the arch transmits the weight to the columns. The experiment is as follows: let a man be placed on a steelyard in the middle of the shaft of a well, then let him spread out his hands and feet between the walls of the well, and you will see him weigh much less on the steelyard. Give him a weight on the shoulders; you will see by the experiment that the greater the weight you give him the greater force he will make in spreading his arms and legs, the more he weighs on the walls and the less he weighs on the steelyard' (B 27 r). The experiment is repeated, and illustrated in Forster III 19 v. Here Leonardo tries to measure this effect, saying, 'If a chimney sweep weighs two hundred pounds, how much force does he exert with his feet and back in the chimney?'

The continuity and breadth of Leonardo's thought and its relevance to the descent of a bird in flight arises from his awareness of a particular function of the feathers at the wing-tip. He answers his question thus: 'The flight of birds has but little force unless the tips of the wings are flexible. This is proved by the fifth of the Elements which says: lateral power checks the descent of objects, as may be seen in the case of a man pressing with his feet and back against two sides of a wall as one sees chimney-sweeps do. Even so in great measure does a bird by the lateral bendings of the tips of its wings against the air where they find support and bend' (E 36 v). In this way Leonardo came to recognise that the bird's wing-tips supply some 'lift'.

Yet another method of obtaining gentle landing is described under the heading 'Percussion – Flight of man'. Here Leonardo asserts, 'Of things that fall in the air from the same height, that will give less percussion which descends by the longer route'. He then makes a series of little drawings of a falling piece of paper swirling to and fro, moving by 'reflected movement from one side to the other as their obliquity changes' (see Figure 6.38).

Leonardo accompanies the sketches with a long involved verbal description which appeals, like Euclid, to various propositions stated in different earlier books (unhappily lost). These state that 'things that percuss the air with a broader part of themselves have less penetrating power in the air, and conversely, the thing that percusses with less breadth penetrates the air more swiftly'. These are applications of his pyramidal law (see Chapter 4, p. 108). Thirdly, he cites the oft repeated law that the heaviest part of a body is the guide of its movement through the air. 'Let *ab* be the heavy substance which though of uniform thickness and weight, being in a slanting position has more of its weight in its front face than any other part'. He then traces the reasons for its oscillating course of descent and comments on the figures of little men attached in the lower four figures thus: 'The air that has the swiftest movement moves most – it follows that the man can descend as shown below. This man will move to the right side as he bends the right arm and extends the left arm; and he will move from right to left by changing the positions of the arm' (G 74 r).

The drawing suggests gliding flight, but it would seem also justifiable to see in it a method of modifying descent obliquely through the air. However, this and the design in Figures 1.25 and 6.36 (pp. 33, 190) clearly anticipate Otto Lilienthal's attempts at flight in 1895, about 400 years later.

The last safety precaution for landing which we will mention consists of an ingenious device produced by Leonardo in his early inventive period. In Figure 6.39 he depicts a long iron bar split in two by a central wedge-shaped rod. 'This instrument', he writes, 'may be used to allow a man to fall from a great height. It is made this way: the man who supposedly falls should hold the iron *ab*. Iron *b* is thin much like your thumb. Iron *a* is 3 times thicker. In

falling, irons *ab* enter through guides *sh* into beams *cn* . . . Therefore as the weight of the falling man is secured to irons *ab*, and as these cannot reach the bottom *rf* because of its narrowness, the percussion perforce will be moderated since it is performed between oblique lines; and it is made slowly since the beams are necessarily spread out, and the more the beams open above at *cn* the more they close at the bottom, *qp*. As they close together they exert force upon a bale of wool which is squeezed by them in the manner of a pincer. Consequently the yielding wool will contribute towards moderating the blow, and thus will prevent the falling man from suffering any harm' (Madrid I 62 v).

This apparatus as applied to the percussion of a plane landing is more succinctly described in Figure 6.40b. Here in the drawing on the right Leonardo represents the force of percussion by a wedge, writing, 'Here I refer to the movement and the beginning of percussion'. On Figure 6.40a (B 89 r) he draws a ladder as a retractable undercarriage to be lowered on landing. On its right is the wedge device which is again drawn in Figure 6.40b besides which he writes, 'Here I refer to the end of the movement and the

Figure 6.39. A shock-absorber for a falling man (Madrid I 62 v).

This ingenious instrument was designed to reduce the impact of percussion of a falling man. Leonardo begins his description thus: 'On the nature of percussion, adducing the fall of Simon Magus and of someone who jumps on to the points of the toes'. Simon Magus was said by legend to have fallen from his car in the sky and broken both his legs. This apparatus is designed to prevent such events. Its mechanism is described in the text. Leonardo also used it to break the descent of a plane.

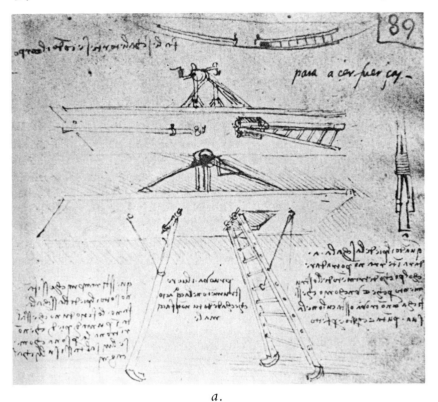

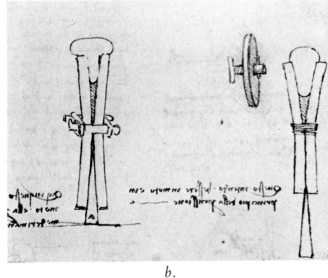

a.   b.

Figure 6.40.

a. Retractable undercarriage and assisted landing devices (B 89 r).

In the upper drawing a ladder and a beam can be seen drawn up and fixed to the underside of the wing. In the lower drawing this apparatus is lowered for landing. At the bottom of each support are claw-like structures intended to slide forward on the ground, so breaking the impact of landing the craft. In the right margin is another device consisting of a wedge inserted between two rods which are bound together. On landing the wedge is forced up between the rods at *a,* so breaking the impact of percussion.

b. The mode of action of the shock-absorber described above is clarified here.

On the right Leonardo writes, 'Here I refer to the movement at the beginning of the percussion'. Beside the left-hand drawing he writes, 'Here I refer to the end of the movement and the percussion; and I demonstrate how the wedge must be situated at *a,* which receives the blow' (Madrid I 63 r).

percussion; and I demonstrate how the wedge must be situated at *a,* which receives the blow'.

Helicopter flight (Figure 1.25a), glider flight (Figure 6.36) and flapping-bird flight – Leonardo studied them all as scientific possibilities. But lack of suitably light materials thwarted his practical efforts.

APPENDIX: LEONARDO'S CALCULATION ON MAN-POWERED FLIGHT

Leonardo's calcuations regarding the weight-lifting power of the muscles of the human body have been confirmed by the flight of Dr. Paul MacCready's Goassamer Albatross across the English Channel in 1979. 'The complete aircraft weighed only 70 lb, half the weight of Bryan Allen, the pilot, and required only 0.3 hp for flight at its optimum speed of 12 mph'. Bryan Allen was 137 pounds in weight, i.e., 171.25 florentine pounds. He was capable, therefore, according to Leonardo, of raising approximately 342 florentine pounds. The flying weight of the Albatross with the pilot was 215 pounds, i.e., 270 florentine pounds. Thus the weight lifted by Bryan Allen was well within that calculated by Leonardo. Even the chain-driven pedal-power used in this feat had been represented in Leonardo's notes by the drawing of a bicycle (CA, 1980 edition, 133 v).

The data regarding the flight of Gossamer Albatross were kindly sent to me by K. W. Clark, Secretary, Man Powered Aircraft Group, The Royal Aeronautical Society, London.

Chapter 7

Leonardo's Anatomical Approach to the Mechanism of the Human Body

As one looks back at the many different radiations of Leonardo's researches, their focus on the anatomy of the human body emerges as almost inevitable. His studies of perspective and mechanics, his interpretations of the circulation of the elements of the atmosphere and of the 'life' of the macrocosm of the earth, all seem to reflect and converge inexorably towards a pyramidal focal point, the body of man.

Leonardo appears to have come to feel this himself. Not only did he make his first sortie into anatomy about 1489, but in the long interval of the following years he also repeatedly reveals his intention of returning to the attack. On at least four occasions he expresses his conviction that the actions of human beings are produced by the 'four powers' (B 63 r; Forster II 116 v; BM 37 r, 151 r and v) with the obvious intention of investigating these actions in further detail. In the interest of brevity only the passages in *Codex Arundel* 151 r and v will be quoted here. 'Force arises from dearth of abundance; it is the child of material motion and the grandchild of spiritual motion, and the mother and origin of weight. Weight is finite, limited to the elements of water and earth. Force is infinite since by it infinite worlds might be moved if instruments could be made by which the force could be generated.

'Force with material motion, and weight with percussion are the four accidental powers on which all the actions of mortals have their being, and their death. Force has its origin in spiritual motion; and this motion flowing through the limbs of sentient animals enlarges their muscles, whence, being enlarged, the muscles become shortened in length and draw back tendons connected to them; and this is the cause of the force of human limbs. The quality and quantity of the force of a man can give birth to other force which will be proportionally so much the greater as the movement of the one is longer than that of the other'. On the following page Leonardo varies this statement, 'Weight and force together with material motion and percussion are the four accidental powers by which the human species in its marvellous and various actions appears in this world to be shown as a second world of nature, since by such powers all the evident actions of mortals have their being and their death'. He goes on to set out separate definitions of move-

ment, gravity and force, ending the page with a further description of how 'the spirit of sentient animals makes the limbs of their bodies move . . . by the shortening of the muscles drawing back the tendons'. In this context he once more asserts that in human limbs 'material movement is born from spiritual' (BM 151 v).

In the same way as Leonardo's experiments on perspective led him to study the anatomy of the eye and brain, so his detailed observations on human movements (outlined in Chapter 6) led him to investigate the anatomy of the human skeleton and muscular systems, and the mechanics of those movements.

Moreover, he had by 1506–1508 developed his theory and practice of the elements of machines in terms of the pyramidal action of the four powers on the four elements. It was Leonardo's conviction that by applying this knowledge to the human body he would be able to explain its feelings and movements. And by virtue of his mastery of perspective he felt capable of describing the human body by illustration rather than words.

For these reasons Leonardo's many discussions about 'demonstrating' human anatomy and physiology centre around two main points: how to illustrate the parts of this human machine like other machines so that their mode of action should be clear to those who look at his drawings; and how to demonstrate the mechanical principles underlying all the living movements from birth to death of this microcosm, the human body.

On two occasions Leonardo set out in detail plans for his books on anatomy, one in 1489, the other about twenty years later, in 1509. Quotation reveals his different outlook on the problems of anatomy at these two different dates.

In 1489 he wrote as follows:

'On the order of the book.

This work should begin with the conception of man and describe the form of the womb, and how the child lives in it, and to what stage it resides in it, and in what way it is given life and food. Also its growth, and what interval there is between one degree of growth and another, and what it is that pushes it out of the body of the mother, and for what reason it sometimes comes out of the mother's belly before its due time. Then I will describe which parts grow more than others after the infant is born; and give the measurements of a child of one year. Then describe the grown-up man and woman and their measurements, and the nature of constitution, colour and physiognomy. Then describe how they are composed of veins, nerves, muscles and bones. This you will do at the end of the book. Then in 4 drawings represent four universal conditions of men, that is joy with different ways of laughing, and draw the cause of laughter; weeping in different ways with their cause; fighting with the different movements of killing; flight, fear, ferocity, boldness, murder, and everything belonging to such events. Then draw labour, with pulling, pushing, carrying, stopping, supporting, and similar things.

Attitudes.

Then describe attitudes and movements.

Effects.

Then perspective through the function of the eye; on hearing I shall speak of music; and describe the other senses.

Senses.

Then describe the nature of the 5 senses' (W 19037 v; K/P 81 v).

It may be noted here that every one of these subjects is developed amongst his anatomical drawings or in his *Treatise on Painting*. This programme, apart from the development of the foetus and the mention of the veins, muscles and bones, is dominated by the outlook of Leonardo the artist—anatomist rather than Leonardo the scientist. The passage describing perspective combines both sides of his nature, for he follows his subject into the physiology of the eye, vision and the other senses. During the following years he analysed the physiology of vision and sensation in detail in such passages as those cited in Chapter 2 from W 19019 r (K/P 39 r) and CA 90 r.

Leonardo's plan for the arrangement of his book on anatomy is significantly different twenty years later. In W 19009 r (K/P 143 r) he writes, 'Arrange it so that the book on the elements of mechanics with its practice shall precede the demonstration of the movement and force of man and other animals, and by means of these you will be able to prove all your propositions'. These words are significantly placed in the middle of a page of inspired drawings of the anatomy and actions of the muscles and tendons of the hand. This view is endorsed on W 19060 r; K/P 153 r (Figure 17.16, p. 361). Here, alongside a drawing of the maternal and foetal circulation under the title 'On machines', he writes, 'Why Nature cannot give movement to animals without mechanical instruments is demonstrated by me in this book on the active movements made by nature in ani-

mals. And for this reason I have drawn up the rules of the 4 powers of nature without which nothing through her can give local motion to these animals. Therefore we shall first describe this local motion and how it generates and is generated by each of the other three powers. Then we shall describe natural weight, although no weight can be called other than accidental, it has been accepted to call it thus in order to separate it from force which in all its operations is of the nature of weight, and for this reason termed accidental weight. And this force is rated as the 3rd power of nature or any natural creature. The fourth and last power will be called percussion, that is, the ending or impediment of movement. And first we shall state that every insensitive local motion is generated from a sensitive mover, just as in a clock the counterpoise is lifted up by a man, its motor. Furthermore the elements expel or attract one another, for one sees water expelling air from itself, and fire, having entered as heat at the bottom of a cauldron escapes from it through the bubbles on the surface of the boiling water. Again, a flame attracts air to itself and the heat of the sun attracts water upwards in the form of moist vapour which afterwards falls down again, condensed as heavy rain. But percussion is the immense power of things which is generated within the elements'.

This important passage containing the germs of many of his physiological concepts necessarily remains almost completely devoid of meaning unless the reader comprehends what Leonard meant by his 'four powers' and their relation to the 'four elements'. It is for this reason that so large a part of this book has been devoted to elucidating the immense amount of scientific work which Leonardo performed in the years between his first suggested arrangement of his book on anatomy and this his second arrangement some twenty years later.

On the page following his comments on the mechanical instruments responsible for the movement of animals Leonardo describes once more the 'Order of the book'. He is concerned with demonstrating by illustration the working parts of the human body as clearly as he has for years been demonstrating the working parts of innumerable machines. This plan is so revealing that it cannot be fragmented without losing vital insights into Leonardo's approach to human anatomy and physiology. He writes:

'This my configuration of the human body will be demonstrated to you just as if you had the natural man before you. The reason is that if you want to know thoroughly the anatomical parts of man you must either turn him or your eye in order to examine him from different aspects, from below, from above and from the sides, turning him round and investigating the origin of each part; and by this method your knowledge of natural anatomy is satisfied. But you must understand that such knowledge will not leave you satisfied on account of the very great confusion that results from the mix-up of membranes with veins, arteries, nerves, tendons, muscles, bones and blood which itself dyes every part the same colour, the vessels emptied of this blood not being recognisable because of their diminished size. And the integrity of the membranes is broken in searching for those parts which are enclosed within them, and their transparency, stained with blood, does not let you recognise the parts covered by them because of the similarity of their blood-stained colour, and you cannot get a knowledge of the one without getting confused and destroying the other.

Therefore it is necessary to perform more dissections; you need 3 to acquire a full knowledge of the veins and arteries, destroying with the utmost diligence all the rest; and another 3 to obtain knowledge of the membranes; and 3 for the tendons, muscles and ligaments; 3 for the bones and cartilages; and 3 for the anatomy of the bones which have to be sawn through in order to demonstrate which is hollow and which not, which is full of marrow and which spongy, and which is compact from the outside inwards and which cancellous. And some are very fine at one part and compact at another, and in one place hollow or filled with bone or marrow or spongy. Thus all these things will sometimes be found in the same bone, and another bone have none of them. And 3 need to be made of a woman, in whom there is great mystery on account of her uterus and its foetus.

Therefore through my plan you will come to know every part and every whole through the demonstration of 3 different aspects of each part, for when you have seen any part from the front with nerves, tendons and veins which arise from the side in front of you, the same part will be shown to you turned to its side or its back, just as though you had the very same part in your hand and went on turning it round bit by bit until you had obtained full knowledge of what you want to know. And so in a similar way there will be placed before you 3 or 4 demonstrations of each part from different aspects in such a way that you will retain a true and full knowledge of all that you want to know about the configuration of man.

Therefore the cosmography of the lesser world [microcosm] will be shown to you here complete in 15 figures in the same order as was used by Ptolemy before me in his Cosmography. And therefore I shall divide the parts as he divided the whole, into provinces, and then I shall describe the functions of the parts from each side, placing before your eyes the knowledge of the whole shape and strength of man in so far as he has local movement by means of his parts.

And would that it might please our Author that I were able to demonstrate the nature of man and his customs in the way that I describe his shape' (W 19061 r; K/P 154 r).

On this same page in the margin Leonardo adds a note in which he applies these principles of anatomical demonstration to a particular part, the hand. Here he describes with his usual diligent detail ten systematic demonstrations of the bones, ligaments, muscles, tendons, nerves, veins and arteries, the tenth being 'the whole hand complete and finished with its skin, and its measurements; and these measurements should also be made of its bones. And what you do from this side of the hand do also from the other

three aspects, that is from the inner side, from the dorsal side, and from the outer side and from the aforesaid [palmar] surface. Thus in the chapter on the hand forty demonstrations will be made and you should do the same for each limb. And in this way you will present full knowledge'.

A few pages later in the same folio Leonardo summarises his anatomical methods, now apparently tested by experience:

'And you who say that it is better to see an anatomy performed than to see these drawings would be right if it were possible to see all these things which are demonstrated in these drawings in a single figure. In this with all your ability you will not see nor obtain knowledge of anything but some few vessels. In order to obtain a true and full knowledge of which I have dissected more than ten human bodies, destroying all other organs and taking away in its minutest particles all the flesh which was to be found around the vessels without causing them to bleed except for the imperceptible bleeding of the capillary vessels. And one single body was not sufficient for enough time, so that it was necessary to proceed little by little with as many bodies as would render a complete knowledge. This I repeated twice in order to observe the differences.

And though you have love for such things you will perhaps be hindered by your stomach; and if that does not impede you, you will perhaps be impeded by the fear of living through the night hours in the company of quartered and flayed corpses fearful to behold. And if this does not impede you perhaps you will lack the good draughtmanship which appertains to such representation; and even if you have the skill in drawing it may not be accompanied by a knowledge of perspective; and if it were so accompanied, you may lack the methods of the geometrical demonstration and methods of calculating the forces and strength of the muscles; or perhaps you will lack patience so that you will not be diligent. Whether all these things were found in me or not, the hundred and 20 books composed by me will give their verdict yes or no. In these I have been impeded neither by avarice or negligence but only by time. Farewell' (W 19070 v; K/P 113 r).

By putting these different arrangements together one obtains a clear idea of Leonardo's approach to anatomy. The earlier plan shows the dominance of the artist intent on depicting the forms of man shaped by time at all ages, his emotional expressions, and proportions based on an understanding of perspective gained from experience. Visual experience itself receives extended anatomical attention.

The later arrangement laid down some twenty years later brings out Leonardo the scientist–engineer, intent on expressing the mechanical principles of the human machine by the diligent application of his knowledge of perspective, now combined with the four powers of nature and the geometry of their mechanics.

As he proceeded with this vast enterprise, Leonardo laid down numerous subsidiary plans to meet the many different kinds of problem which arose. We have already glanced at one such plan, that designed to exhibit 'full knowledge' of the hand. For full knowledge of other parts such as the shape of the cerebral ventricles and heart he used injection methods. Throughout his anatomical investigations Leonardo never ceased to devise plans for demonstrating the perspectival proportions of the parts as the basis for displaying their mechanical actions. His anatomy and physiology continually develops as he endeavours to answer these intelligently planned investigations. As his knowledge increases it gives rise to further questions with further plans for their solution.

Almost by definition such a process was bound to remain continuously unfinished and therefore unfit to be crystallised into that form called a publication. Indeed,

one of Leonardo's last anatomical notes commences thus: 'Definitions of the instruments.
Discourse on the nerves, muscles, tendons, membranes and ligaments'.
Under this heading he writes a page of notes on the mechanical functions of these 'instruments'. The whole page is crossed through with red ink, and in the margin Leonardo makes a further note, 'Define all the parts of which the body is composed, commencing with the skin with its superficial layer' (W 19087 v; K/P 175 v). That this should be nearly his last word on anatomy after some twenty-five years of study producing 120 books on the subject casts a vivid light on both Leonardo's personal outlook and the tragic fate of his unfinished labours.

How much dissecting did Leonardo perform personally? This question has provided good ground for petty scholarly altercation, even to the ridiculous contention that Leonardo

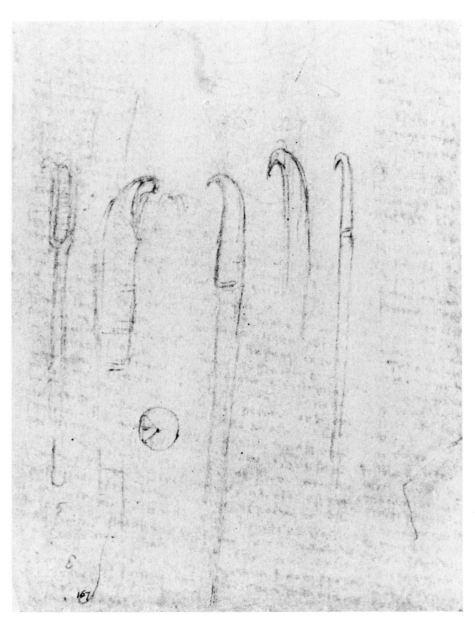

Figure 7.1. Leonardo's dissecting instruments (W 19070 r; K/P 113 v).

On the same sheet as he describes in detailed notes his technique of anatomical dissection Leonardo draws his instruments: his bistouries, hooks and a retractor.

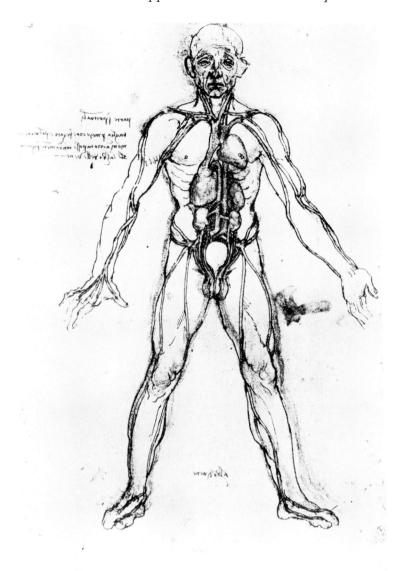

Figure 7.2. The vascular tree superimposed on a drawing
in *Ketham's Fasciculus* (W 12597 r; K/P 36 r).

himself did not perform any dissections – and this after
blindly studying his illustrations of such limb movements
as supination of the hand with flexion of the elbow by the
biceps, depicted with painstaking accuracy by Leonardo for
the first time.

It is significant, too, that Leonardo made drawings of his
dissecting instruments in Figure 7.1. These, interestingly,
are on the same folio as that on which he writes out a
detailed description of his methods for obtaining 'a true and
full knowledge' of the vessels 'to obtain which I have dissected more than ten bodies . . . This I repeated twice in
order to observe the differences'. It is also on this folio that
Leonardo states that he has composed 120 books on anatomy and enjoins himself, 'Get your books of anatomy
bound'.

About the same time (1508–1510) Leonardo listed his
dissections of '3 men finished; 3 with bones and veins; 3

with bones and nerves; 3 with bones alone. These are 12
demonstrations of complete representations' (W 19023 v;
K/P 65 v).

The latest reference to Leonardo's anatomy during his
lifetime comes from the diary of Antonio de Beatis, secretary to the Cardinal of Aragon who visited Leonardo at
Amboise in October 1517. 'This gentleman', writes Antonio de Beatis, 'has written of anatomy with such detail
showing by illustrations the limbs, muscles, nerves, veins,
ligaments, intestines, and whatever else there is to discuss in
the bodies of men and women, in a way that has never yet
been done by anyone else. All this we have seen with our
own eyes; and he said that he had dissected more than thirty
bodies both of men and women of all ages'. This brief
description literally describes the extent of Leonardo's personal dissections.

Since Leonardo mentions the great fourteenth-century

anatomist Mondino by name twice, it is clear that he knew of *Anatomy*, a great textbook written in 1316. This, because it contained the results of some original dissection, remained the classic textbook of practical anatomy until replaced by the *De Humani Corporis Fabrica* of Andreas Vesalius in 1543.

There is much evidence that Leonardo used the Italian version of Mondino's *Anatomy* as a dissecting manual. Indeed, it would appear that in most of his dissections of the abdominal viscera obtained from 'the old man' who died in the hospital of Santa Maria Nuova in Florence, Leonardo followed the instructions for dissection to be found in Mondino's *Anatomy*. This is suggested by the mistakes he makes in representing the abdominal muscles; by his description of the omentum; by his method of removing the intestines; and throughout the text to these dissections Leonardo uses Mondino's terminology. The evidence becomes overwhelming when the blood supply of the stomach, liver, spleen and intestines is illustrated by Leonardo. Brilliant as these are in demonstrating the blood vessels to the stomach, liver and gallbladder, never before illustrated (even down to

such vessels as the left and right gastro-epiploic arteries), all these are mentioned by Mondino, and errors such as that of showing the portal vein joining the splenic vein at the 'middle of its passage' can arise only from Leonardo putting too literal an interpretation on Mondino's guiding words (W 19031 v; K/P 73 v). Later he corrects his error.

That Leonardo used Mondino only as a point of departure is obvious from other dissections, for example, those of the skeleton, muscles and nervous system. But in the matter of the anatomy and physiology of the abdominal viscera his views in large degree remained determined by his dissection of 'the old man' under Mondino's guidance.

The Italian edition of *Ketham's Fasciculus* of medical writings containing Mondino's *Anatomy* elsewhere contains a crude illustration of a Vascular Man which bears a striking resemblance, particularly in posture and facial expression, to Leonardo's Vascular Man W 12597 r; K/P 36 r (Figure 7.2). Moreover, the erroneous uterine 'horns' and other uterine appendages so puzzling a feature of the great double sheet Figure 17.9 (p. 355) are described by Mondino and also illustrated in *Ketham's Fasciculus* of 1493.

Chapter 8

Leonardo's Anatomy of the Eye and Physiology of Vision

In Chapter 2 we described Leonardo's construction of the physics and physiology of perspectival vision. We described, too, how he traced out a perspectival relationship between the object seen and the soul's subjective visual perception and how he constructed a hypothetical process of visual 'experience' extending from the luminous object to the *imprensiva* and *senso comune*. These ideas were worked out relatively early. It was not in Leonardo's character to leave the matter without further later investigation. Unfortunately, this did not lead him further towards a valid theory of optics. One of the reasons for this was his failure to reach a precise picture of the structure of the eye. In this he was hindered by his way of applying his perspectival theory to the physiology of the eye.

We shall first see how this came about and then describe how, in spite of it, Leonardo made important advances in the physiology of vision.

Leonardo's return to the study of the eye is noted for the most part in the Windsor Drawings and in MSS D and K. A number of late optical notes are also to be found in the *Codex Atlanticus*. These all date from about 1508 onwards.

Leonardo's errors arose largely from his discovery of the variations in the size of the pupil with light, as described in *Forster* II 158 v. Here, under the heading 'Perspective', he writes, 'When the eye turns away from a white object illuminated by the sun and goes to a place where there is less light everything there will appear dark. And this happens because the eye that rests on the white illuminated object comes to contract its pupil in such a way that if there were at first a certain visual quantity more than 3/4 of that quantity will be lost, and what is lacked in quantity will also be lacking in power. Although you could say to me that a small bird must see in proportion very little through its small pupils so that the white would appear black, to this I would reply to you that here we are paying attention to the proportion of the mass of that part of the brain devoted to the visual power and to nothing else. To return, this pupil of ours expands and contracts according to the brightness or darkness of its object; and because this expanding and contracting take some time it does not see at once as it emerges from the light into the shade, nor similarly from shade into light. And this deceived me once when painting an eye, and

from it I learnt'. Leonardo wrote this some time between 1495 and 1497.

It will be noticed that Leonardo here attributes the power of seeing to the size of the pupil. This idea persists throughout all his later optical studies. So thrilled was he to discover that the pupil by its alteration in size automatically regulates the amount of light allowed to enter the eye that he ascribed to it the main regulating power of vision.

He even goes so far as to correlate the apparent size of an object with the size of the pupil. 'That pupil', he writes, 'which is largest will see the object largest. This is evident when looking at luminous bodies particularly those in the sky. When the eye comes out of darkness and suddenly looks up at these bodies they at first appear larger and then diminish' (H 88 r). 'Every object we see will appear larger at midnight than at midday . . . This happens because the pupil of the eye is much smaller at midday than at any other time' (H 86 r); the larger its diameter the larger the image.

This false idea persists; it is repeated about 1498 in MS I 20 r and explained in detail in MS D 7 r under the heading 'How the larger pupil sees the object larger'. It is evident that Leonardo never mastered the consequences of his discovery of the variations in the size of the pupil with light, and he failed to observe that the pupil contracts with accommodation.

He found a happier field of observation in night vision, a subject he pursued by comparing the sizes of the pupils at night in man, birds and cats. He was acutely aware that light itself, however dim, is necessary for vision. Animals which see during the night do so by virtue of being able to use less light than the eyes of men. 'I say that sight is exercised by all animals through the medium of light; and if anyone adduces against this the sight of nocturnal animals I would say that this similarly is subject to the same law of nature . . . If you choose to say that there are many animals that prey at night I answer that when the small amount of light sufficient for the nature of their eyes is lacking they avail themselves of the powers of hearing and smell which are not impeded by the darkness and which are far superior to man. If you make a cat leap in daylight among a lot of pieces of crockery you will see them remain unbroken, but if you do the same at night many will be broken. Night birds do not fly unless the

Figure 8.1. The size of the pupil in relation to the visual power (Madrid II 25 v).

In the upper two diagrams the processing station, called by Leonardo the *'imprensiva'*, is represented by the circles labelled *a*. Their proportion to pupil size is emphasized. How two eyes convey more light than one is further illustrated in Figure 8.2. The lowest diagram correlates the size of the pupil with the amount of light at night and during daylight. For Leonardo's comments, see text.

moon is shining either in full or in part; rather do they feed between sunset and the total darkness of night' (CA 90 r).

To emphasise his point Leonardo returns to his concept of the cerebral localisation of vision in different animals. Several such passages are to be found, the most interesting of which are on CA 262 rd, Madrid II 25 v and Figure 8.1 (c. 1504), MS D 5 r, Figure 17.3 (p. 347) (c. 1508), and Figure 8.3 (c. 1513), in which he compares the pupils of man, birds and lions.

His passion for quantitative geometrical expression leads Leonardo to try to express the relation between pupil size and visual power in terms of a circle. The most revealing of these attempts occurs on Madrid II 25 v (Figure 8.1), where he develops his quantitative thesis as follows. Beneath the title 'human *imprensiva*' he draws a circle, *a*, illuminated by two eyes, *b* and *c*. Beside this figure he writes, 'The light seen by one eye has half less power and magnitude as light seen by two eyes'. This statement is clarified by an explanation on Madrid II 24 v (Figure 8.2), where Leonardo continues, 'Again, about the case of two eyes giving more light than one to our *imprensiva;* square *cdef* shall be constructed like a room enclosed on all sides into which the peep-hole [*spiracolo*] *a* will be first made. This will illuminate the entire wall *cd* with one degree of light. Then I will make the peep-hole *b*, which without doubt will double the quantity of the light on the aforesaid wall *cd,* because it has acquired a second degree of light in addition to the first one'.

To return to Figure 8.1, Leonardo now elaborates his assertion about the eyes *b* and *c* as follows: 'Proof; let *a* be the *imprensiva* to which the eye conveys [*conferriscie*] its luminous objects. I say that *b* illuminates by only one degree of light this *imprensiva*. Adding *c* this *imprensiva* receives 2 degrees of light. And since 2 degrees of light are in double proportion to one degree we find this *imprensiva* to be doubly illuminated by two lights, and by a hundred [lights] a hundred times more. That place will be more illuminated which is percussed by the greater amount of light. And likewise it will be less illuminated if it is seen by less light'.

Alongside this passage in Figure 8.1 Leonardo has drawn a diagram which is entitled 'the *imprensiva* of the owl'. Here the *imprensiva* is represented by a small circle labelled *a*, towards which two broad cones of light converge from two larger circles, within each of which he writes, 'nocturnal pupil'. The diagram is deliberately made to illustrate the contrast between the relatively large human *imprensiva* in the drawing above, with the small pupils *b* and *c*. The comparison is carried over into the lowermost diagram. Here the pupils are represented by the three concentric circles, *a*, *b* and *c*. Leonardo labels *a* 'light of the sun'; *b*, he labels 'light of day'; *c* he labels 'nocturnal light which the pupil [*luce*] of the owl receives'. The other eye is represented by the lower set of concentric circles correspondingly labelled *c, e, d*. Leonardo explains these figures as follows: 'From an equal distance a small and powerful light will illuminate a place as much as a very large and weak light which is that much

Figure 8.2. The amount of light conveyed by one and two *spiracoli* (Madrid II 24 v).

This experimental application of the camera obscura to the amount of light admitted into the eye by the pupil is described in the text.

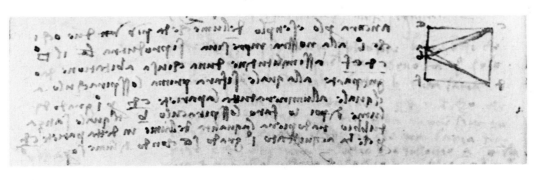

Figure 8.3. 'The eyes of animals' (G 44 r).

This series of small drawings is inserted by Leonardo in a long passage describing the eyes in animals. 'The eyes of all animals have their pupils adapted to increase or diminish of their own accord in proportion to the greater or less light of the sun or other brightness'. He particularly emphasizes that it is the area of the pupil which is affected that is important, not the shape. He points this out in relation to animals 'of the lion species' in which 'the pupil diminishes from the perfect circle to the shape of an oval slit like that shown in the margin'.

It is in this context that Leonardo first shows his awareness of an inverse square form of his pyramidal law. Similarly proportioned circles are drawn in *Codex Arundel* 78 v (see p. 157 and Figure 5.29, p. 154).

weaker as it is larger than the first. Therefore the very small light of the night being large in quantity [area]* will be equal to a small but powerful quantity of daylight. No place is at night completely deprived of light but it seems dark to diurnal animals. Every pupil is continually changing into different sizes in different quantities of light or shade. But the owl, more than any other animal makes the greatest change of all. Men overwhelmed by excess of light almost shut their eyes, cutting off with their eyelids part of the pupil after the pupil by itself has diminished in size as much as it can'.

This correlation between the size of pupil and power of vision is expressed even more clearly quantitatively in *Codex Atlanticus* 262 rd: 'If the eye of the owl has its pupil increased a hundred times in darkness it increases its visual power a hundred times . . . and because things which are equal do not overcome one another the bird sees in the darkness with its pupil increased a hundredfold as it does in the day with the pupil diminished by ninety nine parts in the hundred' (see Figure 8.3). In this same passage he concludes that the geometrical shape of the pupil is not relevant to visual power; it is the size (area) that matters. 'Of the nocturnal animals only the feline species changes the shape of its pupil as that increases or diminishes, but though when at its extreme of diminution this is long in shape, halfway it is oval, and when fully dilated it is circular in shape'.

THE RESUMPTION OF ANATOMICAL RESEARCH ON THE ANATOMY OF VISION

Though those views were derived from his early anatomical researches about 1489, Leonardo retained the views just described to the very end. They were recapitulated in MS G 44 r about 1513. Nevertheless Leonardo returned to further anatomical exploration in this sphere about 1508.

We now find further anatomical verification of his view that the lion has greater visual power than man. 'The eyes of the leonine species have a large part of the head for a receptacle and the optic nerves are immediately joined to the brain. With man one observes the contrary, for the orbital cavities of the eyes are only a small part of the head and the optic nerves are thin, long and weak' (W 19030 v; K/P 72 v).

He goes on to explain how this 'weakness' of human vision is related to the fact that man's pupil varies only by the proportion of two to one, whereas 'nocturnal animals' such as the owl have pupils which vary by as much as a hundred times.

By 1508, when he reviewed and summarised his physiology of vision, Leonardo had discovered the approximate shape of the lateral ventricles of the brain (see Figure 2.27, p. 65) and had labelled this '*imprensiva*'. Moreover, in the geometrical sphere of knowledge he was now aware that the area of a circle is proportional to the square of its diameter (see Figure 5.29) (Euclid, Book XII, Proposition 2). His discussion on the cerebral localisation of vision therefore changes its terms. He writes, 'The nocturnal animal can see more by night than by day. This happens mainly because there is a greater difference between increase and diminution of their pupils than in diurnal animals. For while the pupil of man doubles the diameter at night which means that it is four times the daytime size, the diameter of the pupil of the owl increases 10 times that of the daytime which means it is 100 times that of the day. Furthermore the ventricle located in the human brain called the *imprensiva* is more than 10 times the whole human eye of which the pupil, whence sight arises, is less than a thousandth part; whilst in the owl the pupil at night is considerably larger than the ventricle of the *imprensiva* placed in the brain. Whence in man there is a greater proportion of the *imprensiva* to the pupil, which is not so in the owl where it is about equal; and this *imprensiva* of man compared with that of the owl is like a great hall receiving light through a little hole compared with a little room entirely open. And in the great hall there is night at noon, and in the little open one there is daylight at midnight if the weather is not cloudy. And of this more potent causes will be demonstrated by anatomical dissection of the eyes and the *imprensiva* of animals, that is man and the owl' (D 5 r).

One of Leonardo's extant anatomical drawings corresponds to this concept of the anatomy of vision, that in Figure 8.4. Here are two small sketches of the optic nerves, chiasma and tracts, running from the eyeballs to the brain which they appear to join high up under the olfactory tracts. To the right and below is a sketch of the cerebral ventricles. Both olfactory and optic tracts, as well as the trigeminal nerves, are depicted as entering the unpaired anterior ventricle before it divides into its two lateral horns. The *imprensiva* here is in a position to receive the waves of percussion of

---

*Leonardo often uses the word *quantita* to indicate area in space.

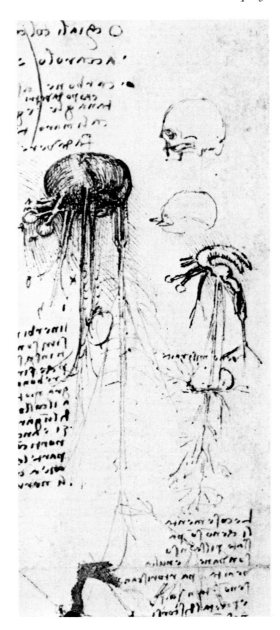

Figure 8.4.   Optic tracts and olfactory nerves ending in the *imprensiva* (W 19070 v; K/P 113 r).

To the left in a sketch of the brain and cranial nerves the optic tracts are shown running to the base of the brain. To the right their destination is shown to be to the anterior ventricle, where the *imprensiva* is located. Here visual, olfactory and other sensory impulses were, in Leonardo's view, processed and relayed to the *senso comune,* located in the third ventricle.

light transmitted from the eyes, and to relay them to the *senso comune* in the middle (third) ventricle. Leonardo located the extrinsic visual power (*virtu visiva*) in the front of the optic nerve. This is percussed by the images (*spetie*) and refers them via the *imprensiva* to the *senso comune* in the same way as do the nerves for smell (olfactory nerves). This whole process, in Leonardo's view, is automated by the action of the 'worm' of the choroid plexus, acting as a

sphincter between the *imprensiva* and the *senso comune* and *memoria* (W 19117 r; K/P 115 r). Here he locates what he calls the 'intrinsic power of vision' (D 10 v), as opposed to the extrinsic power in the optic nerve.

That practical anatomy was much in Leonardo's mind at the time of writing this folio (W 19070 v; K/P 113 r) is shown not only by the long passage describing anatomical investigation and his drawings of anatomical instruments on its recto but also by his reminder once more that 'the book on the science of machines goes in front of the book on physiological functions [*giovamenti*]'; and 'get your books on anatomy bound', he adds.

Evidence that Leonardo's dissections of the eyeball and orbit were incomplete is suggested not only by the gaps in extant drawings but also by his own ingenious suggestion made about 1509: 'In the anatomy of the eye in order to see the inside well without spilling its humour one should place the whole eye in white of egg, make it boil and become solid, cutting the egg and the eye transversely in order that none of the middle portion may be poured out' (see Figure 8.5a). Had Leonardo used this technique successfully, it is unlikely that he would have persisted in drawing the lens as round and central. Either we have lost such drawings or the suggestion was made too late to be used. That his programme for dissection of the eye was in fact never completed is confirmed by a remark made even later (about 1513) in MS G 44 r (see Figure 8.3, p. 203), where he ends his description of differently shaped pupils with the reminder, 'Make anatomies of different eyes and see which are the muscles which open and close the aforesaid pupils of animals' eyes'.

THE CONTENTS OF MS D

About 1508, when he was revising his notes as a whole, Leonardo reviewed and summarised his notes on vision. The small MS D is similar in its structure and brevity to the little *Codex on the Flight of Birds*. In addition to the physiological ideas on vision already described, this manuscript contains passages casting much light on Leonardo's experimental approach to the problem.

Like so many of Leonardo's abortive treatises, this one begins with a startling generalisation, applying his four powers to the power of vision (*virtu visiva*). 'Nature does not give uniform power to the *virtu visiva;* she gives this *virtu* a power which is so much the greater as it approaches its centre. She does this in order not to break the law given to all other powers which become more and more powerful as they approach the centre. One sees this in the action of percussion by a body, and in the supports of the arms of a balance . . . one sees it in columns, walls, pillars; one sees it in heat and in all other natural powers' (D 1 r).

Recalling Leonardo's work on the four powers described in the first part of this book, we realise that he is prefacing his discourse on vision by a statement that the power exerted along the central line of any force such as gravity, percussion, etc., is greater than that along all other more

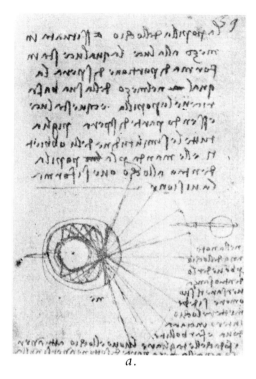

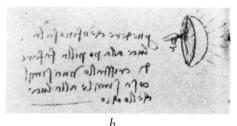

b.

Figure 8.5. Dissection of the eyeball (K 119 r).

a. To the left a hypothetical pathway of rays of light is drawn. It enters through the cornea and pupil and is reflected from the uvea back to the optic nerve; In this sketch it does not pass through the lens. Leonardo abandons this possible pathway.

The small sketch on the right shows a scalpel cutting an eyeball transversely. The eyeball has been previously hardened by boiling in white of egg. The script below is translated in the text.

b. Besides this drawing of the cornea Leonardo writes, 'In order to see the function of the cornea with regard to the pupil have a similar object made of crystal the same shape as the cornea of the eye' (K 118 v).

a.

peripheral lines. Thus the central line of vision is more powerful than any peripheral line (see Figures 2.33 and 8.6).

He then proceeds straight to the question, 'Why Nature made the cornea convex, protruding like part of a ball. Nature made the surface of the opening of the eye [cornea] convex in order to allow surrounding objects to imprint their images at the greatest possible angles' (D 1 r). Hitherto Leonardo had regarded the optic globe of the eye as a system of concentric spheres, as depicted in Figure 2.20 (p. 61). Even in MS D a number of figures show it so. But in Figure 8.7 (p. 206) he shows the curvature of the cornea protruding from the rest of the eyeball, its centre of curvature being on the anterior surface of the lens. This prominence of the cornea is of such importance that Leonardo gives special attention to it in Figures 2.37, 8.5b, 8.11 and 8.13 (pp. 75, 205, 209, 212).

In Figure 2.37 describing how the images of objects come to the eye, Leonardo depicts how 'the eye can see objects both at the sides and behind itself'. 'The eye can see, though it cannot recognise, the movement of two lights on a wall whilst the man who is observing leans with the back of his neck against that wall. This is because . . . the cornea is more prominent than any other part of the eye' (D 8 v). Similar visual fields are shown in Figure 8.11. In Figure 8.5b he adds his skill in model-making to solve this problem. 'In order to see the function of the cornea with regard to the pupil, have a similar object made of crystal the same shape as the cornea of the eye' (K 118 v). He recognises that refraction of a light ray passing from the 'rare' medium, air, to the 'dense' medium of the cornea enables this extraordinarily wide visual field to be comprehended. He tests this experimentally with his crystal model of the cornea (Figure 2.39,

Figure 8.6. Reflection of light from the eyelids (D 9 v).

In these drawings light from a luminous object is shown entering the eye from the front in three parts; the central ray enters straight into the pupil; the upper and lower rays are reflected into the pupil from the lower and upper lids, respectively, with their contained moisture.

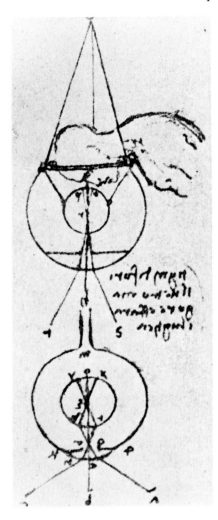

Figure 8.7. The hypothetical experiment of placing the experimenter's eye where the optic nerve should be in a model eye (D 3 v).

As explained in the text, Leonardo proposes to make a model eye, putting his own in the position of the optic nerve under water. The course of the rays of light is drawn in the sketch below. It will be noticed that the rays cross twice, once behind the pupil and once in the lens; all except the central ray.

p. 75) and confirms that only the 'central rays' convey perfectly light and colour. Struggling with the obstinate problem presented by the failure of 'brightness' to conform to the law of perspective with distance, Leonardo proposes in Figure 8.6 its solution by the formation of a watery concave layer between the upper and lower lids, supplementing the central ray by reflecting light into the pupil (D 9 v).

A number of optical experiments are reported in MS D. In Figure 2.16 (p. 56) Leonardo, using a little hole (*spiracolo*) made in a thin sheet of iron, brilliantly illustrates the principle of the camera obscura. 'If you get these images on a sheet of white paper . . . you will see all the objects with their own shapes and colours, but smaller and upside down'. This inversion of the image was his prime reason for postulating that the crystalline lens re-inverted the images which then reached the expanded optic nerve right way up.

He also tested this postulate experimentally in Figure 8.7 by viewing an object through a globe filled with water containing a suspended 'ball of thin glass' to represent the lens. Fitted at the front was the side of a box with a small hole in the middle to represent the pupil; the water filling the whole apparatus represented the 'humours' of the eye, and his own eye was placed in the position of the optic

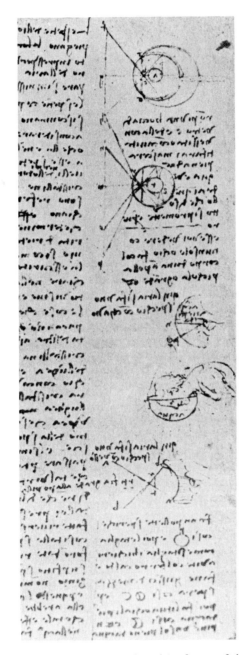

Figure 8.8. Water-lenses placed in front of the eye.

Beneath another drawing of a head held under a convex lens of a bowl of water Leonardo draws a concave lens or glass 'shell' filled with water with an eye behind it. Below this he writes, 'Make small spherical ampullas like this ⟨⟩ and then cut them as glasses are cut in a spiral way with a hot iron and make shells out of each hemisphere thus ◯) ; and then fill your glasses with water thus ⅅ filling one only completely full' (D 7 v).

nerve. 'Such an instrument', wrote Leonardo, 'will bring the images to the eye just as the eye brings them to the visual power'. His diagram shows the two inversions so produced in order to re-erect the image (D 3 v).

On D 7 v (Figure 8.8) Leonardo depicts experiments in which he placed glass 'shells' of various shapes in front of the eye, filling them with water or other clear fluids. He thus constructed the equivalents of convex and concave spectacles. The results are not given.

BINOCULAR VISION

Problems of the judgement of distance interested Leonardo from his earliest investigations of perspective of 1492. We have described previously (pp. 45–49) how he divided monocular judgements of distance into linear perspective, colour perspective and the disappearance of detail

with aerial perspective. He gave equally close attention to distance judgements based on binocular vision. Many experiments were performed on objects placed at various distances from the nose outwards. Objects of all sizes from those smaller than the pupil to mountains in the far distance were systematically observed and considered from the point of view of his theory of the physiology of vision. A few examples of these experiments will illustrate this work.

In Figure 8.9a, beside a clear diagram of binocular vision, Leonardo writes, 'How one and the same object is clearly comprehended when seen with 2 concordant eyes. These eyes refer it to one and the same point inside the head, as appears in *m, n, o, p*. But if you displace one eye [with the finger] you will see one object converted into 2' (CA 204 rb). In Figure 8.9b this experiment is repeated and explained at length (W 19117 r; K/P 115 r). Clear vision is achieved when the pyramidal lines from both eyes meet at

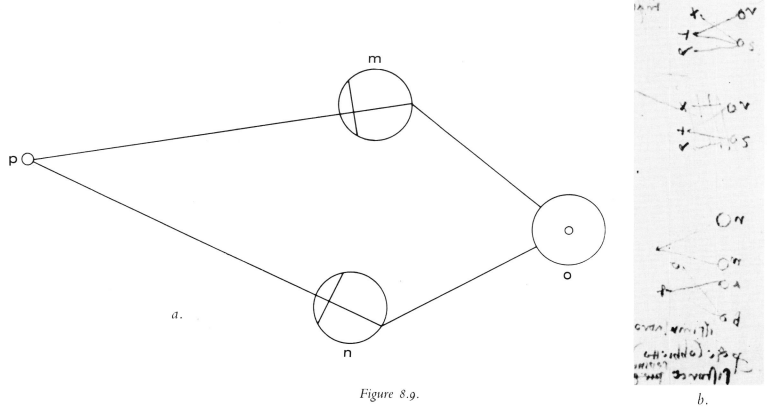

Figure 8.9.

a. Binocular vision; note the convergence of the eyes on the object *p* (CA 204 rb).

Diagram from Leonardo's rough sketch shown in Figure 2.34 (p. 71), in describing which he mentions the effects of displacing the eyeball.

b. The effects of displacing the eyeball.

Below this series of sketches Leonardo refers to the top one as follows: 'Why the object seen by the eye which is turned by being pushed sideways by a finger in any direction makes a movement contrary to that made by the eye. This seems to arise as the figure above shows, that is, if the concourse of the two central lines *rt* and *st* terminate in the angle *t*, and the non-central lines *sx* and *sv* remain apart, and if you push the eye *s* downwards, you displace the master line *st* from its place into which moves the line *sx*. And because the object moves as much towards the line as the line towards the object, therefore moving the line *sx* towards *t*, the object, it appears that this object *t* moves to the position *x* where the line *sx* is' (W 19117 r; K/P 115 r).

The sketches on page 208 show how this complicated figure was constructed.

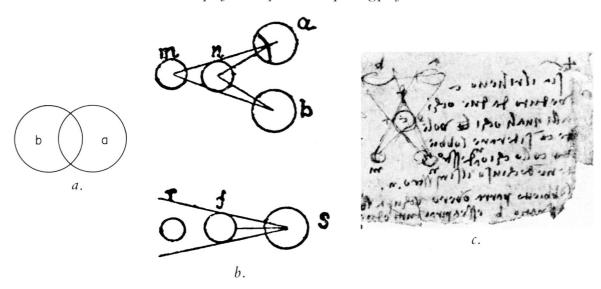

Figure 8.10.   Stereoscopic vision.

a. Overlapping monocular visual fields (I 43 r).

Leonardo becomes keenly aware of the significance of this overlap. In the *Treatise on Painting* (TP I 487) he points out that an object such as a finger held in front of the face can become 'transparent' if one focuses on the background – an experiment one can perform for oneself.

b. The contrast between binocular and monocular vision (A 90 r).

If the object *n* is close to the face the two eyes can see around it, so perceiving the object *m* behind it. With monocular vision, as shown in the lower drawing, this is impossible.

c. The visual fields of each eye shown separately when looking at an object, *c* (W 19102 r; K/P 198 r).

With both eyes open this object takes on the stereoscopic impression of depth as defined by the field *erf*.

the point where the object is placed. 'If the object is placed too near to the eye it cannot judge it well, as happens to a man who tries to see the tip of his nose. Hence as a general rule Nature teaches us that an object can never be seen perfectly unless the space between it and the eye is equal at least to the size of the face' (CA 138 vb).

Leonardo is aware of the effect of overlapping visual fields, saying, 'No opaque body of spherical shape seen by 2 eyes will ever be shown to be of perfect rotundity. Let *a* be the situation of your right eye, *b* the situation of the left. If you close your right eye you will see the spherical body round the centre, *a;* if you close the left eye then the said body will surround the centre, *b*' (I 43 r; Figure 8.10a).

Applying his awareness of the effects of binocular vision to looking at paintings, he explains, 'Why a painting can never appear detached as natural objects do . . . It is impossible that painted objects should appear in such relief as to resemble those reflected in a mirror, although both are seen on a flat surface, unless they are seen with one eye. And the reason is that two eyes see one object behind another as *a* and *b* see *m* and *n; m* cannot exactly occupy the space of *n* because the base of the visual lines is so wide that the second body is seen beyond the first. But if you close one eye as at *s* the body *f* will conceal *r* because the visual line proceeds from a single point, and makes its base in the first body

whence the second of the same size can never be seen' (A 90 r; Figure 8.10b). 'Objects seen with both eyes appear rounder than with one' (H 49 r) sums up this description of stereoscopic vision.

Leonardo's understanding of stereoscopic vision is perhaps best expressed on W 19102 r; K/P 198 r (Figure 8.10c): 'Let the object in relief, *c*, be seen by both eyes [*m* and *n*]. If you will look at the object with the right eye, *m* [*sic*], keeping the left eye shut, the object will appear to fill up the space at *a;* if you shut the right eye and open the left [*sic*] the object will occupy the space *b*. And if you open both eyes the object will no longer appear at *a* or *b* but the space *erf*'.

Leonardo extended his grasp of stereoscopic vision into the field of physiological diplopia (double vision), observing that if you focus your eyes on a near object another object further away will appear double. The simplest example is described in Figure 8.11a, where two eyes, *s* and *r*, are shown focussing with their central lines on a far object, *x*, and a nearer object, *t*. If each eye is closed alternately the central line of each shifts first to one side, *v*, then to the other, *y* – 'The right eye sees with the right adherent line and the left with the left adherent line'. Thus three images of two objects are seen when both eyes are open. This phenomenon is clearly described in Figure 8.11a and elaborated in Figure 8.11b. On this page Leonardo draws sections of the base of the aorta as well as the cornea projecting from the globe of the eye with its far-flung visual field. But of most importance to our present theme is the series of drawings down the right side of the page; the upper three show the two eyes focussing on two objects at different distances; the fourth shows the eyes looking at a string of five objects in a vertical row at different distances from the eyes. Leonardo explains these figures as follows: 'Many things placed one behind the other in front of the two eyes with known distinct distances; all such things will appear double except that which is seen best [i.e., in focus]. And the space interposed between these duplications will seem so much the greater as the objects are nearer to the eye, looking at the farthest one. But if you look at the first nearest one the said spaces will appear so much the smaller as the objects are nearer to the eye' (W 19117r; K/P 115r).

Underneath the sketch containing a string of five objects he writes, 'Here the objects are doubled to the visual power

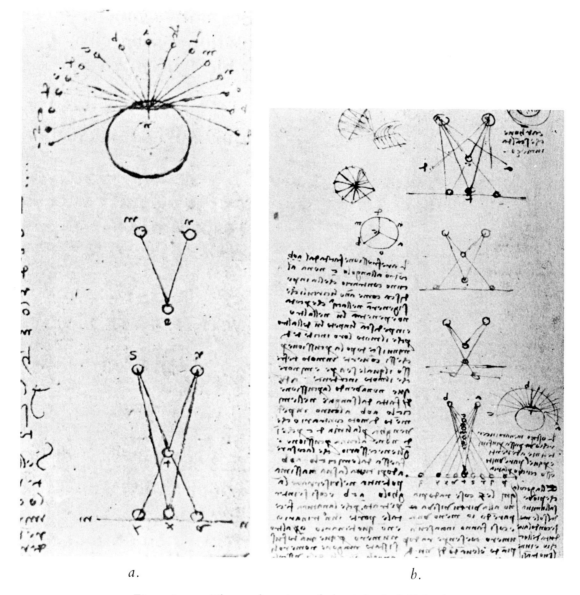

a.   *b.*

Figure 8.11.   The explanation of physiological diplopia.

a. Leonardo explains this drawing as follows: 'If the two central lines of vision concur in the object *x* the subordinate adherent lines *sv* and *ry* will see the object *t* occupy two places on the wall *nm,* that is at *v* and *y*' (D 8 v).

Thus, too, if one focuses on the object *t,* two objects *rx* and *sx,* will also be seen, i.e., two objects will give three images.

b. Having demonstrated the phenomenon with two objects in the upper sketches, Leonardo now increases the number of objects to a receding series of five, concluding that the 'objects are doubled to the visual power, appearing as 10 minus 1' (W 19117r; K/P 115r).

[*virtu visiva*], appearing as 10 minus 1. And they do this for every number, i.e., always doubling and leaving out one from every total sum which results from such doubling. And one begins with 2 which duplicated and 1 taken away from the total leaves 3; therefore the twofold seen by the two eyes will appear threefold. And if you put out 100, double it and take away one, there remains 199, and thus 100 objects being placed one behind the other in front of your eyes will appear as 199. And the cause which gets one taken away, leaving an odd number, is that the angle of the two central lines [of vision] meet together on that one of the objects which is well seen and comprehended and these lines do not intersect after the creation of such an angle. And the 2 lines end on one and the same object; and to the 2 eyes one object does not appear as 2, as happens in objects seen by each of the two-fold visual lines which are not central' (W 19117 r; K/P 115 r). These words checked against his sketch leave no room for doubt that Leonardo describes

physiological diplopia. This experience one can reproduce for oneself by focussing on one finger held in front of the face and noticing that another finger behind it appears double.

SPECTACLES

About the age of fifty Leonardo had to wear spectacles. Naturally he applied his theory of vision to their use. Spectacles in his day were well known; they were divided into two main classes, those for 'the young' (concave lenses) and those for 'the old' (convex lenses). Leonardo's description of his own glasses, those for 'the old' takes us further into the way he thought refraction affected images entering the eye. 'The glasses of spectacles', he wrote, 'show us how the images of objects stop at the surface of such glasses and, bending [i.e., being refracted], penetrate from this surface

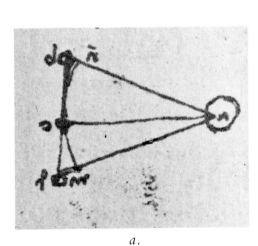

a.

Figure 8.12.   Spectacles.

a. In this drawing *a* represents the eye, *bcd* a straight plane, and *acnm* a curved plane. Leonardo writes, 'Objects do not send simulacra to the eye with their parts in the same proportion as is found in themselves'. On the same page he goes on to discuss 'Whether the species of objects are received by the visual power at the surface of the eye or whether they pass inside it' (see text). Finally he describes 'How the straightness of the convergence of the species is bent on entering the eye' (D 2 r).

b. 'To prove how spectacles help the sight' (CA 244 ra).

This drawing is explained in the text.

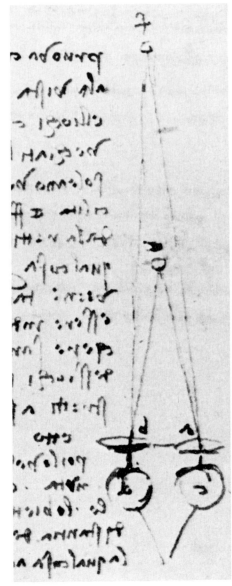

b.

to the surface of the eye so that it is possible for the eye to see the shapes of the aforesaid objects. This is possible because such a surface has in common a boundary between the air and the eye where it divides the albugineous humour from the air. If we affirm that the images of objects are stopped on the surface of the spectacles we are saying that through the spectacles of the old the object is shown much larger than it is in reality, and if it were not for the interposition of the glass between the eye and the object the object would be shown at its natural size. Therefore it is evident that the images, or the convergence of the images of any object which is cut by the interposition of transparent bodies, will be impressed on the surface of such a body and from here a new convergence [of rays] will be regenerated which conducts the images of the object to the eye' (D 2 r; Figure 8.12a).

It is clear that Leonardo saw both the convex lenses of his spectacles and the convex lens of his cornea as transparent, dense, magnifying media, each acting on light in the same way – which indeed they do.

His own personal experience with spectacles would appear to be included in the following analysis of their action: 'To prove how spectacles help the sight. Let the spectacles be *ab* and the eyes, *cd*. Because of old age an object usually looked at with great ease needs to be strongly turned into the straight axis of the optic nerve. By reason of old age the power of turning [the eyes] comes to be weakened so that the eyes cannot be turned without great pain, so that necessarily one is constrained to place the object further away, that is to say, from *e* to *f* where they can see it better but not in detail. Now, by interposing the spectacles, with their help, the object is well appreciated at a distance, *e*. This happens because the object is passed to the eye through different media, i.e., rare and dense; rare through the air between the spectacles and the object, and dense through the thickness of the glass which bends the line *ab* in such a way that the object is seen at *e* as though it were at *f*. Through the comfort of not bending the axis of the eye from that of the optic nerves, and owing to its nearness, the object is seen and appreciated better at *e* than at *f*; and the minute details are very much enlarged (CA 244 ra; Figure 8.12b). Here Leonardo accounts for the mechanism of accommodation by convergence of the visual axes of both eyes and expresses the belief that loss of this capacity is responsible for the presbyopia of old age.

COLOUR VISION

Leonardo concentrated on three aspects of the problem of colour: colour perspective, colour contrast and the subjective or objective nature of colour perception. The underlying theory from which he set out on his observations of colour was closely related to Aristotle's views and to the spurious Aristotelian work *De Coloribus,* which was published by the Aldine Press in 1496.

For Leonardo colour as perceived results from three variables: the colour of the object, the colour of the medium and the sensitivity of the eye. 'Black and white', he writes, 'are not commonly included among the colours because one is the absence of colour, the other the source of colour' (TP I 178). He lists black, white, blue, yellow, green, ochre, purple and red as the eight colours with which a painter has to work. He emphasises that 'nothing ever looks to be its real colour unless the light which illumines it is entirely similar in colour' (TP I 197). And this occurs very rarely because of other lights, shadows and the colour of the medium. Thus colour perspective supplements linear perspective. 'The eye can never arrive at perfect knowledge of the interval between two objects at different distances by means of linear perspective alone unless assisted by colour perspective'. As already noted in Chapter 2, Leonardo reduced this statement to graphic form (see Figure 2.8, p. 49). He also tested it experimentally using the colours of trees, houses or men placed at intervals of 100 *braccia* and painting them on his vertical glass pane. He concluded that 'the second object decreases by 4/5 of the first when it is 20 *braccia* from the first' (TP I 236; CU 77 v). In this instance diminution by linear perspective was confirmed by his colour perspective.

The perspectival diminution of the force or impetus of the waves carrying the '*spetie*' undergoes a loss depending on the resistance or density of the medium through which it travels to the eye. If this is very long and dark, the image progressively loses its light and colour. If the loss of light from white is partial, it becomes some shade between orange, red or green to blue or purple. But 'every impression is preserved for a time in its sensitive object [in this case the eye] . . . and the impression is that of a blow, so the brightness of the sun or other luminous body remains in the eye for some time after it has been seen' (CA 360 ra). 'The eye having looked at light retains some of it; there remain in the eye images of the thing appreciated and they make a less lit place seem dark until the eye has lost the last trace of the impression of the stronger light' (CA 203 ra). The contrast between light and dark after-images is important in Leonardo's view of the physiology of vision. He cites the motion of the firebrand appearing as a circle of light in the darkness and the drop of rain appearing as 'a continuous thread' as examples (CA 360 ra). Moreover, a drop of rain by this same mechanism is 'according to the sight of the eye in its passage appears continuous through as much space as shows all the colours of the rainbow' (W 19117; K/P 115 r). He adds, 'Everything that moves swiftly shapes its passage with the image of its colour' (CA 147 va). All such after-images were seen by Leonardo as evidence that light and colour entered the eye to stimulate vision. Such images could last a definite time, for 'If you look at the sun or other luminous object and then shut your eyes you will see it again in the same form within your eye for a long space of time' (CA 204 ra). Moreover, such after-images were positive or negative; 'When seen by an eye which has recently looked at the body of the sun a dark place will appear sown with spots of light and a bright place with dark round spots' (CA 369 vc).

As a painter Leonardo was primarily concerned with the colours of pigments. He divides them into 'simple' and 'secondary', the latter being produced by mixing the simple colours. For example, 'Green is composed of blue and yellow' (TP I 177). Quite typically he plans to mix these colours systematically in infinite combinations: 'Although the mixture of colours with one another extends to infinity, I will mix each one of them, one by one, then two with two and three with three, following thus through the whole number of colours (TP I 178).

Leonardo's obvious concern with pigments might well have led him to fail to penetrate further into a comprehension of the physiology of colour vision. But he evaded this obstacle by comparing his results from mixing pigments with those from his experiments with the camera obscura. He expresses it thus: 'The intersection of the images [*spetie*] as they enter the pupil does not fuse them one with another in that space where that intersection unites them. This is made evident, because if the rays of the sun pass through two panes of glass in contact with one another, one of these

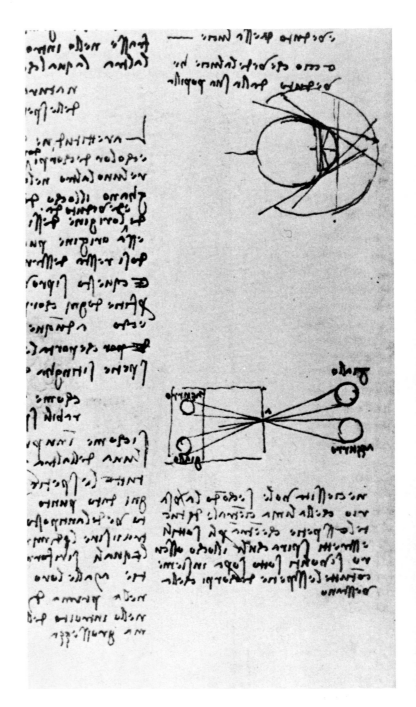

Figure 8.13.   The existence of monochromatic light.

Beside the upper drawing Leonardo writes, 'On the nature of the rays composing the images of bodies and their intersection'.

In the sketch below he draws two sources of light, one labelled yellow, the other blue. The rays from these sources pass through a *spiracolo* representing the pupil of the eye. They intersect but do not fuse, so that the images formed at the back of the camera obscura, representing the eye, are still blue and yellow, as labelled. Leonardo heads this passage, 'How innumerable rays of innumerable images can be reduced to one single point' (W 19152 r; K/P 118 rb).

He here refers to the point of intersection of yellow and blue rays in the *spiracolo* or pupil of the eye. Compare Figures 2.15 and 2.16 (p. 56).

Figure 8.14. The colours of the rainbow.

Much concerned as to whether the colours of the rainbow were objective, or subjectively created by the observer's eye, Leonardo made several experiments. In the top drawing he places his eye near a glass of water, in the glass of which exist minute bubbles. 'These bubbles, although the sun does not see them, generate from one of their sides all the colours of the rainbow'.

Below he writes alongside the drawing, 'How the eye has no share in producing the colours of the rainbow. Place a glass full of water on a window sill in such a way that the rays of the sun strike it from the opposite side; then you will see the aforesaid colours produced in the impression made by the sun's rays which have penetrated through the glass and ended on the floor in a dark place at the foot of the window. And because here the eye is not exerted we can evidently and with certainty say that such colours do not have anything to do with the function of the eye' (W 19150 r; K/P 118 ra).

Thus he proved to his own satisfaction that the different colours of the spectrum exist objectively.

being blue and the other yellow, then the ray that penetrates them does not become tinged blue or yellow but a most beautiful green. And the same thing would happen in the eye if the images of the colours yellow and [blue]* should fuse with one another at the intersection which they make within themselves at the entrance of the pupil. But as this does not happen no such fusion exists' (W 19152 r; K/P 118 rb; Figure 8.13). Compare Figures 2.15 and 2.16 (p. 56).

In Figure 8.13 Leonardo describes the position occurring when two pigments are mixed as the painter mixes them, e.g., yellow and blue make green. In the pupil of the eye, however, no such mixture exists. He goes on to explain the difference in terms of his camera obscura experiments under the title, 'On the nature of rays composing the images [*spetie*] of bodies and their intersections. The straight rays which carry the shape and colour of bodies whence they depart, do not themselves tinge the air, nor can they tinge

each other by contact at their intersection. Indeed they tinge only that place where they lose their existence, because that place sees and is seen by the origin of these rays, and no other object that surrounds that origin can be seen from the place where such a ray when it is cut off and destroyed leaving there the spoil [image] it has carried off. And this is proved by the fourth [book] on the colour of bodies where it says, the surface of every opaque body participates in the colour of its object. Therefore it is concluded that the place, which by means of the ray which carries the images sees and is seen by the origin of that image, is tinged by the colour of that object' (W 19152 r; K/P 118 rb). In these terms Leonardo describes the monochromatic light ray.

The objective existence of monochromatic light was detected by him experimentally by placing a glass full of water on a window sill exposed to the sun's rays so producing 'the colours of the rainbow . . . [they are] produced in the impression made by the sun's rays which have penetrated through the glass and ended on the floor in a dark place at the foot of the window. And because here the eye is not

---

*Leonardo writes 'green' here. The word '*ver*' is crossed out and substituted by '*azurro*' a few lines above, but not here.

exerted we can evidently and with certainty say that such colours do not have anything to do with the function of the eye' (W 19150 r; K/P 118 ra; Figure 8.14).

This observation interested Leonardo by clearly separating 'objective' colour images produced outside the eye from those 'subjective' coloured after-images which he had previously observed 'inside' the eye. Since colour perspec-tive supplied one of the most relevant clues to the objective observation of natural phenomena, this was a vitally impor-tant distinction for him.

Although he reached a concept of light and colour in terms of wave motion and refraction, Leonardo did not get as far as applying either of these concepts to the physics of the wave formation of colour.

Chapter 9

From Sound and Hearing to Voice and Music

Leonardo's approach to the problems of hearing and sound was similar to that of his investigations of light and vision. In both cases he searched for the essential nature of the force, be it light or sound, and then tried to trace its path from its origin in the macrocosm of nature outside to its final destination inside the microcosm of the brain of man.

In one of his early plans for 'The order of the book' of anatomy Leonardo gives a rather strange title, 'Effects' (i.e., that which receives causes) under which he proposes to describe 'perspective through the function of the eye; on hearing I shall speak of music and describe the other senses' (W 19037 v; K/P 81 v).

Thus he makes it clear that he intends to approach the problems of sound and hearing by the same perspectival or pyramidal method as he investigated light and vision.

SOUND

The cause of sound is percussion (see Figures 4.41 and 4.42, pp. 126, 127). 'The blow [percussion] is the end of swift movement caused by force on resistant objects. This same percussion is the cause of all sounds, a breaker and transmuter of different objects, causing again a second movement. It is a transmuter of its effect . . . It leaves its origin by circular movement and is all in all and all in each part' (A 27 v). As an example of such transmutation he shows a bell being struck (or percussed), an event he describes as 'A blow on a resonant object' (A 8 v). And, 'A blow occurring in a resonant body is immediately sensed in the whole of the body' (A 7 v). This movement throughout the bell caused by a momentary blow 'is repercussed in the air, and the air which touches the moved thing resounds' (Triv 36 r). This kind of to and fro movement (oscillation or vibration) Leonardo later calls *'moto ventilante'* fanning movement (see Chapter 4, p. 101–102). By the production of such movements in the air the sound of the bell is transmitted in such a way that a similar bell will respond. 'A blow given to a bell will make another bell similar to it respond and move somewhat. And the string of a lute as it sounds produces movement and response in another similar string of similar tone in another lute; and this you will perceive by placing a straw on the string which is similar to that sounded' (A 22 v). In Leonardo's view the human ear-

drum also picks up sounds in the same way as the second bell or lute-string, from resonating waves of sound in the air.

His employment of a piece of straw to demonstrate resonance at once recalls his similar use of a piece of straw to demonstrate the transverse wave movement produced by throwing two stones into water. Here Leonardo writes, 'Although the sounds that penetrate the air depart by circular movements from their causes, nevertheless the circles moved from their different origins meet together without any impediment and penetrate and pass into one another, always keeping their causes as their centres. Since in all cases of movement water has great conformity with air, I will offer it as an example of the above-mentioned proposition' (MS A 61 r; see Figure 2.18, p. 58). He then describes the experiment of throwing two small stones into water, the spread of circles of waves from these two centres and their intersection without interfering with one another. The waves he describes as 'tremor rather than movement', an observation which he verifies by floating pieces of straw on the surface of the water and observing their movements up and down. Thus, sound spreads in the same way as light.

The clear statement made in MS A 61 r was achieved only after a number of trial runs. This is common with Leonardo when trying to state an important principle, and sometimes his abortive attempts help to clarify his final statement. In *Codex Atlanticus* 373 rb this proposition is incompletely expressed in three versions, two of which say, 'Just as when water and air are percussed, the one by a stone, the other by the voice, you will see the water demonstrate the percussed place by means of its different circles, so the sound of the voice made in the air will be equally heard at equal distances'. Trying again, he writes, 'Just as air percussed by the voice and water percussed by a stone go away in dilating, fleeing movements in circles which demonstrate their cause, since all these circles have their centre at the placed percussed' (CA 373 rb). It will be noticed that all these descriptions, including the drawing in Figure 2.18, are attempts to describe the patterns of observed physical events in geometrical terms. Such movements represent the action of the transmuted 'power' of percussion, and such action though incorporeal in essence, invisible and 'spiritual', by moving corporeal 'instruments' at any time becomes 'material'.

THE ORIGIN OF SOUND

Leonardo makes extensive observations and experiments on the production of sound, and wherever he finds it produced he finds its cause to be percussion. He puts it thus: 'There cannot be any sound or voice where there is not movement from percussion of the air. There cannot be percussion of the air where there is no instrument. There cannot be an incorporeal instrument. This being so, a spirit cannot have either sound, form or force, and should it assume a body it will not be able to penetrate nor enter where the doors are shut. And if any should say that through air collected together and condensed a spirit may assume bodies of various shapes and by such an instrument may speak and move with force, my reply would be that where there are neither nerves nor bones there cannot be any force exerted in any movement made by imaginary spirits. Shun the precepts of those speculators whose arguments are not confirmed by experience' (B 4 v).

Besides explaining his view of the primary source of sound, Leonardo here opens up quite a different question, Can a spirit speak? Here he answers: 'without a material instrument, no'.

As examples of sound-creating 'instruments' Leonardo cites the cannon, which is corporeal with lightning, and hammer on anvil. 'The noise of the cannon is caused by the impetuous fury of the flame repercussed by the resistant air . . . Because fire is rarer than air it follows that the air cannot make way for the fire with the velocity and swiftness with which the fire assails it. Therefore resistance occurs and is the cause of the great din and noise of the cannon (A 32 r). Percussion, it will be recalled, he defines as 'the end or impediment of movement' (A 27 v), and this occurs where any movement meets 'resistance.' Thus, 'If the cannon is moved against the incoming impetuous wind it will be the occasion of a greater thunder by reason of its greater re-

sistance to the flame . . . And if the air be equally dense or rare and without movement the noise will be equally heard around its cause; and it will go on expanding from circle to circle just as the circles of water do when caused by a stone thrown into it' (Triv 18 v).

Here, too, he recognises the effects of blast which are similarly but more slowly distributed in expanding circles (CA 375 rc): 'Where similar instruments [i.e., cannons] are in use the adjacent air will break or move all objects of weak power of resistance', such as large vessels, windows and doors. 'This happens because the air expands and presses itself outwards wishing to escape in all the directions in which movement is possible' (Triv 18 v).

Lightning produces thunder by the same mechanism; its swift 'rare' flame percusses the air in its vicinity rapidly both by its movement and expansion. This in turn percusses neighbouring denser air, so producing the prolonged noise of thunder (K 110 v). These observations are generalized: 'When air enclosed in a space issues forth its percussion made with the air outside makes a noise, as is seen with the bombards, a bursting bubble and the popping which children make with berries pushed into the mouth of a sackbut' (A 31 v). The children's pop-gun is illustrated in Figure 9.1 as an experiment to test this thesis. 'Sound is nothing but the separation of compressed air', he explains. The noise of the hammer percussing an anvil particularly interested him because, 'The time in which the blow is produced is the shortest thing that can be done by man . . . which movement repercusses in the air, and the air which touches the thing moved, sounds' (Triv 36 r). He notices, too, that 'Sound cannot be heard at such close proximity to the ear that the eye does not see the contact of the blow first' (C 6 v). This leads him to realise that 'It is possible to recognise by the ear the distance of a clap of thunder from first seeing the flash of lightning' (A 19 r). 'The more distant the place of percussion the longer the

Figure 9.1.   The sound of a pop-gun (L 89 v).

Leonardo draws a child's pop-gun, showing how a wooden piston with a stopper condenses the air, so ejecting the pellet at the other end. Beside the drawing he writes, 'It is proved by this example how the noise made by a bombard consists solely of the separation of condensed air'.

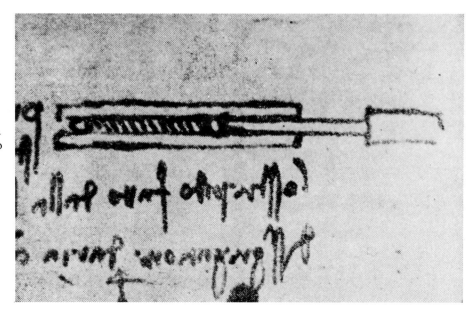

time of its conduction to the ear' (CA 251 rb). He adds, 'Thunder is moved with time in the manner of a wave of water and it makes more noise where it is most impeded' (K 110 v).

The pitch of sound produced by percussion did not escape Leonardo's musical ear. He observes it with regard to the percussion of drums, the plucking of strings, ringing bells and the human voice.

He was very much aware that the pitch of a drum depends on the tension of its skin from experimenting with 'a calf-skin covering a resonant vessel' (A 52). 'Just as the same drum makes low or high-pitched sounds [*boce gravi e acute*] according as to whether the skin is more or less tightly drawn, so this skin differently drawn on the same drum will make different sounds' (Figure 9.2). Each of these is separately illustrated with an ingenuity which shows Leonardo the inventor at his best. On this same page he draws mechanical schemes for altering the tension of the skin of a drum and also drums with different tensioned skins combined into a single instrument so that chords can be percussed on them. Leonardo's ingenious devices have been beautifully analysed in detail by W. Emmanuel Winternitz.[1] They are of particular relevance to hearing since Leonardo, although aware of the eardrum in man, gives no detailed anatomical description of it.

As already mentioned, Leonardo detected the wave movement of vibrating strings by means of his observations of the movement of attached pieces of straw. He also observed that 'If you beat a plank you will see the dust on it collect into little hills' (A 32 v and 71). He observes that if you go on percussing, 'The dust will always pour down from the tip of their pyramid and descend to its base, whence it will re-enter beneath, ascending through its centre to fall back again from the top of the little hill. And the dust will circulate in the right-angled triangle *amn* as long as the percussion continues' (Madrid I 126 v; Figure 9.3).

In these experiments he is using dust as his marker, noting that it flies off from the vibrating part of a plank or rod and settles in pyramidal hillocks at points of rest which are not vibrating – the nodes. The site of the node is even more accurately located by noting the 'circulation' of the dust particles which rise to the greatest height at the point of the pyramid.

'A rod or cord in rapid oscillation appears to be doubled', he writes, and then asks, 'Tell me why the false cord of a lute makes as it oscillates two or three images and sometimes four' (C 15 r). It was a problem that took centuries to answer.

Leonardo relates pitch to velocity of the movement making percussion, not to its wave-frequency. For example, in the context of the movement of birds' wings in the air he states, 'The condensation of the air will be of greater or less density according to the greater or less velocity of the moving thing compressing it, as is shown in the flight of birds, for the sound that they make with their wings in percussing the air is lower or higher pitched according to whether the movement of the wings is slower or faster' (E 28 v). This he

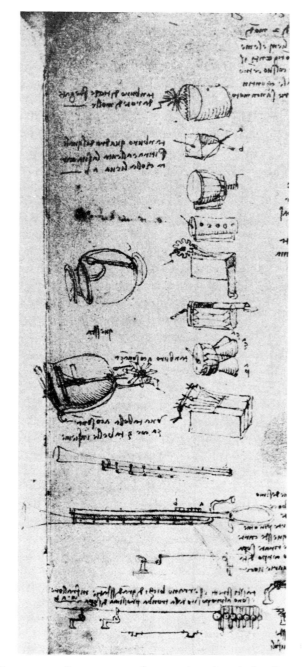

Figure 9.2. Instruments for varying the pitch of sound (BM 175 r).

The top five drawings on the right show a series of drums each with a different mechanism for varying the pitch when percussed. In the top one, wires scratch the drum-skin, producing a continuous sound. In the next below, widening the space *ab* tightens the skin in front, so varying the pitch rapidly at will. In the next drawing, the drum is tightened by a crank at the back. In the next, pitch is altered by blocking a series of holes in the side. Similar variations of open spaces at the sides and end are applied to the two drums beneath. Below these three cones are attached to three strips of drum-skin of different tensions, so producing a harmonic chord. Rectangles perform the same function in the drawing beneath. Trumpets with side-holes apply the variation of pitch in the two lower drawings. Musical pots are shown on the left.

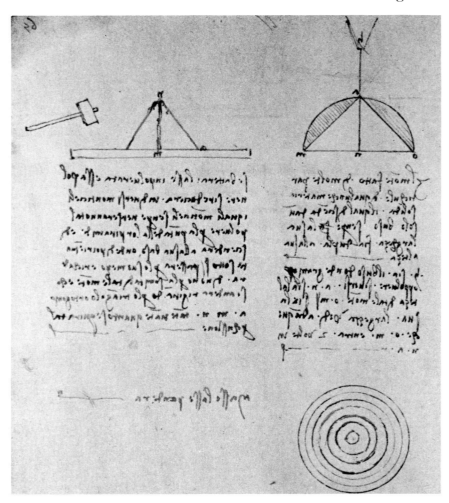

Figure 9.3.   The formation of a pyramid of dust on a beaten board.

Beneath the left-hand drawing Leonardo writes, 'If a board covered with dust is beaten the dust will come together in different little heaps. In these little heaps the dust will always pour down from the tip of their pyramid and descend to its base, whence it will re-enter beneath, ascending through its centre to fall back again from the top of the little hill. And the dust will circulate in the right-angled triangle *amn* as long as the percussion continues'. To the right Leonardo draws the formation of a similar pyramid of dust by dropping dust through the aperture *b*. The 'different little heaps' of dust on the beaten board demarcate the nodes on a vibrating bar. The set of concentric circles illustrate that 'If water enclosed in a vessel is struck at its centre a circular motion will be produced which starting from very small circles will end in very large ones' (Madrid I 126 v).

finds confirmed by his experiments on the wings of flies. Leonardo quotes the then commonly held view that the sound made by flies proceeds from their mouths like the voices of men (BM 257 v) but continues, 'That flies have their voice in the wings you will see by cutting them a little or indeed by smearing them with a little honey in such a way that they are not prevented from flying. And you will observe that the sound made by the movement of their wings is made raucous and the sound will change from high to low-pitch in direct proportion to the degree that their wings are more impeded' (W 19014 v; K/P 148 r). The matter is clinched for him by his answer to his own question, 'Why does a swift wind passing through the same pipe make a high-pitched sound? The wind passing through the same pipe will make a sound so much lower or higher in pitch in proportion as it is slower or swifter. And this is seen in the changes in the sounds made by trumpets or horns without holes, and also in the winds that whistle through little holes [*spiracoli*] and chinks of doors or windows' (E 4 v). He illustrates this theme in his invention of variant of a musical pipe, Figure 9.4. Here he shows one instrument with oblong as opposed to round side holes, and another in which all the side holes are run into one long slit. He writes, 'These two flutes do not change their voices by leaps; rather in the manner which is peculiar to the human voice. And

Figure 9.4.   Two 'flutes' modified to resemble the human voice (CA 397 rb).

The side holes in these two 'flutes' have been modified so that by moving the hand up and down over them a glissando effect can be produced as opposed to the discrete notes of a scale. This Leonardo feels imitates more closely the singing human voice.

this is done by moving the hand up and down as with a coiled trumpet, best of all with a whistle *a*, and you can make 1/8 and 1/16 of the voice or however much pleases you' (CA 397 rb). It will be noticed that the second oblong hole in the left-hand 'flute' is twice the distance of the first from the mouthpiece. This may well coincide with an octave change in his choice of 'making 1/8 and 1/16 of the voice'. That Leonardo persistently uses the word '*voce*' for the sounds produced by this instrument and likens its action to the 'human voice' relates this instrument to the mechanism of the human voice.

HUMAN VOICE PRODUCTION

THE ANATOMICAL APPARATUS

It is desirable to consider Leonardo's anatomy of the vocal apparatus apart from his physiology of its function since his anatomy is brilliantly correct whilst his physiology based on this anatomy, though equally brilliant, is incorrect.

Leonardo was right in realising that voice production involved the integrated function of structures ranging from the thoracic cage, through lungs, bronchi, trachea, larynx, pharynx, nasal and mouth cavities to the teeth, lips and tongue; and he considers all these structures, producing unprecedentedly accurate drawings of them all. Many of these are to be found on one page, Figure 9.5, drawn about 1510. On the left of this page the two bronchi may be seen joining to form the windpipe or trachea, which ascends in front of the gullet (oesophagus) to the thyroid cartilage, just below which is the bulging mass of the thyroid gland. Both this and the thyroid gland suggest the organs of a pig rather than man. Above this drawing is an enlarged version of the upper parts. Here the cricoid cartilage is better shown, and the hyoid with its ligaments appears above the thyroid cartilage. Leonardo has included more of the pharynx and its constrictor muscles. The uvula, prominent in contemporary physiology as an organ which drained off superfluous humours from the brain, is conspicuous and in a further sketch above is shown with a drop of 'superfluous humour' hanging from its tip.

At the bottom of the page, in Figure 9.5, Leonardo has concentrated on the voice box (larynx) with remarkable success. The central sketch shows the larynx from above. In this the aryepiglottic fold with little bulges in it formed by the cuneiform and corniculate tubercles can be clearly seen. To the right of this is a coronal or frontal section of the larynx with the tongue-like epiglottis rising above, labelled *dn*. The vocal cords themselves are clearly seen on each side of the letter *c*, forming a passage for air from the trachea below. On each side of them are what Leonardo calls '*saccules*'. These he labels *a* and *b*. This remarkable anatomical exploration Leonardo describes as follows: 'Air enters and goes out through the mouth [of the larynx] *d;* and when the food passes over the bridge *dn* [epiglottis] some particle could fall through the mouth *d* and pass through *c*, which

would be fatal. But nature has arranged the *saccules a b*, which receive such a particle and proceed to keep it until, with coughing, the wind which issues with impetus from the lungs by way of *c*, eddies and drives out the droplets squeezed by the walls of the *saccules, a b*, out by way of *d*, and so this harmful matter is thrown out of position'.

Down the right border of the page Leonardo depicts how the leaflike epiglottis bends over and prevents food from entering the larynx whilst guiding it into the oesophagus, for which reason, 'One cannot swallow and breathe or give voice at the same time'. In Figure 9.6 Leonardo draws the vocal folds, thyroid cartilage with its central notch, and superior cornua, in undoubtedly human form.

Figure 9.7 shows the hard and soft palate dividing the nasal cavities from the mouth. The whole cavity formed by the pharynx, nose and mouth, Leonardo looked upon as a resonating chamber for the sounds produced in the larynx. This chamber changes its shape continually; for example, the soft palate is mobile, let alone the tongue, cheeks and teeth. The tongue is dissected, and the muscles of the lips carefully exposed in order to analyse their movements.

Taken together, these drawings show Leonardo's integration of the whole mobile anatomical apparatus of speech.

THE PHYSIOLOGY OF SPEECH AND SONG

It is evident that Leonardo attributed sound to the percussion of a 'dense' body on a 'rarer' one. All the examples of sound production previously described, from the thunder of a cannon to the sound of a bell or a flute, are accounted for by percussion. It is therefore not surprising to find that he applies the same mechanism to the production of the human voice. For this reason it is of first importance to him to decide an issue which occupies a prominent heading in Windsor Drawings 19064 v (K/P 157 v): 'Whether the wind which issues [*si fugge*] from the trachea is condensed in its passage or not'. In answer he notes that 'the trachea is narrowed at the larynx [*epigloto*] in order to condense the air which comes to it from the lung for the creation of different kinds of voice [*diversi generationi di voce*] . . . Therefore if the trachea were to be dilated at its upper end as it is in the throat, the air could not be condensed and perform the functions or services necessary to life and to man, that is in speaking, singing and the like'.

The quest is pursued in a long note written about the same time, 1509, in Figure 9.7. Here Leonardo draws on his previous observations on the physics of sound. 'Furthermore', he writes, 'You will describe and draw in what way the function [*ufitio*] of varying, modulating and articulating the voice in singing is a simple function of the rings of the trachea moved by the reversive nerves [from the vagus] and in this case no part of the tongue is used. And this rests on what I have proved before, that the pipes of an organ are not made lower or higher pitched by the mutation of the fistula (that is the place where the sound is produced) by making it wider or narrower, but only by the mutation of the pipe

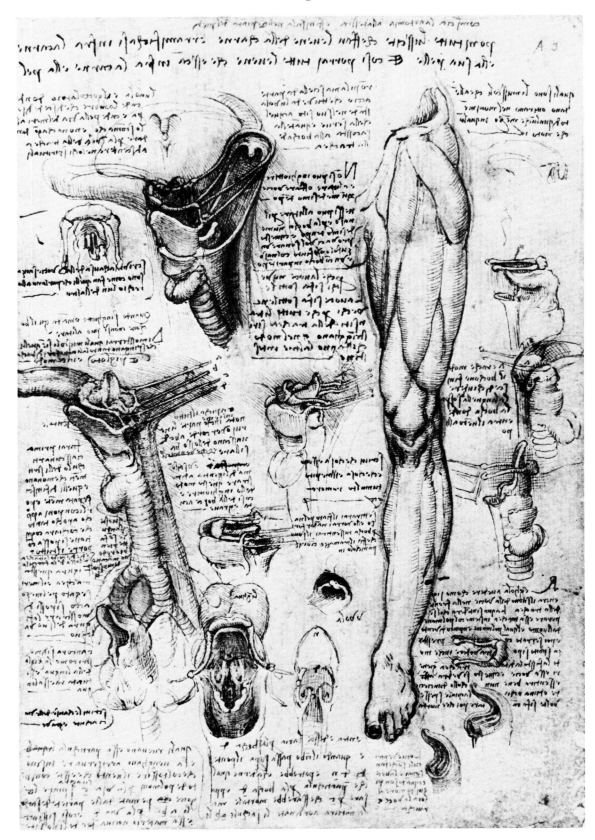

Figure 9.5.   Some of the apparatus for voice production (W 19002 r; K/P 134 r).

On this page Leonardo fits together the individual parts shown at the right (epiglottis and larynx), the vocal cords (below) with a composite picture of the whole apparatus on the left.

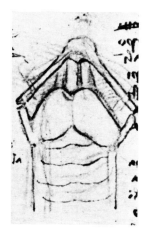

Figure 9.6. The larynx and vocal folds (W 19114 v; K/P 109 r).

In this view of the larynx the vocal cords, the thyroid cartilage and the windpipe are clearly recognisable.

lower-pitched noise then the longer. And on this I shall not expatiate because it is fully treated in the book on musical instruments'.

He then describes how he considers this process accounts for variations of pitch of the human voice. 'The extension and narrowing of the trachea together with its widening and shortening are the causes of the variations of the voices of animals from high-pitched to low-pitched and from low-pitched to high-pitched'. According to his principle of continuity, if the tracheal channel narrows, it will, comparably with water channels, accelerate air flow, and conversely if it broadens the flow of air within, it will slow. Leonardo is concerned that this slowing may be so marked as to fail to raise the pitch of the voice [*achuire la vocie*]. 'But of this', writes Leonardo, 'we shall make an experiment in the anatomy of animals by blowing wind into their lungs and [then] compressing them, narrowing and widening the fistula [*fistola*] which generates their voice (W 19115 v; K/P 114 r).

into wide or narrow or into long or short, as is seen in the extension or shortening of the coiled trumpet. Furthermore, in a pipe of fixed width and length sound is varied [in pitch] by putting into it wind with greater or less impetus. And such a variation [of pitch] does not occur in things percussed with greater or less percussion, as one hears in bells beaten with very small or very large percussors. And the same thing happens in pieces of artillery similar in width but varying in length. But here the shorter makes a greater

Passing through the narrow slit (*fistola*), the air is put into eddies as the result of friction. 'The voice is the movement of the air in friction with a dense body . . . and this friction of a dense with a rarefied body condenses the rare one and so makes it resistant. Moreover a swift-moving rare [body] in a slow-moving rare [body] condense each other in their contact, and make a sound or very great din' (W 19047 v;

Figure 9.7. The formation of vowels (W 19115 r; K/P 114 v).

In the centre above is a sketch of the teeth, hard and soft palate and uvula separating the shaded nasal cavity from the mouth. To the left of this are three sketches of the lips as formed in the pronunciation of the vowels *a, o,* and *u*. The table on the right combines the five vowel sounds with consonants to which the lips and teeth contribute in pronunciation. The muscles and movements of the tongue are analysed below.

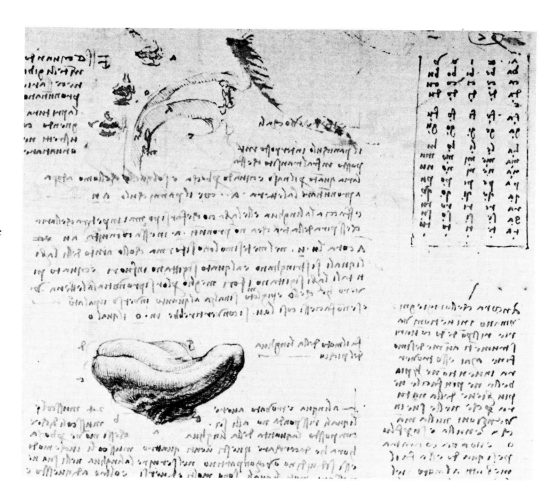

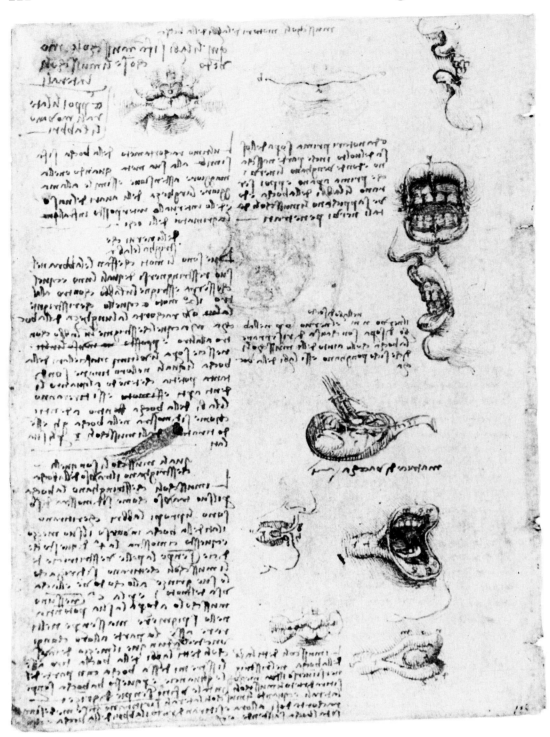

Figure 9.8. The movements of the lips (W 19055 v; K/P 52 v).

The top left drawing depicts pursing of the lips. To the right of it is the opposite movement, the broadening of the lips in a smile – the only study we possess of the famous smile of the *Mona Lisa*. Muscles used in opening and closing the lips are studied in the other drawings. Note, however, that the fourth drawing on the right margin depicts a pregnant cow's uterus.

K/P 48 v). It is at the fissure between the narrowed vocal cords that such friction is maximal. The eddies arising from such movement revolve in the laryngeal sinuses (sinuses of Morgagni) just above, so creating or amplifying sound. 'The two ventricles are those which make the voice sound and when they are full of the humour then the voice is hoarse' (W 19002 r; K/P 134 r). This whole process is very similar to that described by Leonardo on the very next folio (W 19116 r: K/P 115 r), of eddies made by blood passing through the narrow aortic valve. Here too he makes a note about how the motor nerves of the voice and how they act in high, low and medium-pitched voices. When the movement of the air is brief, speech results; when prolonged, it produces the sound of singing, which he sees as comparable to the note of a flute or organ pipe.

Referring to the singing of base notes, Leonardo points out that 'the rings of the trachea having been narrowed, reopen like a spring' and that this their natural size is further augmented by the shortening of the trachea and pharynx 'as do those counter bases [*contra bassi*] who shorten the throat the more the lower they make their base-notes' (W 19068 r; K/P 161 r). The opposite occurs in the production of the high-pitched piping voices of old men. In this case, 'all the passages of the trachea become narrowed in the same way that other entrails are' (W 19002 r; K/P 134 r). One of the underlying anatomical changes in old age, according to Leonardo, was a 'narrowing' of all the tubular structures of the body, the trachea, intestines and arteries being the most important.

To the mobile soft palate Leonardo atributes the production of all vowel sounds. 'The membrane interposed between the passage which the air makes partly through the nose and partly through the mouth [the palate] is the only one which a man uses to pronounce the letter A' (W 19115 r; K/P 114 v). This is labelled *an* in the accompanying drawing, Figure 9.7. He then goes on to assert that 'all vowel sounds are pronounced with the back part of the mobile [soft] palate which covers the epiglottis'. This, too, can be seen in the same figure. On this same page at the right upper corner the vowels are laid out in five columns coupled with consonants. This introduces the subject of the movements of the tongue and lips.

Beneath a drawing of the tongue in Figure 9.7 Leonardo writes a long passage, commencing, 'The tongue is found to have 24 muscles corresponding to the six muscles of which the mass of the tongue that moves in the mouth is composed'. The rest of this passage discusses the intrinsic and extrinsic muscles of the tongue and suggests investigation of their origins, arterial, venous and nerve supply. All the rest of the prolific text scattered over the page describes phonation, discussing with typical thoroughness the parts played by the larynx, soft palate, nose, tongue and teeth in the formation of phonemes. They may be briefly summarised as follows: The letter *a* arises from the soft palate, whatever the tongue and lips may do. The vowel *u* arises from the same place, 'with the aid of the lips which are compressed and thrown somewhat outwards, and the more the lips are protruded the better is the letter *u* sounded; and with this sound the epiglottis rises. If it does not, the sound *o* is produced'. The lip movements of these three vowel sounds are illustrated at the top of the page. All the vowel sounds arise from the soft palate and lips which shape the original sound emanating from the larynx. Consonants are supplied by the tongue and lips. 'The tongue is employed in the pronunciation and articulation of the syllables, components of all words'.

On another page Leonardo describes the muscles and movements of the lips (see Figure 9.8) as follows: 'The muscles which move the lips of the mouth are more numerous in man than in any other animal; and this design is necessary for him because of the many operations in which he continually employs his lips as in the 4 letters of the alphabet, *b, f, m, p*' (W 19046 r; K/P 51 r). Developing this theme, he adds, 'Consider well how through the movement of the tongue with the aid of the lips and teeth the

pronunciation of all the names of things is known to us and the simple and compound words of a language are brought to our ears by means of this instrument. And if all the effects of nature had names, these would extend to infinity' (W 19045 v; K/P 50 v). This theme leads him, however, into a philosophical discourse on the rise and fall of languages rather than any further analysis of the physiology of speech.

THE PERSPECTIVE OF SOUND

The sound of the human voice produced in the larynx constitutes what Leonardo calls a 'sensible' wave motion, and the cavities of the throat, nose and mouth respond to these waves. They shape the sound by a responding 'sensible' motion, that is, resonance, which lasts and does not immediately cease with the cessation of the stimulus. In fact the mouth and nose react in a way similar to that of a percussed bell. Percussion leaves an 'impression' on anything percussed, and 'Every impression is preserved for some period of time in a sentient object . . . I apply the term sensitive to that object which is moved by any impression from what it was before; an insensitive object is one which though moved from its earlier state does not preserve in itself any impression of the thing which moved it. The sensitive impression is that of a blow received on a resonant object such as a bell and similar things, such as the voice in the ear. For if this did not preserve the impression of voices it could not derive pleasure from a solo song' (CA 360 ra).

Thus from the moment that the voice issues from the mouth into the surrounding air it spreads out in circular waves which lose their power directly in proportion to their distance from their source, i.e., in perspectival or pyramidal proportion (see Figure 2.18, p. 58).

The propagation of sound has many characteristics similar to that of light. Not only is its perspective subject to similar proportionate diminution with distance but also its reflection in the form of an echo obeys the rule that the angle of incidence equals the angle of reflection (A 19 v) like a ball thrown against a wall (see Figure 2.13, p. 55). This rule is not taken for granted; it is tested experimentally and elaborated in Figure 4.42 (p. 127) and Figure 9.9. In Figure 4.42 Leonardo depicts a hammer percussing a bell and shows the reflection of sound from the ledge of a wall in front of it. He writes, 'I say that the sound of the echo is reflected by percussion to the ear just as the percussions made in mirrors by the images of objects are to the eye'. One of the most interesting features of his sketch is his representation of the sound of the bell being produced by the percussion of a little hammer; its reflected percussion from the wall and its action on the ear are also shown as little hammers, so revealing his basic idea of the origin and hearing of sound.

This experiment is elaborated in Figure 9.9. Here he shows a man blowing a horn, the sound of which is repeatedly reflected from a series of receding steps placed both in front and behind him. He comments, 'How one should make the voice of the echo so that whatever you may say

Figure 9.9. The formation of echoes (B 90 v).

It will be noticed that the angles of incidence of the sound of the horn are equal to the angles of their reflection to and from the walls.

will be repeated to you by many voices. The voice leaving the man is repercussed and will fly off to the wall above. If there be a ledge on this wall at a right angle the surface above will send back the voice towards its cause' (B 90 v).

Later he generalises his observations and experiments on echoes as follows, 'The sound of the echo is either continuous or discontinuous; it occurs singly or is united; it is of brief or long duration, a finite or infinite sound; immediate or far away. It is continuous when the vault on which the echo is produced is uniformly concave. It is discontinuous when the place which generates it is discontinuous. It is single when it is produced at one place only. It is united when it is produced in several places. It is long-continuing when it goes circling round within a bell or cistern which has been percussed, or some other hollow place; or in clouds wherein the sound extends by degrees of distance in degrees of time, always uniformly growing fainter if the medium is uniform like the wave which spreads itself in a

circle over the sea. Sound often appears to proceed from the direction of an echo, and not from the place where it really arises' (CA 77 vb).

The 'size' or amplitude of sound gave Leonardo's precise mind a lot of trouble. For sound produced from any given object such as a freely suspended bell it was clear to him that the amplitude of sound was proportionate to the power of percussion. However, one sound double another is not heard at twice the distance: 'If it were so 2 men shouting would be heard at twice the distance of one; nevertheless experience does not confirm this' (A 43 r). Moreover, 'If you take ten thousand voices of flies all united together they will not be heard as far as the voice of a man. And if the voice of a man be divided into ten thousand parts no one of these parts will be equal to the size of the voice of a fly' (A 23 r).

Applying these observations to bells he claims that 'If a bell's sound were heard at two miles and it were then melted down and cast again as a number of small bells it is certain that if they were all sounded at one time they would never be heard at as great a distance as when they were in one single bell' (Forster II 32 v).

Though Leonardo from early days thought of sound as emanating from its source in spreading wave-movements, he did not associate the loudness of sound with the amplitude of sound waves produced by the power of percussion. However, his association of the loudness of sound with the power of percussion enabled him to find solutions for the puzzling observations just mentioned. These solutions, as one might expect, took the form of perspectival or pyramidal figures. For sound, as for light, Leonardo described a linear form of perspective and an aerial perspective subject to the density and movement of the air.

The linear perspective of sound is drawn and described in Figure 9.10. Here Leonardo writes on 'The loss of the

Figure 9.10. The perspectival spread of sound (L 79 v).

If the amplitude of a sound, *f,* is twice that of two small sounds, *m* and *n,* as represented by the bases of their pyramids, they will spread only half as far. The lines *ab* and *cd* represent the equivalent of vertical glass planes, as used by Leonardo in his experiments on linear perspective.

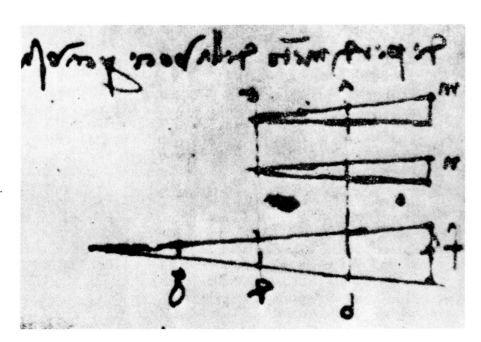

sound and voice by reason of distance'. His simple explanatory drawing consists of three triangles or pyramids, one twice the breadth and length of the other two. The amplitudes of the sounds are represented by the letters *m, n* and *f* at the bases of the pyramids. A vertical line labelled *ab* cuts the small pyramids into two halves, whilst it cuts the large pyramid at a quarter of its length. The line *cd* which passes through the points of the small pyramids *m* and *n* cuts the large pyramid in half. Leonardo explains the diagram thus: 'At the distance *ab* the two voices *mn* are diminished by half; whence, although there are two half voices, they are not as powerful as one whole voice, but only as half a voice. And if an infinite number of halves should reach such a distance they would still only amount to half. And at the same distance the voice *f* which is double *n* and *m* having lost the fourth part of its power, there is left consequently one voice and a half; whence the power at *g* comes to exceed by three times the power which is at the distance *ab* in *m* and *n*' (L 79 v). By this ingenious visualisation of the perspective of sound Leonardo has found an answer to all his problems of little bells and flies' voices not carrying as far as those of big bells and men.

However, such linear sound perspective depends upon uniform density and stillness of surrounding air, and such conditions rarely exist in nature. Even then, Leonardo asserts, 'A small near sound can appear as loud as a big sound far off' (Triv 7 v). 'The ear is deceived by perspective of a sound or voice, which may appear to be moved to a distance and is not moved from its position' (CA 357 vb). And as we have already seen, such 'deception' can also be produced by the straight or eddying movements of air so that eddying winds, for example, can even make sounds travel in a curved path, or by increased density of the 'air', as with clouds, echo or reflect back the sound. Thus aerial perspective of sound modifies its linear perspective grossly and often incalculably. The path of light, on the other hand, is not affected by the movement of air.

THE EAR AND HEARING

ANATOMY

From the anatomical point of view Leonardo had relatively little to draw or write about the structure of the ear (see Figure 9.11). He gave particular attention to the details of the pinna, which catches sound, drawing the helix with its Darwin's tubercle, the antihelix dividing into two crura with the triangular fossa between them, the concha, the tragus and the antitragus. He also takes great pains in locating the ear in relation to both the other parts of the face and head as well as other parts of the body. For example, 'The hole of the ear, the prominence of the shoulder, that of the hip, and of the foot [the external malleolus] are in a perpendicular line' (W 19136–39 r; K/P 31 r). But he does not appear to have explored the external auditory meatus, eardrum or the middle ear. Regarding the internal ear Leonardo remained content with his early description of 1489,

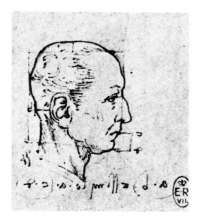

Figure 9.11. The external ear (W 12606; K/P 20 r).

referring to it as 'the hollow porosities of the petrous bone' (W 19019 r; K/P 39 r). In the beautiful drawing of the inside of the skull in Figure 2.25 (p. 64) Leonardo depicts the internal acoustic meatus in the petrous part of the temporal bone; he also shows the auditory nerve on its way to the *senso comune*. From his earliest drawings of the head (e.g., those in Figures 2.20 and 2.21, pp. 61, 62). Leonardo consistently delineated the auditory nerve as well as the optic nerves. Usually both end in the middle ventricle, where he locates the *senso comune*.

PHYSIOLOGY

Aware of the fact that sound, including that of the voice, spreads in circles from its point of origin, Leonardo raises the question whether only part of any sound will reach the ear: 'Whether the whole circle made in the air by the sound of a man's voice carries with it the whole of the spoken word, since the part of the circle percussed into another man's ear does not leave in that ear part of the word; and it is not the whole [circle]' (CA 199 vb). This statement is made clear by the accompanying figure.

Leonardo answers this question by appealing to the analogy of light. 'What has been said', he continues, 'is defined in the case of light. One can say whether the whole light illuminates the whole of a dwelling since part of the dwelling would not be illuminated by only part of the light. If you wish to argue and say that this light illuminates part of the dwelling with part of its light and not the whole, I would cite the example of one or two mirrors set in different positions in such a place that each face of such a mirror will have in itself the whole of the said light. Whence it will be demonstrated that such a light is all in all and all in each part of this dwelling. And the voice does the same in its circle' (CA 199 vb).

It will be remembered how important Leonardo considered the central line of all 'powers' to be. He stressed this on the first page of his codex on vision, MS D 1 r. What he said there he applied also to the power of hearing: 'Nature does not give uniform power to the *virtu visiva;* she gives this *virtu* a power which is so much the greater as it approaches

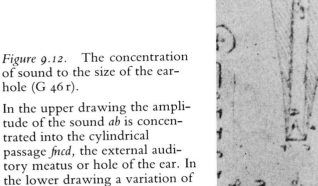
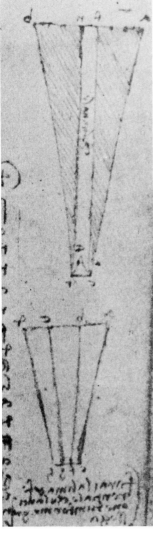

Figure 9.12. The concentration of sound to the size of the ear-hole (G 46 r).

In the upper drawing the amplitude of the sound *ab* is concentrated into the cylindrical passage *fncd,* the external auditory meatus or hole of the ear. In the lower drawing a variation of pyramidal concentration is depicted.

its centre. She does this in order not to break the law given to all other powers which become more and more powerful as they approach the centre. One sees this in the action of percussion by a body'. And, of course, hearing, like seeing, results from percussion. In the case of the ear, however, the approach towards the centre does not have to be carried as far as the central line, as with vision, but only as far as the diameter of the ear-hole, the external auditory meatus. Leonardo illustrates this pyramidal or perspectival concentration of sound entering the external acoustic meatus on at least two occasions, Figures 9.12 and 9.13.

In Figure 9.12 Leonardo explains his drawings. In the top drawing it will be noticed that the horizontal line *ab* is divided into 7 parts. From the central segment of this line *fn* a channel is drawn down to *cd,* the eardrum. Leonardo here uses the horizontal line *ab* to represent the original amplitude of the sound. He writes, 'As many times as the voice *ab* receives into itself the voice *fn* so many times does your ear, *cd,* receive it more powerfully. And it [the ear] receives as many more voices from *ab* than from *fn.* Because in action

ab is almost infinitely greater than *fn* the voice percussing the ear *cd* becomes infinitely greater than if it were percussed from *fn*' (G 46 r).

The same drawing appears again in Figure 9.13 without any relevant text except the words 'On the voice'. However, the small pyramid within the channel marked *cdg* in Figure 9.12 is here brought into greater prominence. In a series of drawings from left to right this little 'pyramid' grows larger until it emerges from a cubical channel drawn in three-dimensional perspective. In the last drawing the pyramid is divided into equal parts in the ratio of 1, 3 and 5. Interpretation must, of course, be speculative, but it would appear that Leonardo here visualises how 'a small near sound can appear as large as a big sound far off' (Triv 7 v) as far as the eardrum is concerned. This corresponds to a similar problem with linear perspective in the visual field. He shows, too, how such a point of origin of sound outside the meatus can be geometrically transmuted into an equivalent cube-shaped portion of the meatus.

At the top of Figure 9.13 is the well-known note, 'Messer Battista dell'Aquila steward in waiting to the pope has my book in his hands'. Beneath this, above the drawings just described, he writes, 'On the voice'. It seems doubtful whether the two entries should be taken together as evidence the Leonardo wrote a book 'On the voice' which was in Messer Battista's hands, though this has been commonly assumed.

The function of the ear, asserts Leonardo, 'is to receive the infinite images [*spetie*] of sounds' (W 19045 v; K/P 50 v). The question of sentient impressions is also discussed at length. After asserting that all such impressions are retained for some time in a sentient object, Leonardo goes on to say, 'The sentient impression is that of the blow received in a resonant object like a bell, and similar things such as the voice in the ear. For if this did not preserve the impression of voices it could not derive pleasure from a solo song, because leaping from the first to the fifth is like hearing two voices at the same time, and thus one hears the true harmony which the first makes with the fifth. But if the impression of the first note did not remain in the ear for some space of time that which follows immediately after the first would appear to be alone' (CA 360 ra).

Leonardo is fascinated by the intervals which form harmonious chords and at one time tried to correlate musical intervals with visual perspective: 'I will make my rule on intervals measuring 20 *braccia* each, just as the musician does with voices, which although united and strung together none the less he makes intervals by degrees, from voice to voice, calling them unison, second, third, fourth and fifth, and so on until names have been given to the various degrees of high and low pitch of the human voice' (BN 2038 23 r). He later abandons this analogy, though obviously reluctantly. The rhythm of musical harmony he describes beautifully as surrounding 'the proportionality of the parts composing its harmony as the contour bounds the members from which human beauty is born' (TP I 39). However, 'Music born from continuous and discrete quantities

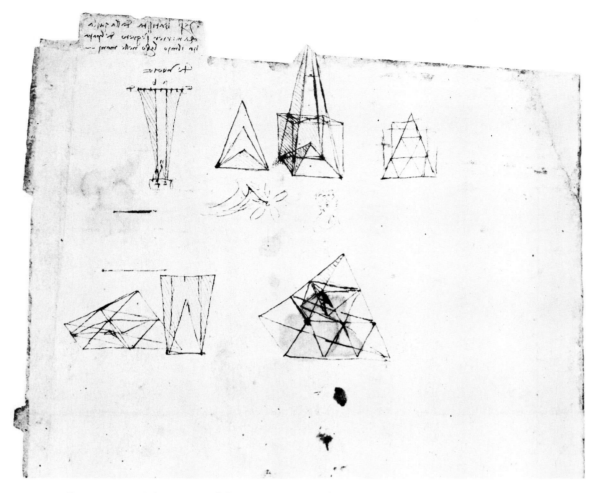

Figure 9.13. The pyramidal concentration of sound entering the ear (CA 287 ra).

Here Leonardo shows how the concentration of sound into the dimensions of the extended auditory meatus can be achieved. He shows, too, how the big sound arising from *cd* can resemble the little sound arising at *n* by a process of three-dimensional pyramidal transformation.

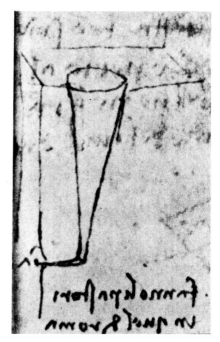

Figure 9.14. Amplification of sound of a shepherd's horn (K 2 r).

Leonardo notes how if a small horn inserted into a rocky mountain at *a* is blown, the sound will be amplified by resonance. He notes that this amplification is made by a conical or 'pyramidal' form.

is devoted to the ear, a sense less worthy than the eye, by which so many various concords of diverse instruments are sent to the *senso comune*' (Madrid II 67 r).

Preservation of '*spetie*' or 'images' in the ear corresponds to the similar retention of 'images' in the eye. Thus resonating impulses in the form of wave-fronts are despatched into the internal ear. This, it will be remembered, is described by Leonardo simply as 'hollow porosities in the petrous bone'. This bone he has beautifully drawn, even to the delineation of the rounded curve of its arcuate eminence (W 19058 r; K/P 42 r). Thus, Leonardo would be very ready to see an analogy between this cavernous, petrous (rocklike) bone and a cavern in the rocky hills of the Romagna. In Figure 9.14 he observed that 'The peasants make large concavities in the mountain in the form of a horn. On one side they fasten a horn, and this little horn becomes one and the same with the concavity already made; whence is made a great sound' (K 2 r). In this way resonance is once more brought into the process of hearing. And amplified impulses of sound will be sent along the auditory nerve, whence after joining other sensations in the *imprensiva,* they find their controlled way to judgement of the soul in the *senso comune*.

REFERENCES

1. Winternitz, W. Emmanuel. 'Leonardo and Music', in *The Unknown Leonardo,* ed. by Ladislao Reti. London, Hutchinson, pp. 110–135, 1974.

Chapter 10

The Senses and the Soul

THE SENSE OF SMELL

Leonardo's personal sensitivity to smell often emerges in his notes. In an early recipe (about 1480) he writes, 'Take rose water and moisten the hands; then take the flower of lavender and rub it between the hands; it is good' (CA 295 ra). A few years later he recommends another recipe, 'If you want to make a stench, take human stools and urine, stinking goosefoot . . . place them together in a glass jar, tightly sealed, for a month beneath manure; then throw it where you wish to make a stench, and break the jar' (B 11 r). How are these scents and stenches created?

Leonardo rarely speaks of the nature of smell as an isolated source of sensation. He almost always compares it with the special senses such as sight and hearing. For example, he discusses at great length whether 'bodies send out from themselves their form'. He asserts that 'the rays from luminous opaque bodies fill all the air round with their similitudes', giving as an example the image of the sun. Then follows another, 'Example. There are many things which send out images [*spezie*] of their form without harming themselves. Musk which always keeps a great quantity of air charged with its odour, and which if it is carried a thousand miles will occupy the air without any diminution of itself' (CA 270I vb and vc). He adds a similar comment about the sound of a bell. Smell, therefore, Leonardo views as a set of forms which comes from bodies in a way similar to light emanating from the sun or sound from a bell. This odoriferous form also is taken up into the air. 'Air on its own is cold and dry and it is void of all matter or vapours; and it changes readily, or to put it more truly, is infused with the nature and image of those things which touch it such as objects. As regards things that touch it, when an odoriferous thing like musk or sulphur or other powerful odour touches it, it is instantly infused into the air; also if a luminous body be placed within it the whole of the surrounding air will be lit up' (A 20 r).

Smells then stimulate the special receptor in the nose in a manner similar to that of light and sound, i.e., by percussion. But there is an important difference. 'Smell', he writes, 'with different odours pleases the *senso comune*. But although these odours give rise to fragrance, a harmony similar to music, nevertheless it is not in the power of man

to make a science of it. And the same applies to taste and touch' (Madrid II 67 r).

THE ANATOMY OF SMELL

Leonardo draws the olfactory tracts in his earliest explorations of cerebral anatomy, as in W 12626 r (K/P 6 r, Figure 2.21, p. 62) and W 12627 v (K/P 4 r). In both of these drawings the olfactory tracts are shown passing back to the middle ventricle, whilst the optic tracts run to the anterior ventricle. The olfactory tracts are faintly outlined in Figure 2.25 (p. 64), passing back from each side of the crista galli towards the *senso comune*, located by the intersection of two lines. Prominent bulges are drawn on the frontal bone. These prominences, which appear in most of Leonardo's skulls, are found again on W 19059 r (K/P 40 r) and W 12602 r (K/P 103 r, Figure 2.28, p. 66). In Figures 2.23 and 2.24 (pp. 63, 64) they are shown to coincide with the inner ends of the frontal sinuses, first brought to light by Leonardo. Evidently he associated these sinuses with the sense of smell; for the *Weimar Blatt*, Figure 2.30 (p. 67), shows the olfactory tracts running to the base of these prominences which Leonardo has labelled '*caruncholi*'. This is a term used by Mondino for the olfactory bulbs. Leonardo uses the same term again in Figure 2.29 (p. 66), which contains his best drawing of the cranial nerves. Here the olfactory bulbs are clearly shown lying on each side of the perforated cribriform plate, the olfactory tracts passing back above the optic nerves and chiasma. Leonardo writes, 'The optic nerves are situated below the nerves of the caruncles. But the optic nerves serve the visual power [*virtu visiva*] and the caruncles the olfactory power [*virtu dell'odorato*]'. We have already noted how in Figure 8.4 (p. 204) the olfactory tracts are drawn by Leonardo running back to the 'anterior' ventricle, where the *imprensiva* is located. From here the waves of this sensation pass to the *senso comune* in the middle ventricle.

Leonardo's association of the sense of smell with the frontal sinuses may well have been the result of following up observations recorded in Figure 17.3 (p. 347): 'I have found that in the composition of the human body as compared with the composition of all animals, it is of blunter and grosser sensitivity as it is composed of less ingenious instruments, and of localised parts less capable of receiving sensory power [*la virtu de'sensi*]. I have observed in the leonine species that the sense of smell is endowed with a part of the brain substance which descends into a very capacious receptacle to contact the sense of smell, which enters among a great number of cartilaginous cells [the ethmoid sinuses] with many passages which lead to the aforesaid brain'.

He continues this passage with a similar comparison between the optic nerves of the lion and man, once more emphasising the surprising relative imperfection of the nerve in man.

TASTE

This sense Leonardo relegated to the group which is useless for science. He compares the multiplicity of sensations received by the different special senses – the infinite varieties of form and colour received by the eye, the infinite mixture of odours received by the nose, etc., and he finishes up by asserting that 'The tongue also perceives an infinite number of flavours, both simple and compound' (W 19045 v; K/P 50 v). Leonardo extends the sense of taste to 'the uvula where one tastes food', he confidently states (W 19058 r; K/P 42 r). Taste buds are indeed found there.

It is interesting to find that although Leonardo abandoned this sense as useless for science, he did not omit taste in a late assertion of his pyramidal law that 'the same power [*virtu*] is so much more powerful as it is more concentrated' (G 89 v; see Figure 4.44, p. 129). Having applied this to light, heat, percussion and other 'powers', he adds, 'So also such qualities as sweetness, bitterness, sharpness, roughness, do the same as has been stated above. And an example of this is shown when any of these is increased in quantity by being mixed with snow or water, which neither gives nor takes away flavour' (G 89 v), and 'water takes up odours and flavours with ease' (A 26 r).

TOUCH

About 1487 Leonardo undertook his only indubitable experiment in vivisection, the pithing of a frog. This provided him with ideas about the pattern of nervous action, which will be described later. Here it is relevant to note that in Figure 10.1, on the same page, in a diagram accompanying the experiment, he labels a cordlike structure, 'the origin of the nerves' and 'the sense of touch' (W 12613 v; K/P 1 r). Until the frog had been pithed it continued to show (reflex) response to touch. Therefore Leonardo gained the impression that touch was centered in this region of the nervous system, i.e., the medulla oblongata and the fourth ventricle, 'at the end of the spinal cord'.

This experience influenced him at a later date when he injected the cerebral ventricles with wax, so outlining their shape (see Figure 2.27, p. 65). For here he must have had this experiment in mind when he wrote, 'Since we have clearly seen that the ventricle *a* is at the end of the spinal cord where all the nerves which give the sense of touch come

together we can judge that the sense of touch passes into that ventricle, since Nature operates in all things in the shortest time and way possible' (W 19127 r; K/P 104 r). Two of his accompanying drawings of the injected ventricles on this page show the fourth ventricle significantly connected with the spinal cord, so illustrating his statement. Yet on the same page he has labelled the lateral ventricles '*imprensiva*', into which he has said all senses pass on their way to the *senso comune,* and it is difficult to see how the sense of touch can pass from the fourth ventricle to the *imprensiva* before reaching the *senso comune* in the middle (third ventricle). Here Leonardo is up against the difficulty of incompatible observation and theory.

Leonardo is interested in the distribution of the sense of touch over the body. 'The sense of touch', he writes, 'clothes all the superficial skin of man'. On the same page he discusses, 'How the 5 senses are the servants of the soul'. He says of touch, 'Touch passes through the perforated nerves and is carried to the *senso comune*. Its nerves proceed spreading out into infinite ramifications in the skin which encompasses bodies, limbs and viscera. The perforated nerves carry orders and sensation to the functioning parts . . . being spread into the extremities of the digits they carry to the *senso* [*comune*] the cause of their contact . . . the *senso comune* is the seat of the soul, and memory is its store and the *imprensiva* is its standard of reference.

'How the sense gives to the soul and not the soul to the sense; and where a sense is lacking which should administer to the soul, the soul in such a life lacks information from the function of this sense, as appears in the case of a [deaf] mute or one born blind' (W 19019 r; K/P 39 r).

Of all parts of the body using touch the hand is for Leonardo the most significant. He early performs the Aristotelian experiment of feeling an object with the fingers crossed, calling it 'An experiment with the sense of touch. If you place your second finger under the tip of the third in such a way that the whole of the nail is visible on the far side, then anything touched by these two fingers, providing that it is a round object, will seem double' (CA 204 va).

Many years later beside a magnificent drawing of the distribution of the median and ulnar nerves to the hand and fingers (Figure 10.2), Leonardo makes a series of comments about the sensation of touch: 'See if you think that this sense [touch] is affected in an organ player at the same time as the mind attends to hearing'. This is a straight reference to his belief that sensory nerve impulses, travelling as waves up a nerve, take time.

Here he recalls the experiment of *Codex Atlanticus* 204 va, done some twenty years before, and presents his answer in terms of anatomy (Figure 10.2): 'Why does one and the same thing touching the side of the finger, *b,* and the side of the 2nd finger *a* seem to be two things; and if it touches at *nm* it seems to be one? It is because *n* and *m* arise from one nerve only whilst *a* and *b* arise from two nerves . . . If one and the same object is touched at *c* and *n* it appears to be two'. In this case one part is supplied by the median nerve, the other by the ulnar. But this solution of the double sensation of touch

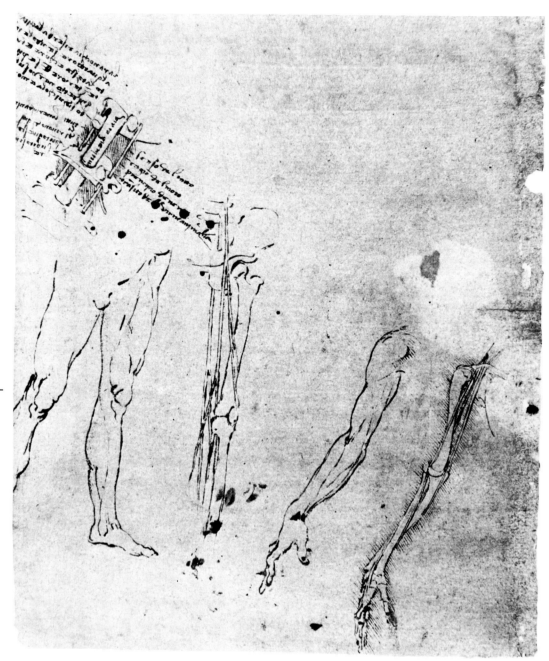

Figure 10.1. Experiment on the spinal cord of the frog (W 12613 v; K/P 1 r).

Top left: the spinal cord is shown emerging from the base of the frog's skull. Along it is written 'generative power'. Two other cords run parallel to the spinal cord. These Leonardo labels 'Sense of touch; cause of motion; origin of the nerves'.

Below right: Leonardo follows three such nerves down the arm corresponding to the ulnar, radial and median nerves. At this early date he has no concept of the brachial plexus.

or pain, ingenious though it is, is not correct. The tips of the fingers concerned are labelled *b, m, n, a, c* in the figures alongside, and the distribution of the median and ulnar nerves to the hand and fingers is wonderfully displayed. Leonardo always paid attention to the effects of injury to parts of the body. Here he notes, 'Here, following a cut in the hand sometimes the sensation and not the movement is lost; sometimes the movement and not sensation. Sometimes it is both sensation and movement' (W 19012 v; K/P 142 v). See Figure 10.2.

He carries this observation further in Figure 10.3. Here in the right-hand margin he draws two fingers from the side, clothed with their flexor and extensor tendons, blood vessels and nerves. The lettering below the finger labels these

structures, each of which receives a comment in the text beneath. Here the letter *c* represents the nerve. Leonardo writes, '*c* is the nerve which gives sensation, etc. This being cut, the finger no longer has sensation even though placed in fire. For this reason skilful nature took care to place it between one finger and the next so that it would not be cut' (W 19009 v; K/P 143 v).

Leonardo also gave separate attention to temperature sense, although it was then usually included under touch. It is worthy of note that in describing the reflection of heat from the surface of a concave mirror he assumes a pyramidal concentration of those rays and intensity of sensation as they converge: 'As many times as this point enters into the surface [of the mirror] so many times will this point

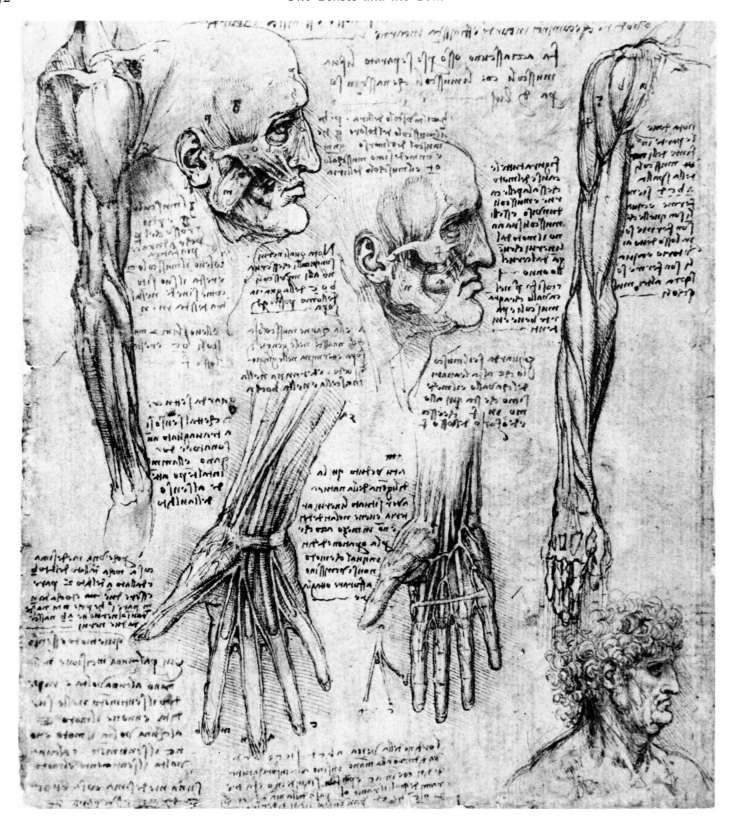

Figure 10.2. Dissections of the nerves and arteries of the hand (W 19012 v; K/P 142 v).

In the left-hand drawing Leonardo makes a meticulously accurate dissection of the terminal branches of the median and ulnar nerves. He relates this pattern to the sensation of touch at the finger-tips. To the right he makes an equally accurate representation of the superficial palmar arch formed by the ulnar artery. Above these drawings are his last dissections of the facial muscles.

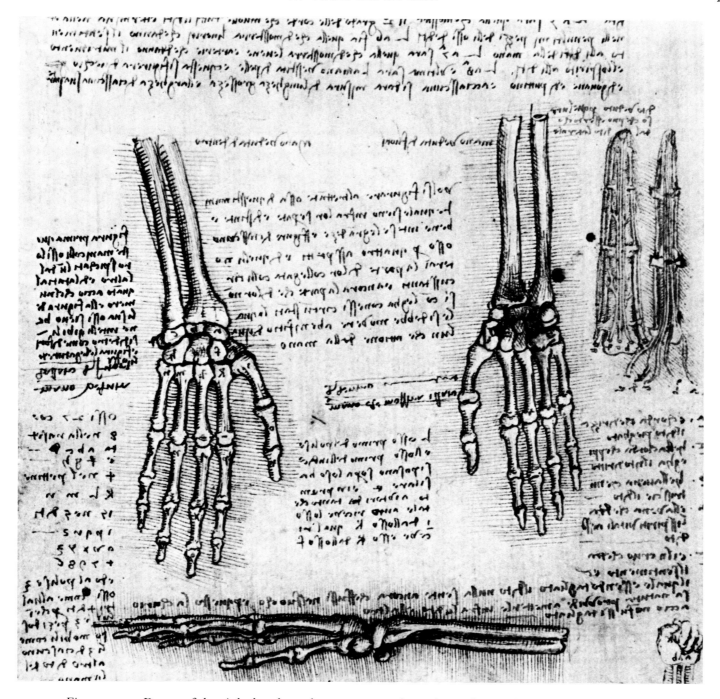

Figure 10.3. Bones of the right hand; tendons, nerves and arteries to fingers (W 19009 v; K/P 143 v).

To the right of unprecedented drawings of the carpus and digits Leonardo draws two fingers clothed by their tendons, nerves, arteries and veins. These are lettered. Below the drawing the letter *c* refers to the 'nerve which gives sensation'. (See also Figure 12.13a, p. 263.)

be hotter than the heat found at the surface of the mirror' (A 20 r; see Figure 2.19, p. 59).

On this same page, Leonardo draws a sketch of a circle of bellows blowing 'cold and dry air' on to a central point in order to obtain a similar concentration of cold sensation. 'Many puffs on the same point make extreme cold', he writes.

PAIN AND TRAUMA

Leonardo's keen awareness of all forms of sensory perception is evident with pain as well as the other sensations and emotions. In his art we find this depicted in the picture of *St. Jerome* (Figure 1.1, p. 10). In his science we find it in his investigation of the damaging effects of percussion on the sensitive human body. His antipathy to vivisection expresses Leonardo's intense aversion to inflicting pain.

The picture of *St. Jerome* is notable for at least two features of the physiology of pain. First, there is his knowledge of the bodily expression of pain; this is particularly vividly displayed in the agonised action of the muscles round the mouth, eyes and forehead. Much anatomical study must underlie this. Expressions of pain are described on BN 2038 31 r in terms of 'wrinkled brows and clenched fists'. And later he describes the medial part of the frontalis muscle as 'the muscle of pain' (W 19012 v; K/P 142 v). The second notable feature of this picture is St. Jerome's chosen method of self-torment by striking his chest with a stone producing the painful dark bruise shown over his heart. Damage produced by percussion, whether by sunlight on the eye, a cannon ball on a fortress wall, or by a stone on the sensitive human body was of absorbing interest to Leonardo.

Pain, in Leonardo's view, arises from bodily 'damage', most commonly produced by percussion. Not only does he consider 'simple' percussion as in the picture of *St. Jerome*, but he also compares this with the effects of 'compound' percussion, i.e., that carried through intervening bodies which percuss one another successivly. 'The hand that holds the stone', he writes, 'when it is struck by a hammer feels only a part of the pain which the stone would feel if it were a sentient body' (A 33 r). 'The body that receives a blow is not injured on the opposite side as it is on the side percussed. Experience shows this when a stone in a man's hand is struck; for the hand holding the stone that is struck is not injured as much as it would be if it actually received the blow' (A 53 v; see Figure 4.43, p. 128). This result he accounts for by an observation many times reported that 'Percussion is very short-lived', its effect ends before it reaches the base of the object percussed, 'for this reason a percussed object like a wooden rod splays out at the top more than lower down. Similarly with the mason who breaks a stone in his hand without damaging his hand' (C 6 v). If there are several or many objects between the part percussed and the percussing force then the pain and damage are progressively diminished. 'Percussion falling through many [divided]

movements is made almost insensible' (CA 373 vb). As an example of this he cites the percussion of the front page of a book. Its many separate pages absorb the percussion until at the back of the book the blow has almost disappeared (BM 82 r). In the human body the joints fulfill a similar function, absorbing the shock of a blow successively, so that if a man jumps onto his toes the percussion of landing may become almost painless by virtue of the spring in his muscles and joints. This has been described in Chapter 6.

In contrast, 'The blow given to any dense and heavy body passes naturally beyond this body and damages everything in surrounding bodies whether they are rare or dense [see Figures 10.4 and 10.5]. Example; if you give a great blow to a rock all the fish beneath or beside it will come to the surface of the water as if dead . . . This blow will naturally pass into the whole quantity of the beaten body, and even into those bodies surrounding it. And if any animals are near these neighbouring bodies they too will be penetrated by the noise and movement so that, not being able to resist, they are left as dead. And their pain is the same as the pain which the hand receives when a stone placed in it is beaten with a hammer' (A 31 r). Destruction of life by percussion is again described in *Codex Atlanticus* 363 vd (quoted in Chapter 4, p. 128).

Pain produced by jumping onto the heels as compared with jumping onto the toes is described several times (e.g., CA 144 ra, p. 129, and 338 ra, in Chapter 6, p. 181), the principle being, 'That which gives more resistance to a blow suffers most damage' (A 36 r); thus, landing on the heels is more noisy and more painful than landing on the toes. In his efforts to prevent such pain Leonardo devised the apparatus drawn in Figures 6.39 and 6.40 (pp. 193, 194), which has been described in Chapter 6, pp. 193–194. He writes about pain in the foot thus: 'On the nature of percussion, giving as examples the fall of Simon Magus and of someone who jumps on to the points of the toes, or receives in his hand, as stone-masons do, bricks thrown from a height. Either, like a man who has pain in the foot so that when he stands up he lets his weight fall on that side; or like someone who puts his feet on something he does not want

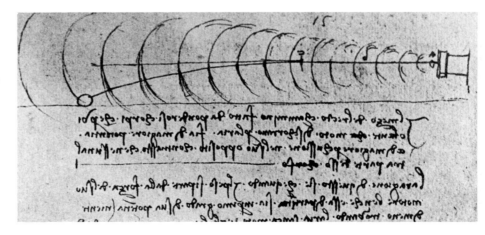

Figure 10.4. Blast waves from a gun and the trajectory of the cannon ball (A 43 v).

Percussion from firing a gun, or from a blow of any sort, spreads in waves. Maximal damage is produced along the central line *abc* of the wave, followed at first by the cannon-ball, which then declines and falls to the ground as gravity exerts its 'natural' power overcoming gradually in pyramidal proportion the 'accidental' force of the initial percussion.

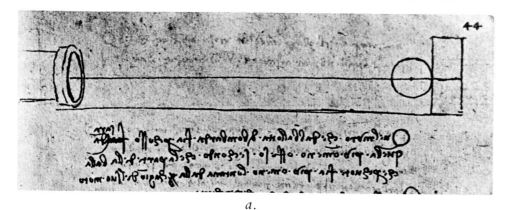

a.

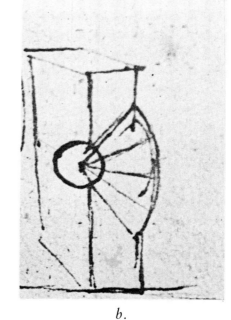

b.

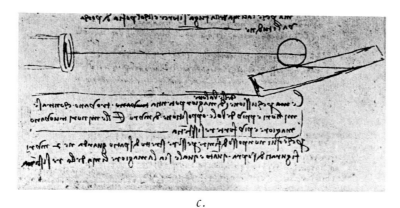

c.

Figure 10.5.

a. Percussion along the central line; damage maximal (A 44 r).

The percussion of any force is maximal along the central line. Here it produces most pain and damage. The resistance of the wall to the cannon ball is maximal.

b. As the ball percusses, its waves of force radiate from this point producing damage along the lines shown in Figure 10.5b. If the wall is replaced by the human body, e.g., head or legs, then pain, trauma or death are produced, whether the medium is air or water, as with the fish cited in the text (A 28 r).

c. How damage is diminished by percussion on an inclined plane (A 44 r).

Here the impact is away from the central line; the cannon ball glances off, and its energy is absorbed in rotation. This principle Leonardo applied to fortresses (see Figures 1.22 and 1.23, p. 31) or the protection of men's bodies.

to tread on. Whence the explanation [*ragione*] how it would be possible to walk placing the bare foot on the point of a knife, or even on an egg without breaking it' (Madrid I 62 v).

Leonardo's curiosity about the pain produced by percussion took him into the observation of human torture. In this content he notes, 'A blow given to the rope by which tortured men are lifted into the air does from this cause double their pain' (Madrid I 180 r). He is referring to the contemporary form of torture by '*colla*' in which a man is hung up by a rope attached to his arms bound behind his back. Elsewhere Leonardo asks, 'What difference is there between the colla being tightened by force and tightening it with a blow?'

(CA 65 va). He illustrates this form of torture in one of his pictographs or rebuses in Figure 10.6.

Figure 10.7, the better-known drawing of the hanged man, Bernardo di Bandino Baroncelli, may well show the effect of percussion on the cervical vertebrae as the result of the sudden end of a long falling movement. The victim's head tilts fatally forward. 'All injury', writes Leonardo, 'leaves pain in the memory except the greatest injury, that is death, which kills memory with life' (H 33 v).

Leonardo realised that the suffering of pain depended upon the sensitivity of the individual as much as on the injury itself. 'Where there is most power of feeling there of martyrs is the greatest martyr' (Triv 23 v). This significant

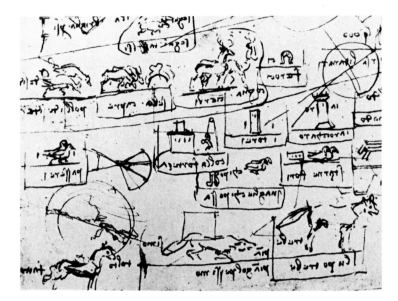

Figure 10.6. Torture by the *'colla'* (W 12692 r).

In the centre of this extract from a page of picture-writing is a small sketch of a man suspended by a rope round a pulley. The rope is attached to his arms, which are bound together behind him. Beneath the figure Leonardo writes the word *'colla'*, upon which he is making a pun in the illustration.

assertion is followed by one of Leonardo's long lists of words, some 400 of them, which significantly enough are related to pain and suffering; *crucified, tormented, suffering* and *percussed* are among them.

Leonardo also observed that the front of the body was more sensitive to pain than the back. He notes, 'In the movement of man, nature has placed in front all those parts which on being percussed cause a man to feel pain; and this is felt in the shins of the legs, in the forehead and nose. And this is done for man's preservation, for if he were not prepared for pain in these parts the many percussions received by such parts would certainly cause his destruction' (W 19141 r; K/P 99 r). He continues this line of thought elsewhere: 'Though nature has given sensitivity to pain to such living organisms as have the power of movement in order thereby to preserve the parts which are liable to be destroyed, living organisms which have no power of movement do not have to encounter objects in front of them. Plants consequently do not need to have sensitivity to pain and so it comes about that if you break them they do not feel pain in their parts as do animals' (H 60 r). He sums all this up in the phrase, 'Pain is the preservation of the instrument' (H 32 v).

Pain results not only from gross mechnical percussion but also from percussion of other 'powers', for example, excessively bright light and heat. 'Pain from bright light on the eye' with the threat of destruction of the organ stimulates him to devise a method 'for seeing an eclipse of the sun without the eye suffering' (Triv 6 v). The mechanism of such pain he describes as due to 'the sudden closing of the

pupil with sudden touching, and friction of the sensitive parts of the eye' (C 16 r). In a pupil dilated in darkness, 'even moderate light is destructive' (F 50 r). Incidentally, this experience of pain from light constituted for him a strong argument for the intramission theory of vision as opposed to the extramission of visual power from the eye to the object seen.

Figure 10.7. The hanging of Bernado di Bandino Baroncelli.

Hanged for his part in the assassination of Giuliano de' Medici, Bandino's body was drawn by Leonardo, who, amongst others, was commissioned to make a portrait. The text beside this sketch consists of details of the colours of the man's clothing. Though no care was taken in those days to ensure a drop long enough to break the neck of the victim, Leonardo's drawing showing the position of the head and body in relation to the rope suggests that the man's cervical vertebrae were fractured by the fall (Bonnat Collection, Bayonnc).

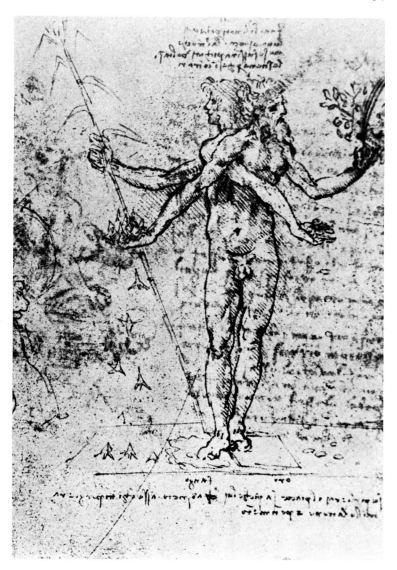

Figure 10.8. Allegorical representation of Pleasure and Pain.

Like Siamese twins, these two are joined together in one trunk but with two heads and shoulders, in which Leonardo contrasts youth and age. These two look in opposite directions. The youth holds a reed symbolic of 'vain and lascivious pleasures' in his right hand. As Leonardo explains, 'In Tuscany such reeds support beds . . . and this reed is useless . . . and the wounds made with it are poisoned'. In his left hand the youth scatters gold. The old man holds a branch of laurel in his left hand, whilst from his right he drops calthrops, sharp pieces of iron which were painful to tread on (Oxford Drawings, Part II, No. 7).

Though pain served to protect the body, in excess it appeared to Leonardo as the 'chief evil' of life. 'The chief good is wisdom', he writes, 'the chief evil is bodily pain. As we are composed of 2 things, that is, soul and body, of which the soul is the better and the body the worse, wisdom is the best and the chief evil belongs to the worse part and is the worst. The best thing is the wisdom of the soul; the worst thing is pain of the body' (Triv 2 v).

At the emotional level Leonardo saw pleasure, as did Plato, as a twin of pain. In Figure 10.8 he portrays Pleasure and Pain as two beings arising from one body. 'Pleasure and Pain', he writes, 'are represented as twins joined together, since there is never one without the other; and they turn their backs to each other since they are contraries. If you choose pleasure know that he has behind him one who will deal out tribulation and repentance . . . They are made growing out of the same trunk because they have one and the same foundation; for the foundation of pleasure is labour with pain, and the foundations of pain are vain and lascivious pleasure' (Oxford Drawings, Part II. No. 7).

THE *SENSO COMUNE* AND THE NATURE OF THE SOUL

One of Leonardo's very rare physiological investigations consisted of pithing a frog. What induced him to do this? Leonardo quotes Livy (Book 27, Chapter 49), where he gives an account of elephants being killed by their drivers by 'driving in a chisel between the ears where the head joins the neck' (B 9 v). This, it appears, was 'the most rapid death that could be given'. Beside the passage Leonardo draws a sketch of the instrument used for the purpose, derived from Valturio's account. This important experiment (not known to have been repeated until 1739 by Alexander Stuart) is described twice. He writes, 'The frog instantly dies when the medulla of the spine is perforated; and previously it lived without head, without heart or internal [organs] intestines or skin. Here therefore appears to lie the foundation of movement and life' (W 12613 r; K/P 1 v).

In Figure 10.1 (p. 231), beside a drawing of the spinal cord issuing from the base of the skull, he writes, 'The frog retains life for some hours when deprived of the head and heart and all the interior organs. And if you prick the said nerve it suddenly twitches and dies'. Below this he adds, 'All the nerves of animals derive from here; prick this and it instantly dies' (W 12613; K/P 1 r).

The experiment appears conclusive. Yet Leonardo's difficulties in interpreting this result are shown by his labelling the spinal cord as it emerges from the skull, 'generative power'. And it is to two cordlike structures alongside this that he gives the labels, 'sense of touch' 'cause of movement', 'origin of nerves', 'transit of animal powers'. It is evident that Leonardo has at this time accepted the Platonic and Hippocratic notion that the semen of man is derived from the white soft substance of the spinal cord. This he illustrates in the coitus figure in Figure 11.2 (p. 244). It is a view he later abandons, but at this stage it gave him obvious difficulties in allocating sensory and motor powers to the spinal cord. He erroneously made these powers pass through the foramen of the transverse processes of the cervical vertebrae.

The experiment, however, astounded Leonardo. It was the first of its kind dealing with the ancient problem of the essential nature of life and death, a problem that even in the twentieth century we have been unable to define clearly. As for the ancients, Plato and Galen had allotted the properties of life in varying degrees to the liver, heart and brain. Aristotle had placed the seat of life uncompromisingly in the heart. And here was Leonardo dealing with a frog, 'deprived of head, heart and all the interior organs' yet retaining life until its upper spinal cord, in the region of the medulla oblongata and fourth ventricle, was destroyed. He was driven to the conclusion, shared today, that ultimate death is brain death. And if the soul is the essence of life, this is where the soul must be located.

But, as we have seen, his later investigations of the special senses of sight, hearing, and smell led him to find their sensory nerve paths converging further towards what we now call the third ventricle. Moreover, when he later found a copy of Mondino's *Anatomy,* in the section on the brain he would read these words, 'in the middle is the *sensus communis* which comprehends the species brought by the special senses, and so the sensitive faculty ends here as do streams at a fountain'. Thus Leonardo's own experience, gained by dissection, was confirmed by authority. Leonardo henceforth confidently located the soul in the *senso comune* in the third ventricle.

What was Leonardo's view of the nature of this soul? Of course, it was in the first place 'spiritual'. This latter word Leonardo defined as a force or power of nature which is 'incorporeal and invisible . . . because of active incorporeal life; and I call it invisible because the body in which it is created does not increase either in weight or size' (B 63 r). Spiritual entities were for Leonardo energy or 'power' entities. These can be converted into 'material movement'. Consider the example of a nerve impulse where 'spiritual movement flowing through the limbs of sentient animals enlarges their muscles. Whence, being enlarged, these

muscles contract and draw back their tendons . . . And from this arises the force and movement of human limbs. Therefore material movement arises from the spiritual' (BM 151 v). This description has a striking similarity to his description of the movement of a point, which 'has location without the occupation of space. It is given in nature; it is movable together with the place within which it resides; the movement of a point describes a line . . . surface is generated from the transverse movement of a line . . . and body is made of movement' (BM 159 r).

Thus Leonardo sees the material body as ultimately derived from spiritual force through movement. For him, just as God is the soul or spiritual force which creates the universe, so the human soul is the spiritual force which creates the human body. Moreover, man's spirit, like that of God, is creative. This is evident from his crafts, skills and works of art, as expressed, for example, in machines, architecture and painting. Indeed, it is only in this creativeness that man differs significantly from animals, 'In fact man does not differ from animals except in accidental things; it is in this that he shows himself to be a thing divine. For where nature finishes in the production of her forms or species, there man begins to make with the aid of nature an infinite number of forms with natural things. And as these are not necessary to those so well adjusted [*ben si correge*] as are animals, they are not disposed to search for them' (W 19030 v; K/P 72 v).

Man's creativeness is enhanced, if not dependent upon, his comprehension of the macrocosm or world of nature. Such comprehension can be obtained only through the senses, through 'Experience, the interpreter between resourceful nature and the human species'. It is experience that 'teaches that what nature works out among mortals is constrained by necessity and cannot operate in any other way than that which reason, its rudder, teaches it to work' (CA 86 ra). Science for Leonardo was clearly a half-way house to creative art. Science was not fulfilled unless and until it was manifested in creative art. And if science is the product of the sensory side of human activity, particularly that of the sight of the eye, then art is the creative motor activity flowing from this knowledge. Leonardo saw this transmutation of energy as occurring in the *senso comune,* where the soul was located. 'The soul apparently resides in the seat of judgement and judgement resides in the place where all the senses meet together, which is called the *senso comune . . .* The *senso comune* is the seat of the soul, memory is its monitor, and the *imprensiva* is its standard of reference' (W 19019 r; K/P 39 r). This early conclusion he never abandons. It is worth noting that Leonardo describes the soul as 'residing' in the *senso comune,* just as he describes the geometrical point which 'has location without the occupation of space' as 'residing in a place'. He conceives both as belonging to the spiritual realm of power or energy – as non-material entities.

For many years Leonardo believed that 'Experience is never at fault, it is only your judgement that is in error' (CA 154 rb). But as time and his investigations progressed he realised that some, if not many, forms of experience did not represent 'real truth'. Quite early he began to realise this limitation of the artists' linear perspective; even more limited were those other sensations which aroused emotions; reactions such as pleasure and pain. With regard to experience from the senses, particularly perspective, Leonardo responded by further and deeper study in an endeavour to explain such anomalies. But pleasure and pain, as we have already seen, he came to look upon as 'evils' which inevitably falsified the soundest judgement, which he called 'wisdom'. 'The best thing in the soul is wisdom . . . wisdom is the supreme good of the soul, that is of the wise man, and nothing else can be compared to it' (Triv 2 v).

As we have seen already, the motor side of the soul expressed itself in the movements of man in terms of emotional expression, gestures and local motion from one place to another. (See Chapter 6, pp. 159–161.) All such movements were, in Leonardo's view, transmitted down nerves from the soul in the *senso comune,* possibly also from the fourth ventricle of memory which at one time he also labelled 'Will' – but on this point, as we have noted, Leonardo seems to have remained vague. In any case, he accepted that motor impulses pass down the spinal cord and emerge through the spinal nerve roots, whence they find their way via the peripheral nerves to the muscles. It is to be noted that Leonardo uses the same word, '*sentimento*', to describe nerve impulses travelling along sensory and motor nerves. He describes motor nerves as 'couriers of the soul which have their origin from its seat, and command the muscles that move the parts of the body at the good pleasure and will of the soul' (W 19088 r; K/P 175 r). Here the 'command' is equated with the word *sentimento,* as it is in describing the nerve-supply of the intercostal muscles: 'And the nerves which give a stimulus [*sentimento*] to these muscles take their origin from the spinal cord which passes down the spine of the back' (W 19047 r; K/P 48 r). Leonardo evidently believed that all nerves were 'perforated' by a small central tube. Along this travelled the nerve *sentimento,* stimulus or impulse, a force created by movement of the soul transmitting a wave front of 'power' to the muscles. But at the muscle a difficulty arose for him. How did this stimulus produce 'enlargement and shortening' of the stimulated muscle in its contraction? Like so many of his mechanistically minded successors, Leonardo sought for a fluid which might inflate the muscles. He thought of the possibility of air 'pressing and dilating the different passages of the brain and so travelling down the perforated nerves' (W 19064 r; K/P 157 r). This proposition entailed answering the question whether the soul, or spirit, can be 'a quantity' or not. If, for example, the soul is a fine air (or *pneuma*) or even a vacuum, it will rarefy the air with which it is mixed. Consequently, this air, made lighter than the surrounding air, will fly 'upwards of its own accord, and will not remain among the air which is heavier than itself. And furthermore as this spiritual power spreads out it disintegrates and alters its nature thereby losing something of its former power' (W 19048 r; K/P 49 r). He clinches his argument in answering his own question, 'Whether the spirit, having taken a

body as air, can move by itself or not. It is impossible' (W 19047 v; K/P 48 v). He asserts, 'The definition of a spirit is a power united to a body, because by itself it can neither support itself nor take any kind of local movement' (W 19048 r; K/P 49 r). Thus body and soul, for Leonardo, are one indivisible entity. 'The soul desires to dwell within the body because without the organic instruments of that body it can neither act nor feel' (CA 59 rb).

THE BIRTH AND DEATH OF
THE BODY AND SOUL

Leonardo emphasised the essential unity of body and soul from the beginning to the end of human life. Of the infant in the mother's womb he says, 'One and the same soul governs these two bodies, and the desires, fears and pains are common to this creature as to all other parts animated by the soul' (W 19102 r; K/P 198 r). During pregnancy 'The soul of the mother which first composes within the womb the shape of man, then in due time, awakens that soul which is to be its inhabitant' (W 19115 r; K/P 114 v). The food of the pregnant woman contributes to the maintenance of her own body as well as the creation of new life within her. 'We make our life', writes Leonardo, 'by the death of others. In dead matter there remains insentient life, which on being united to the stomachs of living things resumes a life of the senses and intellect' (H 89 v).

In the mature adult body, shaped by the soul, Leonardo notes that limbs are moved by levers of the second class. This leads him to express profound admiration for nature's subtle mechanics: 'Nature does not go in for counter weights when she makes organs suitable for movement in the bodies of animals, but she places inside the body the soul, the composer of this body' (W 19115 r; K/P 114 v). His conviction of the possibility of obtaining wisdom through the soul is reiterated: 'Wisdom is the food and true riches of the soul; for so much more noble are the possessions of the soul than those of the body' (CA 119 v).

Disease and age cannot affect the soul, asserts Leonardo. 'The soul can never be corrupted by the corruption of the body; it works in the body like air which causes the sound of the organ, in which when a pipe breaks it ceases to have any good effect' (Triv 40 v). Marvelling at the human body as a wonderful work of nature, Leonardo adds, 'and if this his composition appears to thee a marvellous construction remember that it is nothing as compared with the soul which dwells within that structure. For truly whatever it may be it is a divine thing. Leave it then to dwell in its work at its good pleasure, and let not thy rage or malice destroy such a life. For in truth he who values it not deserves it not. And since it parts from the body so unwillingly I indeed believe that its cries and pain are not without cause' (W 19001 r; K/P 136 r).

Seeing the body as but a transient being, formed and activated by the spiritual powers of the soul, Leonardo accepted its inevitable dissolution into the shapeless chaos called death. He expresses it thus: 'You see that the hope and desire of returning home into the first state of chaos is like the moth to the light. And the man who with constant longing awaits with joy each new spring, each new summer, each new month and each new year, deeming that the things he longs for are ever too late in coming, does not perceive that he is longing for his own destruction. But this desire is the very quintessence, the spirit of the elements, which finding itself enclosed within the soul of the human body is ever longing to return to its Author. And you must know that this same longing is that quintessence inseparable from Nature, and that man is a model of the world' (BM 156 v). 'While I thought I was learning how to live I have been learning how to die', he adds (CA 252 ra).

Chapter 11

The Nervous System

It must not be forgotten that the nervous system supplied Leonardo with that 'experience' upon which he set such great store. It provided his chosen gateway into science. Much of his exploration of the sensory nervous system, particularly that on vision and other special senses, therefore stemmed from a double peak of interest, one of which was early, the other occurred years later. The former, in so far as it was related to his exploration of perspective, vision and the other special senses, we have already described. It remains to integrate his work on the central nervous system as described in the *Anatomical Corpus*.

ANATOMY

Leonardo's advances in anatomical knowledge, and of the central nervous system in particular, can be divided usefully into three stages. The first of these comprised his early explorations from 1487 to about 1495. During this period he was intensely concerned in elucidating the physiology of vision in relation to the artist's perspective and in evaluating the visual sense in relation to the other senses of the body. The second stage dates from a revival of interest in anatomy, probably following the acquisition of Mondino's *Anatomy* in Italian, published in 1493. Leonardo clearly used this as a guide to further dissection of the nervous system as well as other parts of the body. The third stage is demonstrated by his advance in certain anatomical fields far beyond Mondino's work.

In the first period Leonardo obviously had access to human heads for study but dissections of other parts were largely confined to the organs of other animals: monkeys, bullocks, dogs, horses and bears provide the major part of general anatomical illustrations of this period. In the second period Leonardo by no means abandoned such animal dissections, but human organs become frequently illustrated and increasingly accurate. This applies particularly to his studies of the cranial nerves and to the brachial plexus. In the third period his illustrations so far outstrip Mondino's anatomy as to sometimes reach comparison with modern anatomical illustration, particularly with regard to the anatomy of the nerves to the arms and legs. It is interesting to note that none of Leonardo's extant drawings of the brain and cranial nerves have been dated after 1508, and it was between this date and 1514 that Leonardo produced most of his own personal outstanding anatomical contributions. In Volume II of the Windsor *Anatomical Corpus* drawings of the brain are conspicuous by their absence. By this time Leonardo was presumably satisfied with his cerebral anatomy and physiology, though it did not reach the standard of detailed accuracy that he attained in illustrating other parts of the body.

The most progressive leap in Leonardo's cerebral anatomy is illustrated by the difference between the drawings in Figures 2.20, 2.21 and 2.27 (pp. 61, 62, 65). The former show the cavities of the three rounded cerebral ventricles. Then there is the sudden change to a remarkable approximation to the true shape of these ventricles found in Figure 2.27. The reason is explained by Leonardo at the top of the page: 'Make two vent-holes in the horns of the greater ventricles and inject melted wax with a syringe, making a hole in the ventricle of memory [fourth ventricle]. And through such a hole fill the 3 ventricles of the brain. Then when the wax has set dissect off the brain and you will see the shape of the three ventricles exactly. But first put fine tubes into the vent-holes so that the air which is in these ventricles can be blown out and make room for the wax which enters into the ventricles' (W 19127 r; K/P 104 r). Underneath the drawing to the left, in which the great anterior ventricle is labelled '*imprensiva*', the middle ventricle, '*senso comune*' and the posterior (fourth ventricle), '*memoria*' Leonardo writes, 'Shape of the *senso comune* cast in wax through the hole *m* at the bottom of the base of the cranium before the cranium was sawn through'.

The 'hole *m*' is, in fact, not marked in this drawing in Figure 2.27. Instead, the site of the fixture and nozzle of the syringe at the base of the third (middle) ventricle can be clearly seen. The syringe is also to be seen similarly situated beneath the third ventricle in the sketch of the wax-injected ventricles in the lower right corner of the page. Beneath this and to the left Leonardo describes how 'we have clearly seen that the ventricle *a* [fourth ventricle] is at the end of the spinal cord'. He then makes a sketch of the exposed part of the floor of this, the fourth ventricle. Evidently his wax injection has entered the very top of the central canal of the spinal cord, on its way distorting the median fissure in the floor of the rhomboid fossa, a distortion which also can be seen from the horizontal section of all the ventricles in the large drawing above. This sketch shows again the site of injection marked by the letter *m* at the base of the middle (third) ventricle. To the right of this is a pencil sketch of the cerebral cortex now seen from above. The longitudinal sulcus is clear and the arrangement of the other sulci and gyri correspond to those of an ox. The presence of the network of vessels, the rete mirabile, at the base of the brain drawn in the sketch in the central lower part of the Figure 2.27 confirms that this work was indeed done on the brain of an ox, for this rete mirabile is not present in man or horse.

As a result of this demonstration of the basic shapes of the cerebral ventricles Leonardo subsequently depicts them thus. (It may be noted that the wax set before it penetrated to the posterior and inferior horns of the lateral ventricles.) He shows them similarly shaped in Figures 2.28 and 8.4 and on the *Weimar Blatt,* Figure 2.30, dated about 1508 (pp. 66, 204, 67).

These pages also show a progressive knowledge of the anatomy of the cranial nerves. In Figure 2.28 the optic nerves, chiasma and tracts are the focus of attention; the olfactory tracts above can be seen ending in unduly large olfactory bulbs with an exaggerated bony prominence of the forehead in front of them. Three other pairs of cranial nerves, of doubtful identity, are shown running towards the ventricle containing the *senso comune.* The spinal cord continues to be accompanied by the two 'vertebral' cords, first drawn on the occasion of the experiment of pithing the frog (Figure 10.1, p. 231).

Progress in Leonardo's knowledge of the cranial nerves reaches its peak about 1508 in drawings in Figure 2.29 and the *Weimar Blatt,* Figure 2.30 (pp. 66, 67).

Basing his technical dissection on the procedure described by Mondino, Leonardo in Figure 2.29, p. 66, makes a step-by-step exposure of the base of the brain from the front, so revealing the cranial nerves and noting their intracranial course before they perforate the various holes in the base of the skull. With typical thoroughness he improves on Mondino's procedure: 'Such knowledge you will obtain with certainty when you diligently elevate the pia mater little by little commencing from the edges and noting bit by bit the position of the above-mentioned perforations, commencing first on the right or left and drawing that [side] in its entirety; then you will follow the opposite side which will give you information as to whether the previous side was correctly positioned or not. Furthermore it will inform you whether the right side is similar to the left; and if you find differences review the other anatomies to see whether such a variation is universal in all men and women' (W 19052 r; K/P 55 r). The results of such diligence were brilliant, as shown on the two drawings in the top half of this page. Here the olfactory tract and almost all the cranial nerves can be identified. The most noteworthy points are the delicate representation of the perforations of the

cribriform plate of the ethmoid lying between the two olfactory bulbs, *a* and *b;* the clear picture of the optic nerves from the eyeballs to the chiasma and beyond. The oculomotor nerve is easily identified; the other cranial nerves less easily so. Leonardo did not define the Gasserian ganglion and thus shows only what appears to be the ophthalmic branch of the trigeminal nerve. The relation of all these nerves to the anterior and middle fossae of the skull is beautifully shown, with their sites of perforation from the inside of the cranium.

The emergence of these nerves outside the skull is equally brilliantly shown on the *Weimar Blatt* (Figure 2.30, p. 67), where many of the holes made by their perforations are depicted. This sheet continues, as it were, the anatomy of Figure 2.29 (p. 66). Here too the olfactory tracts and bulb with the very prominent bony glabella labelled '*caruncholi*' can be seen, also the optic nerves and chiasma. The ophthalmic division of the trigeminal reappears. The maxillary nerve, seen on the floor of the orbit, pierces the infraorbital canal to emerge on the cheek. Leonardo has found the middle and anterior superior alveolar nerves supplying the upper gums and teeth but fails to locate their origin. The lingual nerve to the tongue and the inferior alveolar nerve to the lower gums and teeth are also shown erroneously, since they apparently arise directly from the base of the brain. This error again arises from his failure to detect the Gasserian ganglion.

In Figure 2.30 in the drawing below and to the right, Leonardo sums up the whole anatomy of the skull, brain and cranial nerves in an 'exploded' view, designed not to show detail but, by separating contiguous parts, to give knowledge of the integrated pattern of these structures in relation to the whole. Here he is obviously showing how the brain fits into the skull, and how the cranial nerves and spinal cord pass out of the skull through their various holes. It will be noticed that even at this date the 'vertebral' nerves still accompany the spinal cord.

Leonardo saw the nervous system (and the cardiovascular system) as integrating mechanisms of the body's function. The nervous system integrated sensory and motor activities. This is expressed not only in his percussion-wave theory of sensory and motor nerve impulses but also in Figure 11.1, which he entitles 'Tree of all the nerves' (W 19034 v; K/P 76 v). This contains three drawings. The two larger drawings show schematically the whole nervous system consisting of brain, spinal cord and peripheral nerves to arms, trunk and legs, one seen from the front the other from the left side. Between these two drawings he writes, 'Tree of all the nerves; and it is shown how all these have origin from the spinal cord, and the spinal cord from the brain'.

Above the drawing he explains the importance of seeing the relation of the nervous system to the whole body. 'The whole body takes origin from the heart in so far as its primary creation is concerned. Therefore the blood vessels and nerves all do likewise. However all the nerves are manifestly seen to arise from the spinal marrow, remote from the heart, and the spinal marrow consists of the same substance as the brain from which it is derived'.

The little drawing between the two larger drawings in Figure 11.1 reveals one of Leonardo's principles of anatomical demonstration: 'In each demonstration of the whole extent of the nerves draw the external outlines which denote the shape of the body'. This almost purely schematic drawing of the central nervous system contrasts sharply

with his attempted drawings of the spinal cord, which appear in Figures 11.2 and 11.3.

The early drawing (Figure 11.2, W 19097 v; K/P 35 v) has all the signs of an attempt to execute a plan suggested in Windsor Drawings 19037 v (K/P 81 v): 'This work should commence with the conception of man'. Here Leonardo visualises conception arising from the act of coitus, according to the ideas of Plato and Hippocrates, in which the

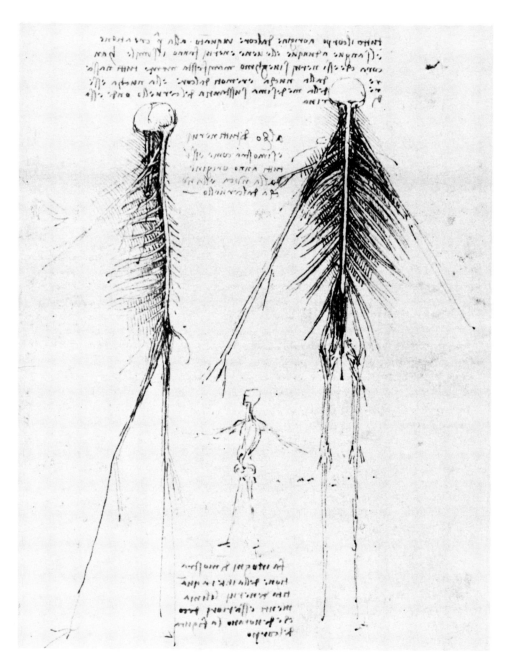

Figure 11.1.   The 'Tree of all the Nerves' (W 19034 v; K/P 76 v).

The object of these drawings is to show how the tree of all the nerves when taken as a whole outlines the shape of the human body. They also demonstrate that the spinal cord and its nerves spring from the brain.

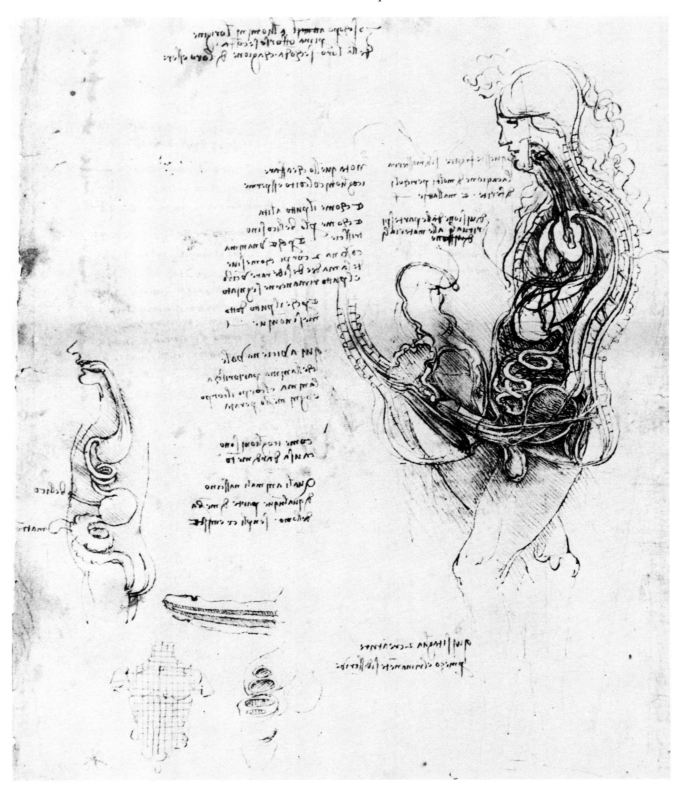

Figure 11.2. The coitus figure (W 19097 v; K/P 35 r).

The coitus figure represents the idea of the spinal cord as the source of semen, as described in Figure 10.1 (p. 231). Here 'nerves' to the penis are shown originating in the spinal cord and carrying sperm to be ejected in coitus. It will be noticed that the female is also given similar 'nerves' to the uterus to carry her seed. The penis of the male, according to this theory, has two channels, one for the seed and one for the urine. Leonardo depicts these realistically in the central drawings below. This idea is found in Plato's *Timaeus*.

To the left is a presentation of visceral anatomy according to the ideas of Aristotle.

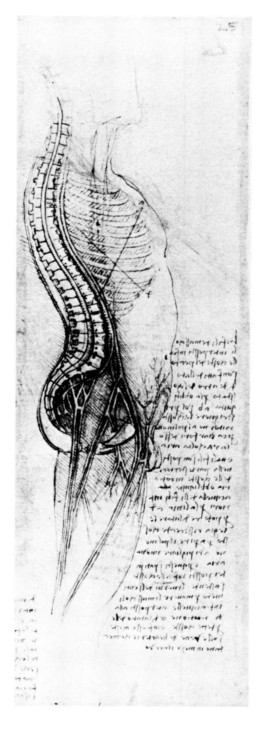

Figure 11.3. Hemisection of the human body (W 19114 r; K/P 109 v).

Here Leonardo uses an imaginary hemisection of the body (reminiscent of a butcher's cleavage of animal carcasses) to summarise his anatomical progress. The curvatures of the spine, the spinal cord and the sacral plexus are shown. Whilst in front of the vertebrae the aorta and inferior vena cava are depicted passing into the pelvis. On the abdominal wall the superficial epigastric veins run down to the saphenous vein which is shown joining the femoral vein.

In the chest the crossing lines indicate the lines of action of the internal and external intercostal muscles.

semen flowed from the spinal cord to the penis. Here the imagined nerves necessary for such transmission are displayed. Plato in his *Timaeus* saw the spinal marrow as containing 'the universal seed stuff' of mortals. 'From the passage of the egress of drink [the penis] . . . is bored a hole into the condensed marrow which comes from the head down by the neck and along the spine, which marrow we termed 'seed'. And the marrow inasmuch as it is animate and has been granted an outlet has endowed the part where the outlet lies with a love for generating by implanting therein a lively desire for emission' (*Timaeus*, 91 B). This drawing of Leonardo follows Plato's description so literally that it is difficult to avoid the conclusion that Leonardo was illustrating what he read in the *Timaeus*.

Erroneous as is most of the anatomy in this drawing in Figure 11.2, it is worth noting that the curvatures of the spine are already correctly shown—an achievement which defeated Leonardo's contemporaries, and even Vesalius.

In Hippocrates Leonardo found confirmation of Plato's view: 'Hippocrates says that the origin of our semen is derived from the brain and from the lungs and testicles of our forefathers' (Forster III 75 r), he notes about 1493. This question of the source of the sperm is succinctly answered by Guy de Chauliac, 'From all the body, and especially from the principal members for the breeding of their vessels as heart, liver and kidneys, and for the cause of delectation the brain hath commutation therein, for the sinews [nerves] which descend from the brain to the bollocks. Thus of all the body it taketh nature, not by quantity but by vigour' (*Questionaries of Chyrurgerie,* 32 v). Guy de Chauliac also describes the penis, as illustrated by Leonardo in Figure 11.2: 'It hath two waies, that is of the sperme and of the urine' (32 r). Other features on this page accord with Mondino's and Aristotle's anatomical views. The figures on this page would therefore appear to represent Leonardo's visual representation of the literature he had read in preparation for his own anatomical explorations (see also Chapter 17).

The idea of a hemisection of the body, so striking a feature of this demonstration of coitus, was common in medieval anatomy and was advocated by Mondino. Sagittal hemisection of the body for anatomical demonstration seems to have appealed to Leonardo for he uses it again on W 19114 r (K/P 109 v; Figure 11.3). The contrast between the two figures strikes one sharply. Whereas the former (Figure 11.2) is composed almost entirely of visualised structures as gleaned from the literature, Figure 11.3 is composed from the results of Leonardo's own dissections. It comprises a demonstration of the two great integrating systems of the body, the nervous and vascular systems. The vertebral column and its contained spinal marrow are not so very different from the earlier drawing. In both, the curvatures of the spine are correctly shown; in both, the spinal cord is incorrectly shown extending down into the terminal coccyx instead of ending at the first lumbar vertebra, but when it comes to the drawing of the sacral plexus of nerves there is a marked difference. Though the pattern is still only approximately correct, it is obviously obtained through painstaking observation, showing the formation of a simplified sacral plexus composed of lumbo-sacral roots, from which springs the sciatic nerve, the largest nerve in the body. Leonardo shows it here (and in other drawings) as dividing high up almost straight away, into its two main branches, the tibial and common peroneal nerves, which run down the back of the thigh. This is a normal variation.

The blood vessels arising from the aorta and inferior vena cava are equally recognisably the result of dissection. The accompanying text is wholly concerned with the action of the external intercostal muscles (*cd*) and the internal intercostal muscle (*ab*), whose lines of action are represented by the two lines over the chest crossing at right angles. Though not drawn here, he has not forgotten the intercostal nerves coming off the spinal cord. He writes, 'Through these [intercostal] spaces interposed between the ribs, the nerves of stimulation [*sentimento*] are extended to move the muscles'.

Another field of anatomy in which Leonardo's progress presents clearly demarcated steps is that of the brachial plexus. In Figure 10.1 (p. 231) we see Leonardo's first attempts to explore this complicated region. Figure 10.1 is the page containing the drawing of the frog's spinal cord. Alongside it are two parallel nerve cords. We have previously mentioned how these cords led Leonardo into error in subsequent sudies of this part of the nervous system. In the frog the sympathetic trunk passes down alongside the vertebrae and spinal cord from the skull to the ansa subclavia sending branches to the spinal nerves. It then loops round the subclavian artery. There can be little doubt that this is the nerve trunk found by Leonardo in the frog, and that he has continued to represent this in other animal forms, e.g., monkeys and man, as passing through their vertebral canals in the neck. This error is only too clearly displayed on W 19040 r (K/P 63 r, Figure 11.4). On this page Leonardo depicts seven cervical vertebrae sectioned in a frontal plane to show the spinal cord running down the centre and two lateral cords running down in the vertebral canals in the transverse processes. Nerve-connections are shown between these cords, the spinal medulla and the peripheral nerve roots in the lower four. This is obviously a hypothetical schema representing what Leonardo at this time felt ought to happen. In the drawing below, the spinal cord and attached nerves have been removed from the bony shell to show his hypothetical schema more clearly. In the upper right corner is a little drawing of the spinal cord along which Leonardo has written, 'Spinal medulla and a nerve which arises from it'. As can be seen, the nerve root is covered by 'The two coats which clothe the spinal cord are the same as those which clothe the brain, that is the pia and dura mater'. Most of the text on this page analyses the structure of the peripheral nerves and their function: 'The spinal cord is the source of the nerves which give voluntary movement to the limbs'. There is no mention of the structure of the brachial plexus here.

On W 19040 v (K/P 63 v, Figure 11.5) progress is recorded. The brachial plexus is shown derived from four roots only, but the peripheral nerves on the back of the arm are identifiable. Particularly clearly delineated are the ulnar and radial nerves on the back of the hand. In the two drawings to the right Leonardo has concentrated on the nerves distributed to the muscles of the arm, represented by small bellies. In the right-hand drawing he has given names to some of these muscles, for example, 'humerus' probably indicates the deltoid, and 'fish of the arm' almost certainly refers to the biceps, which is fish-shaped.

That Figure 11.5 is relatively early is indicated by its position on the verso of Figure 11.4 and a list of 'demonstrations' still to be made which occupies the lower half of this

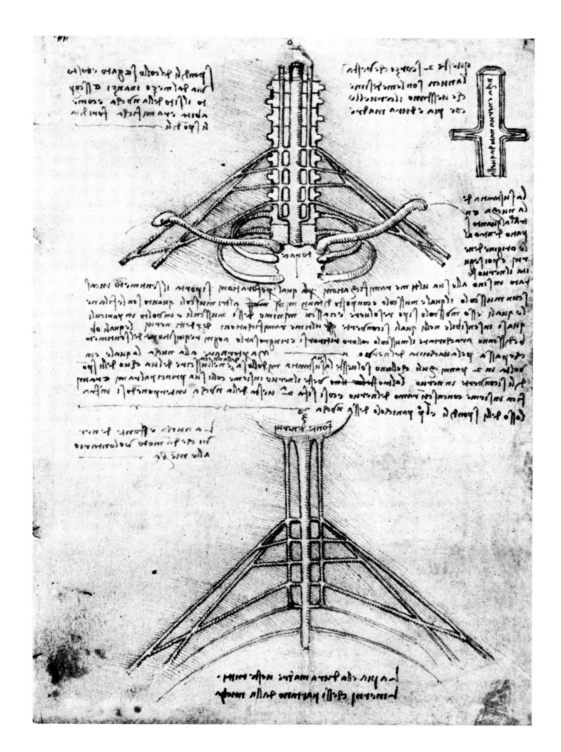

Figure 11.4. A hypothetical scheme of the cervical cord and brachial plexus (W 19040 r; K/P 63 r).

The two drawings represent Leonardo's early conception of the cervical part of the spinal cord and the emergence of its roots to form a vestigial brachial plexus. In the lower drawing the spinal cord and nerves have been removed from the vertebrae. In the top right-hand corner he depicts the origin of a nerve root from the spinal cord surrounded by dura and pia mater.

Figure 11.5. The brachial plexus and
nerves to the arm (W 19040 v;
K/P 63 v).

Marked progress is shown towards an
appreciation of the correct constitution
of the brachial plexus here on the verso
of the page containing Figure 11.4.
Here the plexus is drawn as derived
from four (not five) nerves with the
formation of trunks which later split.
These formations are here only approximately accurate.

page. This includes demonstrations of the bones, nerves
and muscles of the body, but dissections of the viscera of
thorax and abdomen are included.

Leonardo's peak of achievement in analysing the confus-
ing complexity of the brachial plexus is reached on W
19020 v (K/P 57 v, Figure 11.6). Above the main drawing
on this page are the words '*del vechio*', indicating that the
dissection was performed on the 'old man' whom Leonardo
saw die suddenly in the Hospital of Santa Maria Nuova in
Florence. Here the brachial plexus is seen correctly as arising from the union of the anterior divisions of the lower
four cervical nerves and the first thoracic nerve. It is shown
extending from the lower part of the side of the neck to the
axilla. These five nerves are shown uniting to form upper,
middle and lower trunks, which are shown in their turn
dividing and re-uniting into cords. Allowing for the variability of this arrangement in nature, this pattern as detected
by Leonardo can be considered accurate. But at the level of

the cords and their branches Leonardo becomes confused, though the median nerve (*b*) and the ulnar (*r*) are clearly distinguished. However, this whole map of the brachial plexus should be evaluated as a major discovery in Leonardo's exploration of the lesser world, the microcosm of man. Alongside it he remarks, possibly as a result of his experiences in the campaigns of Cesare Borgia, 'Any one of these 5 branches which is preserved from a sword cut suffices for the sensation of the arm'.

To add to this remarkable achievement Leonardo, without verbal comment, illustrates the upper intercoastal nerves, and crossing the necks of the ribs are five thin nerves, the first illustration of the sympathetic chain and formation of the splanchnic nerves, labelled *n*.

How right he was when he exclaimed, 'Do not busy yourself in making enter by the ears things which have to do with the eyes' (W 19071 r; K/P 162 r). But how equally true is it that so many have had their minds' eyes so firmly closed as to be unable to see what was placed before them here.

The elaborate and detailed exploration of the anatomy of the nervous system, of which we have given some examples, served to confirm Leonardo's ideas of its function.

This, in the main, is to convey sensation to the soul and force from it. 'Nature has arranged in man the function of muscles to draw the tendons, and these are able to move the limbs according to the will and desire of the *senso comune,* just like the functions of the officials distributed by their lord in various provinces and cities to represent and carry out out his will in those places. And the official who has on more than one occasion carried out the commission given him by the mouth of his lord will then himself in time do something which does not proceed from the will of the lord. Thus one often sees with the fingers, how, after having with the utmost docility learnt things upon an instrument as commanded by the judgement, they will afterwards play it without accompanying judgement' (CA 119 va).

Leonardo was impressed by the fact that sometimes action without 'the will of the lord' occurs; 'How the nerves sometimes work on their own without the command of any agent or the soul; for this you will see clearly appear in the movements of paralytics and those who shiver benumbed by cold, their limbs such as the head and hands trembling without permission of the soul; which soul, with all its

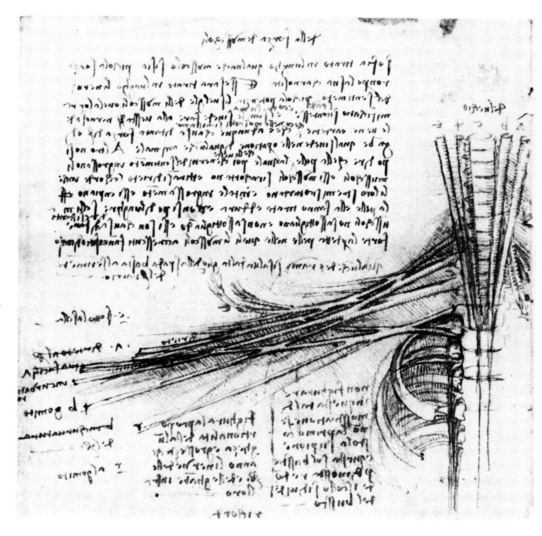

Figure 11.6. The brachial plexus (W 19020 v; K/P 57 v).

Here the brachial plexus in all the complexity of its detail is successfully drawn by Leonardo. Not only has he achieved this, but also below the plexus the upper intercostal nerves are shown, and crossing the necks of the ribs is a series of nerves labelled *n,* which undoubtedly represent the sympathetic chain and formation of the splanchnic nerves. The passage headed 'On the force of the muscles' at the top of this page is quoted on p. 276.

powers cannot prevent these parts from trembling. The same happens with epilepsy or with severed limbs such as the tails of lizards' (W 19019 v; K/P 39 v).

Both sensory and motor sides of the nervous system are linked in his description: 'The perforated nerves carry command and sensation to the functioning parts, which nerves enter into the muscles commanding them to move. These obey and are put into action by shortening for the reason that such swelling shortens their length and pulls them back. The nerves which are interwoven amidst the particular parts of the limbs being infused into the ends of the digits carry to the *senso comune* the cause of their contact. Tendons with their muscles serve the nerves just as soldiers serve their leaders; and the nerves serve the *senso comune* as do leaders their captains; and the *senso comune* serves the soul as the captain serves his lord' (W 19019 r; K/P 39 r).

Chapter 12

The Mechanism of the Bones and Joints of Man

Of all Leonardo's achievements those of his drawings of the human skeleton and muscles have perhaps been accepted as the most outstanding. This aspect of his exploration of the human body, supplemented by his unique capacity for representing movements in visual perspective form, makes his drawings of bones live in a way seldom achieved by modern anatomical illustration. Leonardo's explorations of the microcosm of the human body bear a curious likeness to those explorations of the macrocosm of the earth by his contemporary, Columbus.

As an explorer of a New World, Leonardo was much more akin to Columbus than is commonly appreciated, by reason of his persistent analogy between the microcosm of Man and the macrocosm of the World. 'This earth is the world machine enclosed in the larger air', he wrote (CA 252 rb), and it was as a machine enclosed in the larger air and subject to the same laws that he saw the human body. The bones of the macrocosm were its mountains and rocks. 'As man has within himself bones as supports and protection [*armadura*] of the flesh, so the world has rocks, supports of the earth' (A 55 v). 'If the body of earth were not like the body of man it would not be possible for the water of the sea . . . to rise to the summits of the mountains' (see Chapter 15). And although in later years this analogy changed its form, it did not weaken. For example, nearly twenty years later he discusses 'Minerals which are joined together in continuous ramification' (Leic 18 r) and describes how gold grows in the earth 'with a slow movement and it converts into gold whatever touches its extremities; and note that therein is a vegetative soul which it is not within your power to generate' (W 19045 v; K/P 50 v).

Another strong influence which Leonardo brought to bear on his study of bones was his architectural knowledge. No doubt he viewed with favour Vitruvius's and Alberti's repeated comparisons between buildings and animals: 'The philosophers have observed that Nature in forming the bodies of animals always takes care to finish her work in such a manner that the bones should all communicate . . . So we also should connect the ribs [of the structures] together' wrote Alberti (*On architecture,* Book 3, Chapter 12). Leonardo carries the analogy further, to the consideration

of a cathedral in need of repair. 'A sick cathedral', he writes, 'requires a doctor architect who understands well what an edifice is and upon what rules the correct method of building is based; and whence these rules are derived, and into how many parts they are divided; and what are the causes which hold the structure together and make it last; and what the nature of weight is, and what the desire of force is, and in what manner they should be combined and related; and what effect their union produces' (CA 270 rc).

If this was his approach to the structure of a static, immobile edifice such as a cathedral, how much more complex was his approach through his 'rules' and 'four powers' to the jointed, mobile edifice of the human body? The nature of Leonardo's studies of stresses on beams has been touched on in Chapter 4. He applied the same principles to bones, though in this field he concentrates more on their accurate proportionate illustration according to his perspective principles. These involve many instances where his mechanical studies of weight, leverage and percussion are applied. For example, an experiment on a horse's bone is reported in Figure 4.43 (p. 128) on a page mostly devoted to studies of stress and percussion of beams. Here he carefully notes the site of fracture of 'a horse's bone supported by a straw 1/4 *braccio* long', when percussed. This occurs 'under the blow given from above'. He compares the leverage of muscles on the radius and ulna of a monkey with that of a human arm in Figure 12.1 (see also Figure 6.11), and he illustrates the leverage of the ribs by pulley mechanism in two drawings in Figure 14.2 (p. 292).

Leonardo's famous drawings of the Vitruvian proportions of a man's body first standing inscribed in a square and then with feet and arms outspread inscribed in a circle (Figure 12.2) provides an excellent early example of the way in which his studies of proportion fuse artistic and scientific objectives. The centre of the standing man in the square is at the pubis. It is Leonardo, not Vitruvius, who points out that 'If you open the legs so as to reduce the stature by one-fourteenth, and open and raise your arms so that your middle fingers touch the line through the top of the head, know that the centre of the extremities of the outspread limbs will be the umbilicus, and the space between the legs will make an equilateral triangle' (Accademia, Venice). Here he provides one of his simplest illustrations of a shifting 'centre of magnitude' without a corresponding change of 'centre of natural gravity'. This remains passing through the central line from the pit of the throat through the umbilicus and pubis between the legs. Leonardo repeatedly distinguishes these two different 'centres' of a body, i.e., centres of 'magnitude' and 'gravity' (see Chapter 4).

This kind of study is pursued in many drawings of the body, as on W 19132 r; K/P 27 r (Figure 12.3). Here again the centre of a standing man is shown to be at the pubis. The centre of height of a kneeling man is at the umbilicus; if he bends his elbows, putting his hands to his breast as in praying, the tips of his elbows also come to the level of the umbilicus. The middle of a sitting man is marked by a line 'below the breasts' against which Leonardo has written the word '*mezo*' (middle). All these postures are illustrated later by the movement of articulated bones, which is one reason why Leonardo's skeletal levers seem to move and live (CA 349 rb; see Figure 12.4).

Leonardo, as we have seen, laid great stress on the importance of a mastery of perspective in representing human anatomy. To many of us in the twentieth century the photographic film has exhausted the stimulus from this achievement, but to Leonardo, striving to create 'true' pictures of reality, it was a most exciting union of art and the knowledge we now call 'science'. Without perspective the human machine, like other figures of machines, could not represent 'true knowledge'. This revealed parts that were measurable so that the viewer was given the power of recreating the illustrated measurements of the machine in three dimensions as models.

The majority of notes on Leonardo's anatomical pages consist of suggestions of improvements in his technique of illustration. For example, in relation to bones he writes, 'True knowledge of the shape of any body consists of seeing it from different aspects. Therefore in order to give knowledge of the true shape of any limb of man, the first of beasts among aminals, I will observe the aforesaid rule by making 4 demonstrations of each limb from their 4 sides. And I shall make 5 of the bones, sawing them along the middle and showing the empty space in each of them [the marrow cavity], one of which is full of marrow, another spongy, empty or solid' (W 19000 v; K/P 135 v). Though he usually

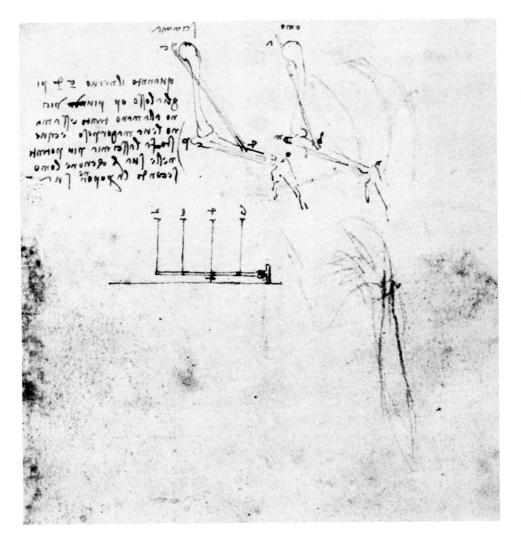

Figure 12.1. Comparison of the leverage at the elbow in man and the monkey.

The right-hand drawing sketches the arm of a man showing schematically the site of insertion of the flexors of the elbow such as the biceps. The left-hand drawing compares this with the situation in monkeys. The comparable powers of leverage are depicted below by the relative forces required to elevate the lever representing the forearm, which has its fulcrum at the end on the right. Leonardo writes, 'The nearer the tendon *cd* which flexes the bone *op* is to the hand so much the greater weight does this hand lift. And this makes the monkey more powerful in its arms than man, proportionally' (W 19026 v; K/P 68 v). See also p. 268.

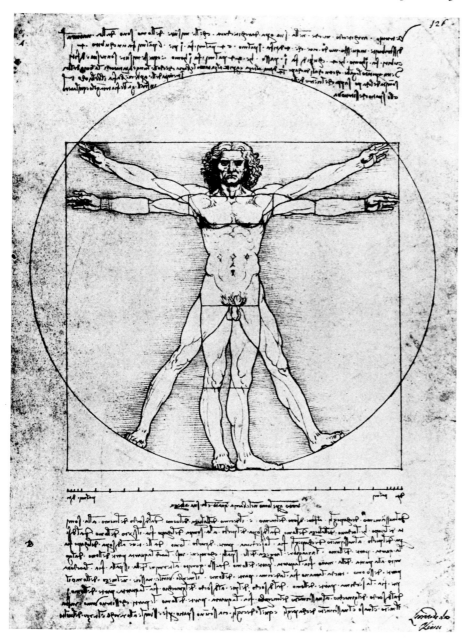

Figure 12.2. Proportions of the human body (Accademia, Venice).

In the accompanying note Leonardo draws attention to the fact that the standing man has his centre of the square at the level of the pubic genitalia, whereas if the arms are raised and the feet parted so that the height of the body is diminished by one-fourteenth, the body can be inscribed in a circle with its centre at the umbilicus, and an equilateral triangle is formed by the legs. Such geometrical transformations were an essenial element of Leonardo's visual language.

diligently draws the views he mentions there are no extant drawings of bones sawn along their length. This is one of the clues strongly suggesting that a number of anatomical sheets have been lost.

He suggests making 'these 4 demonstrations twice; and of this duplication you will make one kind in which the opposite facets of the bones are joined to the corresponding ones, as nature made them; and the other demonstration you will make with the bones separated; and by this means you will see the true shape of the opposite facets of the bones which are joined together' (W 19000 r; K/P 135 r). This is the rationale of Leonardo's unique 'exploded' views of the joints by which he revealed the types of movement and mechanisms that are particular to different joints (Figures 12.5 and 12.6).

His dissections of bones and joints lead him to describe them as follows: 'Bone is of inflexible hardness suitable for resistance; it is without sensation and ends in cartilages at its extremities. Its marrow is composed of sponge, blood and soft fat coated with a very fine membrane. Spongiosity consists of a mixture of bone, fat and blood' (W 19088 v; K/P 175 v). So accurate is this verbal description of compact, trabecular (spongy) bone, bone marrow and the endosteum that one regrets all the more the absence of figures illustrating these points.

Cartilage Leonardo describes as 'a hard substance like, so to speak, indurated tendon or soft bone. It always lies between bone and tendon since it partakes of both substances. It is flexible, unbreakable, and being bent, it straightens itself on its own like a spring' (W 19088 v; K/P 175 v).

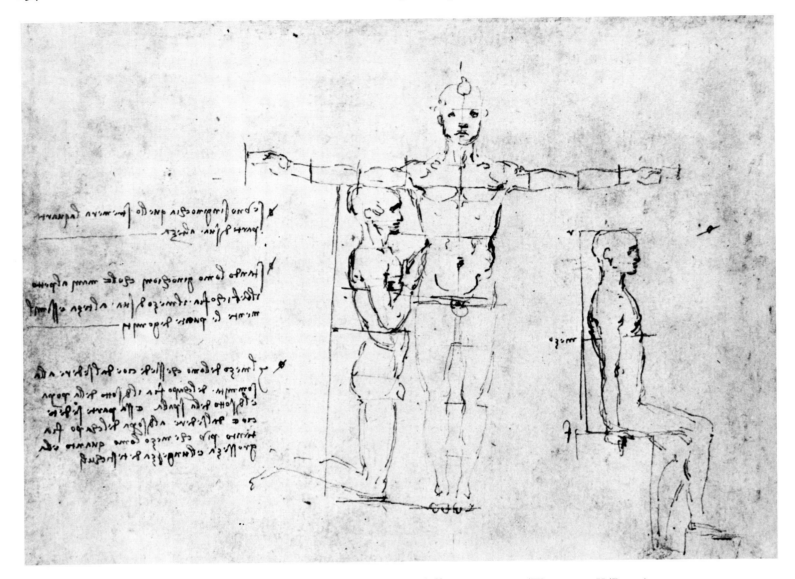

Figure 12.3.   Studies of human proportions in different postures (W 19132 r; K/P 27 r).

Proceeding from the Vitruvian figure, Leonardo compares the proportions of the human body, standing,
kneeling and sitting, paying particular notice to the proportions of the body in each case.

A joint Leonardo describes as the basis of movements thus: 'There are 6 things which come together in the composition of movements, i.e., bone, cartilage, membrane, tendon, muscle and nerve' (W 19093 r; K/P 177 r). And he describes the movement of a joint as follows, 'Tendons are mechanical instruments . . . which carry out as much work as is assigned to them [by muscles] . . . Membranes interposed between the tendons and cartilage are made to join the tendon to the cartilage by a broad continuous junction in order that it shall not break from excessive force . . . Ligaments are joined to tendons and are a kind of membrane which bind together the joints of bones and are converted into cartilage. They are as numerous in each joint as are the tendons which move that joint and the tendons coming to the joint from the opposite side. Such ligaments are all joined and mixed up together, aiding, strengthening

and correcting one another. The muscles of limbs are drawn towards the bone to which they are joined [at their origins] and they draw their tendons behind them together with the limb which is joined to this tendon [at its insertion]' (W 19088 v; K/P 175 v).

Drawings of individual joints show this pattern of construction of joint capsules (see Figures 12.15 and 13.9). A surprising omission is that of the synovial membrane, particularly as Leonardo was acutely aware of the necessity for the lubrication of the moving parts of machines (see Chapter 4, pp. 123–124).

Regarding the exertion of power through joint surfaces, Leonardo notes a special function for force exerted along the 'central line' on W 19000 r (K/P 135 r, Figure 12.6). 'When the line of power of a movement passes through the centre of the junction of movable bodies they will not be moved, but they will be stabilised in their primary straight line, as is demonstrated by *an,* a mover which passes through the centre of two movable bodies, *nm* and *mo,* and stabilises them. But if the line of power of the mover is outside the central line of the two rectilinear movable bodies *cd* and *de* . . . then the further it is from the axis of these moved bodies the more it will bend their straight line into an angle, as does a bow-string'. He applies this principle in Figure 12.6 particularly to the action of sesamoid bones beneath the great toe, and in Figure 12.7 to the knee-joint (W 19008 r; K/P 140 r).

The nearest approach to a drawing of the complete skeleton is found on W 19012 r (K/P 142 r, Figure 12.8). These drawings reveal Leonardo feeling his way into the bony

labyrinth of the microcosm, changing his illustrations as he goes along. The first drawing, that on the upper right of the page (since Leonardo nearly always worked from right to left), shows a thoracic cage with twelve ribs meeting in the middle. He ignores the sternum. In contrast, the last of the three views of the thorax, below and to the left, shows the thorax from the front with remarkable accuracy. Errors, however, are still there; e.g., the sternum is divided into seven segments as reported by Mondino and present in the monkey (which he dissected earlier), and an acromion bone is shown, separately at *a.*

The text in Figure 12.8 consists almost entirely of Leonardo's instructions to himself regarding methods of making his 'demonstrations' intelligible. The most notable

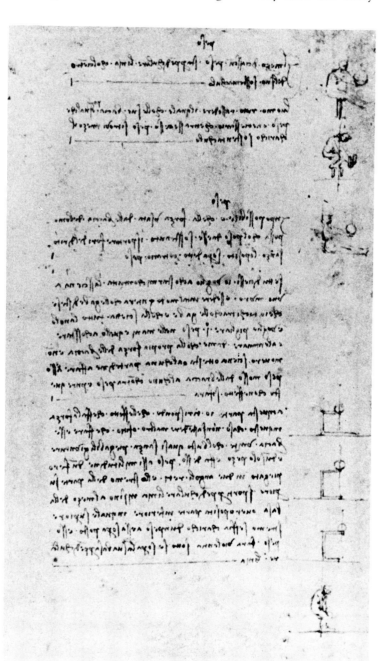

Figure 12.4. Posture related to leverage and weight.

In this series of drawings the relation between a weight held in the hands and the centre of gravity is studied. In a long passage Leonardo writes, 'The middle of each weight is in a perpendicular line with the centre of its support'. Thus the support (or fulcrum of balance) must lie between a man's own centre of gravity and that of the weight. In studying the case of a man sitting with his back to a wall Leonardo states that the arms act as 'one piece, making this piece like a bar of iron bent into two right angles' (CA 349 rb).

Thus, the seated man is converted into a weighted lever system.

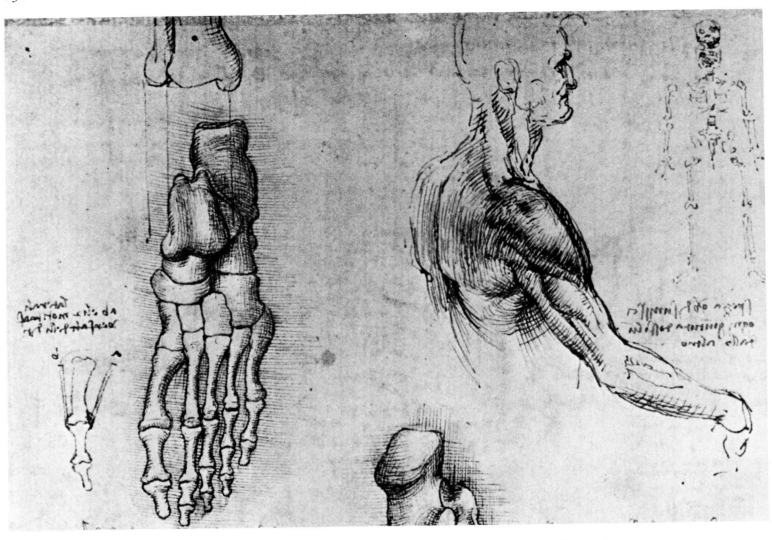

Figure 12.5.   The plan for 'exploded views' of joints (W 19002 v; K/P 134 v).

In the top right-hand corner Leonardo draws a small skeleton. All the joints of the bones have been 'exploded' so that the bones are not in contact. The skull has been divided and turned down at the temperomandibular joint, bizarrely obscuring the neck. The left fibula is sketched in on the inside of the tibia. This curious error is emphasised when Leonardo applies this scheme of anatomical illustration to an otherwise wonderfully accurate drawing of the bones of the left foot.

features of the page are those of the pelvis and legs. For the first time the pelvis is shown correctly shaped, placed and angled in relation to the spine, sacrum and thigh-bones. The kneecap receives special study with its attachments to the ligaments above and below. The right marginal figure gives a beautiful lateral view from the lumbar spine down to the feet. Above this the curvatures of the thoracic and cervical spine are suggested in outline correctly. These drawings convey deep understanding of the statics of the transmission of weight, in the human erect posture. Leonardo makes no comment on them.

The spinal part of this study is carried over to W 19007 v (K/P 139 v, Figure 12.9), whilst the pelvis and lower limb are further studied on Figures 12.7 and 12.10).

Figure 12.9 (W 19007 v; K/P 139 v) provides a very good example of Leonardo's prowess, not only as an anatomist demonstrating the curvatures of the spine but also as an artist putting into practice his principles of illustration from different aspects, as well as using the 'exploded view' of the three cervical vertebrae in the lower left-hand corner. The first drawing on the right shows the whole spine from the front. So beautifully is it drawn that one can see the different curvatures, as shown more explicitly in the lateral view in the centre of the page. In each drawing Leonardo clearly demarcates with letters the cervical, thoracic, lumbar, sacral and coccygeal vertebrae, and in a little table below he enumerates them correctly, making thirty-one in all. At the bottom of the page he illustrates the spine from behind. To the left of this, in a special study of the seven cervical vertebrae, he clearly shows the peg of the odontoid process in position. To the left of this again he gives an 'exploded view' which successfully demonstrates all the complexities of the articular facets of these three vertebrae. The peg of the odontoid process of the axis is now removed from its joint facet on the body of the atlas. The text on this page is notable for the absence of any comment on his beautiful illustration of this extraordinarily subtle mechanism. However, Leonardo again stresses the importance of drawing 'these bones from 3 aspects joined together and 3 aspects separated. And you will do the same then from two other aspects, i.e., from below and above'. These views are absent, presumably on a lost page of his work. That he appreciated the importance of his method is suggested by one of his few outbursts of pride: 'Thus you will give true knowledge of their shapes which would have been impossible to ancient and modern writers without an immense tedious and confused amount of writing and time. But through this very quick way of representing them from different aspects one gives a full and true knowledge of them. And in order to give this benefit to men I teach ways of reproducing and arranging it; and I pray you, O successors, that avarice shall not constrain you to print it in . . .' This tantalisingly unfinished passage has given rise to much speculation, the word *woodcuts* being the most acceptable ending.

The posterior view of pelvis and legs which completes the drawings on Figure 12.10) is found on W 19004 r (K/P 138 r, Figure 12.10), occupying the centre of the page. In a note written between the legs Leonardo remarks that 'This tail-bone [coccyx] is bent backwards when a woman gives birth'. All the remaining drawings are concerned with the movements of the elbow. On the right, pronation of the elbow is shown from the side (top), from behind (below) and from the front (centre) with the elbow extended. Views of the elbow flexed from back, front and side in supination are shown on the lower half of the page. Flexion of the elbow in pronation is drawn from another subject on the centre left. This drawing is particularly interesting, firstly, because both it and its companion on the left margin show the results of a past fracture of the radius and ulna with the remains of callus formation. Leonardo, not knowing this, asks, 'Observe the two prominences at *n* on the inner sides of the two bones of the arm; what function do they serve?' In the same drawing he stresses a point made in the *Treatise on Painting* (CU 104 v), pointing out that 'The arm in extension is shortened by 3 1/2 fingers from the shoulder to the elbow'. This is due to the olecranon process entering the olecranon fossa at the lower end of the humerus as the arm is extended. When the elbow is bent, this process emerges and lengthens the distance from shoulder to the elbow tip. In the *Treatise on Painting* Leonardo says that the difference in full

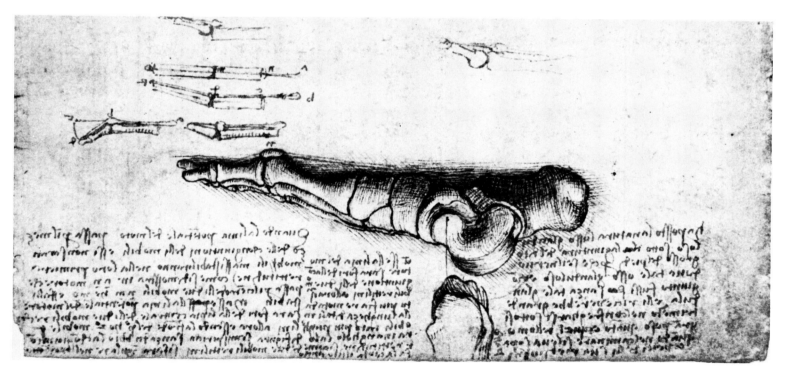

Figure 12.6. Central and eccentric lines of force in the toe joints (W 19000 r; K/P 135 r).

Below a rough sketch of a sesamoid bone on a toe Leonardo draws two jointed rectangular bodies through which a central line of force, *anme*, is shown. When this acts it pulls the surfaces of the joint together. In the drawing below an eccentric force, *bcde*, is shown. This flexes one body on the other at the joint. In the drawings below this he illustrates how such an eccentric force moves the great toe. The side view of the ankle and foot is drawn upside down to show the sesamoid bone at *n*.

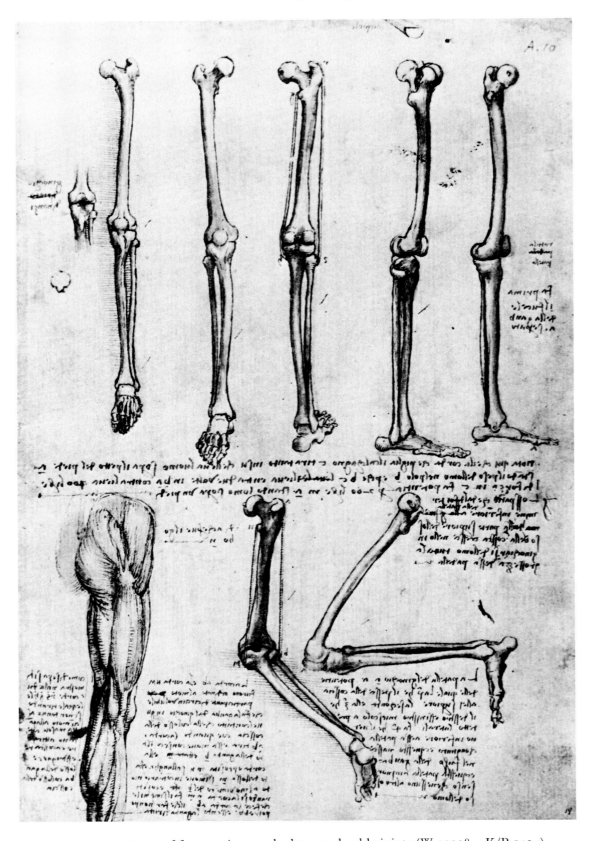

Figure 12.7.   Lines of force acting on the knee and ankle joints (W 19008 r; K/P 140 r).

In the central drawing of the upper row two symmetrical lines of force act on the knee so pulling its surfaces together and stabilising the joint. Below, the same forces acting on the flexed knee are shown to rotate the lower leg at the joint.

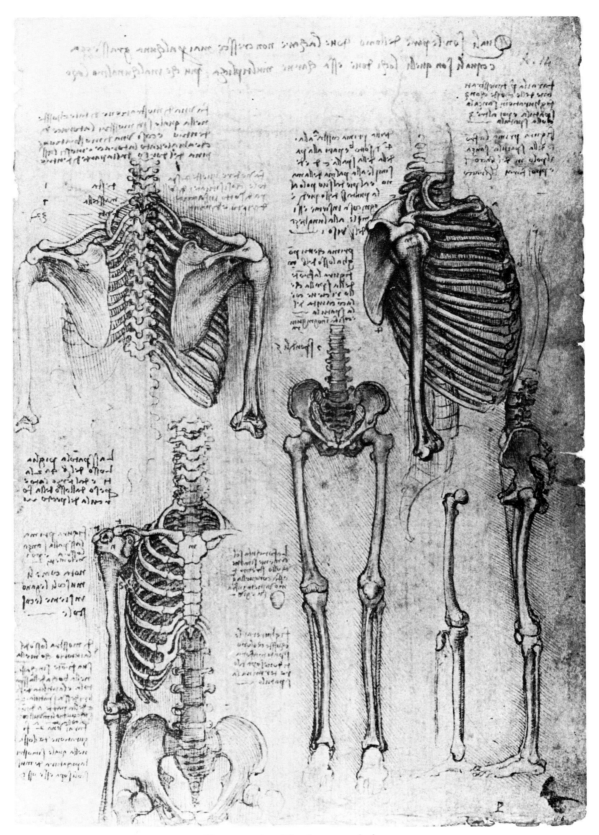

Figure 12.8. The human skeleton.

In this group of drawings Leonardo pays particular attention to displaying the articulated skeleton from the back, front and side in order to give 'true knowledge' of its structure. For the first time the pelvis is shown correctly angled in relation to the spine and thigh bones (W 19012 r; K/P 142 r).

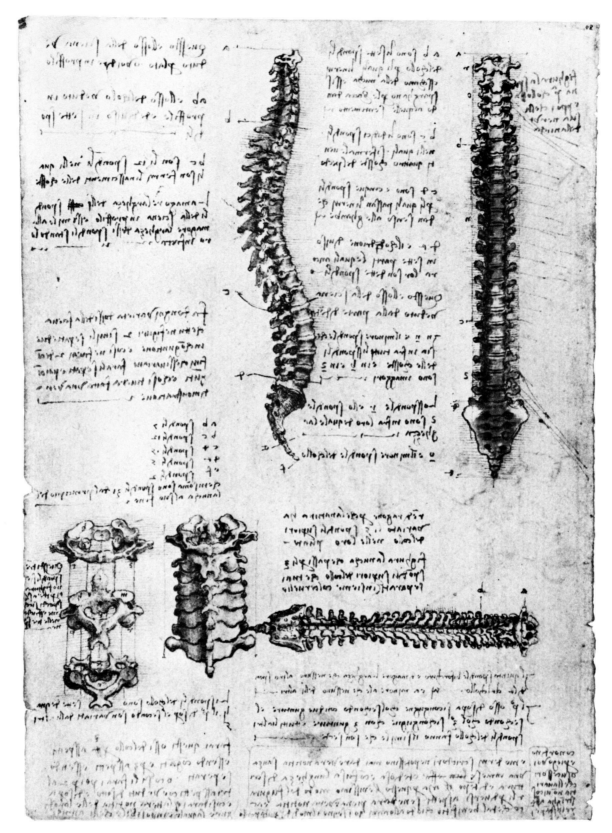

Figure 12.9.   The vertebral column as a whole; the cervical vertebrae in particular
(W 19007 v; K/P 139 v).

The curvatures of the spine from front, back and side are correctly represented. They illustrate the potent flexibility so well shown in man and animals (see Figure 6.18, p. 177). His
use of the 'exploded' view of the joints of the three upper cervical vertebrae is well shown,
but the articulation with the skull is either omitted or lost.

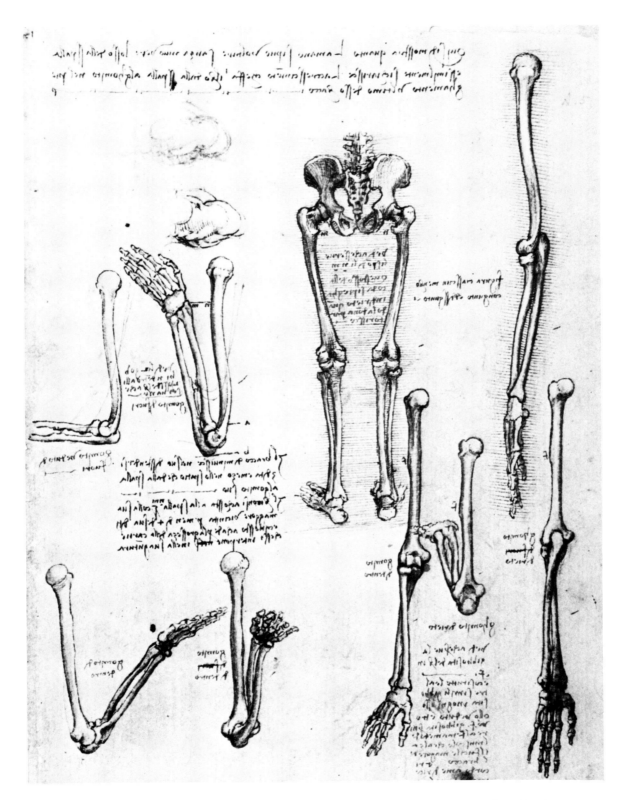

Figure 12.10. The pelvis and legs (posterior view); flexion, pronation and supination of the forearm (W 19004 r; K/P 138 r).

The view of the pelvis completes the series begun in Figure 12.8 (p. 259). All combinations of the movements of flexion and extension of the elbow with pronation and supination are shown. In the two upper left drawings an old fracture of the radius and ulna is illustrated. Leonardo, unaware of this possibility, asks what the bosses on the bones here are for.

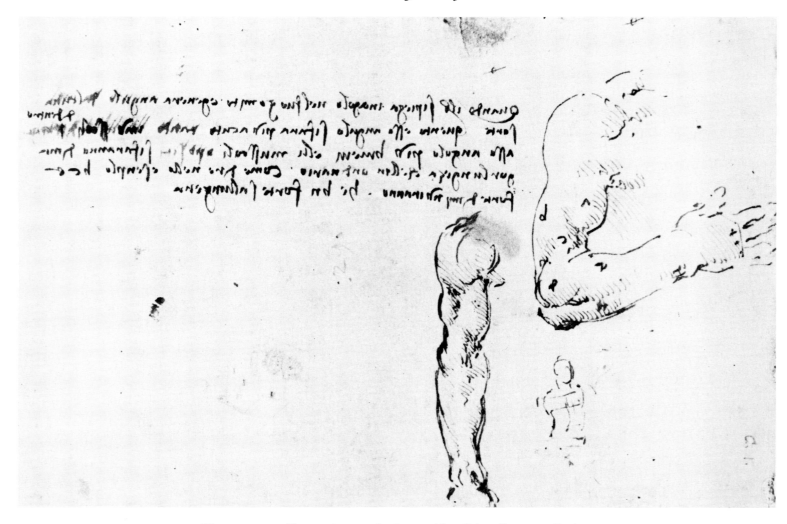

Figure 12.11.   Shortening on the inner side of the elbow on flexion.

Leonardo writes, 'When the arm is bent into an angle at the elbow . . . this angle will become more acute the shorter the muscles inside that angle become. And the muscles on the opposite side [the extensors of the elbow] will become longer than usual' (W 12614 r; K/P 93 r).

This general rule applies to all joints of the body (see Figure 4.9a). He applies it to the knee on Figure 12.7, and in the little sketch here below as well as to flexion of the whole body to one side; also to elongation of the recti abdominis on hyperextension of the spine. The principle rests on the changes produced on bending a straight beam. Only the central line remains unaffected (see Figure 4.6, p. 97).

flexion and extension equals 'one-eighth of the total length' of the arm. He also shows in the drawing on this anatomical page that 'The arm does not approach the shoulder at its closest, *nm,* by less than 4 fingers breadth; and this happens through the thickness of the flesh which is interposed at the [elbow] joint'. Leonardo's lasting interest in pronation and supination is repeatedly evident on other pages. On W 19000 v (K/P 135 v, Figure 13.5, p. 272) it is studied in detail. On W 19118 v (K/P 116 v) he explains why pronation is limited by flexion of the elbow, an observation easily confirmed on oneself.

The leg and foot receive comparable attention on W 19008 r (K/P 140 r, Figure 12.7, p. 258), where both left and right legs are shown in a series from front, back and sides, on standing and kneeling. The first right-hand drawing

shows Leonardo advancing from merely descriptive anatomy to the application of the principle of the lever to movement of the heel (see Chapter 4, p. 108). In this medial view of the leg, the tip of the medial malleolus of the ankle is labelled *b,* the end of the heel, *c* and the ball of the foot, *a.* Leonardo describes the lever system so constituted as follows: 'Note here that the tendon which takes hold of the heel *c* [tendo Achilles] pulls it upwards, so that it raises a man on to the ball of his foot, *a,* the weight of the man being taken at the fulcrum [*polo*] *b.* Because the lever *bc* is half the counter-lever *ba,* so 400 pounds of force at *c* produces a power of 200 pounds at *a,* with a man standing on one foot'. Here he is applying the well-known principle of the lever that the product of weight and distance from the fulcrum is equal on each side. Thus in a man standing on his toes, 200

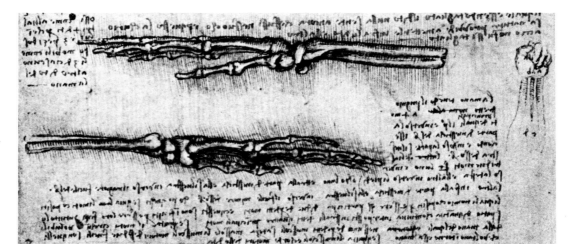

Figure 12.12. Movements at the wrist (W 19009 v; K/P 143 v).

The text on this lower part of the page is devoted to a description of the movements of the hand at the wrist.

multiplied by the distance *ab* equals 400 multiplied by half this distance *bc*. On the left margin of the page Leonardo, having drawn the patella in place, then draws it with the patella tendon cut, the patella turned down so exposing the condyles of the femur and the form of the knee-joint.

In the central drawing of the leg in Figure 12.7 two cords are drawn, one, *gh* from the lesser trochanter to the internal condyle of the tibia, another from the greater trochanter to the head of the fibula. These two cords appear again as *nm* and *ab*, in the drawing of the kneeling figure below. They are muscular powers corresponding roughly to the actions of the vastus lateralis and semimembranosus, the origins and insertions of which he has not yet established. The drawings are made to illustrate the mechanical principle

already referred to in Figure 12.6 (p. 257). Leonardo points out that if the two cords, *rs* and *gh* in the upper drawings, are pulled, 'they give no movement to the leg but only draw and press the bones close together'. This is because their combined power acts along the central axis of the joint. However, if the knee is flexed, as in the lower drawing, 'When the cord *ab* pulls it moves with itself the side of the leg, *b,* and the opposite cord, *nm,* is elongated and the side of the bone *M* is moved in the opposite direction to the movement of *b*'. He then describes the converse movements: 'When the tendon *nm* pulls'. He thus demonstrates how rotation of the lower leg at the knee can be brought about only when the knee is flexed. The vastus lateralis and biceps femoris are dramatically illustrated in the muscle figure to

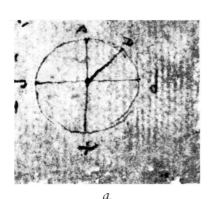

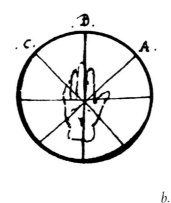

a.   *b.*

Figure 12.13. The four principal movements of the hand.

a. The four principal movements of the hand are represented by the letters *abcd,* which are placed in a circle. Therefore, compound movements such as *e* are infinite (W 19013 r; K/P 144 r).

b. 'That it is impossible to memorise all aspects and mutations of the parts of the body.' This is Leonardo's comment on these two sketches of the movements of the hand in the *Treatise on Painting.* He continues, 'It is impossible for anyone to be able to keep in his memory all the aspects or changes of each part of any animal whatsoever. This we shall exemplify by demonstrating the case of the hand. And because every continuous quantity is divisible to infinity the movement of the eye which looks at the hand as it moves from *A* to *B* moves through a space *AB* which is also a continuous quantity and consequently divisible to infinity. And each part of its movement changes the aspect of and shape of the hand as it is seen, and so it will be as it moves through the whole circle. The hand which is raised up in movement will do the same, that is, it will pass through space which is a continuous quantity' (CU 130 r–130 v).

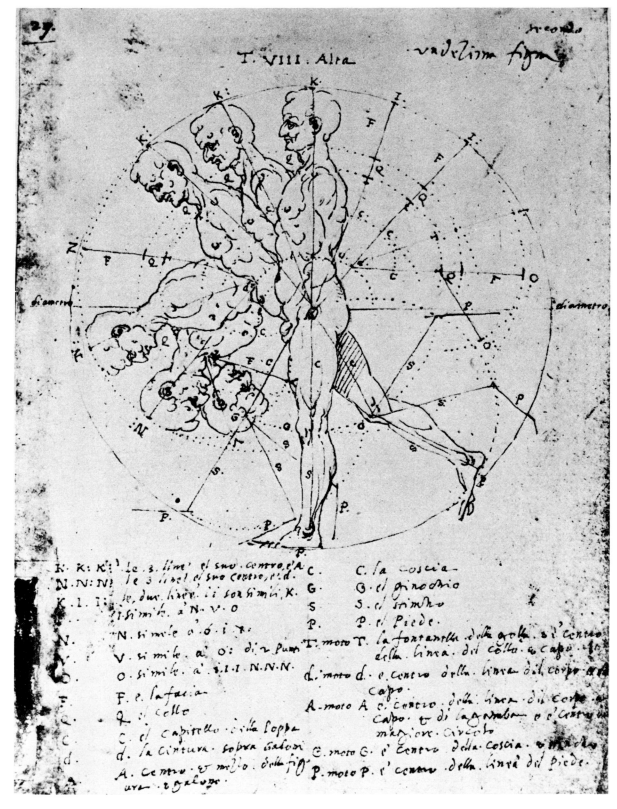

Figure 12.14.   Rotation of the body round the hip joint (*Codex Huygens*, folio 29, figure 13).

Leonardo calls the hip joint the 'pole of a man', '*polo dell'omo*'. Here this concept is illustrated, the trunk and right leg being shown to rotate around the pole marked *A* which is the hip joint. As the man bends forward other poles come into play, e.g., the upper and lower poles of the cervical vertebrae.

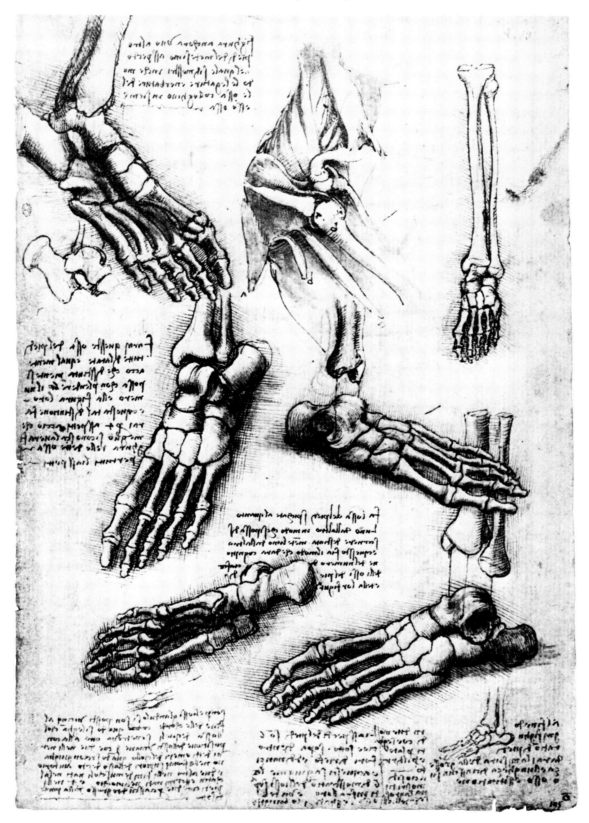

Figure 12.15. The bones of the foot (W 19011r; K/P 145r).

On this page Leonardo applies his principles of 'exploded views' from different aspects. The sesamoid bones are drawn in situ (lower left drawing) and then separate. Each foot, Leonardo notes, is to be accompanied by measurements of its length, breadth and position.

the left, with a muscle outline consistent with semi-membranosus on the internal side of the thigh.

In the remaining drawing in Figure 12.7 of the flexed knee from the medial side Leonardo is again showing the point that he made about the elbow, that when there is lengthening of one side of arm or leg on flexion there is diminution on the other (Figure 12.11): 'When a man kneels', he writes, 'the distance between the lower edge of the patella and the part of the bone of the thigh above increases by the whole size of the patella'. See also Figure 4.9a.

Leonardo's finest studies of the bones of the hand on W 19009 v; K/P 143 v (Figures 10.3, 12.12, pp. 233, 263) were the first to show the bones of the wrist and hand accurately. He displays them from front, back and both sides. The long discussion at the head of one page is concerned with his plan for making eight 'demonstrations of the hand'. Again he emphasises the importance of showing the bones 'together and a little separated'. On the right margin he shows the phalanges covered by the flexor and extensor tendons, arteries, veins and nerves, labelled *e, a, d, b* and *c*, in that order (see Figure 10.3). Below the text describing these structures is a small sketch of the forearm with a closed fist; the long text to its left is entirely devoted to this drawing (Figure 12.12). 'The hand held as a closed fist', writes Leonardo, 'has 4 principal movements', flexion, extension, abduction and adduction. These four movements he calls 'compound' because 'they are infinite, by being made throughout the whole of the space which is a continuous quantity interposed between the said four principal movements'. Four muscles acting thus, Leonardo considers, can always produce an infinity of circumvoluble movements. For this reason he looked for 'four muscles' as sufficient for all the movements of the eyes but did not appear to find them. Four such movements are combined, or 'compounded', to produce 'infinite movements' of the fingers. This is described on W 19013 r (K/P 144 r, Figure 12.13a), accompanied by a diagram. 'Circular movement of the hand showing first the four principal movements, *ad* [flexion] and *da* [extension] and *bc* [adduction] and *cb* [abduction]. And besides these four principal movements one makes mention of the non-principal movements which are infinite'.

A further variation of the same statement appears in the *Treatise on Painting* (CU 130 r) (Figure 12.13b), where the assertion 'that it is impossible to memorise all the aspects and mutations of the limbs' is explained on the principle that it is impossible to memorise infinite quantities that are infinitely divisible. Leonardo speaks in exactly the same terms of the movements of the shoulder: 'The principal movements made by the shoulder joints are simple, that is . . . up, down, forwards and backwards, though one could say that such movements were infinite'. He then describes putting the shoulder firmly to a wall and tracing a circular figure with the arm . . . and a circle is a continuous quantity' (CU 104 r). On W 19005 v (K/P 141 v) he adds the refinement of 'moving the hand with a piece of charcoal around the shoulder, keeping the arm straight so as to make out the circle on the wall'. In the case of the shoulder and the hip joints Leonardo had special reasons for his appreciation of circular movements, for these are 'ball and socket' joints.

On CA 357 vd Leonardo draws figures of what he calls a *'polo universale'*, a universal joint (often called a Cardan joint from the work of Jerome Cardan, who was well acquainted with Leonardo's notebooks). On Codex Madrid I 100 v (Figure 4.16, p. 104) Leonardo describes his universal joint as 'permitting all movements of objects joined to it'. Leonardo uses the word *'polo'* to describe the fulcrum of a balance around which there is rotation in one or many planes. We still talk of the North Pole and the Pole star in a similar sense. *'Polo universale'* applies to as wide a range of rotations as possible in a joint, as in a 'universal joint' of which the hip and shoulder are examples in the human body. Leonardo was well aware that the ball and socket joints exposed by his dissections of shoulder and hip were examples of his universal joints, capable of giving rise to universal circular movements in many planes.

The hip joint, being situated at the centre of the height of the standing body, becomes the axis around which the body rotates backwards and forwards when the legs and feet are firm on the ground as well as the axis of movement of the legs in walking. Leonardo calls it the *'polo dell'omo'*, the pole of man (BN 2038 29 v). It is not surprising that amongst those who knew of Leonardo's work on joint movement there were some who developed this view of human movement systematically and more extensively. Such movements form the essential basis of the first two books of the *Codex Huygens;* the *'polo'* of the hip joint is particularly well illustrated in Figure 12.14.

Leonardo's magnificent drawings of the bones of the foot on W 19011 r (K/P 145 r, Figure 12.15) show all his illustrative devices in action: e.g., different aspects, and bones 'separated from each other' with perfect anatomical shaping. Leonardo pays particular attention to the sesamoid bones drawn at the bottom left-hand corner like two oval peas. These 'glandular bones', as he calls them, 'are always placed near the ends of tendons when they are attached to the bones'. He classed the patella correctly as a sesamoid bone. On this page he shows two sesamoid bones under the great toe. 'Nature has placed the glandular bone under the joint of the great toe', he writes, 'because if the tendon to which this glandular bone is united were to be deprived of it, it would be severly damaged by friction under such weight as that of a man when walking and raising himself up on the joints of his feet'.

It is interesting to note that the excellent drawings of the skull and sinuses by Leonardo were made in the first period of his anatomical researches, and he did not return to make fresh studies of this region in later years. From the text with which he accompanied these drawings it is evident that his main object was to demonstrate the relation of the parts of the cranium to the optic nerves and other special senses. He showed relatively little interest in other details of the structure of the skull. These drawings have therefore been included in Chapter 2, where Leonardo was making such an intensive search to define the basis of sensation, judgement and especially 'experience', in physiological terms.

Chapter 13

Muscles, the Forces of 'This Machine of Ours'

Of all the 'powers' of nature, movement was to Leonardo the most basic and the most fascinating. And of all the powers of the human body, movement, with its changes of shape or 'mutations', was at the very heart of his lifelong activities. Studies of the muscles bringing about geometrical mutations of the human body occupied a great deal of his time. It is no exaggeration to say that some hundreds of pages of his notes are devoted to some aspect of muscular action. Leonardo was, one may say, dissecting the movements of the human body rather than a motionless dead cadaver.

'Force with material motion, and weight with percussion are the four accidental powers with which all the actions of mortals have their existence and their death. Force has its origin in spiritual movement which flowing through the limbs of sentient animals broadens their muscles. Being broadened by this flow these muscles become contracted and draw back the tendons which are connected with them; and this is the cause of the force of the limbs of man' (BM 151 r), so runs Leonardo's description of the relation between the energy of stimulation and change in muscle form.

Leonardo's ambitious programme for acquiring anatomical knowledge of muscles is foreshadowed early: 'That painter who has knowledge of the nature of tendons, muscles and sinews will know very well in the movement of a limb how many and what tendons cause it. And which muscle by swelling, causes the contraction of that tendon, and what tendon is converted into the very fine membrane which surrounds and supports the said muscle' (BN 2038 27 r).

He embarked upon this enterprise intermittently from about 1487. His studies of the muscles of the neck in Figure 13.1 provide a good example of early work. Here he shows the anatomy of the posterior triangle between the sternomastoid and trapezius muscles well. These early drawings suggest a relation to the head of *St. Jerome,* but his explorations of deeper structures in the neck were only partially successful (Figure 13.2).

By far the largest number of notes written on the many pages containing drawings of muscles are devoted to ways of demonstrating them and their actions. These notes are Leonardo's signposts designed to guide him into what it

must be remembered was almost entirely unexplored territory. His objective was the mechanics of the human limbs. Just as he set out to trace the 'real' appearance of objects back to the perception of events in the brain, so he set out to analyse all the activities of man into the movements of the bones and muscles within man's body. 'After demonstrating all the parts of the limbs of man and other animals you will represent the way these limbs work so well from different aspects, that is, in getting up from lying down, walking, running, jumping, lifting and carrying great weights, throwing things to a distance, and swimming. And thus in every act you will demonstrate which limbs and muscles are the cause of the above-mentioned actions' (W 19010 v; K/P 147 v; see Figure 13.3).

He pursued his objective, as usual, with meticulous thoroughness. The mechanical action of muscles depends on their shape. Leonardo classifies muscles by their shapes: those with, and those without tendons such as the tongue and the heart; muscles like straps, e.g., sartorius; and muscles with two or more heads, e.g., biceps and triceps. This knowledge enabled him to find the 'central line' of their action along which their force acts. 'Make a demonstration of the thin muscles by using rows of threads. Thus you will be able to represent one upon the other as Nature has placed them; and thus you will be able to name them according to the part they serve, that is, the mover of the tip of the great toe, etc. And having given such knowledge you will draw alongside this the true size, shape and position of each muscle. But remember to make the threads that denote the muscles in the very same positions as the central lines of each muscle. And thus these threads will demonstrate the shape of the leg' (W 19017 r; K/P 151 r).

He first tried out the method of making cross sections of the leg in which he shows the tibia and fibula surrounded by muscles. Figure 13.4 (W 12627 v; K/P 4 v) shows seven cross sections with a detailed drawing of two sections, one through the quadriceps tendon of the thigh and one a few inches above. Here there is some evidence for identification of the parts. The great belly of vastus lateralis, always shown so prominently in Leonardo's drawings of the thigh, stands out in the drawing to the right with its exaggerated ilio-tibial band running to the lateral condyle of the tibia. In

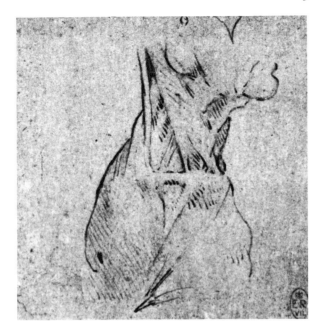 

Figure 13.1.   Superficial dissections of the muscles of the neck (W 12610 and 12611; K/P 17 r and 18 r).

These early dissections of the neck clearly demonstrate the clavicular and sternal heads of the sternomastoid muscle and outline those forming the base of the posterior triangle, which separates this muscle from the front edge of trapezius.

the drawing on the left, the femur has also been sectioned and a whole segment has been 'removed'. Its perspective has been tilted to display the arrangement of the labelled muscle masses. Some of these can be confidently identified, particularly that of the vastus lateralis, marked *e,* the un-labelled rectus femoris, the vastus medialis, labelled *a,* behind which lie the adductor group of muscles (*b* and *c*). However, technical difficulties and the unsatisfactory mechanical information obtained from such sections may have led Leonardo to abandon this method of demonstration, for he does not use it much in later studies.

A technique which Leonardo particularly favours is that of laying the muscles on the bones layer by layer, showing the action of each on the relevant joints. This is the converse of the more obvious method of flaying the corpse and dissecting off the muscle layers from the surface – a method which Leonardo, of course, by no means neglected.

This technique of building up the muscles on their bones of origin and insertion depends obviously on previous knowledge of the bones; it therefore appears only in his later dissections. It is also ideal for demonstrating the mechanical action of muscles – the way in which they bring about leverage. This makes the method doubly attractive to Leonardo. His insight into the value of the technique comes gradually. It begins with the statement, 'Muscles always arise and end in bones adjoining one another. They never arise and end on one and the same bone because nothing would be able to move except [the muscle] itself into a rarefied or dense state' (W 19035 v; K/P 77 v). 'Therefore the joints between bones obey the tendon, and the tendon obeys the muscle, and the muscle the nerve' (W 19019 r; K/P 39 r).

These general statements lead him to practical applica-tion: 'Make for each tendon the bones where they arise and where they come to an end' (W 19015 r; K/P 149 r).

One of the simplest earlier examples of this technique is to be found in Figure 12.1 (p. 252). Here he makes two diagrammatic sketches comparing the leverage of the bi-ceps (its short head, arising from the coracoid process) in flexing the elbow in the monkey and in man. Below he makes a diagram of a lever with its fulcrum at the end, corresponding to the elbow-joint. The proportional dis-tances of elevating forces, from the elbow as fulcrum, are marked out. Here the principle of moments is expressed according to his pyramidal law. At a distance of 2 units 4 units of force are required; at a distance of 4 units 2 units of force are required. This principle of moments is again ex-pressed in MS E 15 r: 'When you want to represent a man as mover of any weight, consider that the movements are to be made through different lines, e.g., from below upwards with simple movement as a man does who stoops down to take up a weight which he lifts up as he straightens himself . . . Here remember that the weight of the man pulls in proportion as his centre of gravity is distant from the fulcrum'.

The progress made by Leonardo in clothing the bones with muscles is superbly shown in Figure 13.5, where in a series of five drawings he studies the mechanism of prona-tion and supination. In the top drawing he concentrates upon the proportionate lengths of the bones and a single overlying muscle, the biceps: 'The tendon of *d* [biceps] is attached midway between the joint of the shoulder and the tips of the fingers'. This is an important observation with regard to mechanical leverage. The main point of the draw-

Figure 13.2. Early exploration of the deep anatomy of the neck (W 12609 r; K/P 3 r).

Here the sternomastoid and trapezius (anterior border) are again clearly drawn. The muscles forming the floor of the posterior triangle have been removed. Identification of the schematic representation of these deeper structures as cords is problematical although some have been labelled by letters.

In the faded sketch to the left the temporalis muscle is well shown being inserted into the coronoid process of the mandible, and preliminary sketches of some of the muscles of expression on the face are made. Below, crossing of the recti abdominis is faintly visible.

Figure 13.3. The actions of agricultural man (W 12644 r).

Studies of the movements and forces of the human body provided the basis for his anatomical explorations. Here it will be noticed that all the actions sketched are agricultural, in contrast with Figure 6.5 (p. 164), where industrial actions are portrayed.

ing is to show the two origins of the biceps, from the glenoid cavity of the shoulder joint and from the coracoid process, the two heads joining together halfway down the humerus. The insertion of the tendon is clearly shown as a protruding band at *d*.

The second drawing in Figure 13.5 consists of an 'exploded' view, in order to show the true shape of the shoulder joint, which can be seen here as a ball and socket joint. The rounded head of the humerus is removed rather far in order to show the relation of the two heads of biceps to the joint. The mode of articulation of the elbow joint and its facets are most beautifully and accurately depicted.

The third drawing shows the two heads of biceps once more at their origin. (This time the pectoralis minor is shown being inserted by two strands into the coracoid process.) The heads of biceps are cut at *b* just before they join the tendon *a*. This is inserted into the radius so that it rotates this bone in such a way as to turn the attached hand so that its palm 'faces towards the sky', as Leonardo describes supination. In this position the radius and ulna (the two bones of the forearm) lie parallel. Pronation, 'with the hand turned towards the earth', as Leonardo expresses it, is illustrated in the next drawing, where the radius is now shown rotated so that it crosses the ulna, turning the thumb through an angle of about 180 degrees. Meanwhile the dis-

tal end of the ulna can be seen to have shifted to the opposite (lateral) side ending at *a*, where the end of the radius was previously. Thus the forearm is slightly shortened on pronation. Leonardo explains these movements as follows: 'The arm which has two bones [radius and ulna] interposed between the hand and the elbow will be somewhat shorter when the palm of the hand faces the ground than when it faces the sky, when a man stands on his feet with his arm extended. This happens because these two bones in turning the palm of the hand to the ground become crossed in such a way that that [bone] which arises on the right side of the elbow goes to the left side of the palm of the hand; and that which arises from the left side of the elbow ends on the right side of the palm of the hand'. In the small diagram alongside he describes how rotation of a line about the centre *a* shortens 'its depth as it becomes more oblique', so explaining the shortening of the forearm.

In the last drawing in Figure 13.5 Leonardo analyses the muscles producing pronation. Here he shows the pronator teres arising from the medial epicondyle of the humerus running like a strap to the radius so rotating it in the opposite direction to the biceps, which is also shown in this drawing. The other pronating muscle, pronator quadratus, is illustrated in the third drawing as a square band joining the lower ends of the shafts of the radius and ulna. In the text here Leonardo does not say a word about either of them, but in the *Treatise on Painting* (CU 119r) he describes this square band as a 'tendon without muscle, the largest in man, from the middle of one arm bone to the other; square in shape, 3 fingers wide, 1/2 finger thick; it keeps the bones together'. He has described pronator quadratus but missed its muscular structure and misinterpreted its action.

Pronator quadratus is drawn and described again in Figure 13.10 (p. 279). Though recognising it now as a 'very

fleshy' muscle, Leonardo fails to see its pronating action. For him it still 'stops the two bones from separating'.

The next stage of clarification of the movements of the elbow is that of representing the action of the muscles by 'threads'. He applies this to the movements of the elbow on W 19103 v, (K/P 196 v, Figure 13.6). Here under the heading, 'The instrumental use of limbs', the biceps is again drawn with its two heads, but the insertion on the radius at *c* shows it clearly 'twisted' round the neck of the bone in such

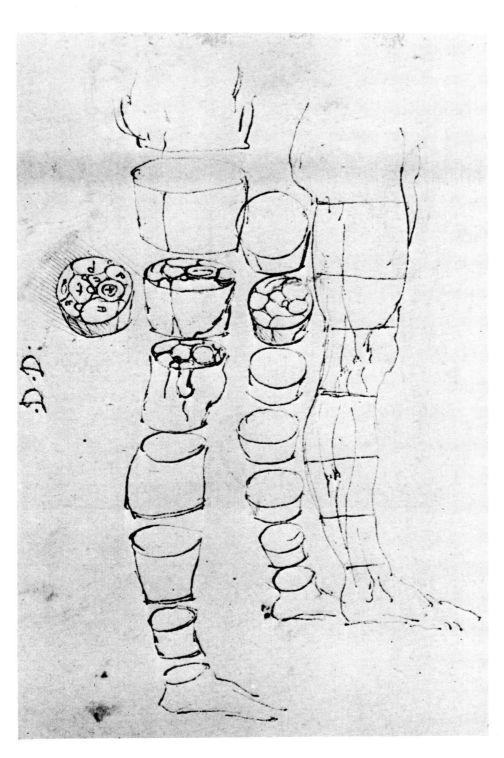

Figure 13.4. Cross sections of the leg (W 12627 v; K/P 4 v).

On the right Leonardo displays the sites of his proposed cross sections. To the left of this he shows them in perspective; in one section he outlines the muscles of the thigh. To the left of this he draws two sets of muscles in rough outline, finally resecting a whole segment in which he labels the main muscles. Note how the sequence reads from right to left.

This method of anatomical illustration was unsatisfactory without fixation of the tissues. Leonardo did not use it much, and it did not become accepted until the nineteenth century.

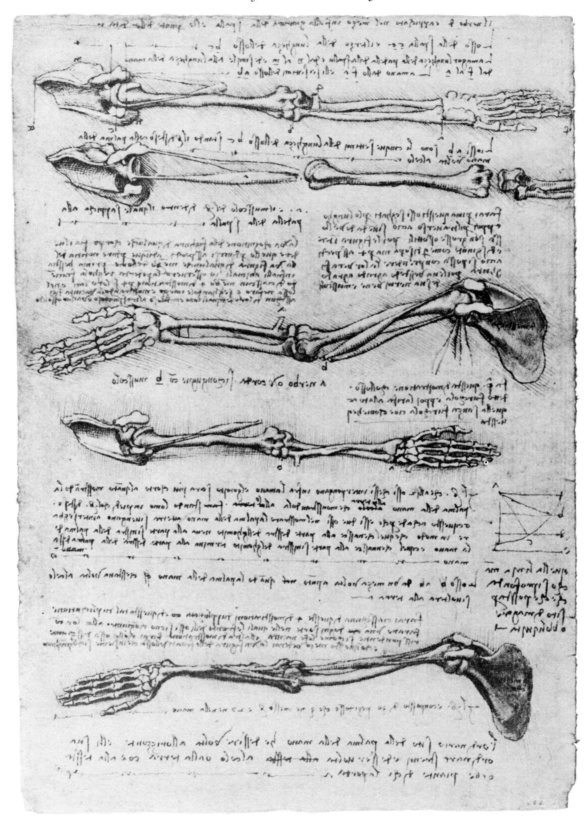

Figure 13.5. How the biceps acts as a supinator as well as a flexor muscle at the elbow joint (W 19000 V; K/P 135 v).

The origin of the biceps at the shoulder is shown with its two heads meeting at *b* to be inserted into the radius at *a* in the third drawing. The manner in which the radius rotates and crosses the ulna carrying the wrist with it in pronation is demonstrated.

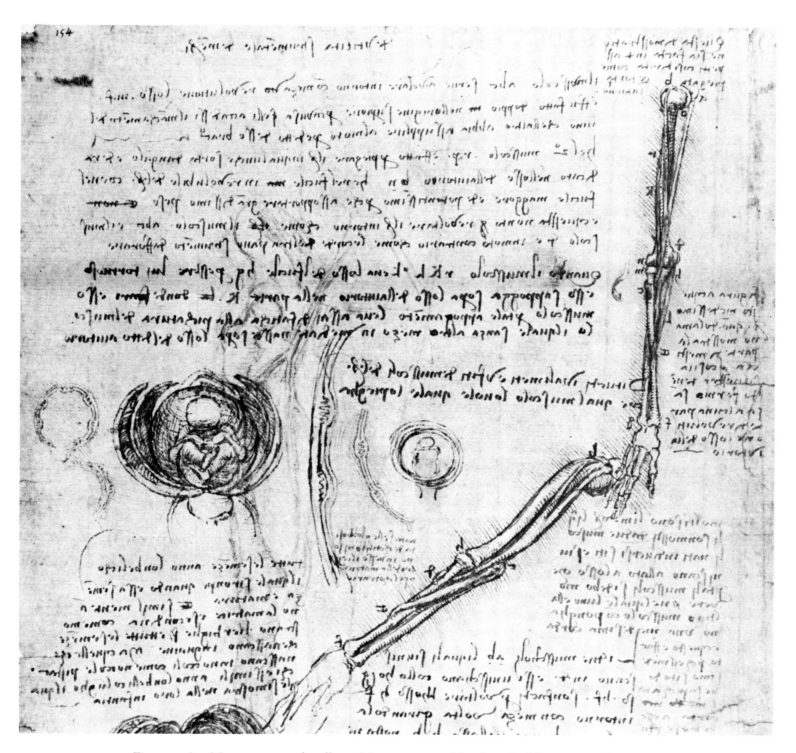

Figure 13.6. Movements at the elbow joint represented by threads (W 19103 v; K/P 196 v).

In a page largely devoted to the development of the foetus Leonardo produces his most successful schematic illustration of the muscles producing flexion, supination and pronation of the elbow. In the upper drawing the insertion of biceps is shown at *c* wrapped round the neck of the radius, all ready to twist the bone back from its pronated position into supination. In the lower drawing pronator teres, *ce,* is shown ready to perform the opposite action.

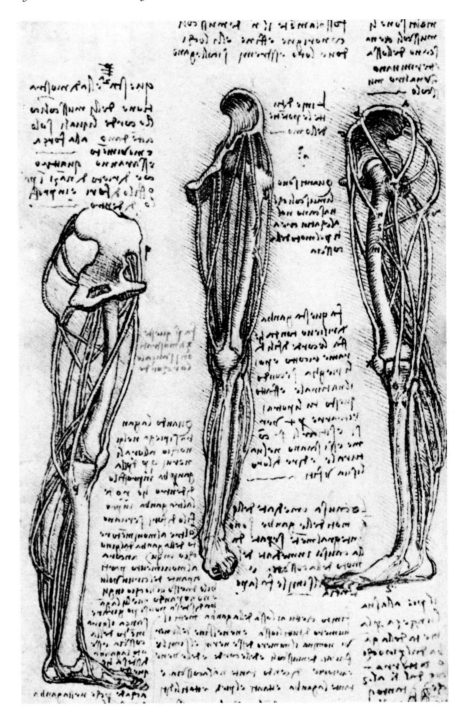

Figure 13.7. Muscles with their central lines of force represented by copper wires (W 12619 r; K/P 152 r).

By constructing a model composed of copper wires in place of the central lines of force of the muscles Leonardo hopes to be able to demonstrate the actions of these muscles.

a way that its action of rotating the radius and with it the hand into a position of supination with palm 'facing the sky' is unmistakable. In this drawing he also represents with a thread the brachialis muscle, which he describes as follows: 'And the 2nd muscle [rk] is made to bend the arm to any angle whatever; it arises from the humerus *bn* and goes into the unrotatable bone of the forearm [the ulna]. It is very powerful because it has to support a very great weight; and it cannot rotate the arm like the muscle *abc* [biceps] and the muscle *de* [pronator teres] is in contrary movement like the

cords of a trephine, an instrument for drilling'. All these structures are drawn and described again in the lower drawing with the additional remark that 'These two said muscles [biceps and pronator teres] have been arranged by the Author to be able to turn the hand to the front and the back without having to rotate the elbow of the arm'. He finishes this magnificent analysis of the movements of the elbow by bringing it back to everyday life thus, 'This rotatory movement is highly necessary in feeding oneself, since the fingers as they take food turn their dorsal side to the mouth; and when they put food into the mouth they are turned in the opposite direction in such a way that their tips, together with the food, are directed towards one's mouth'.

As he builds up the muscles from their bony foundations or dissects down to their deeper layers from the skin, Leonardo reveals a complex pattern, one which he can neither interpret mechanically nor demonstrate without the further technical innovation here used: 'Before you shape the muscles make in place of them threads which demonstrate the positions of these muscles; the ends of which [threads] will terminate at the centre of the attachments of the muscles to the bones. And this will give a clearer knowledge when you want to draw all the muscles, one above the other. And should you do it in any other way your drawing will be confused' (W 19003 v; K/P 137 v).

He makes a special point of drawing 'threads' and not just 'lines'. 'When you have drawn the bones and you wish to draw on this the muscles joined to these bones, make threads in place of muscles. I say threads and not lines in order to know which muscle passes above or below another, which cannot be done with simple lines' (W 19009 r; K/P 143 r). 'Every muscle moves the limb to which it is joined along the line of the filaments of which the muscle is composed' (W 19015 r; K/P 149 r), and 'Every muscle exerts its force along the line of its length' (W 12636 r; K/P 111 r). Thus, from anatomical demonstration he moves to the geometry of muscular movement. Below a drawing of the arm he writes, 'Remember to draw the origins of the two tendons *ab* by uncovering the muscles which use them. And you will do the same for all muscles leaving each one on its own upon the bare bone, so that not only does one see its origin and end, but so as to demonstrate how it moves the bone to which it is attached, and of this the scientific reason is to be given by means of simple lines' (W 19005 v; K/P 141 v).

By these stages Leonardo reached his goal of converting the vast complexity of man's muscular system into geometrical patterns of forces conveyed along the central lines of muscles.

This achieved, he turned to his principle of creating a model of the muscles and joints. On W 12619 r (K/P 152 r, Figure 13.7) he makes a model of what he calls at the top of the page, 'The lines of all the powers of man'. Leonardo describes this model in the following terms: 'I am making only the number of muscles at their origin, and the places where their extremities are attached'. Above the drawing on the left he writes, 'This is the 2nd demonstration of the muscles with their tendons; only of those which are concerned with force and movement. And there will be four [demonstrations], that is, from behind, from the front, in profile from outside and in profile from inside'.

The fact that there are only three drawings on this page indicates that the fourth, missing, demonstration has been torn off. Indeed, at the bottom of the centre of the page Leonardo writes, 'These four legs are to be on one and the same sheet of paper so that one may better understand the positions of the muscles and be able to recognise them from several sides'. The fourth, missing, view is obviously that from behind.

Between the central and right-hand demonstrations Leonardo describes the nature of this undertaking: 'Make this leg in full rounded relief [i.e., like a statue] and make the tendons of tempered copper wires and then bend them according to their natural form. After doing this you will be able to draw them from 4 sides, and to place them as they exist in nature and speak about their functions'.

The interpretation of these drawings is hazardous for various reasons. Many of these 'lines of force' do correspond with identifiable muscles, or muscle groups such as the gluteals, the rectus femoris, the gastrocnemius-soleus group, but a number do not. For example, there does not appear to be any recognition as a group of the four parts of the quadriceps. Instead tensor fasciae latae (a muscle which always catches Leonardo's attention) is only partly understood and seems to come into relation with a 'line of force' acting through vastus lateralis, the ilio-tibial tract and sometimes the lateral hamstring. Sartorius finds happier representation. Again he notes that when both lateral and medial groups act, 'the extended leg cannot rotate without rotating the thigh, and this occurs because in the extended leg the above-mentioned tendons . . . pull the leg straight on to the knee joint'. When the knee is bent the tendons inside and outside the knee joint 'further serve movement, taking part in rotation of the leg from the knee downwards' (W 12619 r; K/P 152 r). (Compare Figures 12.6 and 12.7, pp. 257, 258.)

One notices that when Leonardo had reduced his problem to 'lines of force' acting on fulcra (joints), he tended to find muscles where his mechanical theory led him to expect them, as well as where his dissection revealed them. Nevertheless most of his 'lines' were obviously diligently revealed anatomically for he makes a 'Reminder; To make certain of the origin of each muscle, pull on the tendon issuing from this muscle in such a way as to see the muscle move, and its origin on the ligaments of the bones' (W 19017 r; K/P 151 r).

MUSCULAR CONTRACTION

Leonardo repeatedly searched for the nature of muscular action. 'What is it', he asks, 'that increases the size of muscles so rapidly? They say it is moving air [*vento*]. Where does that go to when the muscle diminishes so quickly? Into the nerves of stimulation which are hollow? That would indeed be a great wind'. He goes on to recall seeing 'a mule which was almost unable to move through fatigue from a long journey under a heavy load, which on seeing a mare, suddenly its penis and all its muscles swelled up so that it multiplied its force so much and acquired such a velocity that it overtook the course of the mare which fled from it, and which was obliged to obey the desires of the mule'. 'Suppose that it is the air in the nerves, then what air coursing through the muscles brings them so much hardness and power at the time of the carnal act?' (W 19017 r; K/P 151 r).

Leonardo was sceptical about air or *pneuma* in the cerebral ventricles constituting the source of power of the soul (see Chapter 10, p. 239). He was sceptical too of moving air or 'wind' as a source of muscle or sexual potency. In his hands the nebulous, traditional theory of the psychic *pneuma* or 'animal spirit' is converted into what could be better termed a 'pneumatic theory' only to be rejected.

Having rejected the *pneuma* or air as the moving force of muscles, Leonardo examines the structure of the muscle in detail in search of a more adequate theory. On the page which contains his remarkable drawing of the brachial plexus (Figure 11.6, p. 249), he embarks on an attempt, illustrated in Figure 13.8, to answer this vexed question. The page is headed, 'On the force of the muscles'. 'If any muscle', he writes, 'be pulled out lengthwise, a small force will rupture its flesh. And if the nerves of stimulation be pulled lengthwise a small force tears them from the muscles where their ramifications are interwoven, spread out and consumed. The same thing is seen to happen to the sinewy coat of the veins and arteries which are mingled with these muscles. What then is the cause of so much force in the arms and legs which is seen in the actions of any animal? One cannot say other than that it is the membrane which invests them, for when the nerves of stimulation widen the muscles these muscles shorten and draw after them the tendons into which their ends are converted; and in such enlargement they fill the membrane and cause it to pull and become hard. And the muscles cannot lengthen unless they become thin, and by not thinning they are the cause of resistance and of giving strength to the aforesaid membrane within which the swollen muscles perform the function of a wedge' (W 19020 v; K/P 57 v). Thus he returns to his concept of 'spiritual force' acting as a form of energy through the nerves or fleshy muscle fibres, 'transmuting' their shape, 'enlarging' or broadening them and shortening their tendons; thus space acquired equals space left (p. 100).

The maintenance of this 'wedge' (isometric contraction) depends largely on the amount of resistance from surrounding muscles or fat. This theme is developed in the *Treatise on Painting* (CU 117 v): 'Those who are inclined to be fat increase a good deal in force after their early youth because the skin is always drawn over the muscles; but such persons are not too dextrous and agile in their movements; and because their skin is tight they have great general power infused into all their limbs. From this it comes about that those who lack such a disposition of the skin help themselves by wearing their clothes tight around their limbs and bind themselves with various bandages so that when their muscles are condensed they may have places where they can be pressed together and supported. But when fat people become thin they are much enfeebled because the deflated skin is left slack and wrinkled; and when the muscles do not find support they cannot condense nor make themselves hard, whence they remain of little power'.

Having reviewed Leonardo's methods of anatomical re-search on muscles and his theory of their action, a brief survey of his achievements in exploring the different regions of the body is indicated.

Leonardo's descriptive anatomy comes across in his drawings. There can be no adequate substitute for the detailed inspection of their intelligent, rich content. This, indeed, is how he wished his anatomy to be communicated: 'Do not busy yourself in making enter by the ears things which have to do with the eyes' (W 19071 r; K/P 162 r).

Leonardo's anatomy of the muscular system varied considerably in quality with the different regions of the body. He was most successful with his anatomy of the arm, particularly with the elbow and hand; slightly less so with the foot and lower leg; least successful with the trunk and thigh. Such a judgement is, of course, made from the extant notes, and there is good reason to believe many have been lost. To some extent it corresponds to his areas of interest. But time was his greatest enemy. As he recognised, 'I have been impeded neither by avarice nor negligence, but only by time' (W 19070; K/P 113 r).

THE ARM AND HAND

Leonardo's technique of dissection at progressively deeper levels is nowhere better demonstrated than in his drawings of the shoulder and arm. On W 19013 v, K/P 144 v (Figure 13.9) he draws three dissections of the arm at three different levels. At the top of this page he prefaces his drawings with the comment, 'And you who want to demonstrate with words the figure of man in all aspects of his limb-formation do away with such ideas, because the more minutely you describe the more you will confuse the mind of the reader . . . Therefore it is necessary to make a drawing of it as well as describe it'.

The most superficial dissection is drawn at the top right corner, where the deltoid and pectoralis major are clearly shown; below these the superficial muscles of the arm and forearm appear. In the second drawing of the series the deltoid is shown lifted off; beneath it pectoralis major and latissimus dorsi are shown, muscles which bring the arm down with impetus as in hammering or percussing. Pectoralis major particularly interested Leonardo since he knew of its power in birds and considered its possible power in the flight of man.

The third dissection (bottom left) in Figure 13.9 shows the pectoralis major lifted off to expose pectoralis minor running up to be inserted on to the coracoid process of the scapula, as well as the two heads of biceps. In the right margin Leonardo makes a cord diagram of these muscular forces.

Leonardo's dissections of the hand on W 19009 (K/P 143 r, Figure 13.10) also show progressively deeper dissections. The relationship between the tendons of flexor sublimis digitorum and flexor profundus digitorum obviously attracted Leonardo's attention by virtue of their mechanical ingenuity. Between the two main drawings of these ten-

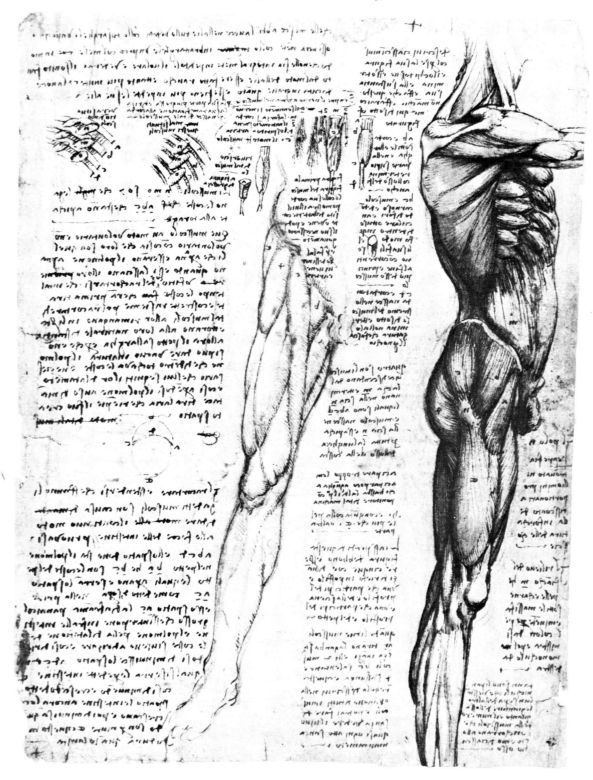

Figure 13.8. The structure of muscles (W 19014 v; K/P 148 v).

The set of small drawings to the left of the shoulder of the large figure on the right margin shows Leonardo's concept of the sinewy origin, rounded muscle belly and cordlike tendon, *cd*. To the left of this figure he draws a small lettered figure of nerve, artery and vein entering a muscle. Two further little diagrams showing the cut muscle appear further to the left. All these drawings illustrate his investigation of the structure of muscle as described in Figure 11.6 (p. 249). The muscles of the trunk and thigh are seen in the right-hand figure. This powerful muscular figure is described on p. 282 of the text.

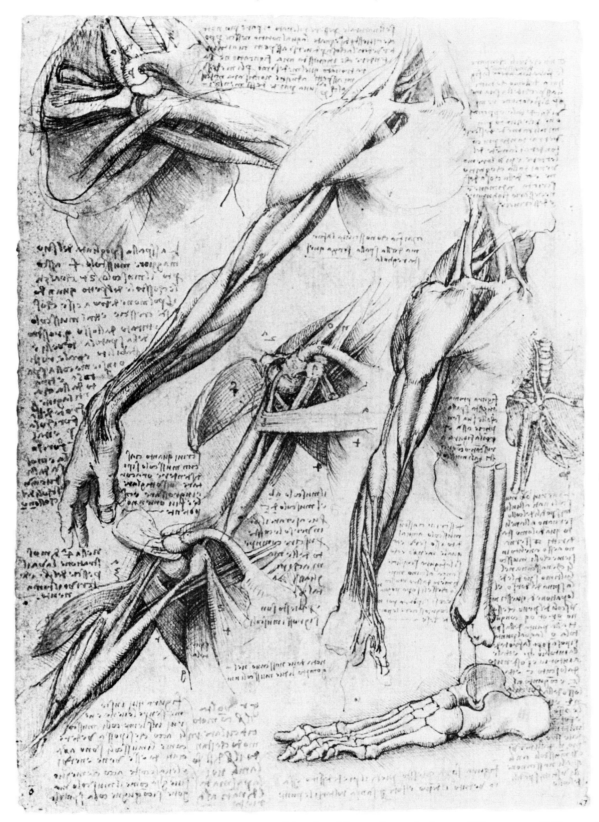

Figure 13.9.   Demonstration of the anatomy of the shoulder and arm in layers (W 19013 v; K/P 144 v).

The note at the top stresses the futility of using words for demonstrating anatomy. Three dissections, numbered '1st', '2d' and '3d', are shown from the right upper corner to the lower left. Each reveals a deeper layer of muscles. In the right margin is a cord diagram explaining the forces exerted by each muscle group. Special stress is laid on pectoralis minor as raising the ribs on inspiration. Top left is a deep dissection of the shoulder; bottom right, a beautiful exploded view of the ankle joint and foot.

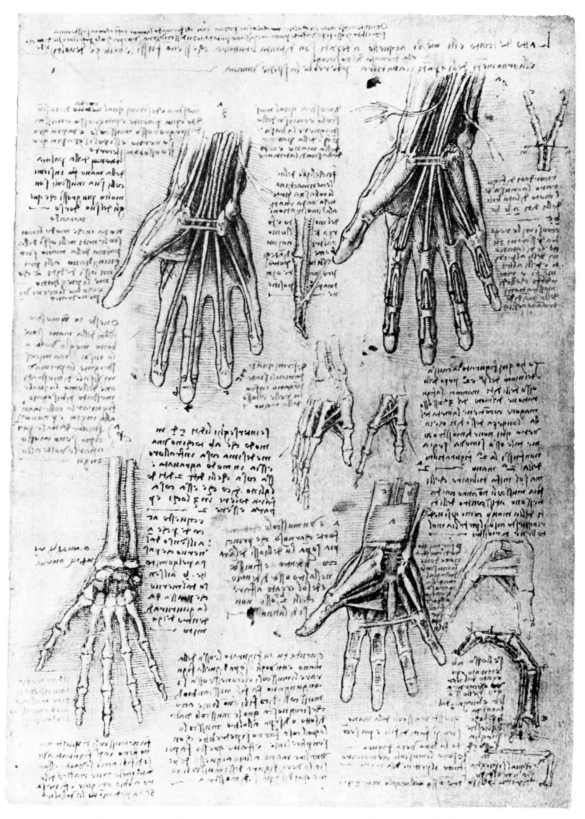

Figure 13.10. Dissection of the hand in layers (W 19009 r; K/P 143 r).

A series of dissections of the hand shows deeper layers from the tendons of flexor digitorum sublimis (top right) down to the bones (bottom left). Between the drawings of flexor digitorum sublimis and profundus is a small drawing showing how the latter tendon pierces the former. Around this drawing Leonardo writes a note emphasising the importance of mechanics to understanding the forces of man.

dons at the top of the page he draws a finger with these two tendons especially clearly displayed. Around this small drawing he writes a note saying, 'Arrange it so that the book on the elements of mechanics with its practice shall precede the demonstration of the movement and force of man and other animals; and by means of these you will be able to prove all your propositions'.

THE LEG AND FOOT

Leonardo made a great number of studies of the superficial muscles of the leg. Many appear to have been made about the time he was concerned with the painting of *The Battle of Anghiari* (Figure 1.26, p. 34) and correspond to his comment, 'Strong nudes will be muscular and thick, those of less strength will be sinewy and thin' (W 12632 r; K/P 8 r). Certainly he often draws grossly hypertrophied leg muscles.

Of greater anatomical significance is a study of kneeling in W 19037 r (K/P 81 r, Figure 13.11). Here he adds muscles to the kneeling skeleton of Figure 12.7 (p. 258). The anterior thigh muscles are distinctly divided into the sartorius and tensor fasciae latae, both arising from the anterior superior iliac spine. The sartorius is here shown being inserted into the upper part of the medial surface of the tibia. Behind it appear the tendons of semitendinosus and semimembranosus (the internal hamstrings) inserted into the medial tibial condyle. Their action is shown to rotate the lower leg at the bent knee.

W 19017 r (K/P 151 r, Figure 13.12) is a large double-page study of the leg muscles concerned in raising up (ex-tending) the toes. A beautiful and minutely accurate drawing of the way the extensor tendon is inserted into the back of the great toe is drawn in the lower right-hand corner. The muscle bellies of the short extensor muscles on the back of the foot, labelled *a, b, c, d,* are superbly shown. It is on this page that Leonardo takes issue with his preceptor: 'Mondino says that the muscles which raise the toes of the feet are found on the external side of the thigh'. The fact that Leonardo denies this confirms that he used the Italian translation of Mondino's *Anatomy* since in this it is said that the digits of the feet are raised by muscles, '*liquali sono nela cossa* [thigh] *nelaparte silvestre*'. In the Latin version Mondino refers to the side of the *lower* leg. The mistake therefore made by the translator into Italian deceived Leonardo.

The sole of the foot is less artistically but almost as searchingly drawn on W 19019 r (K/P 147 r) and W 19006 v (K/P 146 v). They are accompanied by the typical comment, 'The hand does in relation to the arm just as the foot does in relation to the leg' (W 19010 r; K/P 147 r). Obviously the great complexity of 'lines of force' which Leonardo found to be exerted on the joints of the body made it very difficult for him to apply his simple rules of the lever to many of them. He does however apply his mechanics to the weight of the body standing on the foot normally both on the heel (Figure 13.13a) and on tip-toe (Figure 13.13b). On W 19144 r (K/P 102 r, Figure 13.13a) Leonardo describes the weight-bearing foot in terms of a balance in equilibrium: 'The fulcrum *a* [ankle joint] is that [place] where a man balances his weight by means of the tendons *mn* [tendo Achilles] and *op* [dorsiflexors of foot] which are to the shank of the leg above the said fulcrum as the shrouds

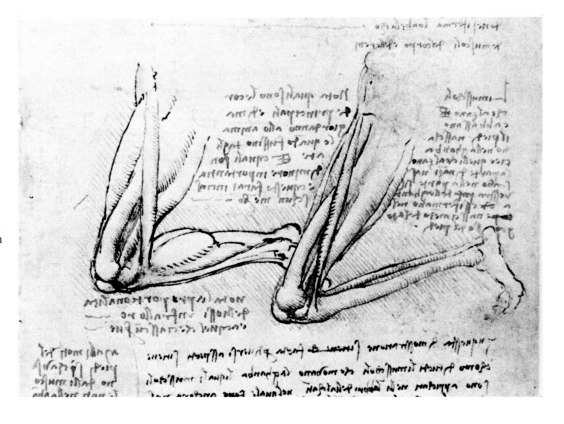

Figure 13.11. The act of kneeling (W 19037 r; K/P 81 r).

The schematically drawn muscles of the thigh on the right are more fully filled in on the left. These muscles are added to the kneeling bones studied in Figure 12.7 (p. 258).

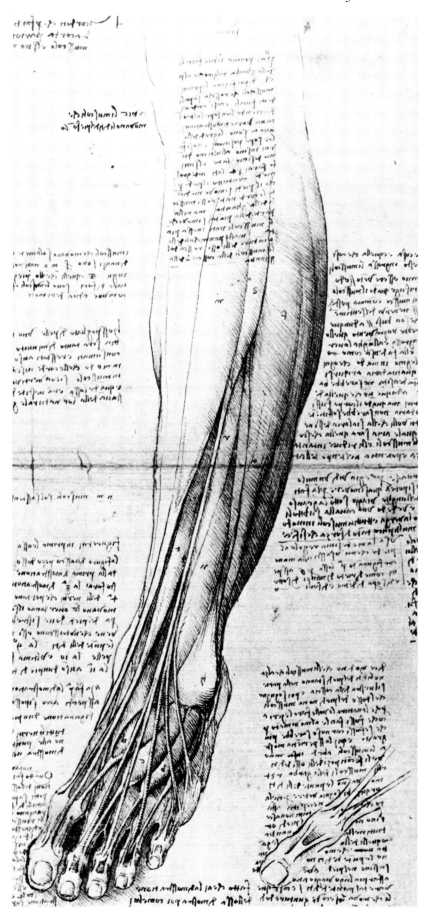

Figure 13.12. The anterior muscles of the leg (W 19017 r; KP 151 r).

This large drawing of the muscles in front of the leg was made to demonstrate the action of the extensor muscles of the toes. The drawing is noteworthy for its accuracy, particularly in the depiction of extensor digitorum brevis on the dorsum of the foot. In the lower right-hand corner the extensor hallucis longus is finely shown.

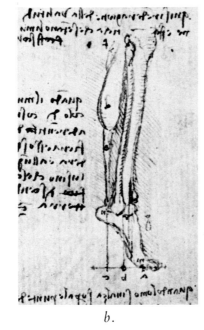

a.

b.

Figure 13.13.

a. The ankle joint and its balancing forces (W 19144 r; K/P 102 r).

The shank of the leg is likened to the mast of a ship; the muscles in front and behind balance it like the rigging of a mast.

b. The forces involved in raising the heel off the ground (W 19010 v; K/P 147 v).

The pull of the tendon of gastrocnemius on the heel at *n* is measured by the increased weight on the ball of the foot at *a* about the fulcrum *b*. Compare Figure 12.7.

are to the masts of ships. And one pound of his weight which a man throws to any side away from this fulcrum *a* will carry weight to the opposite tendon, as will be demonstrated in its place, namely, on the power of the limbs'. This he does in Figure 13.13b.

In Figure 13.13b (W 19010 v; K/P 147 v) he draws the line of force of the gastrocnemius passing down through the tendo Achilles (marked *o*) to the heel-bone (calcaneus) marked *n*, and on to the ground (*c*). The central line of weight passes between tibia and fibula through point *v* to *b* on the ground; the ball of the great toe (*m*) now rests on the fulcrum (*a*). Leonardo explains this drawing at length: 'Here is given the causation of the strength and work which muscles experience'. In summary, he points out that when the calf muscles are relaxed the heel meets the ground at *c*. When a man raises himself on tip-toe the calf muscles feel half the weight of this man standing on one foot. This is shown by the balance *abc,* of which the fulcrum is at *b* from which *a* and *c* are equidistant. With the fulcrum under the ball of the foot at *a* the central line or fulcrum of natural weight at *v* cannot descend because the calf muscle is pulling upwards at the heel, *n*. So the central axis of the leg at the

ankle joint 'supports a weight of 200 lbs of which 100 of natural weight is at *m* or *a* and 100 of accidental weight is at *n*. Both powers are directed upwards because *m* wants to go to *g* and *n* is pulled upwards by the calf of the leg'. These two studies are developed to explain the movement of jumping (compare Figures 6.26b and 12.7, pp. 182, 258).

THE TRUNK AND THIGH IN MAN

The most impressive anatomical drawing of the trunk and thigh is to be found on W 19014 v (K/P 148 v, Figure 13.8, p. 277). The external oblique muscle of the abdomen (*m*) surges upwards to interdigitate with the forward-pointing fingers of serratus anterior on the ribs in a dramatic and accurate confrontation. Latissimus dorsi also runs upwards across serratus anterior, converging into a tendon disappearing under the outline of the upper arm. Below the line indicating the crest of the illium things are not so satisfying. The letters *a, b, c, d* and *r* indicate muscle masses comprising tensor fasciae latae and gluteus medius (*b* and *c*) divided by Leonardo's mechanical lines of force into two parts; *d* provides a surprisingly inadequate figuration of gluteus maximus, the main muscle of the buttocks, because Leonardo appears to consider it as terminating near the greater trochanter. He notes that the 'muscle of the back of the thigh and that of the buttocks' make a bigger variation in their extension and contraction than any other muscle in man (CU 118 v). This is because of their stretching on bending down and contracting when standing up, a movement made around the hip, the transverse '*polo*' or axis of the whole body.

All these muscles he considers concentrate their action round the prominence *n* which is the greater trochanter, and this in its turn marks the hip joint, which Leonardo calls the '*polo*', pole or fulcrum of the body, 'which is always found in the well-proportioned man opposite the fork of the thighs'. From this point the vastus lateralis is drawn descending to the level of the patella. Tensor fasciae latae (*a*) is drawn joining it and the ilio-tibial tract (*h*). The letter *r* marks the sartorius arising with tensor fasciae from the anterior superior iliac spine. The view from the front shows these two muscles more clearly, labelled *c* and *d*, like an inverted *V* with the rectus femoris emerging from between its arms. Leonardo pays much attention to these muscles because he regards them as the only flexors of the hip joint; he does not describe psoas major or iliacus.

It is evident that Leonardo himself is amazed at the wonderful complexity and beauty of the mechanics of the human body revealed by his exploration, for here, in one of his unexpected outbursts (just below the note on the buttocks) he exclaims, 'He who considers this too much, let him subtract from it; and he who considers it too little let him add to it; and he for whom it is sufficient let him praise the first composer of such a machine'.

Superficial muscle configurations of exaggerated shape ('Anghiari' type) are drawn in Figure 13.14. This drawing,

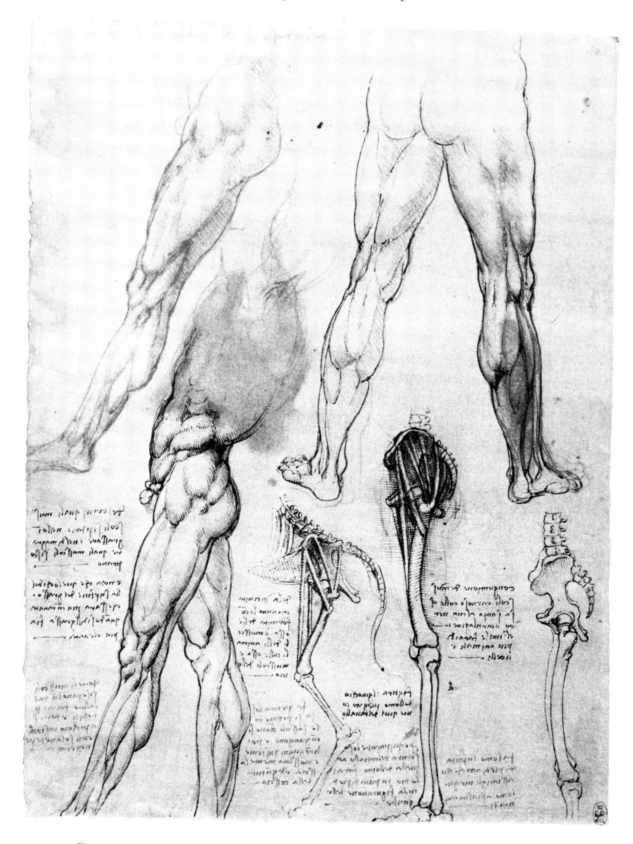

Figure 13.14. W 12625 r; K/P 95 r. The superficial muscles of the leg; exaggeration of muscle outlines was replaced by threads to represent their forces. Comparison of the leg of a man with that of a horse. See text.

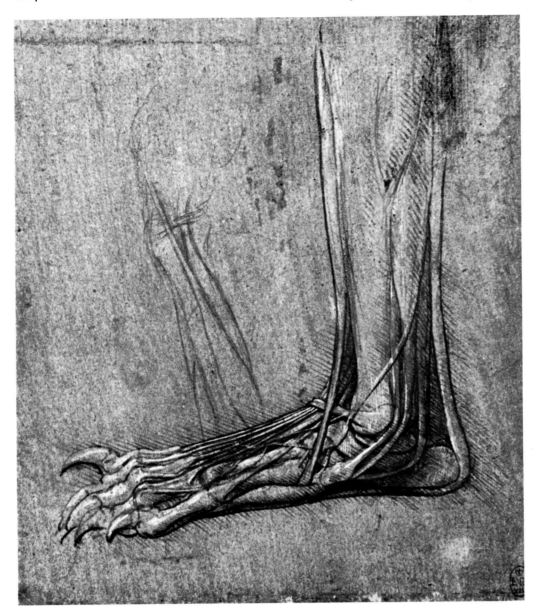

Figure 13.15. Comparative anatomy; the bear's foot (W 12372 r; K/P 12 r).

An early study in Leonardo's phase of purely descriptive anatomy, this bear's foot shows the annular ligament of the ankle joint binding down the dorsiflexor tendons. This feature Leonardo does not demonstrate in his drawings of the human ankle.

however, is most noteworthy for Leonardo's advance into the field of comparative anatomy. The central lower drawing shows the hind leg of the horse with some of the hip muscles represented by his lines of force or 'threads'. The glutei and the sartorius with tensor fasciae latae as flexors of the hip joint can again be easily recognised. They may be compared with the neighbouring study to the right which represents the bony configuration of the human skeleton with similar muscles shown as threads. Perhaps the most striking feature of these drawings, however, is the representation of the bones of the horse's lower leg and foot, showing full appreciation that compared with the leg of man the horse stands on the tip of its toe.

Leonardo showed keen interest in comparative anatomy at this time. He remarks, 'You will draw for comparison the legs of frogs which have a great similarity to the legs of

man, both as to their bones and their muscles. Then you will follow with the hind-legs of a hare which are very muscular and have distinct muscles because they are not encumbered by fat' (W 12631 r; K/P 89 r). Though these comparative drawings are not extant, there are a number of drawings of the anatomy of horses, bears (Figure 13.15), birds, bats, monkeys, dogs, pigs and, of course, oxen. Though these are not all done in order to compare their anatomy with that of man, they do lead Leonardo to the general conclusion that 'All terrestrial animals have similar limbs, that is muscles, sinews and bones, and nothing is varied except in length and size, as will be demonstrated in the Anatomy' (G 5 v). As usual, concentrating on movement, he adds, 'Note the bendings [flexion] of the joints, and in what way flesh increases on them in their flexion and extension. And of this most important knowledge write a

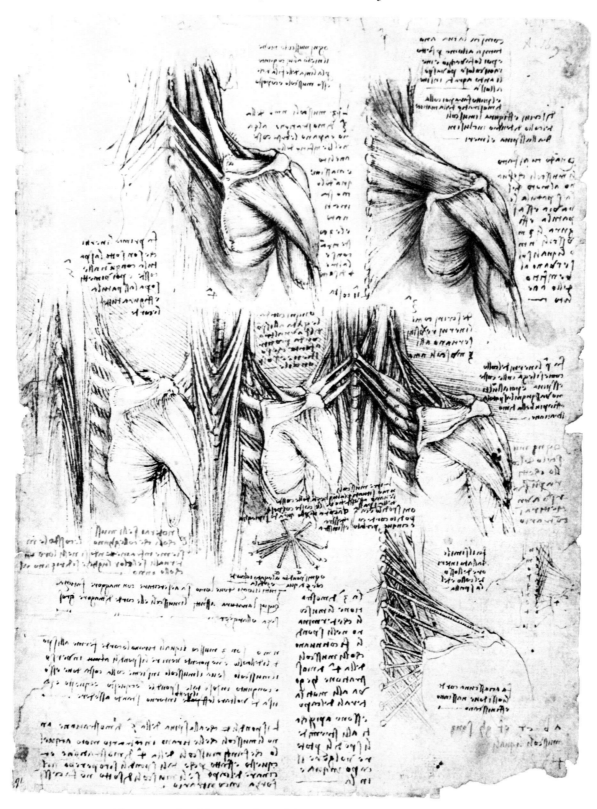

Figure 13.16. The muscles of the spine (W 19015 r; K/P 149 r).

In these drawings Leonardo illustrates his concept of symmetrical pairs of spinal muscles attached to the vertebrae, some pulling upwards some downwards. Acting together, such pairs stabilise the joint of the spine acted upon. Acting separately, they flex the spine. The unilateral geometrical pattern of these forces is shown in the cord drawings in the lower right corner.

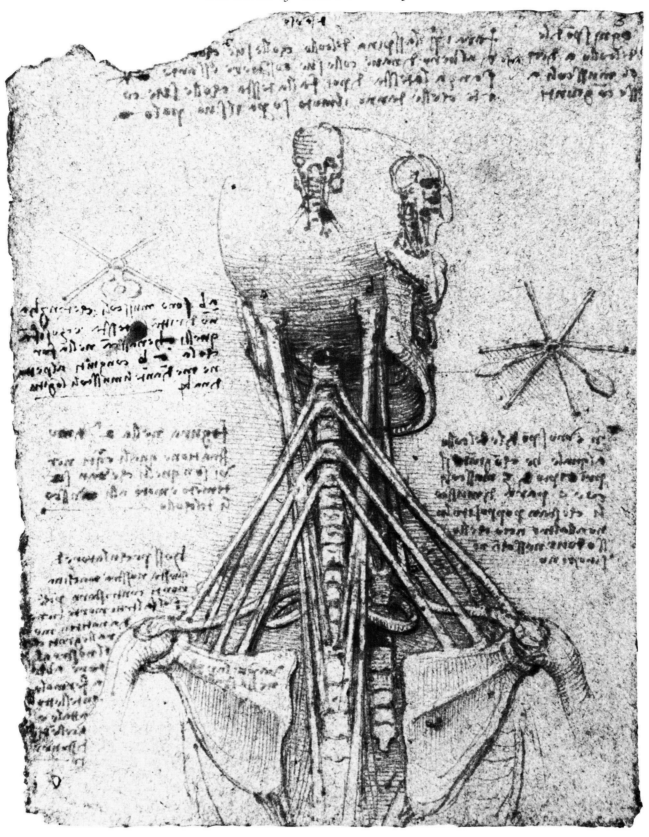

Figure 13.17.   The muscles of the spine   (W 19075 v; K/P 179 v).

The muscles of the cervical spine are here represented entirely as threads or ropes analogous to the mast of a ship with rigging. Some are attached to the skull. The more central forces stabilise and the more lateral forces bend the neck.

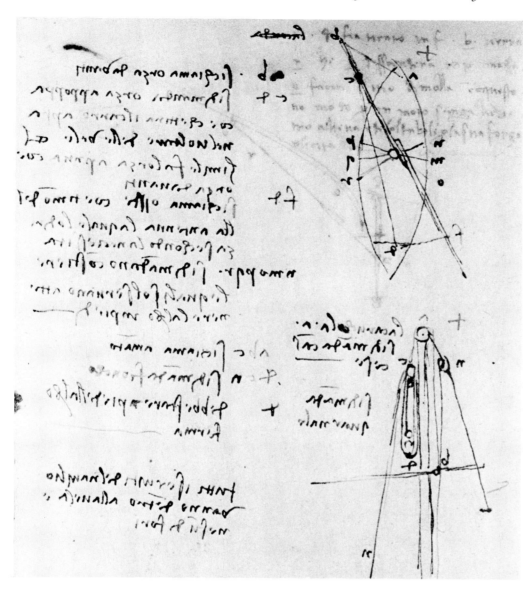

Figure 13.18. Studies of ship's rigging (Madrid II 121 v).

Leonardo's proclivity towards making the analogy between the spinal column and a ship's mast is based on his experience of sailing. On this page he sketches and names the different parts of the rigging. The central and lateral parts are clearly differentiated in these drawings. Compare also with Figure 14.1.

separate treatise on the description of the movements of four-footed animals, among which is man who likewise in infancy moves on all 4 feet' (E 16 r).

The muscles of the spine presented Leonardo with a complexity which he schematised into a mechanical pattern. They are most thoroughly studied on W 19015 r (K/P 149 r, Figure 13.16). In a series of drawings Leonardo (proceeding from right to left) makes progressively deeper dissections, ending at the lower right corner with his final idea of the geometrical pattern of the central lines of muscular forces. Against the top right-hand drawing he explains, 'When you have made the muscles which serve the movement of the scapula lift off this scapula and draw the 3 muscles *nmo* [serratus posterior superior] which serve only to help breathing'.

The next level of dissection in Figure 13.16 is in fact to the left, where these three muscle slips are drawn but not labelled. The levator scapulae arising from the point *a* runs

upwards and inwards to *b* in the neck. In the right-hand drawing of the central row the three muscle-bands *nmo* find clear representation arising from the ribs beneath and passing upwards and inwards to the spine of the lower cervical and two upper thoracic vertebrae. Leonardo attaches very great importance to this muscle, repeatedly drawing and describing it. We shall consider this under the movements of breathing in Chapter 14. Let us continue now with the series of dissections on this central row. The middle drawing is designed to illustrate how a series of paired muscles arise and pull upwards on the vertebral spines of the upper thoracic vertebrae on both sides. In the drawing further left he shows a similar series of paired muscles pulling downwards on the same vertebral spines. He explains the drawings thus, 'Every muscle in the neck which pulls in one direction has another which pulls in the opposite direction'. And in between the drawings he writes, 'Each tendon pulling on a vertebra has another tendon acting through the spinous process which supports that vertebra'. His concept of the principle underlying the muscles of the spine is now expressed in the cord diagram. Here the elaborate but logical pattern of the tendons attached to each vertebral spine on one side is laid out. It will be noticed that in this diagram the seventh cervical vertebra is stabilised, its tendons exerting the counterforce postulated. To the left of this diagram another small drawing depicts a vertebral spine with ten tendons attached like the rays of a star. Underneath this Leonardo writes, 'Each vertebra is bound to 10 tendons, that is, 5 on this side, 5 on that'. Thus movement of any part of the spine is based on a mechanism which can fix neighbouring vertebrae either from above or below. If this fixed or stabilised region moves up or down the spine whilst the rest of it moves, it will produce a wave of related movement. If the movement is one of rotation, the different parts of the body will face in different directions and present a shape like that so movingly rendered in the turning figure of the *Leda*. If the stabilised and/or moving area also travels along the spine longitudinally, the rippling, 'serpentine' movement of horse, cat or dragon emerges, as drawn in Figure 6.18 (p. 177) and W 12363.

The two movements thus based were of special interest to Leonardo; their links with that of the ribs with respiration we shall describe under the respiratory system; and that of the movements of the head and neck.

Leonardo returns to the movements of neck and skull about 1513. At this time he was in Rome, where he was not allowed to pursue his human anatomical dissections. The anatomical basis of his mechanistic physiology in the drawing on W 19075 v (K/P 179 v, Figure 13.17) is much diminished. Leonardo here develops his analogy between the sta-

bilising of the cervical spine and the rigging of the mast of a ship, derived from his experience of sailing (see Figure 13.18). He seems to be continuing the theme expressed in Figure 13.16, for his opening remark is, 'Every vertebra of the neck has ten muscles joined to it'. He continues at the top of the page, 'You will first make the cervical spine with its tendons like the mast of a ship with its side-rigging, without the head. Then make the head with its tendons which give it its movement on its fulcrum [*polo*]'. The crude sketches render a rather pathetic illustration of what he has in mind. On the right margin eight (not ten) muscle-tendons are shown converging onto a spinous process. Beneath it Leonardo describes, '3 pairs of muscles opposite one another, so that the bone where they arise shall not be broken'.

The central drawing in Figure 3.17 consists of his beloved lines of force embodied in a cord diagram showing an almost entirely hypothetical set of muscles supporting the 'mast' of the cervical spine (consisting incidentally of ten, not seven, vertebrae). Two muscles marked *a* and *b*, possibly reminiscent of semispinalis, are 'muscles which keep the head straight, and so do those, *cb*, [sternomastoid] which arise from the clavicle and are joined to the mastoid [*pettine*] through longitudinal muscles'.

Below this ghost of his anatomical prowess Leonardo writes, 'O observer of this machine of ours let it not distress you that you give knowledge of it through another's death, but rejoice that our Author has established the intellect in such an excellent instrument'. This and the drawing constitute Leonardo's anatomical apologia, for it was in Rome that he protested, 'somebody impeded my anatomy; defaming it before the Pope and likewise at the hospital' (CA 182 vc).

Chapter 14

The Ebb and Flow of Respiration

Leonardo's studies of breathing evolved from a relatively literal analogy between the macrocosm of the world and the microcosm of man to a refined mechanical consideration of the respiratory phases of inspiration and expiration with recognition of residual air. His physiology of respiration, as of other vital functions, cannot be appreciated without considering his earlier speculations and seeing how they were transformed into those final closer approximations to his perpetual goal, 'true knowledge' of the process.

In 1492 Leonardo writes, 'Man has within him a pool of blood wherein the lungs as he breathes increase and decrease, so the body of the earth has its ocean which also increases and decreases every six hours with the breathing of the world' (A 55 v). This analogy between the 'tidal' movement of air and 'tidal' seas remains with him. It is evident some twenty years later when he writes in a suggested arrangement of a Book on Water, 'Whether the flow and ebb [*flusso e reflusso*] arises from the moon or the sun; or is it true that it is the breathing of this terrestrial machine' (Leic 17 v). An even more literal interpretation of this analogy is expressed about the same time in the assertion that 'The body of the earth is of the nature of a fish, a sea-monster [*orca*] or sperm-whale, because it breathes water instead of air' (CA 203 rb).

The thread of continuity between these various statements lies in the fact that they are all examples of movements of oscillation, of wave-movements, pendular movements, or '*moto ventilante*' (see Chapter 4). All of these terms are used almost interchangeably by Leonardo, but the most relevant to the tides of water and air are '*flusso e reflusso*', which he constantly looks upon as one movement pattern. We shall translate this euphonious Italian phrase by the clumsier English term *ebb and flow*. This is inept to the extent that it fails to recognise the reflection of an incident movement inherent in Leonardo's own term.

The problem presented to Leonardo by breathing therefore consisted of two parts: first, what is the mechanism of the ebb and flow movement of the chest and lungs; and second, what is the function of this tidal flow of air into and out of the lungs? At first he talks of the lungs 'pressing on the ribs and these are pushed outward', and of air leaking into the pleural cavities (W 19034 r; K/P 76 r). But later Leonardo concludes that all movements and forces within the body originate from muscles. In what would appear to be an early confused attempt to represent the intercostal muscles Leonardo describes the movements of breathing thus: 'These muscles have voluntary and involuntary movements, for they are those which open and close the lung . . . [by moving the ribs] as the chest expands. And since no vaccum can occur in nature, the lung which is in contact with the ribs internally necessarily follows their dilation, so opening the lung like a bellows drawing in air to fill its generated space' (W 19014 v; K/P 148 v).

Finding no help with regard to the nature of movements of the chest in the work of Mondino, Leonardo turned (as advised by Mondino) to Avicenna's chapter, On the Organs. In Figure 14.1 (W 19015 v; K/P 149 v)* he makes a drawing of serratus posterior superior which exactly fits Avicenna's description. This says, 'The fourth pair [of muscles] which expands the chest arises from the seventh cervical vertebra and the first two thoracic vertebrae and is inserted into the ribs of both sides. These are the muscles which expand the chest'.[1] Leonardo himself, discussing the vexed question as to whether muscles attached to both ribs and spine act on the spine or ribs states, 'One treats of man according to the instrumental method', and explains, 'It happens almost universally that muscles do not move the part on which they are established but move the part to which the tendon arising from the muscles is joined, except that which raises and moves the ribs in serving respiration'. Around this drawing he adds, 'All these muscles are elevators of the ribs, and elevation of the ribs is dilatation of the chest, and the dilatation of the chest is expansion of the lung, and expansion of the lung is attraction of the air which penetrates through the mouth into the increased capacity of the lung'. But he notes too, 'These same muscles hold the spine of the neck straight when the power of the lower muscles arising from the pelvis prevails, these terminating

---

*This page is the verso of that on which the muscles of the spine are studied (Figure 13.16, K/P 149 r). The continuity of the thought is obvious.

Figure 14.1. Elevation of the ribs in breathing (W 19015 v; K/P 149 v).

Above, Leonardo draws the serratus posterior superior muscle, which consists of three bellies arising from cervical and upper thoracic vertebrae and being inserted near the angles of the ribs.

Below, he represents one belly of the muscle schematically in order to consider its dynamics, asking, Does it raise the ribs, stabilise the spinal column, or do both according to circumstances? To the right of this sketch he draws a diagram of a mast of a ship with central and lateral ropes attached. Compare Figure 13.18 (p. 287).

on the ribs. When these exert force they act to resist and support the roots of those muscles which hold the neck straight'.

Below this passage in Figure 14.1 are two drawings; that on the right represents the mast of a ship with 'cords' attached, *ab* and *ac* diverge sharply whilst *am* and *an* run down close to the mast. This diagram is transmuted into anatomical terms in the neighbouring drawing. Here the cervical spine is shown; a muscle belly like that of serratus posterior superior is shown attached, arising from what would appear to be the first rib, and a more divergent cord runs from a cervical vertebra to the lateral curve of the same rib. The analogy is clear. Leonardo writes, 'Those ropes more easily prevent the fall of the mast which are joined to the ends, and which converge to their junction on the mast at a greater angle'. And beneath the 'anatomical' drawing he asserts, 'And that rope has less power of pulling down the mast which is joined to this mast by more unequal angles'. He refers here to the direction of lines of force of muscles; the most powerful of all would be that acting on the mast 'between equal angles', i.e., at a right angle; less powerful is that acting at an angle of about 45 degrees, (*ac*); and least powerful is that acting at about 5–10 degrees, as do *am* and *an*. The comparison goes for the imagined muscle acting along the line *ad* as compared with the 'muscle' *ac*. As we have already seen, this analogy between the neck and its muscles and the mast of a ship with its stays remained potent with Leonardo until the end of his life as an anatomist (see Chapter 13). These diagrams afford a good example of the mechanical principles which guided him in his anatomical exploration of muscles.

Leonardo now turns to the centre of the page in Figure 14.1 to elaborate the significance of these three drawings. The top paragraph is headed, 'On demonstrating how the spine is set on the neck'. He emphasises how the muscles arising from the ribs and running to the spine have a double action, 'that is, they support the spine by means of the ribs, and they support the ribs by means of the spine'. And in just the same way, 'the ropes of a ship support its mast, and the same ropes bound to the mast also support in part the sides of the ship to which they are attached'.

The second paragraph is headed, 'On the method of drawing the origins of the movements of any limb'. It constitutes one of his many discussions on the technique of anatomical representation.

The third paragraph is headed, 'On the muscles established on the ribs drawn above'. It constitutes a fascinating discussion as to the action of the muscle bellies of serratus posterior superior, in which he refers to the two drawings on the right in support of his conclusion that these muscles act more strongly in stabilising the spine than in elevating the ribs. In this context it is of interest to quote from *Gray's Anatomy*,[2] which says of this muscle, 'By their attachments it is clear that this muscle could elevate the ribs but experiments in dogs indicate that it is not a respiratory muscle. Its role in man is not clear'.

For Leonardo, however, the serratus posterior superior remained of first importance in elevating the ribs. He again describes the mechanics of this action in detail on W 19061 v (K/P 154 v, Figure 14.2). Here the three familiar muscle bellies (labelled *anm*) are again seen running upwards and inwards to the spine. Below are two drawings in which Leonardo illustrates the action of these muscles by means of figures of a pulley and a lever. In the upper the pulley cord pulls the rib, divided into four parts, through the arc upwards from *e* to *c*. Clearly it rotates on the fulcrum of the vertebra at *b*, so that it comprises a lever of the second order. The mobile advantage of this type of leverage is illustrated by the fact that the arc through which the rib-end moves, *ec*, is 'four times more than at *a*. Consequently dilatation of the lung acquires height by raising up this lid formed by the upper parts of these ribs'. The same point is made in the lower diagram, where the muscle is represented by *rs;* 'as it raises the rib at *s* its end rises from *e* to *b*, four times the distance'. Here, however, Leonardo also illustrates the 'enlargement' forwards as well as upwards 'from the vertical line *fd* to the vertical line *ac*.'

But serratus posterior superior can act only on three ribs. What about the others? In the same previously cited passage of Avicenna Leonardo found a misleading clue for another similar muscle, serratus anterior. Avicenna describes this vaguely as expanding the chest and arising 'from under the scapulae'. It was just what Leonardo was looking for. Once more he falls into the trap of illustrating 'lines of force' in the form of erroneous anatomy. Extending the series of serratus posterior superior downwards, he depicts serratus anterior as arising 'under the scapula' from the vertebrae, and not from the scapula itself. He now has a complete set of muscular forces for dilatation of the lungs in all three dimensions. 'The three upper muscles [serratus posterior superior] are constituted for elevation, and on their pulling they raise upwards 3 ribs to which they are attached, drawing with them the other lower ribs, whereby they are opened, dilated and acquire capacity . . . But the elevation of all the ribs by these upper muscles is not enough unless the ribs are also widened and dilated by the lower muscles. To this dilatation are dedicated the six lower muscles [serratus anterior]. These in their pulling move the flexible cartilages placed at the ends of the ribs. Thus we have found out what opens and raises the ribs in breathing, and overcomes the power of the constriction and contraction which the lateral muscles of the diaphragm exert when the said diaphragm flattens out its dome and increases downwards the space into which the lung expands on filling with air' (W 19067 r; K/P 160 r).

In this discussion the intercostals have not been mentioned. They come under the category of muscles for both expansion and contraction of the chest – according to both Avicenna and Leonardo. They receive repeated attention, always with prominent lines marking the line of action of the external intercostals downwards and forwards, thus serving to raise the ribs, and in a line at right angles, down-

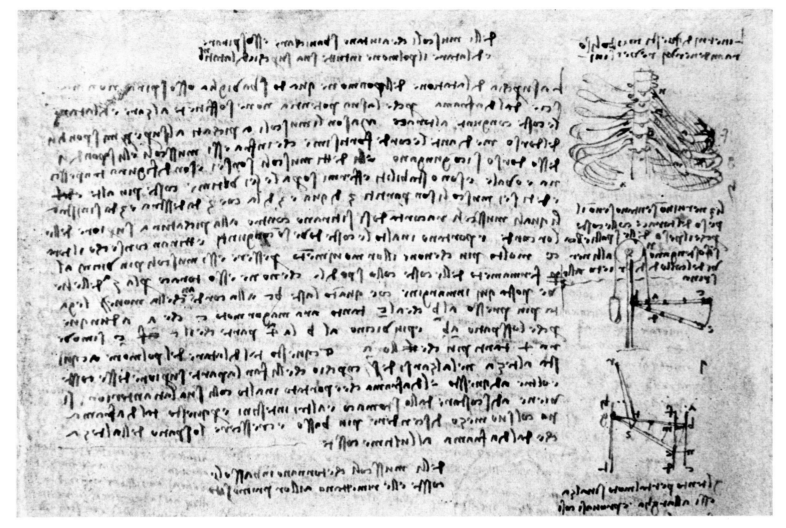

Figure 14.2. The leverage of the elevation of the ribs (W 19061 v; K/P 154 v).

The top drawing is a sketch of the spine and ribs from the back showing muscles descending obliquely from the spine to the ribs. The middle drawing represents this action on a rib in terms of a weight hanging on a pulley, the rope of which, *na,* is fixed to the rib at *a* or *d.* Leonardo explains that since the muscles are attached nearer to the vertebrae than is the rib cage, 'the nearer this cord is attached to *b* than *c* so much the greater will be the movement of *c* than *a'.* And since the distance *ab* is a quarter of the distance *bc,* '*c* will move 4 times more than *a'.* And such movement will elevate the whole chest in breathing.

At the same time the chest will increase in size from back to front, as shown in the lowest drawing, where dilatation of the chest is represented by the space between the two vertical lines *ac* and *fd* made by the movement of the rib *ge* through the arc *eb* up to *gb.*

wards and backwards to show the line of force of the internal intercostals serving to depress the ribs. In Figure 14.3 (W 19044 v; K/P 47 v) we see all the above-mentioned muscles acting on the chest. The right-hand drawing shows the fibres of the internal intercostal muscles between the ribs, their axis being indicated by the line *fn.* The middle drawing shows the fibres of the external intercostals running at right angles to the internal intercostals, a fact made emphatically clear by Leonardo's line of force, *mn,* forming a right angle with the line *fn.* Above, Leonardo draws attention also to the serratus posterior superior, saying, 'The 3 muscles [*opq*] which pull the ribs up we shall call pullers'. The third left-hand drawing is particularly interesting since here he adds the serratus anterior, at this time seen as a long belly *ab* lying on the chest wall under the scapula with five short anterior slips inserted on to the ribs far short of their cartilages. Above this drawing he writes, 'To the five muscles *cdefg* [slips of serratus anterior] created for the dilatation of the chest we give the name, dilators'.

The passage below the drawings in Figure 14.3 describes the action of the intercostal muscles 'dedicated to dilatation and contraction of the ribs. These two contrary movements are arranged for collecting and breathing out the air in the lung which is enclosed in the rib-cage. The dilatation of

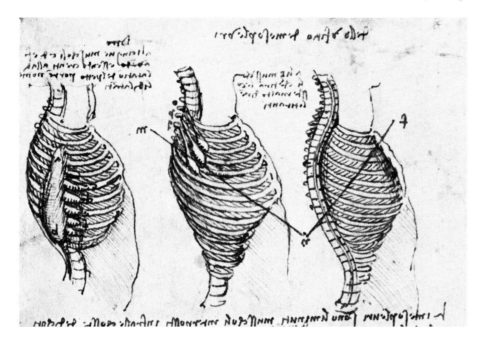

Figure 14.3. Muscles of the chest wall used in breathing (W 19044 v; K/P 47 v).

On the right the internal intercostal muscles are shown. Their line of force on contraction is indicated by the line *fn*. To the left the contrasting line of action of the external intercostal muscles is drawn, *mn*. At the back of the chest the three bellies of the muscle serratus posterior superior are shown. These are limited to the upper ribs. To supply the remaining ribs with similar muscles Leonardo draws digitations of serratus anterior, *ab,* labelled *cdefg*. It is to be noted that he makes this muscle arise from the back of the ribs erroneously.

Figure 14.4. The diaphragm; the motor of both the air and food (W 19065 v; K/P 158 v).

At the top Leonardo draws the diaphragm relaxed into a dome, *abc;* the oesophagus pierces it. Below he shows the diaphragm at two levels, *nmf* and *ngf;* the space *a* representing the degree of its descent. Corresponding to this an equal space is created by the movement of the abdominal wall from *fhs* to *fcs*. As the diaphragm rises again to its previous position, *nmf,* so the abdominal wall returns to *fhs*. They are in equal and reciprocal motion. They thus fulfill Leonardo's concept of geometrical transmutation; space acquired equals space left.

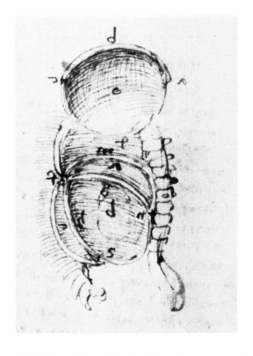

these ribs arises from the extrinsic muscles of the ribs which are placed at the obliquity *mn* with the help of the three muscles *opq* [serratus posterior superior], which by pulling the ribs upwards with great power enlarge their capacity in the way one sees done by the ventricles of the heart. But the ribs . . . would not be able to descend by themselves if a man lay down if it were not for the intrinsic muscles which have an obliquity contrary to that of the extrinsic muscles which extends along the line *fn*'.

With regard to the diaphragm Leonardo found Mondino much more useful. In fact his description of the functions of the diaphragm paraphrases Mondino's. 'The functions of the diaphragm are four. The first is that it is the origin of the dilatation of the lung by which the air is attracted. 2nd that it presses on the stomach covered by it and drives from it digested food into the intestines. 3rd that with the aid of the mirach [abdominal wall] it squeezes the intestines and expels the superfluities. 4th that it separates the spiritual from the natural organs', i.e., the thorax from the abdomen (W 19064 v; K/P 157 v). All these four functions of the diaphragm Leonardo discusses at great length. Those concerned with the abdomen we shall return to later. But Leonardo emphasises that all four of them 'are produced by one and the same cause, its [the diaphragm] relaxation and contraction' by which it ascends and descends within the chest.

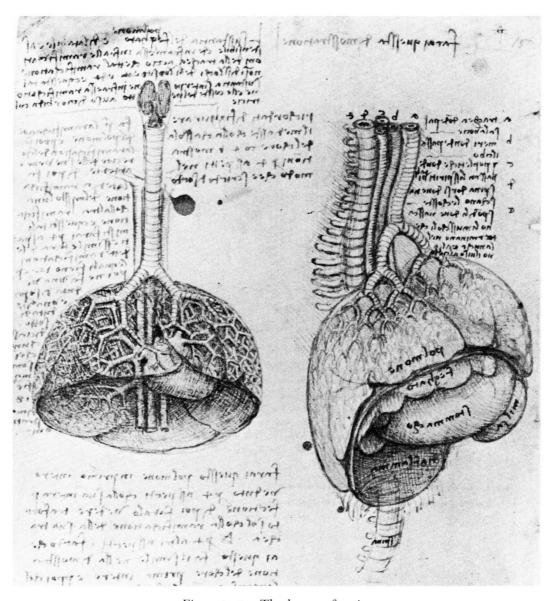

Figure 14.5. The lungs of a pig.

This early drawing was designed to obtain a general picture of the main viscera. In the note (top right) Leonardo lists the viscera illustrated, e.g., trachea, oesophagus, vertebrae, etc., and in the drawing he labels the viscera, lung, liver, spleen, stomach, spine. He describes the lung 'substance' as being 'interposed between this ramification [of the trachea] and the ribs of the chest like a soft feather bed' (W 19054 v; K/P 53 v).

After a number of pages of indecisive discussion of this difficult problem Leonardo reaches a definite and clear-cut concept of the action of the diaphragm (W 19065 v; K/P 158 v); see Figure 14.4. The whole movement is beautifully expressed by the two drawings on the right margin (here enlarged) The upper drawing illustrates the hollow dome of the raised diaphragm *abc* as seen from below. The lower drawing shows the chest and abdomen in sagittal section, the spine on the right and the abdominal wall on the left. The diaphragm divides the 'spiritual' organs of the chest from the 'natural' organs of the abdomen. Both the diaphragm and abdominal wall are represented by two separate 'threads' indicating their movements. Leonardo's explanation in the text alongside is very detailed and clear. It is headed, 'How the diaphragm of itself has only one movement'. 'The diaphragm on its own has only one movement. This is that which makes it withdraw from the lung when the lung follows behind it. The second movement [of the diaphragm] arises elsewhere; it is that which makes it go back to the lung when it flees back. And that which puts it to flight is produced by the abdominal wall. Thus the natural movement which is generated in the diaphragm *nmf* is demonstrated by the movement of *nmf* and *ngf*. This would leave behind it a vacuum in the space *a* unless the lung filled it up by its enlargement as air fills it. At the same time the abdominal wall *fhs* withdraws to *fcs* and the intestines *b* which are expelled from the space *a* escape into the space *c*; that is, when the intestines at *a* descend to *b*, those at *b* descend into the space *c*. Now the slight curvature of the diaphragm *ngf* cannot by itself increase into the former greater curvature *nmf* because the function of muscles is to pull and not to push. Hence if it is to occupy the position *nmf* it is necessary that it should be helped to make its curve by another curvature which is drawn back and straightened into a lesser curve. This the abdominal wall will do. This having been pushed forwards from *fhs* to *fcs* by the straightening of the diaphragm, will presently return back, straightening itself, to *fhs*. And it will drive the diaphragm at *ngf* to the position *nmf*. And so these two contrary motions act like a flux and reflux made by the diaphragm against the abdominal wall and then by the abdominal wall against the diaphragm'.

Thus Leonardo comes to describe the diaphragm as 'the motor' of both air and food: 'Of the two motors of the food and air within the human body, the first was the diaphragm . . . subsequently the straightening of the abdominal wall restored to the diaphragm its previous roundness'. Once more he describes the alternate movements of descent of the diaphragm with rounding of the abdominal wall alternating with the straightening of the abdominal wall and the rounding of the diaphragm, adding, 'and so it will act successively throughout life with this succession of flux and reflux' (W 19065 v; K/P 158 v).

Before we turn to the pulmonary aspect of breathing, it is worth taking note of Leonardo's view of his achievement in his study of the movements of respiration. He has discovered the forms of the bones of the spine and ribs, with

their joints, and reduced them to a multiple lever system with fulcra. On these he has laid an approximately correct pattern of muscles, reduced to 'lines of force' acting on the lever system. He has, with great difficulty, analysed the complex movements of the diaphragm from a relaxed curve to a tense straight line by its muscular action with descent within the thoracic cage and related this to reciprocal movements of the abdominal wall.

All these he has integrated into a movement pattern of the chest and abdomen, that of ebb and flow (*flusso e reflusso*), the 'tidal' movements of inspiration and expiration of air. In the case of the abdomen he has demonstrated a physiological example of geometrical transmutation which fulfils his favourite mathematical dictum that 'If what is taken away is then replaced, nothing is lacking'. This principle was described in Chapter 5 and is evident from inspection of drawings in Figures 5.21 and 5.27 (pp. 148, 152). In short, he has step by step brought his anatomy and physiology of breathing into the form of geometrical transmutation, that formal picture of causation which he considered to be 'true knowledge' of any natural process. It remains to follow Leonardo's studies of the tidal air involved in inspiration and expiration, his prototype of *moto ventilante*.

We have already noted in Chapter 3 the pervasive influence of the macrocosm–microcosm analogy on Leonardo's studies of breathing. They were not easily abandoned. In *Codex Arundel* (1500–1505) Leonardo attempts to make a comparison between the respiratory rate of a man and that of the earth. His arithmetic, here very shaky, is concerned with cube roots. 'The cube root of a man is one *braccio*', he writes, 'and the root of the earth is 21 million miles; therefore the cube of a man is 1 *braccio* and the cube of the earth is 441 million million million. A man making one inspiration and expiration takes four harmonic tempi, of which there are 1,080 to the hour. And if the earth had to breathe in and out in a time proportionally greater than man, the time would be [blank space]. This is to be tested by comparing proportionally the quantity of the earth with that of man'. The interest in this passage lies not in its fanciful arithmetic but in Leonardo's attempt to push his analogy between macrocosm and microcosm into the quantitative field. Incidentally he calculates the respiratory rate of man here at about 4 to the minute, a figure which reflects the absence in his day of any good timepiece.

This effort does not stand alone. Some years later, about 1508, he asks if 'The attraction of the water which the earth would make in 12 hours with ebb and flow could show us the size of the lung of the earth'. 'If we say that the lung is half a *braccio* squared, that is 1/8 of a square *braccio*, and a man breathes 270 times per hour, how great will the lung of the earth be which breathes once in 12 hours?' He then attempts with his faulty arithmetic to answer this unanswerable question, using his pyramidal law in the form of the 'rule of three'. 'I will say this; 12 times 270 makes 2,940. It follows that the lung of the earth is 2,940 times greater than the lung of man', and he proceeds into fantastic calculations which it is pointless to follow. It is to be noticed

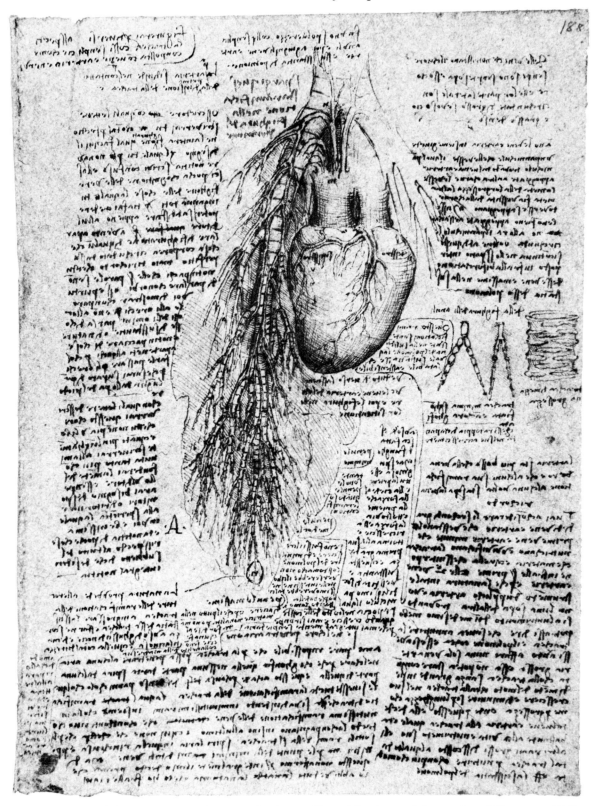

Figure 14.6. The left lung and heart from the back; small ramifications of the trachea (W 19071 r; K/P 162 r).

In this lung of an ox the bronchial vessels are shown. The lobes of the lung are demarcated, its substance described as enclosed by the pleura. At the bottom an encrusted cavity is drawn, presumably tuberculous or hydatid. To the right the trachea and two small forked bronchi are drawn. With inspiration that on the left is shown widened as to both its calibre and its fork.

that Leonardo uses no units of volume in these calculations; this was a concept which apparently he never mastered.

This passage, however, defines the breakdown of his long-held literal belief in the analogy between earth and man on the grounds that 'first, the earth's surface would have to move too much; secondly, that so much water would have to be taken in into the earth that the seas would lose their water'; thirdly, such enormous movements of water 'would make a very great wind which would issue from the earth in the six hours of flow, and the same very great wind would return in the next 6 hours . . . such *flusso* and *reflusso* continuously in the air would go and return by the same path'. And then Leonardo leaps from the use of numerical speculative analogy to his own experience, describing 'A tiny hole [*spiraculo*] in a cask in which a man is enclosed'. 'It is well known', he says, 'that in holding a light in front of this hole it makes no movement because the man in breathing out diminishes his chest and the air fills up the vacuum left by his chest; and in drawing in the air again his chest enlarges and draws in the air which was driven out. But the earth does not move like this chest' (CA 260 ra).

Leonardo devoted apparently little attention to the anatomy of the lung; no indubitable dissection of a human lung is extant. He introduces himself to pulmonary anatomy by dissecting a pig's lung (Figure 14.5). The great majority of his pulmonary studies are late; they were made when he was prohibited from human anatomy in Rome and consequently used animal, mostly bovine, material for his work on the heart and lungs (Figure 14.6).

He notes the incompleteness of the tracheal rings of cartilage, saying that they 'do not join [posteriorly] for two reasons: the one because of the voice and the other to give room for the food between themselves and the bone of the neck' (W 19055 v; K/P 59 v). It will be remembered that he considered elevation of the pitch of the voice to be due to narrowing of the trachea with a consequent increased velocity of air flow through a fixed laryngeal slit (see Chapter 9). He illustrates also in Figure 14.6 (W 19071 r; K/P 162 r) his conviction that the 'trachea minima', that is, the small bronchi, dilate in inspiration and narrow in expiration. 'The ramification of the trachea', he writes, 'divides into the most minute branches together with the most minute ramifications of the vessels. This [lung] in itself is dilatable and extensible like tinder made from a fungus. And if you press it it will yield to the force of compression, and if the force is taken away it increases again and returns to its original size. This substance is clothed by a very thin membrane [pleura] which is applied to the spaces between the ribs opposed to it. When it [the lung] enlarges it is never ruptured because it is never completely filled with air. Nor is it ever emptied of residual air'.

On the same page he depicts the bronchial arteries (see Chapter 15). On this page, too, he divides his drawing of the lung into three parts, correctly labelled, 'Smallest lobes [*penule*]' 'middle sized lobes' and 'largest lobes', divisions which are clearly demarcated in the lung of an ox.

Here, too, in Figure 14.6, Leonardo asks, 'Whether air penetrates into the heart or not'. He answers, 'To me it seems impossible that any air can penetrate into the heart through the trachea because if one inflates it, no part of the air escapes from any part of it. And this occurs because of the dense membrane with which the entire ramification of the trachea is clothed. This ramification of the trachea as it goes on divides into the most minute branches together with the most minute ramification of the vessels which accompany them in continuous contact right to the end. It is not here that enclosed air is breathed out through the fine branches of the trachea and penetrates through the pores of the smallest branches of such vessels. But concerning this I shall not wholly affirm my first statement until I have inspected the anatomy I have in hand'. Thus in Leonardo's view the cooling of the blood in the lung takes place by contact. No air penetrates directly into the pulmonary veins and heart.

Leonardo's late discussions of respiratory movements revert to the action of the diaphragm and the muscles of the abdominal wall. In a note headed 'On the maximum dilatation of the lung', he writes, 'The maximum dilatation of the lung arises from the greatest shortening of the diaphragm and the greatest elongation of the transverse muscles of the abdominal wall. But during this time the muscles interposed between the ribs of the chest [the intercostals] are dilated and the muscles which clothe the ribs below the nipples [serratus anterior] towards the flanks contract and draw behind them the ribs in such a way as to dilate them' (W 19086 v; K/P 178 v).

In summary Leonardo's views on the action and function of the lungs remain somewhat ill-defined. He makes a pioneer attempt to analyse the mechanics of the muscular forces which bring about dilatation of the thoracic cavity. His difficulties in identifying the muscles concerned were great. That these limitations should be viewed sympathetically is denoted by the fact that they still exist today. In any case, one sees him progress from the idea that the lungs play an active role to that of their purely passive expansion 'like a bellows'. His description of the diaphragm as 'motor both of the food and air' reflects the vital importance he attached to that muscle.

Leonardo adheres to the Galenic view that the function of the lungs was to cool the blood. He departs from this view on the question as to how this cooling was brought about. He can find no communication between the trachea with its fine ramifications and the blood vessels of the lung. Therefore cooling of the blood must be brought about by contact only. Misled by his proclivity for seeing reflected movement (*flusso* and *reflusso*), ebb and flow, as a basic pattern of physiological movement, he considers that ebb and flow of blood takes place through the pulmonary veins. He does not consider the possibility of it taking place through a pulmonary circulation.

Cooling of the blood prevents the blood from being made too hot within the heart itself. Should overheating take place in the heart, it would be 'suffocated'. Alternatively, if the vital spirit were not generated, life would be destroyed. The nature of this vital spirit, in Leonardo's view, is indicated by his assertion that 'Where air is not received in the right proportion by flame no flame can live, nor can any terrestrial animal'. 'Where flame does not live no animal that breathes can live; excessive air kills a flame, a temperate amount of air nourishes it' (CA 270 ra).

REFERENCES

1. Avicenna. *Canon of Medicine,* trans. by M. H. Shah, in *The General Principles of Avicenna's Canon of Medicine.* Karachi, Naveed Clinic, 1966, p. 89.
2. *Gray's Anatomy,* ed. by Roger Warwick and Peter Williams. 35th edition, London, Longman, 1973, p. 516.

Chapter 15

The Heart, the Most Powerful Muscle

The heart is a principal muscle of force, and it is much more powerful than the other muscles (G 1v).

Two main differences distinguished Leonardo's researches on the heart from those carried out on the bones and other muscles of the body. The first was that the movements of the heart had no external manifestation; they were therefore of no interest to the artist. The second lay in the fact that the heart was an organ primarily concerned with the movement of blood; here it presented a particularly fascinating hydrodynamic challenge involving Leonardo's researches on the movements of water. It will be rightly suspected that Leonardo had a long, troubled path to follow before he reached the significant conclusion expressed in the subtitle of this chapter. We will briefly review these stages in this chapter.

Leonardo's investigations of the heart followed his usual procedure. First he acquainted himself with the views of his available 'authorities'. As we have seen, his main anatomical authority was Mondino, through whom Leonardo became acquainted with the views of Galen. There is little evidence that Leonardo had direct knowledge of Galen's works except towards the end of his anatomical career, about 1510, when he worked with the brilliant young Galenist, Marc Antonio della Torre, for about a year before the young man died of plague. Thus the influence of Galen and Avicenna came mostly through Mondino as far as Leonardo was concerned. In all these three authorities the teleological outlook is strongly expressed, i.e., that the Creator made no useless or ill-constructed part in the body.

Although the substance of the heart looks like a muscle, Galen had insisted that it is not really so; the heart is in essence too noble and 'entirely different' from muscle. He considers the heart under the heading, 'Respiratory organs', so emphasising his idea that its most important function is to draw air into itself from the lungs in much the same way as the lungs draw in air from outside, i.e., by dilatation or diastole. Thus the pulmonary veins contain air and a little blood which is drawn into the left ventricle where it meets with subtilised blood pressed through minute pores in the septum of the heart from the right ventricle. Here the 'vital spirits' with heat are generated. When the left ventricle contracts in systole the blood, heat and vital spirits are distributed all over the body, and some regurgitates through the mitral valve and is expelled from

the lung as smoky vapour. Meanwhile the cool air drawn in by the lungs passes through the pulmonary vein and keeps the heart from overheating (see Figure 15.1).

Blood itself is made from digested and absorbed food, in the liver by the natural spirit, whence it distributes nourishment throughout the body by the venous system. Some of this blood is drawn in by diastole of the right ventricle of the heart through the tricuspid valve. When the right ventricle contracts, its contained blood flows in three directions: most of it goes through the pulmonary artery to nourish the lungs; a small amount passes through the minute pores in the interventricular septum into the left ventricle, where the blood and *pneuma* create vital spirits; the third portion regurgitates back through the tricuspid valve to the vena cava.

Thus Galen and his many successors, like Avicenna and Mondino, saw the heart as consisting of two ventricles with four valves: the mitral admitting and regurgitating blood and air into the left ventricle; the tricuspid admitting blood into the right ventricle; and the efficient pulmonary and aortic valves preventing regurgitation of the blood expelled by ventricular systole into the pulmonary arteries and aorta, respectively. Avicenna made an important erroneous variation on this theme by exaggerating Galen's 'minute pores' into a third 'middle or interventricular ventricle' in the septum (Figure 15.2). He also supported Galen's idea that the veins of the body arise from the liver and not from the heart. This was rejected by Mondino, who took Aristotle's view that the veins arise from the heart.

What of the atria in this scheme? These Galen saw as dilatations at the ends of the great vena cava and the pulmonary vein, which were filled with superfluous blood. During the powerful movement of diastole, when blood was sucked into the ventricles, there was a danger that not only the valve rings but also the veins themselves might be sucked into the ventricles. To prevent this the auricles would throw their content of blood into these vessels at the critical moment.

Around the heart is the pericardium, which contains a little fluid. Galen described the coronary arteries and vein, as did Avicenna, but Mondino contented himself by mentioning only the coronary vein.

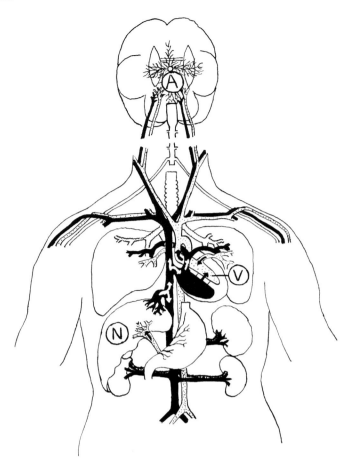

Figure 15.1.   Galen's three types of Spirit, or Pneuma.

N = Natural spirit, created by the liver governing nutrition and distributed through the veins.

V = Vital spirit, or pneuma, created in the heart from air absorbed by the lungs and distributed through the arteries.

A = Animal or psychic spirit, created in the brain from arterial blood, and distributed through the 'hollow' nerves.

Figure from F. N. L. Poynter and Kenneth D. Keele, *A Short History of Medicine* (London, Mills and Boon, 1961), figure 1.

Mondino being Leonardo's closest guide, it is natural that where there are points of difference between these authorities he supports Mondino's view. For example, after human dissections he concludes that the main vein, the vena cava, arises from the heart (W 19028 r; K/P 70 r); he accepts Mondino's description of the 'middle ventricle' as consisting of 'many small cavities', calling it a 'sieve' (W 19116 v; K/P 115 v). But more significantly, Leonardo performs his latest and most searching dissections of the heart according to Mondino's technical advice (W 19080 r; K/P 170 r; see Figure 15.3).

'Cut the heart' writes Mondino, 'first on the right side and begin from its point in such a way that you will not touch the opposite wall but cut along the side of the middle ventricle, and you will at once see the right ventricle and inside it two orifices'. Leonardo's little sketch in Figure 15.3

depicts the incisions into the heart advocated by Mondino in the left-hand drawing, under the heading 'first'. In the drawing to the right of this he shows the opened ventricles and inside the right two orifices. 'When thou has seen this', writes Mondino, 'cut the left ventricle in such a way that the wall of the middle ventricle remains in the middle, and you will immediately see the cavity of the left ventricle, the wall of which is denser and thicker than that of the right'. This incision is depicted by Leonardo in the drawing to the right, labelled 'second'. In the two drawings below on the left margin of Figure 15.3 Leonardo has opened out the two ventricles from the midline and demonstrated the interventricular septum. This, according to Mondino, is called the 'middle ventricle'. Leonardo calls it here 'the sieve', showing it from both right and left sides. To the right of this group of sketches he makes two transverse sections of the heart, one 'through the base' and one 'below the base'. Thus in this little group of drawings he reveals his plan for detailed investigation of the anatomy of the heart.

This group of drawings explaining Leonardo's anatomical method raises a paradoxical question. They are made on a sheet of notes dated about 1513. Why was Leonardo planning methods of dissecting the heart when he was over sixty years old? The answer to this question is both simple and revealing. This was Leonardo's third time round. He had already made two attacks on the problem of the heart and

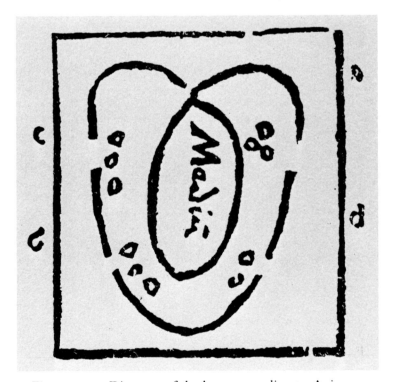

Figure 15.2.   Diagram of the heart according to Avicenna.

The 'middle ventricle' along which the word *Media* is written constitutes a large cavity between the left and right ventricles. The valve orifices are lettered, the bicuspid mitral valve being distinguished from the tricuspid. Figure from Ioannes Adelphus's edition of Mondino's *Anathomia* (Strassburg, 1513).

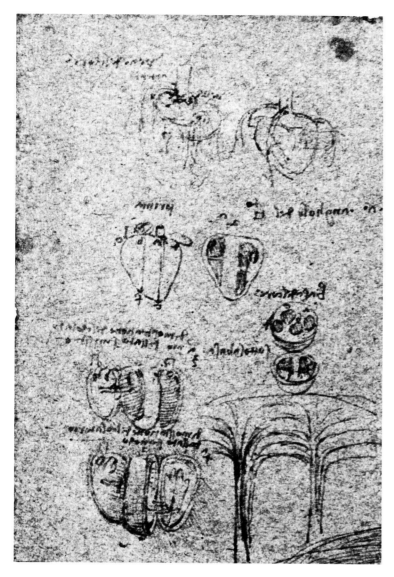

Figure 15.3. Leonardo's method of dissecting the heart (W 19080 r; K/P 170 r).

Beneath two rough sketches of the coronary vessels six diagrams illustrate dissection of the heart. The first, labelled 'prima,' on the left, defines a rectangle, *abcf,* on the front of the heart. Next to *a* is a little *x.* Leonardo writes, '*a* is the angle of the rectangle', thus indicating his line of incision to the left of the ventricular septum. And *bf* is the corresponding line of incision to the right of that septum. In the second drawing the heart is shown opened along these lines. In the drawings below on the left the lateral walls are opened out, first to show the right ventricle with its moderator band; in the lowermost, to show the left ventricle and its mitral valve, the structure of which is drawn in enlarged detail to the right. Above this are two cross sections of the heart, one through its orifices at the base, the other lower down. All these features are elaborated in other drawings of the heart; they illustrate Leonardo's plan of attacking the problem.

circulation; he had already become dissatisfied with the re-sults of both. Now he was preparing for a third. This, indeed, is a typical example of the pattern of his lifetime of research on so many problems. He tackled them all not just once but several times, at different periods of his life. This he did when his observations and experiments in one sphere, for example, mechanics, hydrodynamics or geol-ogy influenced his conclusions in another sphere, such as physiology. His work on the heart and movement of the blood forms a very clear example of this spiral shape of Leonardo's progress in science – and incidentally adds to the apparent confusion of his notes.

Let us look very briefly at his two previous views on the heart and circulation. The key to all three stages of these investigations lies in his firm conviction that the heart was a source and distributor of body heat and nourishment and that heat itself was the 'vital spirit' of life. Thus the question he was continually asking was, How is the body heat cre-ated and distributed by the blood?

In the early period of his anatomical researches Leonardo

concentrated, as we have seen, on the nervous system. At that time he did very little anatomical investigation of the heart. He was rightly concerned with gaining some idea of the general lay-out of the various organs in the body. In Figure 16.1 (p. 328), for example, there is a small figure of a man showing the windpipe leading to the lungs which overlie the heart, and the oesophagus is seen piercing the diaphragm to pass to the stomach and coiled intestines be-low. All these organs are clearly labelled. The lungs and heart are located in the chest, the 'spiritual site', and the diaphragm separates them from the "natural" organs, e.g., the liver and intestines below. On another page from about the same time, alongside a fine drawing of the veins of the face, Leonardo makes the surprising assertion, 'I find that veins perform no other function than to heat; just as nerves and such things have to give feeling [or stimulation]' (W 19018 r; K/P 41 r).

How do the veins or blood vessels of the body carry round this heat? Leonardo's first answer is by a 'power' derived from outside, from the giver of life to the macrocosm of the world as well as to the microcosm of man – the sun. The sun is the source of all heat, giving life to the world, to plants and man. It elevates the 'humours' including the blood of man to heights where they cool, condense and fall down again in the form of rain to the central heart, earth, or sea only to be again raised up in perpetual circulation. In particular this kind of circulation occurs in plants, as exhibited by the fall of sap from the cut branches of vines to be reabsorbed from the soil by their roots (see Chapters 3 and 4 and Figure 3.6, p. 86, where the development of these ideas is described). The most elaborate and perhaps latest visual statement of this idea is in Figure 15.4. Here the heart is made to take the place of a nut or seed from which the plant grows under the influence of the sun.

To us this idea may seem very crude, but it has to be looked upon as an attempt to account for the circulation of the blood in man by purely physical principles without calling in the various kinds of hypothetical 'spirits' which until then had invariably prevailed. And it cannot be forgotten that in the eighteenth century the opponents of the Harveian circulation still based their case on the circulation of sap in plants.[1] However, the theory became increasingly unsatisfactory to Leonardo himself through the years, though it seems to have prevailed with him until about 1508.

In his second round of cardiovascular research Leonardo saw the heart as a stove housing a central fire, the heat of which was conveyed throughout the body by the vascular system. This view coincided with a change in his outlook on the source of the heat of the world, the macrocosm, in relation to its flowing waters, its seas and rivers. This he applied to human physiology in the *Codex Leicester*. First he uses the dialectical form of debate to refute his old views, treating these as if they were the views of an 'adversary'. He writes, 'If you should say that the heat of the sun draws the water up high from the mountain caverns to their summits just as it draws up the open lakes and seas in the form of vapour to form the clouds, the answer is that if the heat should be the cause of drawing the origins of rivers to the peaks of mountains, then where there is greater heat there the veins of water should be greater and more abundant than in cold regions, but we see the contrary' (Leic 3 v). This passage is accompanied by a drawing of the sea sending its 'veins' down into the centre of the earth under the mountains (Figure 15.5). From this centre other 'veins' rise up to the peaks of the mountains above. Below he writes, 'The level of the sea exhaled from the fires which exist at the centre of the body of the earth'. And in the closely written text nearby Leonardo likens the movement of water here to that being distilled in an alembic or retort, where 'the water drops are composed of the moist water-vapour which percusses the top of the alembic'. In *Codex Leicester* 28 r he brings this new principle into line with the movement of blood in animals, saying, 'The ramifications of the veins of water are all joined together in this earth, as those of the blood in other animals. And they are in continual circulation, and thus vivified, they perpetually wear away the places in which they move . . . The heat of the fire generated within the body of the earth heats the waters pent up within it in the great caverns and other places; and this heat

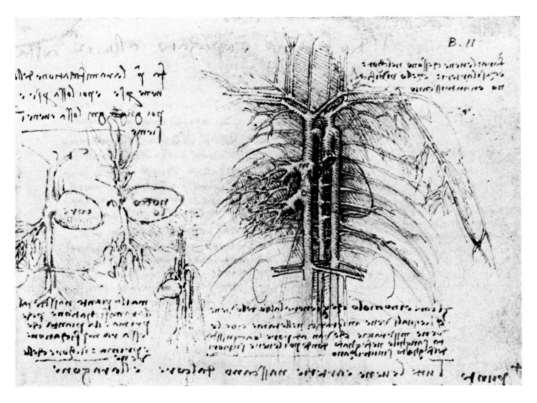

Figure 15.4. The heart likened to a nut producing roots and branches; three drawings of the heart as a nut (W 19028 r; K/P 70 r).

Written within the central sketch is the word *nut;* from it sprout root and branch. Written within the left-hand drawing is the word *heart;* from it sprout the inferior vena cava and hepatic veins (the roots), with the superior vena cava and aorta representing the ascending trunks of the vascular tree. This scheme is transferred to the human body in the drawing on the right.

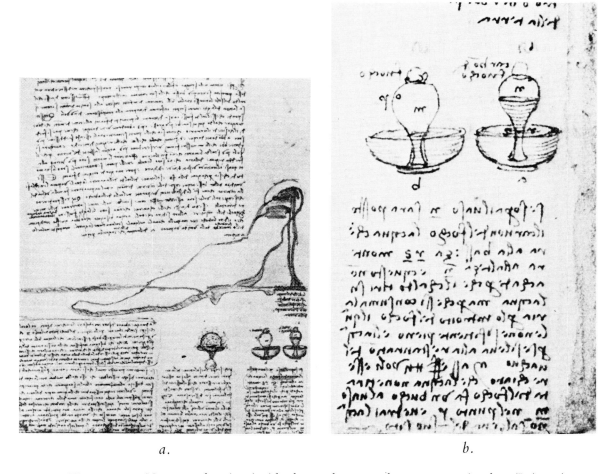

Figure 15.5. How combustion inside the earth causes the waters to circulate (Leic 3 v).

a. Here Leonardo likens the earth to a retort. To the right of the page he shows water at sea-level being vaporised and rising by 'veins' to the top of a mountain, whence it flows by surface vessels (rivers) back to the sea; this part of the drawing in the original manuscript is now much rubbed and faded. Below, he depicts this process occurring in three flasks. Two of these are shown enlarged in *b*.

b. Here he demonstrates that the water rises, not because it is drawn up by the heat of a 'burning coal' inserted at the top, but by burning; 'the air is consumed'. This he proves in the experiment on the left by making a small hole in the inverted vase *m* at *p*. In this case, 'you will see that the water will not be raised'.

causes the water to boil, pass into vapour and be raised to the roofs of the said caverns . . . where coming upon the cold it is immediately changed into water as one sees happen in a retort, and falls down again forming the beginnings of rivers'.

Leonardo applies this analogy to the macrocosm of the world and the microcosm of man: 'The earth, we may say, has a spirit of growth; its flesh is the soil, its bones are the successive strata of rocks which form the mountains, its blood the veins of waters. The lake of blood that lies around the heart is its ocean. Its breathing is the increase and decrease of blood in its pulses, the ebb and flow of the sea. And the heat of the soul of the world is fire which is infused throughout the earth; and the dwelling place of its nutritive spirit is in the fires in divers parts of the world which are breathed out in baths, sulphur mines and volcanoes like Mount Etna in Sicily and many other places' (Leic 34 r). Leonardo brings the analogy of veins of the earth and man ever closer when he writes of 'superifiical and deep veins', the deep veins being nearer to the burning sulphur get heated first (Leic 11 r), and when he refers to 'live veins ramifying in the tissues of the body of the earth', all being joined to the sea (Leic 11 v).

The first reference to a central fire in cardiac physiology occurs in the context of considering a 'kind of bellows for a stove' (Figure 15.6). Here the mechanism of inlet and outlet valves is analysed: 'One which opens outwards serves to send out the heat, which after this movement cannot ever

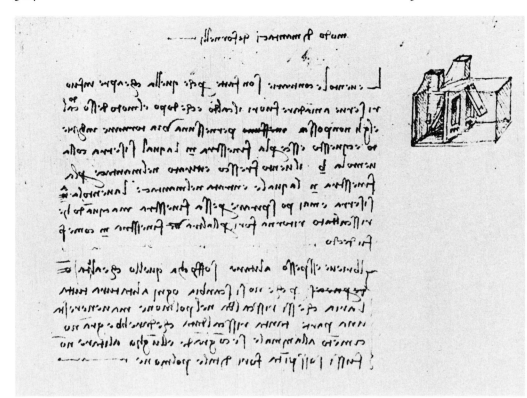

Figure 15.6. The heart as a stove (BM 24 r).

Leonardo here draws a furnace or stove with an inlet valve at *n* and an outlet valve at *m*. These open into separate chimneys. The outlet chimney, labelled *b*, shows also the position of the outlet valve moved from the opening at *m*. Writing about this stove, Leonardo makes it clear that he looks upon it as the heart, and the chimney as the trachea. In Figure 15.7 he draws inlet and outlet passages from the lungs into the heart. These too he labels *m* and *n*.

turn back. This [hot air] goes out through the opening *m*, which is closed by the valve *b*. Fresh air is drawn in through the valve into the bellows through the entrance *n*. When it has entered the bellows the valve is closed and it can never be blown out through this window, *n*; but when it has been heated it can return outside through the other opening, *m*'. Leonardo now reveals that he is considering this 'stove-bellows' as a cardiorespiratory apparatus. He writes, 'Short and frequent breathing suffocates that person who breathes, because each breath does not change all the air in the lung that has been heated. But a part of it remains there so greatly heated that it would do great damage to an animal unless big and long breaths were to drive it out of the lung' (BM 24 r). This 'bellows' type of stove with its inlet and outlet valves thus represents the heart. The lungs get rid of the smoky vapours, acting also as bellows. The trachea is the chimney of the heart acting as a stove. Moreover, Leonardo, contradicting his first view, finds confirmation of this second theory: 'The heart in itself is not the beginning of life but it is a vessel made of dense muscle . . . and it is of such density that fire can hardly damage it. This is seen in the case of men who have been burnt in whom though their bones have been burnt to ashes the heart is still bloody inside. And nature has made it so resistant to heat so that it should be able to resist the great heat which is generated in the left side of the heart through arterial blood becoming subtilised in this ventricle' (W 19050 v; K/P 59 v). Leonardo may well have heard the Florentine tale that after his burning, Savonarola's heart had been found intact amongst the burnt ashes of the rest of his body.

In Figure 15.7 (W 19104 v; K/P 107 r) Leonardo makes a drawing of the heart and lungs fitting this concept of the heart as a stove with inlet and outlet orifices, labelled *m* and *n*, as in Figure 15.6. These may be seen in the drawing on the right; they mark the entrance into a two-chambered heart of two passages from the left lung. Similar passages run from the right lung towards the entrance of the right ventricle. These vessels from the lungs are again seen entering the heart in the drawing on the left side of the page. Leonardo explains these passages in the lung thus: 'The moving air [*vento*] reflected from the lung into the heart cannot enter the heart unless there is an outlet [from the heart]. Therefore two passages are necessary. Of these, one which when the lung sends air into the concavity of the heart at the same time sends air out of the trachea; and a second passage through which air issues from the heart and returns together with the other air which escapes through the said trachea'. These two passages are marked *m* and *n*.

Thus both ventricles of the heart are likened to the 'stove-bellows' fed by air from the lungs in Figure 15.6. The mixture of air from the lungs and blood percolating the septum from right to left ventricle is now likened to a candle, the blood forming the tallow and wick which is burnt in the air. This analogy is developed at great length in Figure 15.8b and on W 19045 r (K/P 50 r).

On CA 270 ra (Figure 15.8b) Leonardo is concerned with the nature of flame. He makes a detailed analysis of a candle flame, studying the relation between the material burnt, the wick, the blue, yellow and red parts of the flame, the creation of smoke, etc., and above all the shapes, movements and powers of flame. Of great significance is his discussion of its relation to air, part of which he asserts is 'consumed' by the flame. 'The light [of the flame] generates a vacuum and the air hastens to fill that vacuum', he writes. It thus percusses the flame giving it its pyramidal shape (Figure 15.8b; compare Figure 3.9, p. 90). 'Excessive wind kills flame, temperate wind nourishes it', and he adds, 'Where flame does not live no animal that draws breath can live' (CA 270 ra). Elsewhere on the same page he asserts that 'The flame which is too condensed immediately dies.

Where a flame does not receive proportionate air no flame can live nor can any terrestrial or aerial animal'. Thus Leonardo recognises that combustion 'consumes' part of the air and is also concerned with the life of animals. This analogy between the candle flame and life is brought into the heart (W 19045 r; K/P 50 r). Here he applies the same principle in a most impressive analogy entitled, 'How the body of an animal continually dies and is reborn'. 'The body of anything that is nourished continually dies and is continually reborn, for nourishment cannot enter except

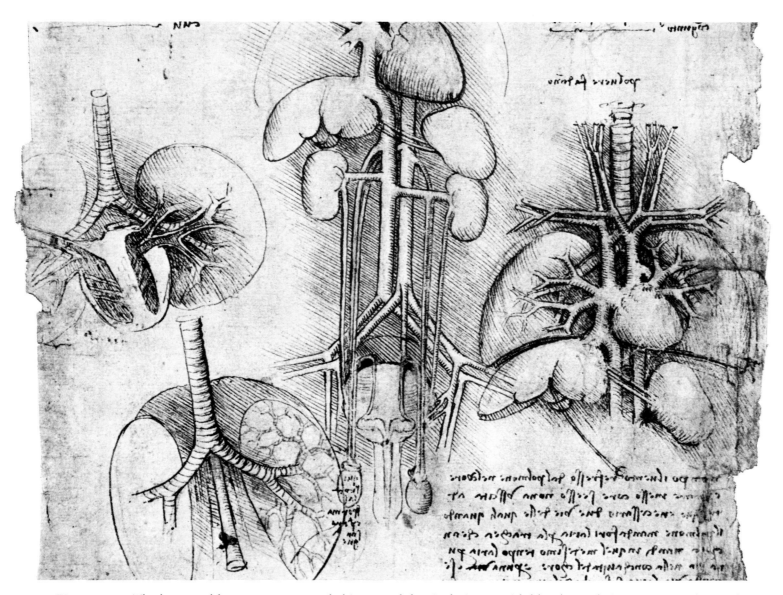

Figure 15.7. The heart and lungs as a stove and chimney; abdominal viscera with blood vessels (W 19104 v; K/P 107 r).

On the right the lungs are drawn with bronchi entering the heart. The left lung sends two passages, marked *m* and *n,* into the heart, one of which, Leonardo explains, acts as an inlet, the other as an outlet for the products of combustion in the heart. Compare Figure 15.6. This construction is drawn more clearly in the upper left-hand drawing, where it is shown to apply to both left and right ventricles.

In the central drawing the liver, spleen and urogenital organs are shown with a schematic rendering of their blood-vessels. Most noteworthy is the drawing of the testicles, the ductus deferens and seminal vesicles with their ejaculatory ducts opening into the posterior urethra.

a.

Figure 15.8.

a. Fire in the macrocosm of the Earth; the flame of an exploding volcano.

This is probably a copy by Melzi of a Leonardo drawing (W 12404 r).

b. The consumption of air by a candle flame (CA 270 va).

The different parts of a candle flame are analysed. The shape of the eddies of air which are attracted to it at the bottom are partly consumed and rise upwards as smoke. For a period this constituted Leonardo's conception of what occurred in the heart of the microcosm of man.

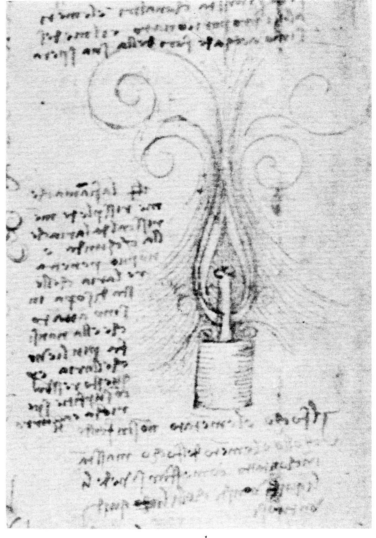

b.

into those places where past nourishment has been exhausted; and if it has been exhausted it no longer has life. Unless therefore, you supply nourishment equal to the nourishment which has departed, life fails in its vigour; and if you take away this nourishment, life is totally destroyed. But if you supply just as much nourishment as is destroyed daily, then as much life is reborn as is consumed, just as the flame of candlelight is made from the nourishment from the humour given to it by the candle. And this light is itself also continually renewed by swiftest succour from beneath by as much as is consumed above in dying, and in dying the bright light is converted into murky smoke. And this death continues as long as the smoke continues, and its continuity is equal to the continuity of nourishment. And at each instant the whole light dies and it is completely regenerated together with the movement of its nourishment; and its life furthermore, receives from it its flux and reflux, as is shown by the to and fro movement of its tip. The same thing happens in the bodies of animals by means of the beating of the heart which generates a wave of blood through all the vessels which continually dilate and contract. Dilatation

occurs on the reception of superabundant blood and diminution on the departure of the superabundant blood received. This the beating of the pulse teaches us when we touch the aforesaid vessels in any part of the body with the fingers. But to return to our intention. I say that the flesh of animals is being continually re-made from the blood which is continually generated for their nourishment. And that this flesh is destroyed and returns through the mesaraic [mesenteric] arteries and passes into the intestines where it putrifies in foul and fetid death as their expulsions and steams show us, like the smoke and fire given as a comparison' (see Chapter 16).

The same mechanism illustrated in Figure 15.7 is again described on W 19112 r (K/P 105 r). Once more in this early drawing of the heart Leonardo describes in detail how the heart's movement is one of 'opening and closing. And it cannot open without drawing into itself air from the lung which it immediately breathes out again into the lung'. 'Many are the times that the heart attracts to itself the air which it finds in the lung and returns this to it, heated, without this lung collecting any other air from outside'

(W 19112 r; K/P 105 r). In both drawings vessels from the right lung run straight into the heart. The walls of both ventricles are shown bound together by cross-bands to prevent their over-dilatation. Later Leonardo reduces their number to one, the true moderator band.

Combustion, in Leonardo's view, whether in a stove, a candle flame or the heart of an animal, consumes a part of the air, and live flame, and having consumed it, dies unless renewed. This part of the air nearly 300 years later was called oxygen.

This, Leonardo's second round of cardiovascular investigations, may possibly have had historical repercussions. The idea that the heat of the body was created by a burning flame in the heart remained alive for one or two centuries. For example, Robert Burton in his *Anatomy of Melancholy*, in the fifth edition, 1651, the first to be published after his death, writes about the heart thus: 'The left creek hath the form of a cone and is the seat of life, which, as a torch doth oil, draws blood into it, begetting of it spirits and fire; and as fire in a torch, so are spirits in the blood; and by that great artery called aorta it sends vital spirits over the body, and takes air from the lungs by that artery which is called venosa'.[2] And William Harvey, in his *Second Disquisition to John Riolan Jun. in which Many Objections to the Circulation of the Blood are Refuted* (1649), specifically denies the idea that 'the heart, as some imagine, is anything like a chauffer [stove] or fire, or heated kettle, and so the source of the heat of the blood'.[3] Almost certainly Harvey was ignorant of the fact that he was referring to the man who 'imagined' and developed such an idea.

Leonardo himself soon became dissatisfied with this concept of the heart as merely the centre for a flame for body heat. The idea did not give sufficient recognition to the mechanical work performed by this 'most powerful muscle in the body'. He appears to have entered quite quickly upon the final, prolonged third round of his investigation. In this he brought to bear the full weight of his work on heat-production by friction (see Chapter 4) and hydrodynamics. He also made a more detailed dissection of the different parts of which the heart is composed (See Figure 15.3). From 1513 he was in Rome; dissection in the hospital was denied him, so he took a pluck of the heart and lungs of an ox from the abattoir and worked on them. The results of these studies were entered into various notebooks. It would appear, too, from his plan of cardiac dissections that he still referred to Mondino's *Anatomy*, though Leonardo very soon departs from any information to be found in that author's work.

LEONARDO'S FINAL CONCLUSIONS ON THE MOVEMENT OF THE HEART: 'THE HEART IS A MUSCLE'

While he was concentrating on the muscular system in general Leonardo came to the conclusion that the heart, too, contrary to Galen's opinion, is a muscle. 'The heart is a principal muscle of force, and it is much more powerful than the other muscles. I have written of the muscles which descend from the base to the apex of the heart and the position of the muscles which arise from the apex of the heart and go up to the base' (G 1 v, Figure 15.9). And he realises that defining the heart as a muscle involves finding veins, arteries and possibly nerves 'like those of any other muscle'. 'The heart is a vessel made of dense muscle vivified and nourished by the artery and vein as are other muscles' (W 19050 v; K/P 59 v). It is to be noted that arteries containing 'vital spirit' were considered to 'vivify', and veins to 'nourish', organs with their contained blood. This was traditional Galenic terminology.

The possible nerve supply of the heart he finds to be the vagus, or 'reversive nerve', as he calls it (see Figure 16.10, p. 336). 'Follow up the reversive nerves as far as the heart

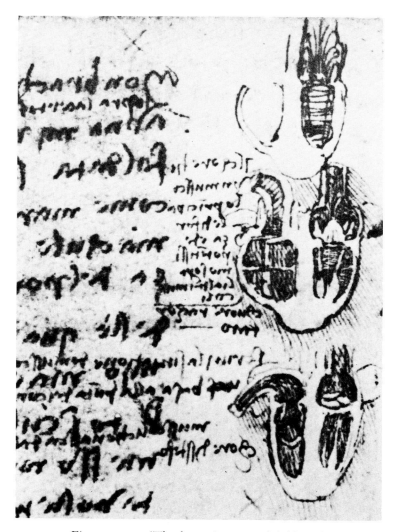

Figure 15.9. 'The heart is a muscle' (G 1 v).

It will be noticed that the atria are shown imperfectly, but blood is expelled from them. In the second drawing the right atrium is contracting, so expelling its blood into the right ventricle, which is dilated. At this time Leonardo thought that the left and right ventricles contracted alternately.

and see whether these nerves give movement to the heart or whether the heart moves by itself. And if such movement comes from the reversive nerves which have their origin in the brain you will make it clear how the soul has its seat in the ventricles of the brain, and the vital spirits have their origin in the left ventricle of the heart' (W 19112 r; K/P 105 r).

THE CHAMBERS OF THE HEART

In the view of all Leonardo's authorities, Galen, Avicenna and Mondino, the heart consisted of two chambers, the ventricles, into which ran the main blood vessels. The vena cava carried blood to and fro into and out of the right ventricle, and the pulmonary vein brought air and a little blood to and fro between the left ventricle and lung. At their mouths, as they entered their respective ventricles, each vein became dilated into little 'ears' or auricles which stored blood for emergency use.

Leonardo, of course, took over this idea through the first two phases of his investigation of the heart, as is shown by his drawings. But in the third phase he came to realise that the heart consisted of four chambers, not two. The upper chambers he called 'upper ventricles' or 'auricles'. We now call them the atria. The lower chambers he called 'lower ventricles'. He had discovered that the atria are not just dilatations of the ends of veins but part of the heart itself. 'The heart has four ventricles, i.e., two lower ventricles in the substance of the heart and two upper outside the substance of the heart, and of these two are on the right and two on the left' (W 19062 r; K/P 155 r; see Figure 15.10).

With this discovery the whole of Leonardo's interpretation of the action and function of the heart changed. 'The auricles of the heart are antechambers of the heart', he writes (G 1 v), Why were they made? 'The auricles of the heart were made in the shape of dilatable purses only to receive the percussion of the movement which the blood makes driven violently out of the ventricles, both right and left, when they contract themselves' (W 19074 r; K/P 166 ra). The idea that when the ventricles contract some of their blood goes back into the atria was fundamental to Leonardo's new idea of the mechanism of the heart. It was also at the root of his failure to reach that concept of circulation of the blood which William Harvey discovered some 120 years later. Nevertheless Leonardo's momentous discoveries about the heart took him far from the prevalent views of Galen and more than half-way to discovery of the mechanism of the heart beat.

The reason for attributing this contractile function to the 'upper ventricles' is illustrated and described in Figure 15.10. 'The heart has four ventricles, that is two lower in the substance of the heart and two upper [atria] outside the substance of the heart; and of these two are on the left . . . The upper ones [atria] are separated by certain little doors, or gateways of the heart, from the lower ventricles . . . The upper ventricles continually make a flux and reflux of the blood which is continually pulled or pressed by the

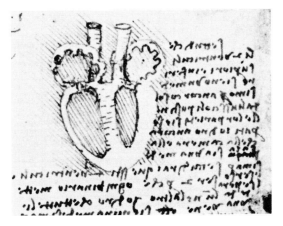

Figure 15.10. The discovery of the atria of the heart (W 19062 r; K/P 155 r).

This drawing clearly distinguishes the atria, called by Leonardo 'upper ventricles', from the 'lower ventricles' of the heart. Note that the atria are given crenated pectinate muscles, indicating that they possess contractile power to expel blood into the lower ventricles.

lower ventricles from [or into] the upper. And since these upper ventricles are more suited for driving blood out of themselves than in atracting it, Nature has so made it that by the closure of the lower ventricles which close on their own, the blood which escapes from them, is that which dilates the upper ventricles. These, being composed of muscle and fleshy membranes, are able to dilate and receive as much blood as is pressed into them; and also capable, by means of powerful muscles, of contracting with impetus and driving the blood out of themselves into the lower ventricles, one of which opens when the other closes . . . And so by such flux and reflux made with great rapidity the blood is heated and subtilised, and is made so hot that but for the help of the bellows called lungs, which, by being dilated draw in fresh air, pressing it into contact with the coats of the ramifications of the vessels, refreshing them, the blood would become so hot that it would suffocate the heart and deprive it of life'.

At first sight this 'flux and reflux' from auricle to ventricle seems rather futile. What did Leonardo think was achieved by it? This question he answers in a discussion about the right ventricle. 'The right upper ventricle is necessary for the flux and reflux which is generated by means of this ventricle. Through this flux and reflux of the blood, and the impetus of its movement from one ventricle into the other, one ventricle drives it out and the other receives it; and that which has received it drives it back again into that which previously drove it out . . . And so between revolving up and down successively it never ceases to flow through the cavernous recesses interposed between the muscles which contract the upper ventricles. And the whirling round in diverse eddies, and the friction which it makes on the walls, and the percussions in the hollows, are the cause of the heating of the blood, and making it from thick and viscous to subtle and penetrative suitable for flowing from the right to the left ventricle through the narrow porosities of the wall interposed between the right and left ventricle' (W 19063 v; K/P 156 v). He explains that the movement of the percussed ventricle 'helping the natural reflection somewhat by its contraction, makes the reflux of blood swifter in returning into the ventricle from which it was first driven . . . And if you should say that the percussion which the blood makes descending to the bottom of the lower ventricle is greater than that which percusses the cover [*coperchio*] of the upper ventricle, because one movement is natural [gravity] and the other is not, here one replies that liquid in liquid has no weight except by the amount percussion generates it'.*

Now the point of this 'flux and reflux' becomes clear. Leonardo is applying his 'four powers' of movement, force, weight and percussion, with impetus and friction, to the movement of the blood within the heart. So far the raison d'être has been the creation of the heat of the body, but he goes on to emphasise that ultimately this flux and reflux will not interfere with the flow of blood through the heart. 'This movement of the blood behaves like a lake through which a river flows which acquires as much water from one side as it loses from the other. The only difference is that the movement of the blood is discontinuous and that of the river flowing through the lake is continuous, and from this lack of such flux and reflux the blood would not be heated and consequently the vital spirits would not be generated and therefore life would be destroyed' (W 19063 v; K/P 156 v).

For all these statements Leonardo has found justification in previous investigations of hydrodynamics. That of the flow of water into lakes, for example, is to be found in *Codex Leicester* 32 v. It begins, 'Given an equal volume of water in an equal time even though rivers may vary in length, breadth, slant, depth . . . or slowness, swiftness, or where the waters fall perpendicularly or rise in eddying flood . . . will not prevent the equal entrance of one being equal to its exit' (see also CA 287 rb, where the principal of continuity is described, and Chapter 4, Figure 4.10).

THE AURICLES (ATRIA) AND VENTRICLES

Leonardo saw no need to modify the Galenic view of a two-chambered heart during the first two phases of his cardiac investigations. Therefore in previous drawings atrial chambers do not appear. Hypothetical vessels from the right lung enter a common chamber at the junction of the superior and inferior vena cava. It is not until Figure 15.10 and Figure 15.11 that he draws indubitable figures of the atria. The latter drawing in which the left atrium is

outlined as well as the right is marked by two asterisks, a symbol used by Leonardo to indicate the importance of the drawing. Yet strangely enough, the ample text with which the rest of the page is filled contains not a word about the heart but is devoted to the vices of mankind, particularly gluttony.

It is in relation to the right 'upper ventricle' that Leonardo refers to blood being drawn out of the lower, 'as much as that quantity which was driven out of the right into the left ventricle' (W 19063 r; K/P 156 r). He thus raises the question of cardiac input and output. But beyond commenting that it must be 'a great weight' of blood, he does not definitely answer this question.

Whilst the muscles of the atria are represented by a crenated line (see Figure 15.10), those of the 'lower ventricles' receive much more detailed attention. In Figure 15.12 (K/P 166 rb) Leonardo puts into practice the idea of making transverse sections suggested some twenty-five years earlier. Here, following the drawings downwards, he draws transverse sections through both left and right ventricles, the right being distinguished from the left by its comparatively thin wall. Between the two is the interventricular septum, labelled by Leonardo '*cholatorio del chore*', sieve of the heart. Below this he illustrates the left ventricle only, in diastole with prominently bulging papillary muscles. Below and to the left of this drawing both ventricles are drawn in systole, i.e., contracted. Alongside Leonardo notes how 'the [papillary] muscles in the cavities of the ventricles by their excessive density prevent the complete closure of the right and left ventricles'. Below this he draws a longitudinal section of the heart showing the papillary muscles in both left and right ventricles. They are connected by simple chordae tendinae to the mitral and tricuspid valves (*cd*). He writes, '*ab* are the gates which open outwards [i.e., the aortic and pulmonary valves] whilst *c* and *d* are gates opening inwards [i.e., the mitral and tricuspid valves]'. He still holds that the papillary muscles 'prevent the heart from dilating excessively when it reopens because . . . the passage of this [blood], owing to its velocity and too much friction, would make it catch fire'. He also holds that the mitral and tricuspid valves (*c* and *d*) close in ventricular systole so that the lower ventricles placed beyond such gates receive the impetus of the blood. 'Their dilatation is the reason why the percussion which the impetus of the escaping blood makes in them is not too powerful'.

The bottom right drawing in Figure 15.12 explains this statement about impetus. The beam of a balance, *ab,* is shown in stable equilibrium on a fulcrum, *c*. This is displaced through an angle, *cde,* so that its right end is elevated and its left end lowered. The diagram illustrates the action of 'impetus'. Leonardo explains, 'Impetus makes accidental lightness and weight'. He shows this in the movement of the beam of the balance. Its right end by moving upwards acquires 'accidental lightness', the left end by moving downwards through the angle *cde* acquires 'accidental weight'. In both cases by their movement they acquire 'im-

---

*In *Codex Atlanticus* 188 vb Leonardo has demonstrated how water 'weighs in air but not in water'. And he asserts, 'No element has in itself gravity or levity . . . [These] arise from movement of the element in itself' (BM 204 r). Both statements underlie these assertions about blood.

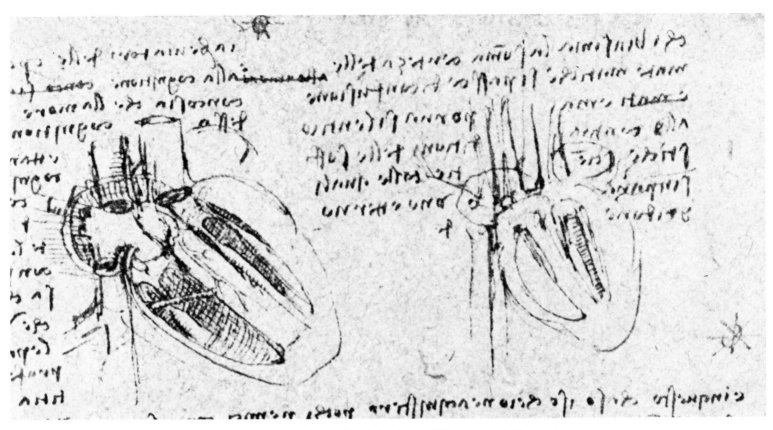

Figure 15.11.   The atria of the heart (W 19083 r; K/P 173 r).

In this drawing the right atrium is shown with the superior and inferior venae cavae entering it, as also does a pulmonary vein from the right lung. The left atrium is also outlined.

petus'. Reference to the drawings in Figure 4.1 (p. 94) will show how acceleration due to gravity results from the addition of 'accidental gravity' to natural gravity, whilst the deceleration of an object thrown upwards is due to the diminution and loss of the impetus of accidental lightness. On the other hand, fire rises through air with increasing velocity by virtue of the increasing impetus of 'natural lightness' in air. This diagram therefore illustrates the different kinds of impetus (accidental lightness or weight) occurring in blood percussed upwards from lower to upper ventricle compared with that percussed downwards when 'natural' is added to accidental gravity or weight.

Each of the 'lower' ventricles of the heart receives detailed attention. Here Leonardo uses his own technique of anatomical display in Figure 15.13. 'Before you open the heart inflate the ventricles of the heart beginning from the aorta, and then tie them up and consider their quantity. Then do similarly with the right ventricle and right auricle, and thus you will see the shape of that which was created to dilate and contract itself' (W 19119 r; K/P 116 r). The beautiful results of this technique are best shown in Figure 15.13. Here (upper left) the right ventricle is shown looking upwards from the apex past the moderator band, called by Leonardo '*catena*' (band), to the three cusps of the tricuspid valve attached to papillary muscles by threadlike chordae

tendineae. The drawing to the right shows the trabeculae carnae, fleshy muscle fibres separated by hollow pits in which Leonardo sees the blood eddying round. At the bottom of the page is a small sketch showing how the moderator band lies in the right ventricle. So well is this shown that some have advocated that this band be named after Leonardo da Vinci.

In the text Leonardo once more concentrates on the heat created by blood eddying round between the trabeculae carnae: 'Not finding any angular impediments it makes easier revolution in its revolving impetus. Thus it becomes heated with as much more heat as the motion of the heart is more frequent. Thus it sometimes creates so much heat that it is suffocated; and I once saw a man who burst fleeing from his enemies, and he poured out sweat mixed with blood through all the pores of his skin'. Though the diagnosis of Leonardo's observation is far from clear, the analogy between the formation of sweat and the condensation of water vapour does emerge clearly.

THE VALVES OF THE HEART

The valves of the heart had been accurately described by Erasistratus about 250 B.C., so it might be thought that Leonardo had little to add in this field. A glance at the

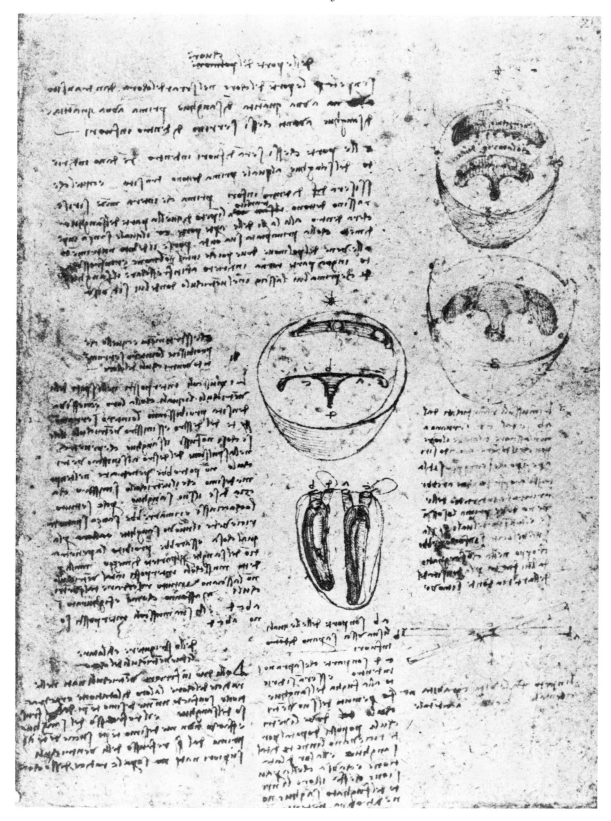

Figure 15.12. Transverse sections of the heart and papillary muscles (W 19073 r; K/P 166 rb).

The way the papillary muscles prevent complete closure of the ventricles in systole is shown in the middle drawing. The faint diagram of a balance in the lower right corner describes its oscillation in stable equilibrium which exerts alternate impetus of 'gravity' or 'lightness' corresponding to the passage of blood upwards or downwards in the heart.

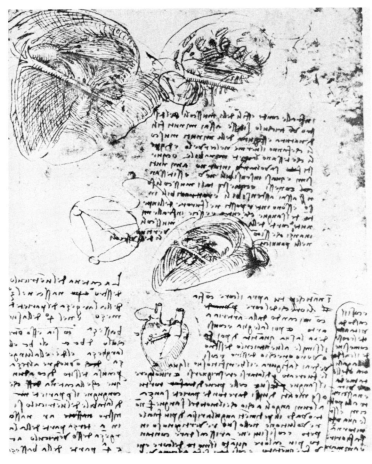

Figure 15.13. The right ventricle, papillary muscles and moderator band (W 19119 r; K/P 116 r).

The moderator band is given marked prominence in these drawings. Its exact location in the heart is demonstrated in the lowermost diagram.

drawings of the tricuspid valve in Figure 15.14 (W 19078 v; K/P 165 v) reveals how wrong such a supposition would be. The top figure shows the three cusps of the tricuspid valve opened out in such a way that the incomplete cusp on the right fits on to the incomplete cusp on the left of the straight line representing the valve ring. Leonardo's instructions written above this drawing are, 'Cut out these three muscles with their cords and [valvular] membranes, and then join them together in the way in which they are placed when the right ventricle is shut; and then you will see the true shape of the [valvular] membranes, and what they do with their cords [chordae tendineae] when they shut'. He has devised a three-dimensional model of the valve (which can easily be made from his instructions) to show not only the structure of the tricuspid valve but also its movements in action. The shape of the closed valve is illustrated immediately below. The remaining drawings, particularly those on the right margin, consist of detailed studies as to the exact mode of attachment of the chordae tendineae, as seen from above and below the valve cusps.

In Leonardo's view all the valves at the beginning of

ventricular systole permit a short phase of regurgitation between the closing valve cusps until their edges have met to produce complete occlusion. He dissociates the papillary muscles attached to the chordae tendineae from the mass of contracting ventricular muscle. He considers that these papillary muscles contract during ventricular diastole at the same time as the atria contract. The papillary muscles, therefore, open the valve cusps (mitral and tricuspid) and facilitate the expulsion of blood from the atria into the lower ventricles.

Normal tricuspid regurgitation is limited, but if the chordae tendineae get impeded by thick short fibres (of fibrin), these 'wrap themselves round the chords of the valve membranes' so that the valve cannot shut well and a great part of the blood escapes through the badly closed valves into the right upper ventricle instead of passing through the interventricular septum to form vital spirit. Therefore, the spirit is insufficient in old people, and they 'often die speaking' (W 19062 v; K/P 155 v).

The pulmonary valve receives little attention from Leonardo; probably because he noted that its structure is identical with that of the aortic valve, to which he allotted unprecedented study. The pulmonary valve is best depicted in Figure 15.25 (p. 321) and in a cross section of the base of the heart showing all the valvular orifices in Figure 15.15, where the mitral valve is also prominently displayed.

The mitral valve finds wonderful representation on the same page (W 19080 r; K/P 170 r) as that on which Leonardo noted down his method of dissecting the heart, Figure 15.16. Here in the middle of the page the vaulted arches of the chordae tendineae rising into the curved closed cusps, as seen from below in systole, present to the observer a dramatic illusion of standing in a cathedral. Below Leonardo portrays the *H*-shaped cleft presented by the closed mitral valve as seen from above, and to the right of this it is again shown, closed, in relation to the aortic, pulmonary and tricuspid valves in cross section. Leonardo gives but little attention to the mechanism of this valve which because of its two cusps was considered by Galenic tradition (as voiced by Mondino) as 'two little gates which close imperfectly' since they draw in air from the lung and return sooty vapours to it. Leonardo, however, has by now abandoned this theory and describes and draws in a note in Figure 15.20 (p. 317) the mitral valve cusps as 'shut with complete and perfect closure', a 'perfect shutting' which is brought about by the same causes which close lock gates (see Chapter 1, p. 28).

The aortic valve particularly attracted Leonardo's attention. All his studies of this valve show him focusing his previous studies of the 'four powers of Nature', in particular those on the movements of water, on the problem. For example, in Figure 15.17 there are vivid drawings of the effects of eddies on the banks from a sudden widening of a river, effects which he transfers to the sudden widening of the aorta distal to the aortic valves.

Leonardo's studies of water had been particularly intense about 1508, thus coinciding with his anatomical studies. He

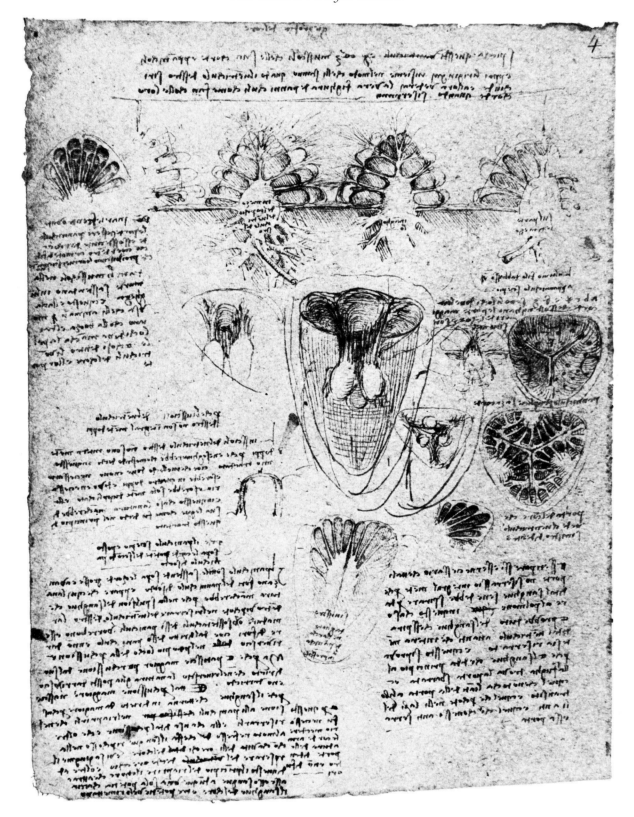

Figure 15.14. The tricuspid valve (W 19078 v; K/P 165 v).

Above, the valve is shown with the valve ring unrolled so that one end can be joined to the other to reform the intact valve. Below this, the papillary muscles are shown with the valve closed. In the right margin the valve cusps are drawn from above and below to show the mode of attachment of the chordae tendineae.

had devoted to them all his investigational skills. He had studied these movements through extensive observations of rivers. He designed special floats as markers which by means of their different specific gravities floated at different levels in flowing water (see Figure 5.7b, p. 136, where a set of such floats is drawn). He describes them as 'objects carried by the current of water between its surface and the river-bed' and explains how they are shaped to achieve this end. It is significant that these notes are written upside down on a page, the recto of which is headed, 'Objects which support themselves between the surface and bottom of water', and is entirely devoted to the study of water.

Observations on water-currents Leonardo had submitted to much experimental exploration, using his favourite method of constructing models in which he could produce controlled variations in such factors as quantity and direction of flow. These models are to be found in Figures 5.12

Figure 15.16. The mitral valve (W 19080 r; K/P 170 r).

The ventricular septum has been turned to the left in the upper left-hand drawing so exposing the papillary muscles, chordae tendineae and cusps of the mitral valve from the side and beneath. Its architecture is elaborated in the drawing beside it. Below this central drawing the valve is shown from above, closed, and in a cross section of the base of the heart.

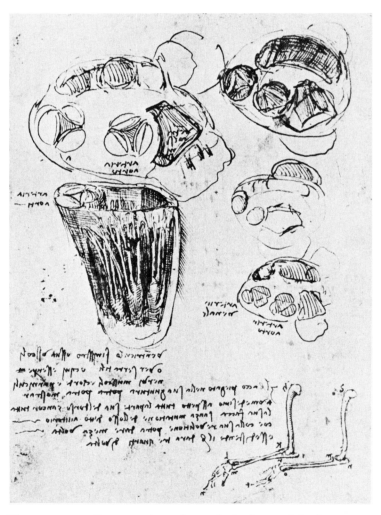

Figure 15.15. Cross sections through the base of the heart; mitral valve (W 19118 v; K/P 116 v).

In these drawings the words *arteria aorta* stand for aorta, and *arteria venale* for the pulmonary artery. The left central drawing depicts the mitral ring and aorta. Below, Leonardo's fascination with pronation and supination manifests itself.

and 5.13 (p. 141). All had glass fronts with opaque backgrounds against which he could observe the movements of such markers as grass seed, bits of paper and coloured inks. It was in this way that he analysed the reflections and eddy formations of water falling into a tank from a narrow pipe – a situation analogous to that of the inflow of blood through the aortic valve into the aorta.

These aortic studies involved detailed studies of the shape of the valve cusps (Figure 15.18). They showed him that the orifice of the open valve was triangular. He applied his hydrodynamic studies to the effects of such a shape on the flow of water from a pipe (Figure 15.19, W 19117 r, K/P 115 r) noting that whether horizontal or vertical, the flow was greatest at the centre of the stream where there was least friction from the walls. He then draws the shape of the out-

Figure 15.17. Eddies produced by sudden widening of a canal (F 91 v).

On widening one bank one set of eddies is created; top drawing. On the widening of both banks two sets of eddies are created, whilst the central stream continues in a straight line, unless, as in the lowest drawing, the turbulence involves the whole width of the canal.

Figure 15.18. The cusps of the aortic valve (W 19079 v; K/P 169 v).

The valve cusps have been removed and drawn from above and below, first closed, then partially and fully open. The wavy edges of the opening cusps and the triangular shape of the orifice between the open cusps are clearly shown.

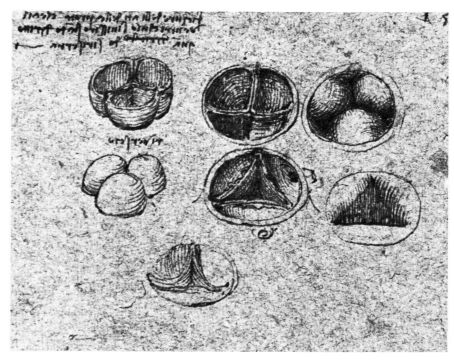

flow through the triangular orifice, making three distinct eddies returning back as they percuss the stationary blood already in the aorta, in the form of expanded 'hemicycles'. The eddy then 'percusses the concavity of the base of each

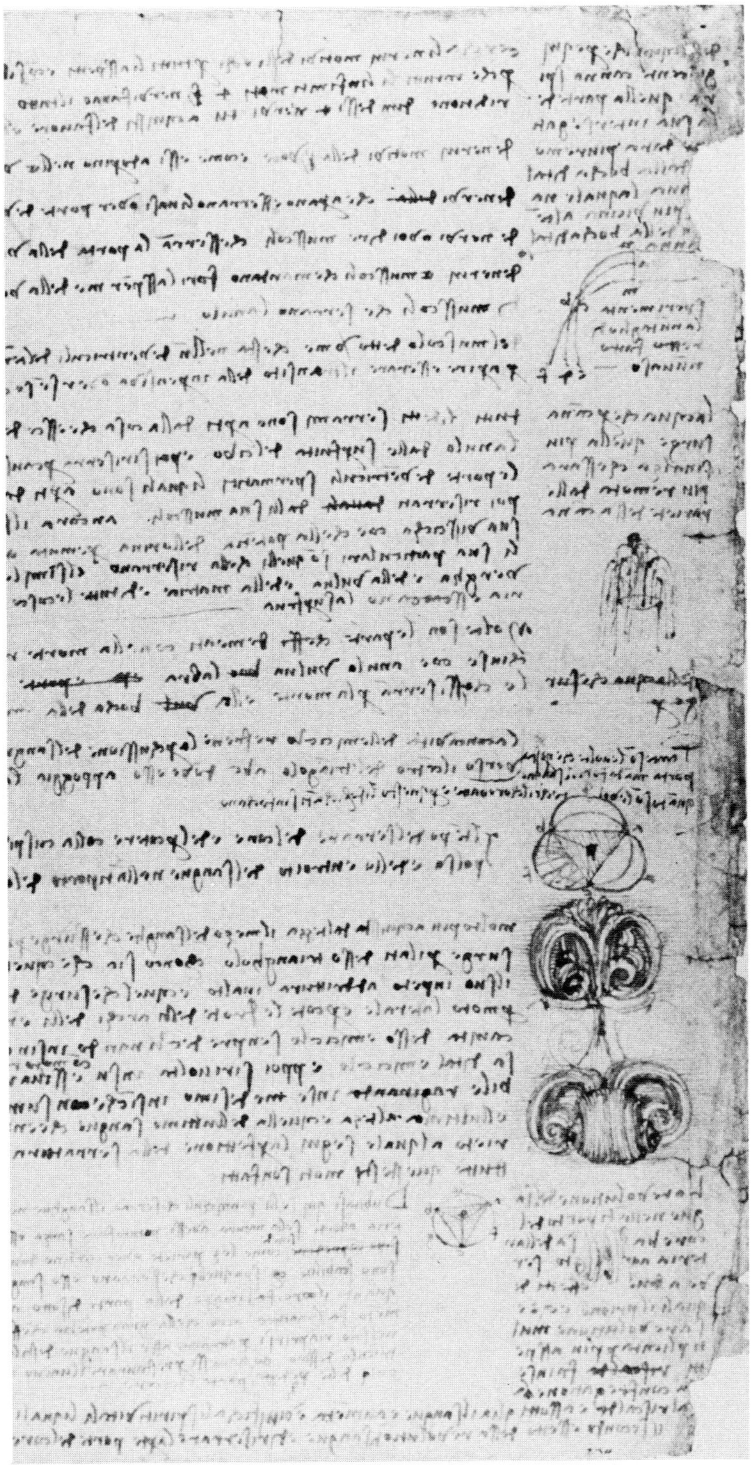

Figure 15.19. The flow of blood through the aortic valve. (W 19116 r; K/P 115 r).

Above are drawings of the flow of water from a pipe (compare with Figure 5.12). Below are drawings of the flow pattern of blood through the triangular aortic orifice.

hemicycle and then goes upwards with reflected movement'. This final upward movement is that which is 'followed by the perfect shutting made by the gate of the heart'. Thus the three valve cusps are visualised as meeting and closing in a vertical plane like that of praying hands, rather than in a horizontal plane. The dilatations so formed at the base of the aorta, from which the coronary vessels arise, are called by Leonardo the 'hemicycles'. They have since been called the sinuses of Valsalva.

The central, most powerful part of the blood stream ejected through the aortic valve produces the pressure-wave throughout the aorta and other arteries which Leonardo recognises to be the palpable pulse. Leonardo approached this problem as he did that of the flow of water on a bigger scale, by making models of the aorta (Figures 15.20 and 15.22). Inside an outline sketch of the aortic root on W 19082 r (K/P 171 r, Figure 15.20) he writes instructions for making a wax cast of the part, over which a hollow cast of gypsum is to be constructed, lined with glass, and then broken. On this same page are other drawings of designs for models of the valve orifice. Earlier, in Figure 15.22 (W 19117 v; K/P 115 v) the detailed drawing of the wavy outline of the opened valve cusps gives irrefutable evidence of him having observed the movements of the aortic valve cusps. Here, too, the movements of the eddies of blood are drawn with the valve open and closed. Indeed, Leonardo wrote opposite the parts of an aortic model in Figure 15.22, 'Make this trial in a glass [model] and move in it water and panic grass seed'. There can be no doubt that Leonardo used the same method of observing the movements of valve cusps in actual speciments of the aorta as well as in his models (see Figure 15.23).*

Retrograde flow of blood in the aorta at the end of systole has been confirmed. Leonardo's actual configuration of the currents of such flow are very probably correct though still technically beyond experimental demonstration.

THE NATURE OF THE HEART BEAT

When Leonardo first approached cardiac physiology he naturally accepted the Galenic view of the heart as an organ which sucked air into itself to form the vital spirits, much as the lungs sucked air into the chest, as well as blood from the veins. In fact, Galen deals with the heart as a respiratory organ. The beat of the heart, visible and palpable on the chest wall, was therefore attributed to its dilation or diastole.

We are now in a position to see that Leonardo altered all this. He came to realise that the essence of cardiac action was its contraction in systole whereby its contained blood was 'percussed' and expelled from the heart along the arteries, so producing the pulse, and also to the lungs along the pulmonary artery. He did not, however, eliminate the path

*Leonardo's drawings of the aortic valve cusps on K/P 115 v, on the left of Figure 15.23, are here compared with the author's cine-photographs of these valves (right).

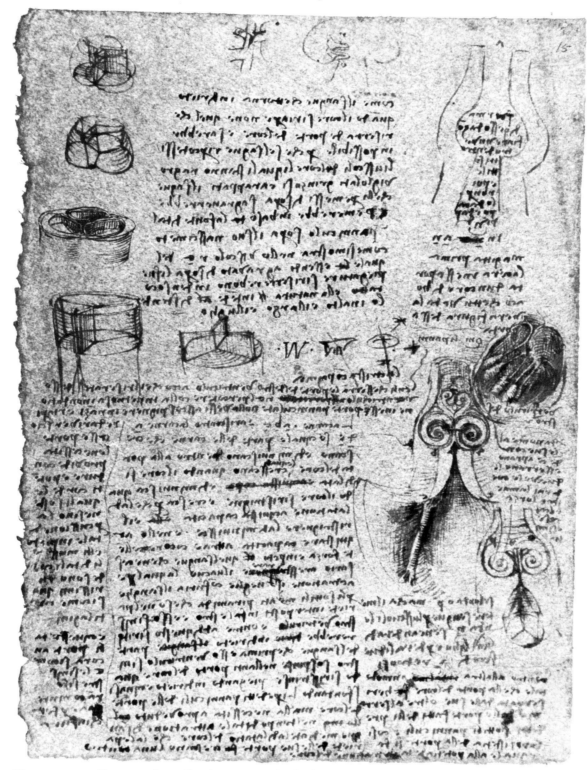

Figure 15.20. Construction of a glass cast and of other models of the aorta (W 19082 r; K/P 171 r).

Within the drawing, top right, instructions for making a glass cast of the aorta are written. The drawings on the left pursue this objective. Below are schematic diagrams of closure of the aortic valves and the waves of blood travelling up the aorta from its percussion in the left ventricle.

On the right margin is a drawing of several papillary muscles with attached chordae tendineae proceeding to two valve cusps. Leonardo writes, 'Give names to the chordae which open and shut these two sails'. Once more he appeals to his experience of sailing boats for a mechanical analogy. Compare Figure 15.21. The pattern of eddies of blood passing into the aorta is developed from the studies illustrated in Figures 5.17 and 15.17.

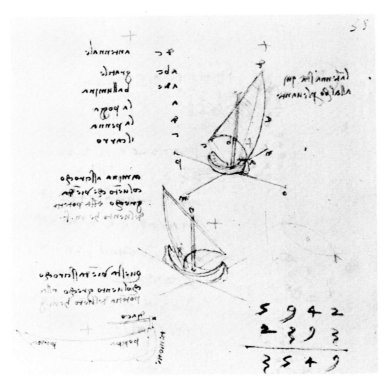

Figure 15.21. Studies of sailing boats, their sails and rigging (Madrid II 35 r).

The closure of the cardiac valves was likened to the billowing sails and attached ropes, here listed and named. Compare Figure 15.20.

Figure 15.22. Movement of aortic valve cusps in systole and diastole; construction of model of the base of the aorta (W 19117 v; K/P 115 v).

On the right, top, are diagrams of bicuspid, tricuspid and four-cusped aortic valve. Below this are the open aortic valve with wavy cusps and studies of the positions of the aortic valve cusps in different phases of systole and diastole. Note that in diastole the cusps are placed perpendicularly, not horizontally.

At the top left are stages in the construction of a model of the base of the aorta.

through the porosities of the interventricular septum from right to left ventricle. Both Galen and Leonardo state that these are invisible and that blood 'sweats' through them. Nor did Leonardo eliminate passage of blood outwards through the veins. His suggested ebb and flow between atria and ventricles was based on his belief that the heart created the heat of the body and that this function was as important as that of mechanical movement of the blood.

If the heart 'percussed' its blood, what was the relation between ventricular movement and this percussion of the blood? Leonardo went to the abattoir to find the answer. Here he observed the slaughter of pigs by piercing their hearts with a skewer. He writes, 'they pierce its right side and heart at the same time with a skewer [*spillo*] pushing it straight in. And if this skewer pierces the heart when it is elongated the heart shortens during the expulsion of blood and draws its wound upwards with the point of the skewer; and as much as it raises the point of the skewer inside, so much it lowers the handle of the skewer outside' (W 19065 r; K/P 158 r). When the heart refills, the converse happens. 'This I have seen many times and I have observed the measurements, having allowed the instrument to remain in the heart until the animal was cut up' (Figure 15.24a).

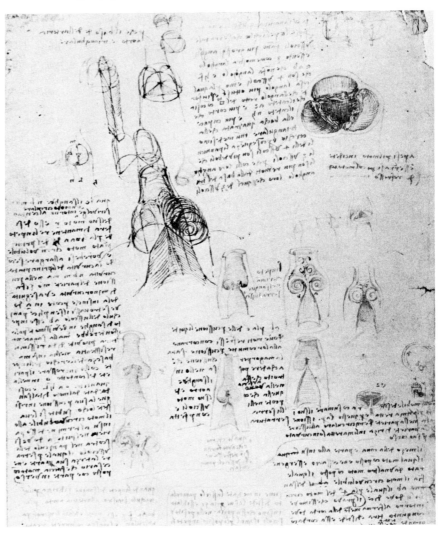

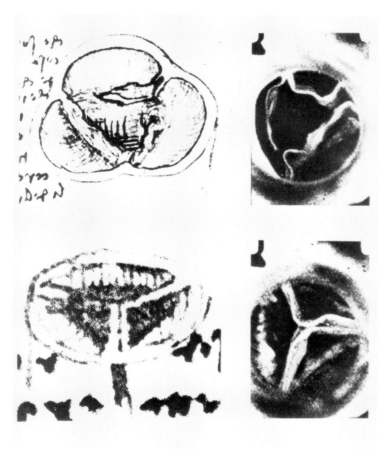

Figure 15.23. Aortic valve cusps open and closed.

On the left are two drawings of the aortic valve cusps drawn in Figure 15.22. On the right are two cinematographic pictures of the aortic valve cusps open and closed.

Leonardo is here putting into experimental service the principle of the lever, which he specifically studied about 1493, comparing the force and movement of the ends of a lever and its counterlever (BM 214 r; see Figure 15.24b). From these measured movements he came to appreciate that the heart shortens during systole and lengthens in diastole – just the opposite of Galenic traditional teaching. 'The flux of blood generated in the heart through the two upper and lower ventricles is the cause of the nourishment making discontinuous motion [pulses] in all parts of the human body nourished by the blood. This is clearly manifested in all the umbilical cords of infants who die in the uterus together with the mother, and even after they have been born' (W 19066 r; K/P 159 r).

He reaches the first correct interpretation of the cardiac impulse on the chest wall in asserting, 'The time of the shutting of the heart [systole] and of the percussion by its apex against the chest wall, and of the beating of the pulse, and of the entrance of the blood into the gateway of the heart [aortic orifice] is one and the same' (W 19997 r; KP 115 r).

It is difficult to believe that these assertions (and others supporting them) were written by a man who was ignorant of the circulation of the blood as we know it, yet this is true. Leonardo had broken through the Galenic tradition of cardiac action, but not the Galenic tradition of the movement of the blood through the body, for he discovered neither the pulmonary nor the systemic circulation of the blood.

THE CORONARY VESSELS

In searching for the blood supply of the most powerful muscle in the human body Leonardo explored the coronary arteries and veins. The mouths of the coronary arteries are drawn in his detailed dissection of the aortic valve cusps (Figure 15.18, p. 315). The pattern of their distribution in the ox heart is beautifully shown (labelled *a* and *b*) in the right-hand heart drawn on W 19073 v; K/P 166 v (Figure 15.25), the pulmonary artery having been cut away to display their main trunks. In this drawing the three cusps of the pulmonary valve are also shown. The same sheet is entirely devoted to views of the coronary arteries as seen from in front and behind the heart. The coronary sinus and great cardiac vein are also clearly shown.

'THE TREE OF THE VESSELS'

In demonstrating the blood vessels of the body Leonardo made his starting point the works of Mondino and Avicenna. But whereas Mondino deals with the anatomy of the body regionally, Avicenna in his *Canon* deals with the different systems; therefore, Leonardo found useful information there. In describing blood vessels, however, there is often confusion between arteries and veins. Avicenna, for example, begins his description of arteries with the words, 'Pulsating vessels are arteries', and later, 'Non-pulsating vessels are veins, all of which originate in the liver'.[4] Leonardo does not always extricate himself from this confusion; he uses the word '*vene*' sometimes indiscriminately to describe either arteries or veins. In such cases the word 'vessel' is used in quoting him. It is worth noting that Leonardo's description of blood vessels, with the exception of the coronary artery and vein and the blood supply of the lung, were made from relatively early dissections, probably none after about 1510.

Like Avicenna, Leonardo was anxious to see the various systems of the body as a whole. His 'Tree of the nerves' (Figure 11.1, p. 243) is accompanied by a drawing of the 'Tree of the vessels [*vene*]' (Figure 15.26), which bears all the marks of being derived from Avicenna, for this unusually crude and unattractive drawing shows the vena cava arising from the liver and dividing into its superior and inferior branches from its 'origin'. The human body in which this 'Tree of the vessels' is incorporated bears a remarkable resemblance to the venesection man in *Ketham's Fasciculus,* at that time the most popular textbook of medicine.

Later, under the influence of Mondino, Leonardo draws the vena cava arising from the heart which he compares to a seed giving rise to its stem and roots (Figure 15.4, p. 302). 'The heart is the nut which generates the tree of the veins',

a.

a. Experimental demonstration of the movement of the heart (W 19065 r; K/P 158 r).

Changes in the heart are shown 'when they kill pigs in Tuscany where they pierce the hearts of the pigs with an instrument called a *spillo*'. Leonardo saw this skewer-like instrument as a lever with its fulcrum on the rib as it pierces the chest wall into the heart. As *a* is carried to *b* in the heart so the handle *f* is carried down to *g*, in the top drawing. 'I will not expatiate further', he writes, 'because a complete treatise on such movements has been made in the 20th On the Powers of Levers'. This work is lost, but the principle is explained in Figure 15.24b, written about 1493.

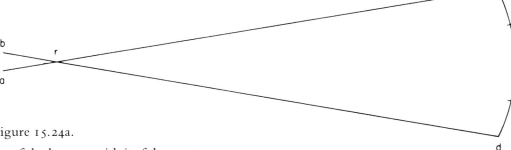

b.

b. Diagram of the lever system used in Figure 15.24a.

Here Leonardo shows how the movement of the lever *ca* with its fulcrum at *r* moves four times as far from *c* to *d*, as does the point *a* to *b*. Should *a* represent the point of the *spillo* in the heart, its movement would be four times as great outside the chest as made by the handle *cd* (BM 214 r).

Figure 15.24.

he writes, and he is at pains to explain that a plant never arises from its roots. 'This is observed by experience in the sprouting of the peach which arises from its nut as is shown above'. From Avicenna he takes the general rule that veins and arteries nearly always accompany and mutually support one another. Leonardo depicts this in several drawings of the vessels. Becoming aware of this error later, he corrects it, in Figure 15.27, alongside an accurate representation of the superficial epigastric and pudendal veins draining into the great saphenous vein. This in turn is shown passing into the femoral triangle to enter the femoral vein. Leonardo here reminds himself, 'Remember to note where the arteries leave the company of the veins and nerves' (W 19113 r; K/P 110 r).

One of his most brilliant original observations consists of showing the effects of time on blood vessels, in Figure 15.28. Above a straight set of branching vessels from a babe two years old he writes, 'youth', and above a tortuous ramification of vessels to its left he writes, 'age'. Beside the drawings he heads a section, 'The nature of the vessels in youth and age'. 'In so far as vessels become old so their straightness is destroyed in their ramifications and they become so much the more flexuous or serpentine with thicker coats as their age increases with the years'. This theme of arteriosclerosis with age is developed at length on W 19027 v (K/P 70 v). Here he ascribes these changes to over-nutrition of the arterial wall by reason of its contact with the nourishing blood. This leads to narrowing of the vessel

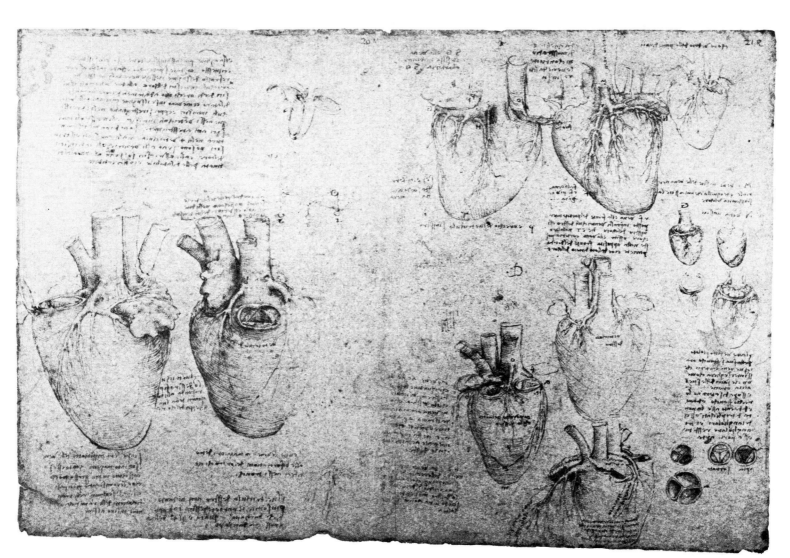

Figure 15.25. The heart, coronary arteries and veins (W 19074 v–19073 v; K/P 166 v).

These drawings of the coronary vessels in the ox are very accurate, increasingly so if one reads them from right to left, as Leonardo probably drew them. The small sketches on the right margin illustrate how these vessels crown the heart, explaining their nomenclature.

Having decided that the heart was a muscle, Leonardo felt bound to search for arteries and veins supplying it, as they supply all other muscles.

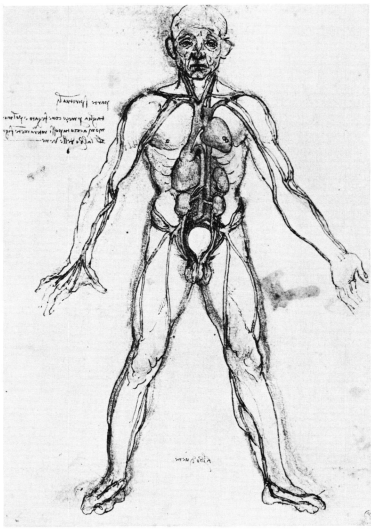

Figure 15.26. 'Tree of the vessels' (W 12597 r; K/P 36 r).

This early drawing of the main blood-vessels of the body is notable for its crudity. Here the liver is the site of origin of the venous system. The heart appears to be fed by a branch from the superior vena cava. The inferior vena cava sprouts from the liver in a strange manner, the traces of which are still evident in Figure 15.7 (p. 305).

lumen; then 'corruption of the blood' is followed by complete obstruction of the blood flow. This process particularly affects small peripheral vessels where they are narrowest. Thus 'the skin dies for lack of sustenance', as do other organs and tissues.

This 'slow death' Leonardo considers to be a normal change wrought by the 'movement of time' from youth to age, whereby 'old people fail and die without fever when they are of great age'. In it he found the cause of death of the 'old man' upon whose body he performed many dissections in his search for the cause of his 'sweet death' whilst chatting to Leonardo sitting by his bed.

It may be noted here that the term *arteriosclerosis* was not introduced until 1833 by Lobstein, who incidentally, like Leonardo, saw the process as due to 'unnatural nutrition of the arterial wall'.[5]

As an example of one of Leonardo's finest vascular dissections, that on W 19051 v (K/P 60 v, Figure 15.29) will reward a more detailed inspection than there is space to give here. It must suffice to draw attention to three points. On the right he draws the detailed anatomy of the hepatic vein as it drains into the inferior vena cava in two distinct locations, the upper consisting of three large venous branches, the lower of smaller veins. In the upper drawing, the coeliac

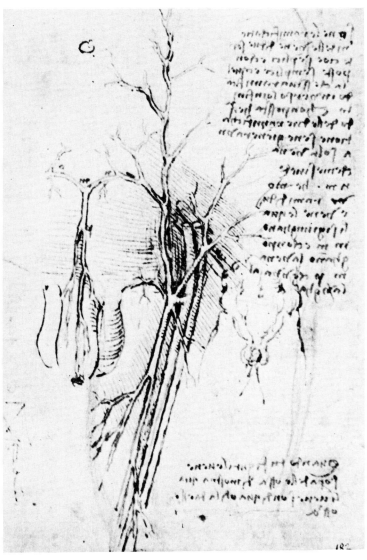

Figure 15.27. The long saphenous vein passing through the saphenous opening (W 19113 r; K/P 110 r).

The long saphenous vein is seen joining the femoral vein after receiving the superficial external iliac, superficial epigastric and superficial external pudendal veins (front view of Figure 11.3, p. 245).

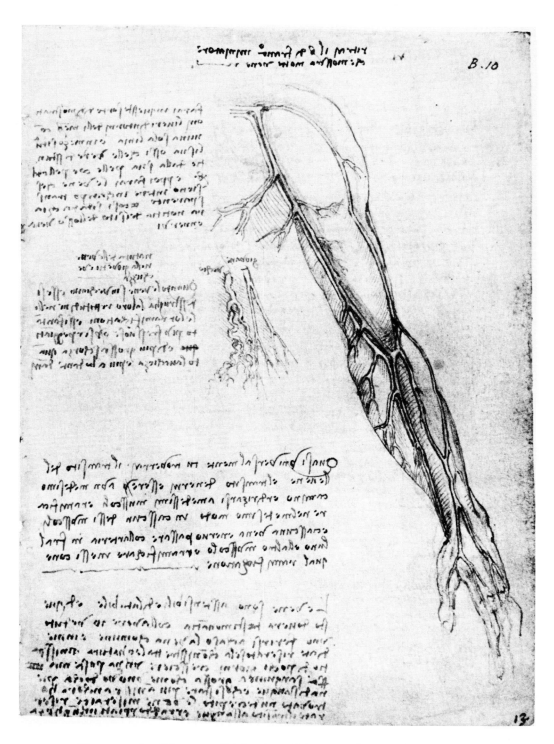

Figure 15.28. The superficial veins of the arm; the vessels of 'youth' (right); the vessels of 'age' (left) (W 19027 r; K/P 69 r).

Leonardo draws most of the veins in the arm accompanied by arteries. The cephalic vein is an exception. The comparison between the straightness of the vessels in youth and their tortuosity in age occupies three pages of Leonardo's script. These constitute the first description of arteriosclerosis.

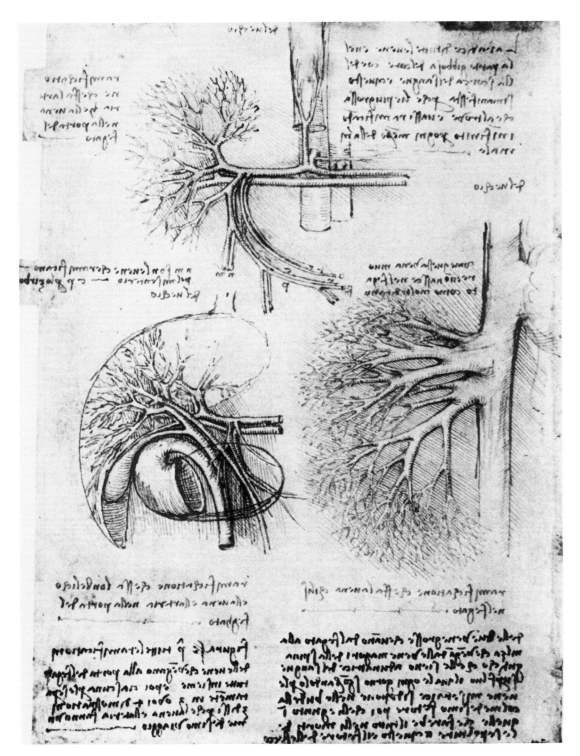

Figure 15.29. The vessels of the liver (W 19051 v; K/P 60 v).

The hepatic veins draining into the inferior vena cava are superbly shown on the right. Above, the coeliac axis and hepatic artery is demonstrated. Below and to the left, the hepatic arterial tree is combined with the portal venous branches and the biliary tree.

Comparison with Figure 15.26 (p. 322) gives some idea of Leonardo's progress.

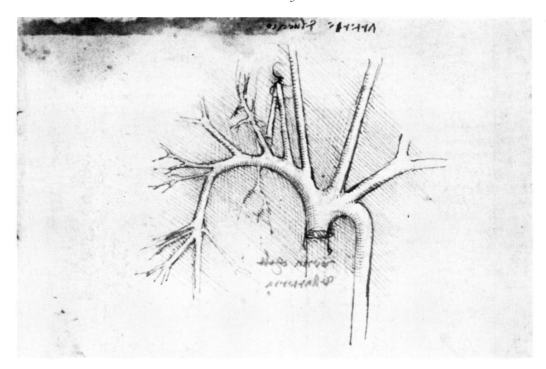

Figure 15.30. The human aorta (W 19050 r; K/P 59 r).

This drawing is the first known representation of the human aortic arch and its branches.

axis arising from the aorta is shown giving off its splenic, hepatic, left gastric and phrenic branches. Below to the left, ramifying into the concave surface of the liver, Leonardo draws the portal vein accompanied by the hepatic artery, the hepatic duct and the remnant of the ligamentum teres, with the gallbladder and cystic duct snugly embedded in the under surface of the liver. The right gastric and the gastroduodenal artery also find recognition.

It will be noticed that Leonardo's drawings of the aorta almost always are schematic or derived from animals rather than humans. The outstanding exception is to be found in Figure 15.30, where the human aorta is for the first time drawn correctly with its branches.It is that of the 'old man'. On W 19025 v (K/P 67 v) Leonardo's particular interest in the anatomy of the hand is reflected in a beautiful study of the ulnar artery and the superficial palmar arch alongside the distribution of the median and ulnar nerves (see Figure 10.2, p. 232).

Incomplete as this outline of Leonardo's anatomy of the blood vessels must be in comparison with the detailed account given in previous works, i.e., *Leonardo da Vinci on the Movement of the Heart and Blood,*[6] and the *Corpus of the Anatomical Studies in the Collection of the Queen at Windsor Castle,*[7] we cannot leave this subject without mentioning his description of the pulmonary vascular supply. This was one of his latest attempts to unravel the complex patterns of the blood vessels. It should be remembered that Leonardo uses

the term *trachea* to describe both the trachea and the bronchi. On W 19071 r (K/P 162 r, Figure 14.6, p. 296) he makes a memorandum, 'You have to consider the second order of veins and arteries which cover the first very small vessels [bronchial veins and arteries] which nourish and vivify the trachea . . . and why Nature duplicated artery and vein in such an instrument, one above the other, for the nourishment of one and the same organ. You could say that the trachea and the whole lung had to be nourished, and if this had to be done with a single large venous artery [pulmonary artery] this could not be joined to the trachea without great impediment of movement which the trachea has in its increase and decrease in length as well as breadth. Wherefore for this reason she [Nature] gave such veins and arteries to the trachea sufficient for its life and nourishment; and the other big branches she separated somewhat from the trachea [bronchi] in order to nourish the substance of the lung more conveniently'. In the accompanying sketch he draws the small bronchial vessels supplying the substance of the bronchi in contrast with larger pulmonary vessels supplying the distal lung substance. To perform this function, 'The vein is always over the artery; and the artery is more refreshed by the moving air of the trachea than the vein because the artery has more need [of refreshment], being hotter' (W 19072 r; K/P 163 r). Thus does Leonardo describe one of his most important discoveries, that of the bronchial arteries and veins.

REFERENCES

1. Gardiner, J. *The Circulation Vindicated*. Moxon, 1702, p. 11.
2. Burton, Robert. *Anatomy of Melancholy*. 5th edition, London, Cripps, 1651. Part I, section 1, subsection 4.
3. Harvey, William. *Second Disquisition to John Riolan Jun.*, in *The Works of William Harvey*, ed. by R. Willis. London, Naveed Clinic, 1847.
4. Avicenna. *Canon of Medicine*, trans. by M. H. Shah, in *The General Principles of Avicenna's Canon of Medicine*. Karachi, Naveed Clinic, 1966, pp. 112, 117.

5. Lobstein, J. F. *Traite d'anatomie pathologique*. 1829–1833. Vol. 2, p. 550.
6. Keele, Kenneth D. *Leonardo da Vinci on the Movement of the Heart and Blood*. London, Harvey and Blythe Ltd., 1952.
7. Keele, Kenneth D., and Pedretti, Carlo. *Leonardo da Vinci Corpus of the Anatomical Studies in the Collection of Her Majesty the Queen at Windsor Castle*. New York, Harcourt Brace Jovanovich, 1978–1980. Vols. I–III.

Chapter 16

Digestion and Nutrition

Leonardo's approach to the problems of digestion and nutrition followed a similar path to that of the other functions of the body. He first consulted the authoritative texts available to him, visualising their verbal descriptions into illustrative sketches. In Figure 16.1 Platonic and Aristotelian patterns of the viscera are found. Aristotle describes the jejunum as lying between the upper and lower stomachs after which the gut becomes narrower and convoluted, ending in a straight portion running to the anus.[1] All this is in Leonardo's drawings.

With regard to the digestive organs he made a particularly close study of the works of Mondino, who dealt with the abdominal organs at length. Leonardo would appear to have followed Mondino's instructions in his anatomical exploration. From this approach he made his own contributions in anatomical illustration. It will be remembered that the physiological views of Mondino were largely derived from Galen, whose works he often cites. Thus Leonardo's views on the function of organs were also often fundamentally Galenic.

From the point of view of the conventional artist of the fifteenth century (as well as the artist of the twentieth century) the organs of the abdomen had no appeal. They were not related to the external appearance of man's form and were therefore irrelevant. This was far from Leonardo's point of view. For him the world of nature within the microcosm of man's body was as full of fascination and beauty as the world of the 'terrestrial machine' in the macrocosm without. Repugnant as 'flayed corpses quartered and flayed and horrible to behold' were to his sensitive eyes and nostrils, he could not resist the desire to explore their internal structures and functions, for here, as with the respiratory system, he could bring to bear his capacities as a bioengineer as well as an artist.

The opportunity for this exploration appears to have occurred by chance. Leonardo was so impressed with the event that he described it in detail. The passage in Chapter 15 (pp. 321–322) comparing the state of the arteries in young and old people is repeated in the story of this important incident in Leonardo's anatomical studies. He writes,

'The artery and the vein in the old which extend between the spleen and the liver generate so thick a coat that it closes the passage of the blood which comes from the meseraic [mesenteric] veins through which the blood passes to the liver and the heart and to the two greater veins [venae cavae] and consequently through the whole body. And these veins as well as thickening their coat grow in length and become twisted like a snake, and the liver loses its humour of the blood which was carried there by the vein. Whence this liver becomes desiccated and like congealed bran both in colour and substance, so that when it is subjected to the slightest friction its substance falls away in tiny particles like sawdust leaving behind the veins and arteries.

And the bile ducts and the vein from the umbilicus which enter the liver through the gate of the liver [porta hepatis] remain wholly deprived of the substance of the liver like millet or broom-corn when their grains have been pulled off.

The colon and other intestines become greatly contracted in the aged; and I have found stones in the veins which pass under the clavicles of the chest. These were as large as chestnuts, of the colour and shape of truffles, or dross, or clinkers of iron. These stones were very hard, like clinkers, and had formed sacs attached to the said veins, like goitres. [These were calcified tuberculous glands.]

And this old man a few hours before his death told me that he was over a hundred years old and that he did not feel any bodily deficiency other than weakness. And thus while sitting on a bed in the hospital of Santa Maria Nuova in Florence without any movement or other sign of distress he passed away from this life.

And I made an anatomy of him in order to see the cause of so sweet a death. This I found to be a fainting away through lack of blood to the artery which nourished the heart, and other parts of the body below it which I found very dry, shrunken and withered. This anatomy I described very diligently and with great ease owing to the absence of fat and humours which greatly hinder recog-

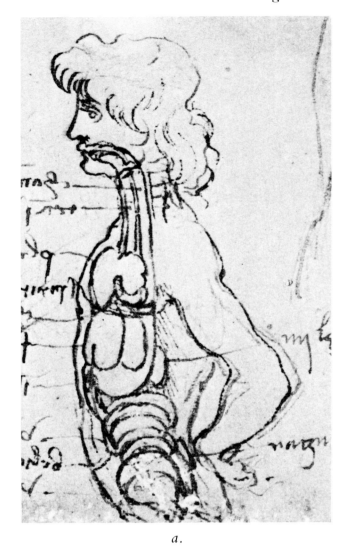

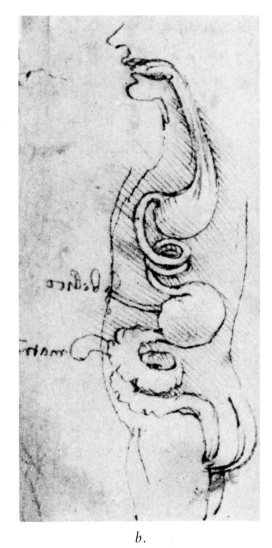

a.   *b.*

Figure 16.1.

a. Drawing of the main viscera according to Plato (W 12627 r; K/P 4 r).

The way the lungs enclose the heart to keep it cool and the separation of the thoracic from the abdominal organs by the diaphragm are described in Plato's *Timaeus*. Leonardo emphasises the position of the diaphragm, labelling it *'spirituale'*.

b. Drawing of the intestine according to Aristotle (W 19097 v; K/P 35 r).

The upper stomach connected by the small intestine with the caecum, called by Aristotle the 'lower stomach', is here visualised in Leonardo's drawing.

nition of the parts. Another anatomy was of a child of 2 years, in which I found everything contrary to that of the old man' (W 19027 v; K/P 69 v).

ANATOMY OF THE ALIMENTARY TRACT

When Leonardo made his proposed dissection of the old man he commenced with the abdominal cavity. This was the custom in those days since putrefaction commences in this region and there were then no methods of preserving the body. Thus Leonardo set to work on his exploration of

the abdomen on the body of this old man and produced a series of twenty or thirty drawings describing all the stages of his progress. Many of these are to be found under the title *'del vechio'*, the old man, so that we know they were made from this particular body.

How was Leonardo, an amateur in such matters, to set about the task of revealing the complex geography of the abdominal organs of man? Here he turned to the current textbook of anatomy, that of Mondino, recently translated into the Italian language, and followed the procedure there described, step by step. Beginning with the muscles of the abdominal wall he penetrated into the peritoneal cavity, exposed the omentum, then the intestines, finally removing the intestines themselves to show the roots of the mesenteric attachments to the spine, the blood vessels therein, and organs such as the kidneys which lie behind the peritoneum in the retroperitoneal space. This beautifully systematic exposure was described by Mondino, and Leonardo can claim little credit for the procedure itself. But, as so often happened when Leonardo used someone else's work, it became transformed into something almost unrecognisable. This happened to Mondino's *Anatomy* in Leonardo's hands, for out of it appeared a superb example of Leonardo's aphorism, 'Do not busy yourself in making enter by the ears things which have to do with the eyes, for in this you will be far surpassed by the painter' (W 19071 r; K/P 162 r).

At first, setting too much store on the work of his mentor, Leonardo copied his mistaken version of the muscles of the abdominal wall. In several early drawings he draws the oblique abdominal muscles as Mondino describes them, 'with their cords crossing after the manner of an X', as in Figure 16.2. Later he corrected this error but finally reverted to it again in order to express his mechanical views.

The rectus abdominis muscle Leonardo divides in Figure 16.3 into four clear-cut segments, or tendinous intersections, on the principle that 'where there is great length of movement there it is necessary to divide the mover into several parts'. He is thinking of the kind of movement made by arching the spine backwards 'as is seen to occur in those contortionists who bend themselves so far backwards that they join their hands with their feet'. This movement, of course, stretches the rectus abdominis muscle maximally. See also Figure 16.11.

Behind the rectus abdominis Leonardo illustrates the transversus abdominis muscle (lower drawing in Figures

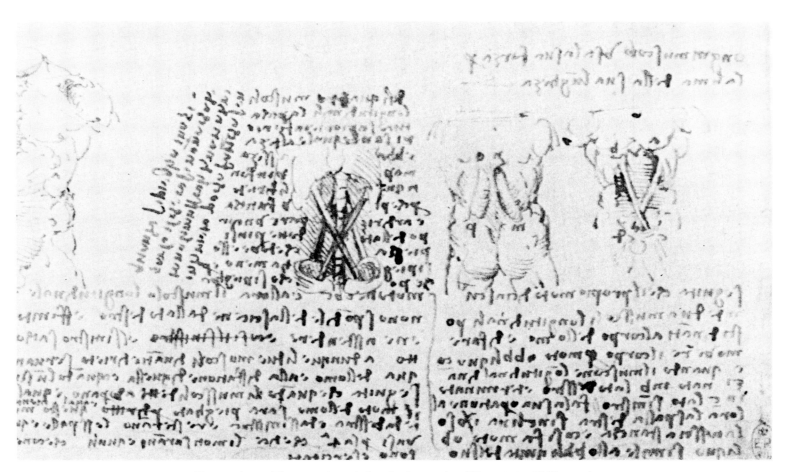

Figure 16.2. The crossing abdominal muscles (W 12636 r; K/P 111 r).

Here Leonardo depicts the anterior abdominal muscles according to his rule, 'Every muscle uses its force along the line of its length', written at the top of the right side of this page. He adds, 'The proper movement of each of the two longitudinal muscles placed in front of the body of man is to make the body move in an oblique movement'. Here he draws the movement rather than the muscles, adding those of the sacro-spinalis posteriorly.

That he was aware of the anatomical structure of the recti and transverse abdominal muscles is shown in Figure 16.3. It so happens that this drawing of the lines of force corresponds to the depiction of abdominal muscles in Pietro d'Abano's *Conciliator,* a fourteenth–century work known to Leonardo.

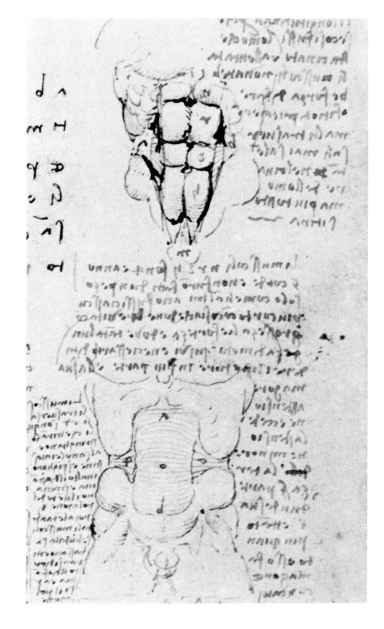

Figure 16.3. The recti and transverse abdominal muscles.

Leonardo describes the rectus abdominis (top) as follows, 'The muscles *nrsh* are 4 and they have 5 tendons . . . Where there is so much length of movement it is necessary to divide the mover into more parts'. Acting together, the recti bend the body straight down; acting separately, they bend the body obliquely, as drawn in Figure 16.2.

Of the transverse muscles of the abdomen (lower drawing) he writes, 'The transverse muscles *cd* are those which constrict and raise the intestines, push up the diaphragm and drive out the air from the lungs' (W 19032 r; K/P 74 r).

16.3 and 16.13). 'The transverse muscles', says Leonardo, 'constrict and raise up the intestines . . . and push the diaphragm upwards'. Having drawn the superior epigastric arteries descending from the 'pomegranato' (the xiphisternum as named by Mondino), Leonardo opened the peritoneum so revealing a beautiful picture of the stomach, liver and spleen – all labelled – and the apron of fat forming the omentum covering the intestines (Figure 16.4). The ligaments, remnants of blood vessels running up to converge on the umbilicus, are residua of the foetal circulation which Mondino repeatedly mentions. The prominent ligament containing a large umbilical vein is a feature of the 'old man's' cirrhosis of the liver, the nature of which, of course, Leonardo did not suspect (W 19039 v; K/P 61 v).

The next stage of dissection was to remove the omentum, so demonstrating the coils of small and large intestines. 'These are better understood if you inflate them', he writes (W 19019 v; K/P 39 v). Leonardo then makes a special drawing, Figure 16.5, of the stomach, liver and spleen. The liver is shown crumbling away as previously described. The large vein running from the spleen to the liver is joined about the middle of its course by the portal vein, as described by Mondino. It will be noticed that the spleen is unduly large, another feature of cirrhosis of the liver. In the lower right-hand corner of the page is a drawing of the 'monoculus' (Mondino's term for the caecum). But what Mondino did not mention was the appendix to the caecum. This is drawn by Leonardo and marked *n*. Below this sketch he writes, 'The auricle [appendix] *n* of the colon *nm* is a part of the monoculus [caecum] and this acts by contracting and dilating so that superfluous wind does not rupture the monoculus' (W 19031 v; K/P 73 v). This is the first known recognition of the appendix, and Leonardo gives it the function of accommodating excessive wind, much as it was thought that the auricles of the heart accommodated excessive blood.

Leonardo draws the gallbladder in Figure 16.6 (W 19039 r; K/P 61 r), where it is shown supplied by the hepatic artery, from which the gastroduodenal and its right gastro-epiploic branch passing to the greater curvature of the stomach and omentum are clearly indicated. The plane in which the omentum hangs down from the stomach in front and the mesocolon behind are recognised, though the actual

foramen of Winslow is not drawn. The 'separation' of the 'branny' particles of liver tissue left Leonardo with a magnificent picture of the vasculature of the liver and the biliary tree, as previously mentioned in Chapter 15 (see Figure 15.29, p. 324).

Next, still keeping to Mondino's instructions, Leonardo removed the intestines in order to expose the structure and pattern of the mesentery (Figure 16.7; W 19020 r; K/P 57 r). In the top drawing the superior mesenteric vein is shown joining the splenic vein to form the portal vein, here more accurately located than in the diagram below and to the

Figure 16.4. The contents of the abdomen (W 19039 v; K/P 61 v).

In this drawing the liver, stomach and spleen are labelled. Leonardo's concern to show exactly how the abdominal organs are placed is revealed in his note, 'Remember to show how high the stomach is above the umbilicus and close to the xiphisternum, and how close the spleen and heart are to the left breast'. The omentum is shown shrivelled back to the level of the umbilicus, as it is 'in the old'.

To the left is a study of the superior epigastric vessels anastomosing with the inferior epigastric vessels.

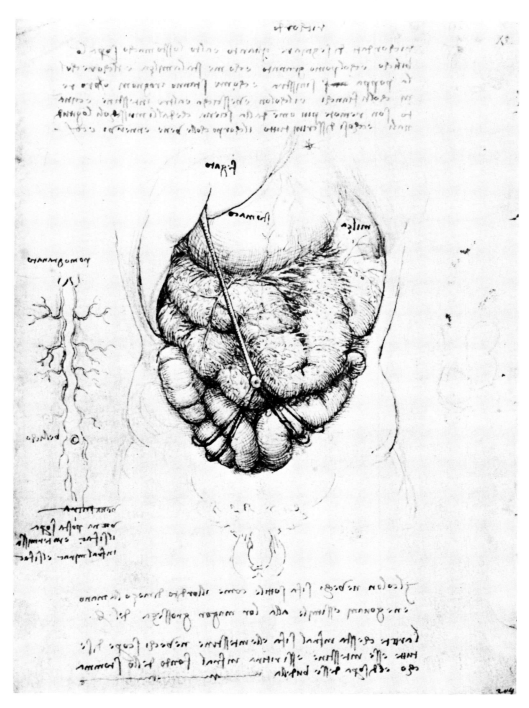

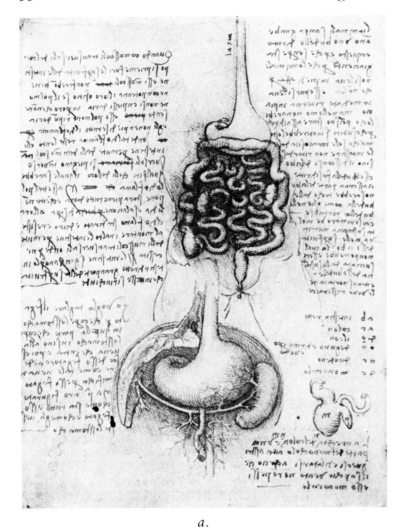

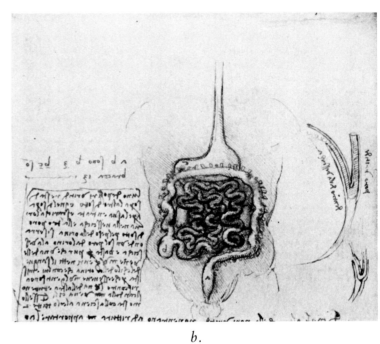

a.   *b.*

Figure 16.5

a. The stomach, intestines, liver, spleen, caecum and appendix (W 19031 v; K/P 73 v).

In the upper drawing the omentum has been removed, exposing the pattern of the intestines. Six different regions are lettered and listed in tabular form on the right margin, beginning with '*ab* straight intestine', i.e., rectum. The other letters are not easy to see even in the original drawings. They define colon, ileum, jejunum, duodenum and caecum. The caecum with its appendix is drawn again, enlarged, on the right. The appendix, labelled *n,* is described as 'adapted to contracting and dilating so that superfluous wind should not rupture the caecum'.

The lower central drawing depicts a large spleen, splenic and portal veins, and a crumbling liver as found in the 'old man' who died with cirrhosis of the liver.

b. Another drawing of the stomach and intestines (W 19031 r; K/P 73 r).

Here the duodenum takes its normal human path; the coils of small intestine are differently arranged; the caecum, *b,* has no appendix; the colon has appendices epiploicae. The kidneys can be seen lying snugly inside the ascending and descending colon. The text on this page is concerned with the passage of urine.

right, where in making his plan of the mesenteric vascular tree Leonardo keeps more closely to Mondino's erroneous description.

In the last of this series of dissections, Figure 16.8 (W 19028 v; K/P 70 v), Leonardo has removed the mesentery and exposed the retroperitoneal blood vessels and kidneys.

After a couple of false starts on the upper half of this page he finally reaches a beautifully accurate configuration in the lower right corner – accompanied once more by a diagram of the tortuosity of aged blood vessels. This same visceral and vascular pattern is transferred to the great double sheet, Figure 17.9 (p. 355), which Leonardo appears to have designed as a demonstration of all the abdominal viscera for teaching purposes. We shall return to it in Chapter 17.

This anatomical series constitutes Leonardo's main exploration of the alimentary tract. He supplemented it with further investigation of both its upper and lower ends, i.e., of the jaws, oesophagus and stomach at the upper end, and the rectum and anus at the lower end. With this knowledge of structure he embarked on a mechanical physiology of the alimentary tract which was very ingenious. Unforunately, it was also largely erroneous.

LEONARDO'S PHYSIOLOGY OF DIGESTION

Three factors have to be taken into consideration before submitting an account of Leonardo's physiology of digestion. First and most relevantly, Mondino, his main authority, nowhere mentions peristalsis in any part of the alimentary tract, and this in spite of Galen's description of this type of intestinal movement in his work *On the Natural Faculties*. Secondly, Leonardo eschewed vivisection in his physiological investigations. This method of investigation was later to become vital to the solution of many physiological problems, including that of the circulation of the blood. Leonardo's failure to use experimental vivisection provides

an example of his keen sensitivity to the suffering of living animals. This overcame his scientific curiosity. He continually expresses in his fables and tales the greatest sympathy for all living creatures; so much so that the thought of men eating animal flesh was repugnant to him. He sacrificed his science to this sensitivity.

The third important factor lay in Leonardo's own basic conception of the physiology of the 'microcosm' consisting essentially of the expression of the 'four powers' acting on the four elements. Ignorant of peristalsis, he was confronted with the problem of accounting for the movement, digestion and absorption of food products through the stomach and along the gastrointestinal tract by pressure gradients created by various combinations of these 'powers'.

He began well in his consideration of the process of mastication. In the action of the teeth and jaws Leonardo found a mechanical problem after his own heart. In Figure 2.24 (p. 64) Leonardo draws one of those beautiful skulls in which he concentrates all his skill in the art of perspective. The front half of the right side of the face is cut away to show that the orbital cavity and the maxillary sinus (one of his discoveries) are about the same size and depth 'and end under the *senso comune* in a perpendicular line'. His dissec-

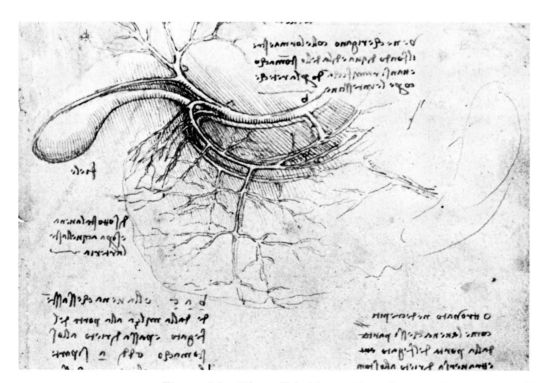
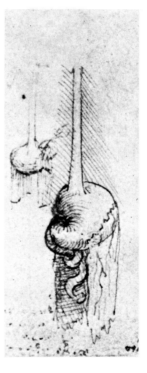

Figure 16.6. The gallbladder and hepatic artery (W 19039 r; K/P 61 r).

The hepatic artery is represented by *bac*. At *a* the gastroduodenal artery branches off. At *c* the cystic artery passes along the lower edge of the liver to the gallbladder. Thereafter the hepatic artery can be seen spreading out into the liver.

On the right the stomach is shown with the omentum hanging down like an apron in front, and the mesocolon behind.

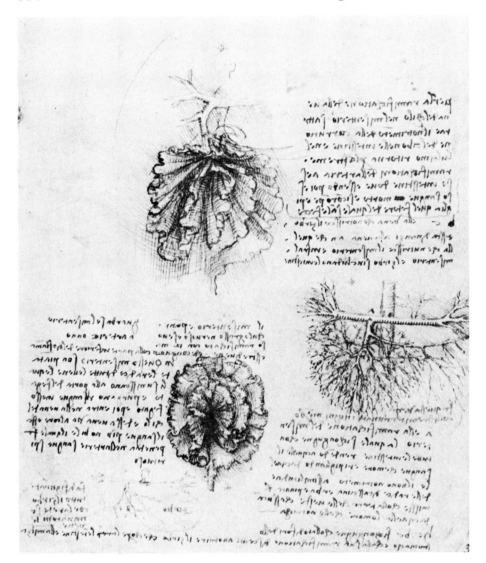

Figure 16.7.   The mesentery and portal venous system (W 19020 r; K/P 57 r).

The leaves of the mesentery are shown with the portal vein emerging to join the splenic vein in the upper drawing. To the right the vascular pattern is depicted in more detail. Leonardo writes, 'In this mesentery are planted the roots of all the veins which are united at the gate of the liver'.

tion of the jawbone leads him to study the different shapes of the teeth and their roots – the first accurate study of these features.

Having obtained knowledge of the movements of the lower jaw on the temperomandibular joint, Leonardo correlated the forms of the teeth with mechanical leverage under the heading, 'On the kinds of teeth and their position and distance from the fulcrum of their movements', in Figure 16.9. 'That tooth has less power in its bite which is more distance from the centre of its movement. Thus if the centre of movement of the teeth is at *a,* the fulcrum of the jaw [temperomandibular joint], I say that the more distant such teeth are from that centre *a,* by so much the less is their power of biting. Therefore *de* is less powerful in its bite than the teeth *bc*. From this follows the corollary which says, that tooth is the more powerful the closer it is to the centre of its movement or the fulcrum of its movement; that is to say, that the bite of the teeth *bc* is more powerful than that of the teeth *de*. Nature makes those teeth less able to penetrate the food and with blunter points which are more powerful. Therefore the teeth *bc* will have their points so much the

blunter as they are moved by greater power. Therefore the teeth *bc* will have their points so much the blunter in proportion to the teeth *de* as they are nearer to the fulcrum *a* of the jaws *ad* and *ae*. For this reason Nature has made the molars with large surfaces suitable for masticating the food and not for penetrating or cutting it up. And she has made the teeth in front for cutting and penetrating and not suitable for masticating food, and she has made the canines between the molars and the incisors' (W 19041 r; K/P 44 r).

The lips, tongue and pharnyx also received study in relation to the movements of food in the mouth and swallowing. The passage of a bolus of food over the epiglottis has already been described in relation to his study of voice production (see Figure 9.5, right margin, p. 220). Here the oesophagus is shown running down behind the trachea; here also Leonardo encounters the problem of finding the motive force which propels it.

In Figure 16.10 (W 19050; K/P 59 v) Leonardo makes a drawing of the vagus nerve descending beside the trachea and oesophagus, supplying both with twigs as it runs on to the surface of the stomach. Alongside on the right he illus-

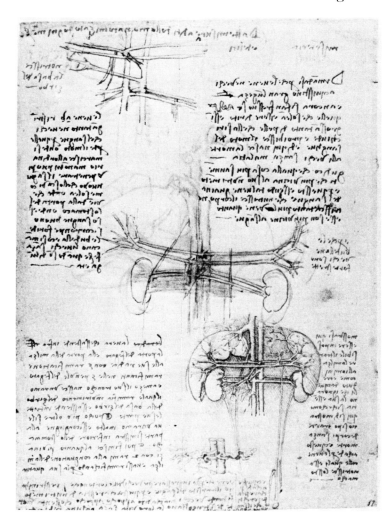

Figure 16.8. Abdominal blood vessels (W 19028 v; K/P 70 v).

The upper sketch of retroperitoneal blood vessels is exploratory. A much improved study succeeds it below. With the completion of his anatomy of the viscera in the abdomen Leonardo makes a drawing which includes the liver, spleen and kidneys.

Most of the notes on this page, however, are concerned with the little drawing of tortuous vessels at the bottom. He writes, 'Vessels which by the thickening of their coats in the old restrict the passage of blood; and through this lack of nourishment the old fade away, little by little in slow death'.

Figure 16.9. Leverage of the jaws and the kinds of teeth (W 19041 r; K/P 44 r).

Leverage was one of Leonardo's favourite examples of his pyramidal law. Here he draws a pyramidal diagram to illustrate the leverage of the jaws in biting, and the relation of the power of the bite to the position of the teeth on the lever of the jaw. The fulcrum of both upper and lower jaws is at the temperomandibular joint, *a*. The shapes of the teeth are drawn in Figure 2.24 (p. 64). The explanation of the diagram is written in the text alongside.

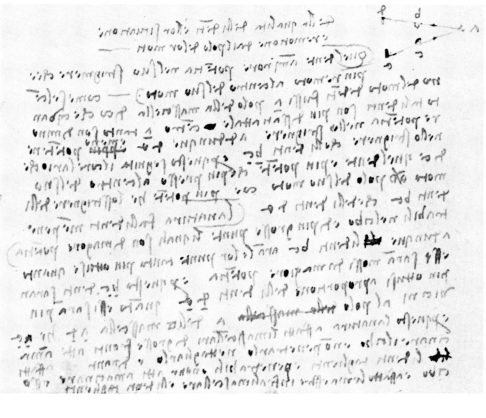

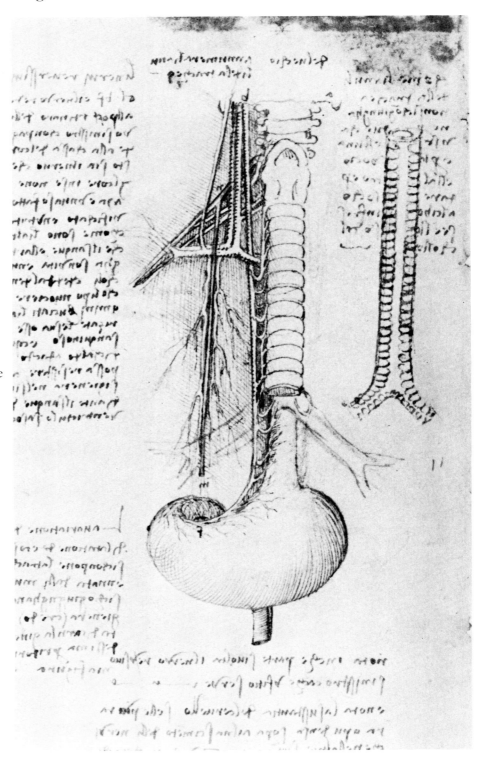

Figure 16.10. The vagus nerve supplying the trachea, oesophagus and stomach; the back of the trachea showing 'how the rings are not joined' (W 19050 v; K/P 59 v).

trates, 'How the rings of the trachea are not joined [at the back] for two reasons; the one is because of the voice and the other is to give room for the food between themselves and the bone of the neck'. Food, he suggests, is forced down the oesophagus by the muscles of the tongue and pharynx above. And dilatation of the oesophagus below to contain the food bolus is obtained through this gap in the posterior rings of the trachea. He suggests also that the anterior ver-

tebral muscles in the neck may squeeze the oesophagus so that food is pushed down into the stomach (W 19068 r; K/P 161 r). In spite of drawing nerve-twigs to the oesophagus he persists in not allotting to it any active muscular propulsive power. He adopts the same attitude to the stomach. 'If you should say that it is the longitudinal muscles of the stomach which are for drawing down the food, and the transverse muscles for retaining it, I shall reply that the whole intestine and everything that acts by dilatation and contraction has transverse and longitudinal fibres, as is seen in the weaving of cloth. And this is done in order that no transverse, longitudinal or oblique power shall be able to break or tear it' (W 19064 v; K/P 157 v).

The alternative motive force is described thus, 'The stomach does not move by itself in the expulsion of the chyle, but is moved by something else, that is by the flux and reflux of the spiritual motion which the diaphragm has in relation to the abdominal wall'. That is to say, when the abdominal wall draws in, the diaphragm relaxes, and when the abdominal wall relaxes the diaphragm contracts (W 19064 v; K/P 157 v). Thus we return to the concept of the diaphragm being 'the motor of the food and air', mentioned previously (see Chapter 14). It will be recalled that on this same page Leonardo allotted to the diaphragm 'four functions', 'The second of which was to press on the stomach which is covered by it, and to drive out of it digested food into the intestines. The 3rd, that it should squeeze and help compress the intestines together with the abdominal wall to drive out the superfluities'. He then goes on to explain that the expulsion of air from the lungs is not brought about by the active motion of the diaphragm but by expansion of air contained in the intestines previously condensed by the descent of the diaphragm at the same time as they are, in turn, compressed by the contraction of the abdominal wall. The lung above the diaphragm is 'composed of a very light substance and its passages stay open, hence air escapes from the lung with ease'.

This concept is discussed at great length. It is summarised on the following page (W 19065 v; K/P 158 v): 'The flux and reflux of the two powers created by the diaphragm and the abdominal wall are those which compel the stomach to produce intermittent expulsion, i.e., impetuses, separated by periods of rest at short intervals of time, during which food is driven out and then retained by the stomach. And it cannot return to the stomach because the pylorus is closed at the time of the flux made by the power reacquired from the diaphragm . . . And because the abdominal wall first contracts from below, it drives the wind from below upwards. Therefore the coils of the duodenum press on the base of the stomach and they compress one another in such a way as to close the pylorus'.

In Figure 16.5a (p. 332) Leonardo divides the alimentary tract into its main parts: oesophagus; stomach; rectum, marked *ab;* colon, marked *ac;* ileum, marked *do;* jejunum, 'so-called because it is empty', marked *on;* duodenum, marked *nr;* and monoculus (caecum), *dc.* Some of these letters are not easily seen.

For Leonardo the stomach was the organ of digestion, the process being completed there. The duodenum he most often draws curling round the back of the pylorus of the stomach, an appearance which is seen in the pig. It is also sometimes seen with right lateral shift of the human stomach filled with a barium meal. On two occasions Leonardo draws the more usual shape of the stomach and duodenum, once it is put in as an alternative form (W 19029 r; K/P 71 r),

and once, in Figure 16.5b (p. 332) it is clearly and unmistakably defined. In this particular drawing Leonardo also represents the appendices epiploicae of the colon, and we are told that 'the colon measures 3 *braccia*', i.e., 6 feet, and the small intestine '13 *braccia*', i.e., about 26 feet.

Leonardo allots no special function to the duodenum. He fails to illustrate its relation to the pancreas. The small intestine as a whole is allotted the function of absorbing the digested nutriment. 'Animals without legs', writes Leonardo, 'have a straight bowel . . . because they always stay horizontal, and because an animal does not rise upon its feet, since it does not have them, and even if it does rise up it immediately returns to the flat position. But in man this would not take place since he stands so erect, and so the stomach would empty itself immediately if the tortuosities of the intestines did not slow the descent of the food. And if the bowel were straight, each part of the food would not come into contact with the bowel as it does in tortuous intestines. And much of the nutritive substance would thus remain in the superfluities of the food which could not be sucked up by the substance of these intestines and transported in the mesaraic [portal] vein' (W 19031 r; K/P 73 v).

Nutriment so absorbed into the portal system is carried to the liver. 'The liver is the distributor and dispenser of vital nourishment to man. The bile is the familiar or servant of the liver which sweeps away and cleans up all the dirty superfluities of the food which remain after it has been distributed to the parts of the body by the liver' (W 19019 v; K/P 39 v). Leonardo here tacitly accepts the Galenic view that chyle absorbed from the intestines is converted into blood in the liver, whence it is distributed to the whole body. Yellow bile is an 'impurity' collected by the gallbladder. Black bile goes to the spleen. Aqueous excess is sent to the kidneys.

According to Leonardo all the heat of the body is created in the heart. Here he differed from Galen's opinion that the liver creates its own heat by coction within itself. For Leonardo his newly defined hepatic artery is a most suitable channel through which the heat of the heart is carried to the liver.

In a similar way Leonardo disagrees with 'coction' in the intestines arising there spontaneously. He finds the mesenteric arteries to be the conveyors of heat to them also, and in so doing they perform another most significant part in the processes of absorption and excretion. In a section 'On the nourishment that makes putrefaction', he writes, 'I say that the extremities of the mesenteric vessels which attract into themselves the substance of the food enclosed in the intestines, are enlarged by the natural heat of man, because heat separates and enlarges whilst cold congregates and constricts. But this would not be enough unless to this heat there were added the foetor formed by the corruption of the blood returned by the arteries to the intestines, which blood acts in the intestines not otherwise than it does in those buried in tombs. This foetor enlarges the viscera and penetrates into all their porosities, and swells and inflates bodies into the shape of casks' (W 19053 v; K/P 56 v).

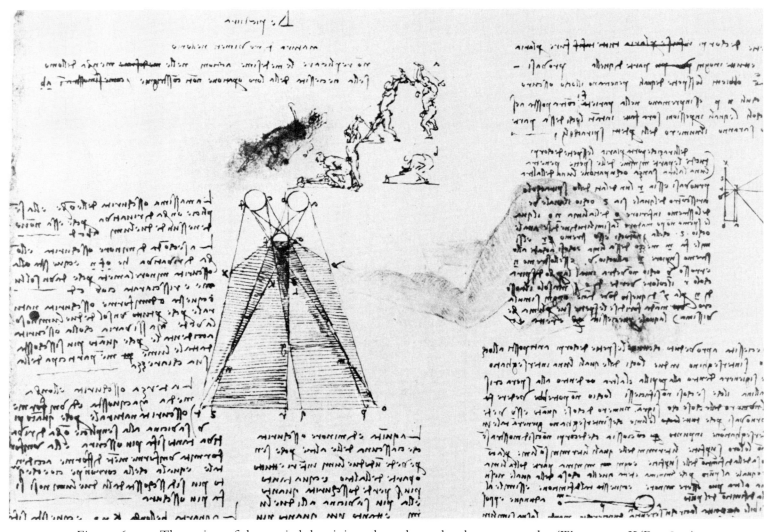

Figure 16.11.   The action of the recti abdominis and quadratus lumborum muscles (W 19149 v; K/P 118 va).

In the series of five drawings at the top of the page Leonardo depicts the most powerful action of 'the longitudinal muscles of the body which were made to flex the body forward'. The analysis commences with the drawings labelled *a,* where there is hyperextension of the man holding the hammer. This enables the flexors to exert their full range of power before finishing their percussion at *b.* This last drawing, *b,* should be compared with the large drawing obscured by script in the middle of the page. Compare with Figure 16.12.

The drawing beneath is a study of the 'degrees' of shadow produced by two lights, *ab* and *cd,* shining on a body *ef.* Compare with Figure 2.11. Figures on the right show how 'images intersect at points without occupying space'.

It will be appreciated that his physical theory of intestinal absorption serves a double purpose in Leonardo's physiology. Not only does it provide him with a theory of absorption of nutriment, but at the same time it explains the origin of faeces. Faecal excrement is formed not only from food residues but also from 'superfluous blood'. The amount of the blood in the body was a subject of great concern to Leonardo. He describes blood formed in the liver and passing through the heart as being 'of great weight' but never progresses to an attempt to measure this 'great weight', as did Harvey. 'Superfluous blood', as he calls it, is blood left in the vascular system after the tissues and organs have taken up all they need (W 19045 r; K/P 50 r; see Chapter 15).

This excess, with the breakdown products of tissues, he believes is transported by the mesenteric arteries to the intestine, where it is excreted in a putrefied form, creating flatus and faeces.

Thus Leonardo describes another form of circulation of the blood. Whereas previously he had thought of blood circulating by virtue of the heat of the sun drawing it upwards to be cooled and then falling back from the cold upper regions of the atmosphere or the cold extremities of man, he now describes a circulation whereby blood formed from the nutriment, absorbed by the intestine and perfected in the liver, circulates through the body. The 'superfluities' left over after its nutritional function has been performed are returned back to the intestine. The fundamental idea of flux and reflux underlay both concepts of such 'circulations'. His basic idea was sound, but both of Leonardo's attempts to describe its details in human physiology were wrong.

The expulsion of the 'superfluities' from the body as excrement was a problem which the sensitive Leonardo did not shirk. As might be expected, he accounts for the expulsion of wind and stools by the production of a pressure gradient. This is the result of two processes: (1) the contraction of muscles surrounding the abdominal cavity; and (2) increased pressure within the bowel itself.

According to Mondino, the longitudinal muscles of the abdominal wall, the chief of which is the rectus abdominis, by their contraction raise intra-abdominal pressure. This Leonardo cogently denies. In a paragraph entitled, 'For what use were the longitudinal muscles of the body made', Leonardo writes, 'The longitudinal muscles of the body [e.g., recti abdominis] were made to flex the body forwards with unalterable power in opposition to the extension of the body [see Figure 16.11]. Their cooperating muscles [synergists] are the lumbar muscles [quadratus lumborum] which lie on the internal side of the spine, and their antagonists are the greater muscles of the backbone' (W 19086 v; K/P 178 v).

On the same page, Figure 16.12, Leonardo asserts, 'The longitudinal muscles are not used in the expultion made from the intestines because the more a man bends himself forwards the more these muscles relax in such a way that they diminish by half their length, as is seen to be done by

Figure 16.12. The act of defaecation (W 19086 v; K/P 178 v).

In the three sketches at the top of the page Leonardo illustrates how in the act of evacuation of the intestines the recti abdominis cannot be used since they are relaxed, a state shown by their wavy folds. In the other drawing he illustrates the quadratus lumborum muscle, labelled 'lonbi', showing how it assists the rectus abdominis in flexing the spine, i.e., in bending forward, as in percussing with a hammer. Compare with Figure 16.11.

those who want to generate great power in this expulsion [see the accompanying drawings at the top of the page]. These people shorten by half these muscles which are compelled to relax and so remain unused without any strength. Therefore these muscles will not exert such a force'.

On the contrary, 'The transverse muscles compress the intestines, but not the longitudinal, because if it were so a man who bends down and relaxes these muscles would not have the power to perform the office of compressing them. But the transverse muscles themselves never relax when a man bends down but rather they contract' (W 19032 r; K/P 74 r; see Figure 16.3, p. 330). He describes the relation between the rectus abdominis and transverse abdominal muscles in the upper drawing of Figure 16.3, saying, '*ab* [the transversus abdominis] are the innermost [layer] of latitudinal muscles and the membranes, into which they are converted, pass at right angles under the longitudinal muscles, *nm* [rectus abdominis]'.

In the lower drawing Leonardo has removed the two recti abdominis muscles, exposing the membranous sheath (*ab*) which covers the transverse abdominal muscles. He describes the situation thus: '*ab* is all gristle which with the sifac [peritoneum] arises from the fleshy muscles *cd* [transversus abdominis]. These muscles enter under the ribs and are the latitudinal muscles, and they arise from the bone of the spine and they alone are those which press out the superfluities of the body'. He later adds to this note in small writing, as if it were an afterthought, 'The transverse muscles *cd* are those which on contraction, squeeze and raise the intestines and push up the diaphragm and drive the wind out of the lung; afterwards then on relaxation of these muscles, the bowels descend, the diaphragm contracts back and the lung is opened'.

In another drawing (Figure 13.8, p. 277) of these transverse abdominal muscles Leonardo explains a feature which is constantly found in all his drawings of the trunk of the human body, saying, 'The transverse muscles are those which cause the compression of the intestines, and the proof of this is, that when the intestines are compressed from such a cause, there are corners of soft fleshy swellings in the flanks'. As an example, such a swelling is very evident, lettered *m,* in Figure 13.8 where Leonardo writes, 'The bulging flank at *m* is of skin and thin flesh, but it is made to project by the colon on the left and by the caecum on the right'.

As we have already seen, Leonardo at one time believed that the diaphragm played a prominent part in increasing intra-abdominal pressure and so expelling the stool. In Figure 16.13, in a passage accompanied by a drawing containing the transverse abdominal muscle alone, Leonardo modifies this view: 'Demonstration how the diaphragm is not used in squeezing out the superfluities of the intestines; breathing clearly not being prevented during the time that superfluities of the intestines are being expelled' (W 19110 r; K/P 191 r). In the drawing alongside he shows the right side of the body bent forwards with relaxed anterior abdominal wall. The diaphragm is still raised into a slight dome, but the transverse abdominal muscle fans out as it comes forward from the spinal region to meet its fellow on the opposite side. The caecum and colon can both be clearly seen beneath it. On the following page he further explains this conclusion: 'If the diaphragm pushed and closed the intestines between itself and the longitudinal muscles, and compressed these, it would follow that the diaphragm would itself follow the motion of descent of the intestines', and this it does not do. Thus he reaches the conclusion that 'The diaphragm does not always augment the expulsive power of the intestines because, although the lung has been emptied of its air, this expulsion does not usually fail to function' (W 19109 r; K/P 186 r).

The second motive force performing this 'expulsive function' is that of the wind contained in the intestine. Of its presence Leonardo cites plenty of evidence. He has noted that 'the duodenum and jejunum, being the first to be occupied by the softer material, are those which empty first, being nearer to the stomach . . . thus the intestine called jejunum is always full of wind' (W 19066 v; K/P 159 v). He has noted, too, that the colon contains wind, even suggesting that the function of the appendix is 'to contract and dilate so that excessive wind does not rupture the caecum' (W 19031 v; K/P 73 v). He has also commented on the production of foetid air from the putrefaction of the superfluous blood excreted into the gut. Thus if the diaphragm remains taut and hard and resists the upward ascent of the intestines compressed by the transverse muscles, their content of wind and faeces will be expelled by a mechanical action similar to that of an air-gun. Air condensed by movement against resistance, as by a piston in a cylinder, increases in pressure. When such air swiftly emerges and percusses the air outside, whether it be from a bombard, a

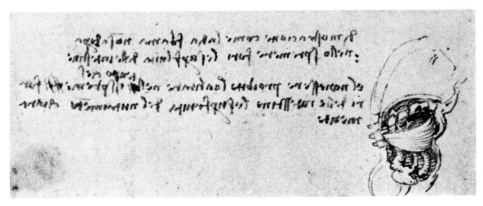

Figure 16.13. The transverse muscle of the abdomen evacuates the intestines (W 19110 r; K/P 191 r).

This is shown fanning forward from left to right. The rectus abdominis is relaxed in waves and the diaphragm is relaxed in its curved dome.

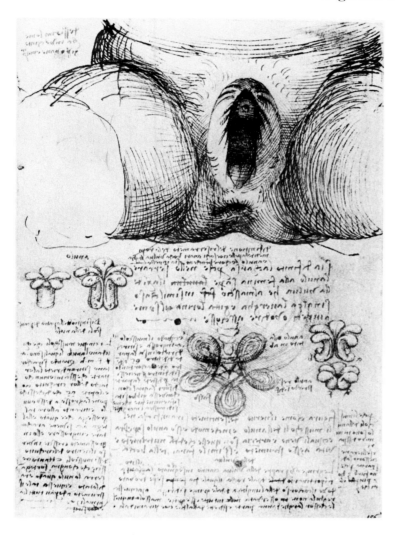

Figure 16.14. Sphincter mechanism (W 19095 r; K/P 54 r).

Beneath the drawing of the vulva and anal sphincter Leonardo draws five sketches of the sphincter mechanism. The largest in the centre of the page he himself labels 'false'. The other two pairs of drawings show the sphincter open and closed. The drawing on the left is lettered and the passage quoted in the text applies to this. Using his general rule that when a muscle contracts or shortens it increases proportionally in width, he asserts that the muscle *oc* shortens, and 'its enlargement increases towards the centre of this circle of muscles . . . so that it shuts the anus'.

bursting air bubble or a pop-gun (see Figure 9.1, p. 216), it makes a sound, 'By disintegration of condensed air' (A 31 v and L 89 v). Thus the ejection of human faeces is often accompanied by noise.

For such 'condensation' of air to build up pressure, and for the prevention of faecal leak or incontinence, there must be, says Leonardo, something that closes the anus—a sphincter. What is this sphincter mechanism? He discusses this problem in relation to the anus in Figure 16.14 (W 19095 r; K/P 54 r), and the sphincter mechanism in general in W 19117 r (K/P 115 r). He broaches this problem by a rather remarkable statement, 'Let the cause be defined why

in the closure of the anus in the female the lips of the vulva open. And in the male in similar circumstances the penis is erect and ejects urine or sperm with impetuses, or as one may say, in spurts'. This observation is written in Figure 16.14 beneath a drawing of the external female genitalia in which the vulva are shown wide open and the anus closed, its site being denoted by converging 'wrinkles'. Beneath this drawing interspersed among a number of notes is a variety of geometrical figures all purporting to show possible mechanisms of the anal sphincter. Leonardo explains these flower-like patterns in the text beneath the lettered drawing. 'Definition of the action of the muscles of the anus', is his heading. 'The five muscles which close the anus are *adfmn*. When they shorten they draw back the part which lies in common circular contact, e.g., the part *oc* which composes the thickness of the anus. Then by pulling on this thickness which is equal to distance *oc,* it shortens and widens, and such enlargement increases towards the centre of this circle of muscles to such a great extent that it closes the dilated anus with great force. And all animals employ such an instrument. When muscle *a* swells up it pulls behind it the inner part *oc*. Thus when this is shortened internally the outside part is necessarily distended. This projects with a convex prominence, as is demonstrated in the margin'. The contrasting shapes of the five anal muscles with the anus 'open' and 'shut' are, however, better shown in the two drawings on the right-hand margin.

Written immediately beneath the large top drawing in Figure 16.14 Leonardo reveals how his thought proceeds from this particular instance to the general problem of sphincter action. Here he writes, 'Definition of the closing of six gaps in the skin; that is the eyes, nostrils, mouth, vulva, penis and anus – and the heart, although this is not in the skin'. He generalises from the vulva as follows, 'The wrinkles of cracks in the vulva have taught us the position of the gate-keeper [sphincter] of the citadel which is always found at the confluence towards which the length of the wrinkles is directed. But this rule is not observed in all wrinkles, but only in those which are broad at one end and narrow at the other, that is pyramidal'.

This subject is taken up again on the page where he is describing the ejection of blood from the heart into the aorta by impetus at each percussion of systole (W 19116 r; K/P 115 r). Here, too, he refers to investigations on the motor nerves of the voice, of the eyes, and 'On the nerves which open and close the vessels or gates of the spermatic ventricles; on the nerves or one should say, muscles, which close the gate of the bladder; on the nerves and muscles which eject sperm with such violence; on the muscles which close the anus; and on the muscle called "vermis" which lies in one of the ventricles of the brain'. 'All the said sphincters', he asserts, 'are opened by the object which issues out of that place closed by them; as the anus is by the superfluities of food, and are then closed by the actions of these muscles. The gates of the spermatic ventricles do likewise, these are opened by the impetus of the compressed sperm and then closed again by their muscles. Furthermore, the urine does the same thing at the gateway of the bladder, that is to say the power of the compressed urine opens this gate and its particular muscles are those which shut it again. And the same is found at the mouth of the penis, of the vulva and of the womb and of all those parts which receive necessary things and expel the superfluous'. It will be noticed that Leonardo omits from this list a sphincter which he studied in considerable detail, the pyloric sphincter of the stomach.

NUTRITION

'Our life is made through the death of others: in dead things life deprived of sense remains, which reunited to the stomachs of living beings resumes sensitive and intellectual life' (H 89 v).

Leonardo's studies of the distribution of blood to the whole body led him to see in this example of his 'rule' that 'The movement of a liquid made in any direction proceeds in its original wave as far as the impetus given it by its prime mover lives in it' (W 19045 r; K/P 50 r). The impetus of blood ejected through the aorta is divided; part is reflected into the aortic 'hemicycles' and whirls round in such a way as to close the aortic valve cusps (see Chapter 15), part carries on through the blood in the aorta and out to the peripheral vascular tree. In both cases, in Leonardo's view, its impetus 'is slowly consumed' and in the end ceases. Thus blood is brought to a standstill by the resistance it meets with in the peripheral organs and tissues; and there the tissues take what they need for nourishment and rid themselves of the dead residue.

As we have seen, 'The body of anything which is nourished, continually dies and is continually reborn, for nourishment cannot enter except into those places where past nourishment has been exhausted; and if it has been exhausted it no longer has life' (W 19045 r; K/P 50 r). He then points out that if this nourishment is insufficient 'then life fails', and if you supply just as much nourishment as is consumed, 'then as much life is reborn as is consumed', and he likens life to the renewal of the flame of a candle (see Chapter 15). 'Equal nourishment or equal cause generates equal effects' (G 17 r) is the 'rule' behind this.

Although the first part of this passage applies to the heart and its movement of blood through the arteries and veins, he ends by saying, 'But to return to our intention, I say that the flesh of animals is being continually re-made from the blood which is continually generated for their nourishment; and that this flesh is destroyed and returns through the meseraic [mesenteric] arteries and passes into the intestines where it putrefies in foul and foetid death, as their expulsions and steams show us, like the smoke and fire given as a comparison' (W 19045 r; K/P 50 r).

On this basis Leonardo accounts in general for healthy maintenance of body weight; increase of weight and size with growth; and wasting with disease or starvation. All food, he emphasises, has once been alive. Dead bodies which have had life pass into the bodies of living animals (CA 145 ra). But he goes further with regard to two dietary components, salts and fat.

THE CONSERVATION OF SALTS

It will be noticed in the preceding quotation he refers to expulsions (the ashes) of animal excreta. We have glanced at Leonardo's concept of '*flusso e reflusso*', of nutrition from the dead body of animals into live animals when flesh and superfluous blood return back to dead, senseless earth and water and are consumed again – a kind of circulation of matter between live and dead forms. This he applied to 'salts' and water as well as 'flesh'.

Leonardo appreciated that 'the quantity of urine shows the quantity of blood that is produced and goes to the kidneys but is first [in part] obliged to pass through the valves of the heart' (W 19069 r; K/P 199 r). He is also aware that 'Salt may be made from human stools burnt and calcined, made into lees and dried slowly at a fire [i.e., evaporated to dryness]. And all the excrements produce salt in a similar way, and these salts when distilled are very strong' (W 12351 v). Later he emphasises, 'Salt is in all created things; we may learn this from passing water through ashes and the residue of things that have been burnt, and from the urine of animals, and the superfluities which issue from their bodies, and the earths into which by corruption all things are changed'. He adds this penetrating comment, 'The human race has perpetually been and will be consumers of salt; and if the whole mass of the earth were composed of salt it would not suffice for human food. For this reason we are forced to conclude either that the forms [*spetie*] of salt are everlasting or that the salt dies and is reborn together with the men who consume it. Since experience teaches us that it does not die, as is shown from the fact that fire does not consume it, and from water becoming more salt in proportion as it is dissolved in it, and from the fact that when water evaporates the original quantity of salt always remains; there must needs pass through human bodies as urine, perspiration or other superfluity as much salt as is brought every year into the cities' (G 48 v and 49 r).

This awareness of the conservation of salt and other forms of flesh which pass through human bodies by 'flowing' in and 'flowing' out as waste is part of Leonardo's analogy between the macrocosm of the world and the microcosm of man.

We see this idea again expressed in relation to water. 'There are many rivers through which the whole element [of water] has passed and returned to the sea'. 'It [water] is transmuted into as many natures as are the different places through which it passes . . . health-giving, harmful, laxative, astringent, sulphurous, salt, sanguineous . . . Now it is the cause of life and then of death, now it nourishes and then does the contrary, now it is salt and then without saltness . . . With time everything is transmuted' (BM 57 r). Thus earth, air, fire and water are transmuted in passing through man's body.

FAT DISTRIBUTION IN MAN

Leonardo's concept of nutrition and metabolic 'transmutation' arose from the geometrical maturity derived from his friend Luca Pacioli. His interest in fat distribution had quite different origins. It arose from his artistic concern with the shapes and forms of infants and is first voiced about 1492.

In comparing parts of the body at different ages he raises the problem of 'How little children have their joints opposite to those of men, in their size. Little children have all their joints thin whilst the spaces between one joint and another are thick. This happens because the joints are only covered by skin without any other flesh. And this skin has the nature of sinew which encircles and binds together the bones, and the fat fleshiness is found between one joint and the other, enclosed between the skin and bone. But because the bones are bigger at the joints than between them, the flesh as a man grows up, loses that superfluity which existed between the skin and the bone, whence the skin is drawn more closely to the bone and the limbs become more slender. But since there is nothing over the joints but cartilaginous and sinewy skin this cannot dry up, and not drying up cannot diminish. So, for this reason children are slender at the joints and fat between the joints, as is seen in the joints of the fingers, arms and shoulders which are slender and dimpled, while in man on the contrary all the joints of the fingers, arms, and legs are big; and wherever children have hollows men have protrusions' (BN 2038 28 v).

With growth from infancy into youth the pattern of limbs changes: 'In the flower of youth the skin is drawn as tight as possible provided the body is not fat or corpulent, and has reached its full height. Later through the exercise of the limbs the skin increases over the articulations of the

joints, and so when the limbs are extended the skin grown over the joints becomes wrinkled' (CU 118 r).

'Muscular men have thick bones and are short, thick, and lack fat, for the fleshy muscles because of their growth are drawn together and there is no room for the fat which would otherwise be interposed between them' (CU 116 r). 'Although fat men are short and thick in themselves like the muscular men mentioned above, they have slender muscles but their skin covers a great deal of fat that is spongy, empty or full of air; and so fat men are supported better by water than muscular men whose skins are full and within whom there is less air' (CU 116 rv).

The relation between the distribution of fat and muscular power claims Leonardo's attention, particularly in relation to the formation and maintenance of the 'wedge shape' assumed by contracting muscle bellies. This has been described in Chapter 13. Here Leonardo is quoted saying that 'Those who are inclined to be fat increase a good deal in force after their early youth'. Here again he emphasises also the importance to muscular strength of a 'tight skin' (CU 117 v). This inverse relation between fatty infiltration and the size and strength of muscles is suggested in his analysis of the nature of fat. In describing how fat fills the intervals between muscle-bellies, thus altering the curvature of the intervening cutaneous hollows, Leonardo points out that 'So that the skin cannot descend into such an angle Nature has filled it with a small quantity of spongy or viscous fat, or you might say with minute vesicles full of air which are condensed or rarefied according to the growth or rarefaction of the substance of the muscles' (Figure 16.15).

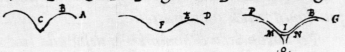

Figure 16.15. The distribution of fat (CU 117 r).

Leonardo writes, 'Of the muscles of animals. The concavities between muscles ought not to be such that the skin looks as if it covered . . . two rods separated a little from such contact. With the skin hanging in an empty space with a broad curvature as at *F* it should be like *I* placed over the spongy fat between angles such as the angle *mno* which originates from the edges of the contact of the muscles'.

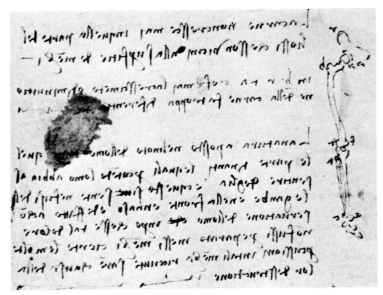

Figure 16.16.   The distribution of fat (W 19141 r; K/P 99 r).

Opposite this little skeletal diagram Leonardo writes, 'At *b r d a c e f* increase of flesh [fat] never makes much difference'.

These are all places where bony prominences exist throughout adult life.

In his dissection of the human body Leonardo became keenly aware of the value of finding a lean body without much fat. His comments about the body of the 'old man' on which he performed an anatomy 'with great ease because of the absence of fat' confirm this. And it was about this same time that he wrote, 'On the human figure' thus: 'Which part of a man is that where in getting fat never increases its flesh? Which is that part which in the wasting of a man never wastes too obviously? Among the parts which fatten, which fatten most? Among the parts which waste, which wastes most? In powerful men of force which muscles are of greatest size and most prominent?' And on the opposite half of the page under the heading, 'On painting', he asks, 'Which muscles are divided up in old age or in youth waste away? Which are the places in human limbs where the flesh never increases whatever the fatness, nor does the flesh decrease on account of any degree of leanness? What is sought for in this question is to be found in all the superficial joints of the bones, such as the shoulder, elbow, knuckles of the hands and fingers, knees, ankles and toes and similar parts which will be mentioned in their places' (Figure 16.16).

Leonardo usually answers his own questions, and frequent references to the distribution of fat do occur in his anatomical notes. For example, in Figure 16.4 (p. 331) he notes that 'The net [omentum] which lies between the sifac [peritoneum] and the intestines in the old uncovers all the intestines and is drawn back between the bottom of the stomach and the upper bowels'. And so it is drawn in Figure 16.4. And he writes at the top of the page containing the very muscular figure of a man, 'Note how the flesh increases on the bones in growing fat and how it diminishes as one becomes lean (W 19032 v; K/P 74 v).

Leonardo notes the distribution of fat between muscles. Under the heading, 'Nature of muscles', he writes, 'In proportion to the greater or lesser lengths of the tendons of muscles so will a man be fatter or leaner in flesh. The flesh in a lean man is always drawn back towards its origin from its fleshy part. And acquired fat spreads towards the origin of the tendon'. The practical significance of leanness to his anatomy is made very evident here by Leonardo's remark, 'Make a demonstration here with lean and thin muscles in order that the space which arises between one and another makes a window to demonstrate what is found behind them; as in this figure of a shoulder made here with charcoal' (W 19014 r; K/P 148 r).

In the following injunction he is bearing both artistic and scientific considerations in mind. 'See how the muscles in the old and lean cover or clothe their bones; and besides this note the rule as to how these same muscles fill up the superficial spaces which are interposed between them; and observe those muscles whose contours are lost to sight from the very least fatness. And in many cases several muscles make one single muscle in getting fat; and in many cases in growing lean or old one single muscle divides itself into several muscles . . . And do the same in a child from its birth up to the time of its decrepitude, through all the degrees of its age . . . And in all these you will describe the mutations in the limbs and joints which get fat or lean' (E 19 v and 20 r).

Leonardo expresses his view of the transmutation and conservation of the matter of the 'elements' in its continual flowing through the macrocosm and microcosm. And since for him 'natural points' or atoms are composed of 'mathematical points' of incorporeal energy, or 'spirit', he sees also a conservation of energy in the physical world. The body of man, as of other material things, is for him but a transient form of geometry, as transient a form as a candle flame; 'it continually dies and is reborn'. 'Dead things reunited to the stomachs of living things resume sensitive and intellectual life'. 'Time transmutes everything'.

This concept of man's nutrition and metabolism, the building up of live bodies and their destruction, is based on Leonardo's geometry of transmutation, applied to the four 'elements' and 'powers' on which he founds his approach to both physical and biological phenomena. In the case of man it is reflected in his persistent theme of '*flusso e reflusso*' which in modern terms can justifiably be translated into anabolic and katabolic phases of metabolism, with 'sensitive and intellectual life' appearing at the peak of their transmutation.

REFERENCES

1. Aristotle. *On the Parts of Animals,* trans. by W. Ogle. London, Kegan Paul Trench Co., 1882. Vol. III, p. 14.

Chapter 17

The Urinary and Reproductive Systems

Leonardo gave relatively little attention to the urinary tract. As with his exploration of the alimentary tract, he followed Mondino's instructions for anatomical procedure and thereby with his drawings transformed Mondino's confused verbal description into a far more intelligible visual form, designed to 'enter by the eyes'.

Again as with the alimentary tract, neither Mondino nor Leonardo suspected that urine is conveyed from the kidney to the bladder by peristaltic movement. Mondino says that the urine is filtered from the blood and 'drips down' the ureters into the kidney. Leonardo cannot accept this from his experience of experiments with water and suggests an alternative hydrodynamic mechanism.

ANATOMY OF THE URINARY TRACT

We have already seen Leonardo's exposure of the whole urinary tract in Figure 15.7 (p. 305). Here Mondino's instructions are followed literally. The left kidney is shown lower than the right because the spleen on the left, says Mondino, 'is lower', and Leonardo here draws the renal veins, omitting the arteries; the whole urinary tract with the spermatic blood vessels running to the testes is presented as Mondino advocates.

Then, as so often happens, Leonardo departs from his authority, adding the ductus deferens and seminal vesicles. On the attached sheet (W 19098 v; K/P 106 v; see Figure 17.1) he demonstrates many directly observed anatomical details of the pelvic organs from the left side; the spermatic vessels descending to the epididymis, and the rising curve of the ductus deferens over the side of the bladder to expand into the seminal vesicles, are again dramatically shown. The bladder (probably artificially inflated with air or water) rises up in the pelvis. Its supply of blood vessels is clearly shown crossing the tube of the rectum which lies snugly ensconced in the curve of the sacrum. From the seminal vesicles the ejaculatory ducts can be seen running forward to empty into the single-channelled urethra, just distal to the neck of the bladder.

Further progress is depicted on W 19054 r; K/P 53 r (Figure 17.2) under the heading, 'Demonstration of the bladder of man'. The progressive aspect of these illustrations of the bladder lies in the depiction of the ureter in the two drawings on the left. The upper shows their site of entry into the bladder wall; the little diagram shows in imaginative detail the ureter penetraing obliquely, forming a small valvular flap, *s*. The middle and right-hand drawings represent views of the blood vessels of the bladder. The small artery encircling the entry of the ureter at *n* in the right-hand drawing is a delicate piece of anatomy that could be depicted only after direct observation. Yet the whole pattern of the vascular supply of the bladder is founded on an obscure and erroneous passage in Mondino's *Anatomy*. Here Mondino refers to two sets of vasa spermatica, one to the epididymis is approximately correct and was verified by Leonardo. But another ascends into the bladder wall becoming more fibrous as it ascends. It would appear that these 'vessels' are confused with the ductus deferens itself or with the branch of the superior vesical artery which follows the ductus deferens and commonly anastomoses with the testicular artery. Thus, in Leonardo's drawing he has made the arteries and veins of the bladder appear to ascend from below supplying a network of vessels in the bladder wall.

Leonardo's best drawing of the kidney, Figure 17.3 (W 19030 v; K/P 72 v), is inscribed with the instruction, 'Cut it through the middle and depict how the passages of the urine are closed and how they are distilled'. The sectioning of the kidney is recommended by Mondino, who, like Leonardo, saw the kidney as a 'purifying sieve' to take off 'aqueous superfluities'. Leonardo uses the term *distillation* of urine, looking upon its formation as similar to that of sweat.

THE PHYSIOLOGY OF URINARY EXCRETION INTO THE BLADDER

The Galenic concept of the oblique entry of the ureter through the bladder wall forming a valve mechanism by which regurgitation of urine is prevented was accepted by Avicenna and Mondino. Leonardo does not accept it. His account is based on his hydrodynamic experiments. He writes, 'Entrance of urine into the bladder. Urine after leaving the kidneys enters the channel of the ureters and from them passes into the bladder near the middle of its height. It enters the bladder through a small perforation made trans-

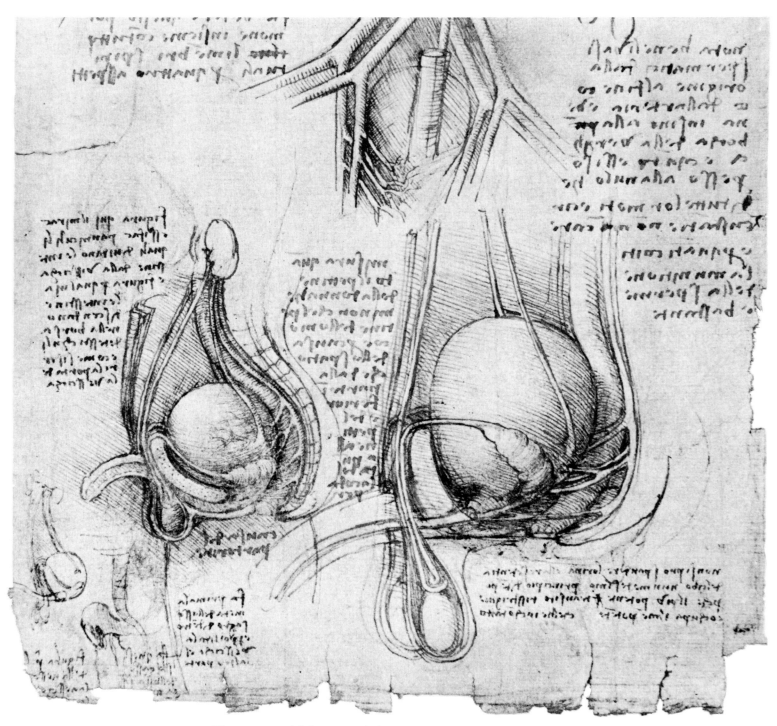

Figure 17.1.   Male urogenital organs (W 19098 v; K/P 106 v).

Amongst the many striking features of these drawings is Leonardo's constant representation of the bladder always as full. In the passage on the left margin he explains the neighbouring drawing as follows: 'Draw here the abdominal wall and peritoneal membranes which divide the intestines from the bladder. And draw the route along which the intestines descend into the purse of the testicles [the scrotum]. And how the gate of the bladder is shut'.

Evidently he looked upon the presence of a hernia as commonplace.

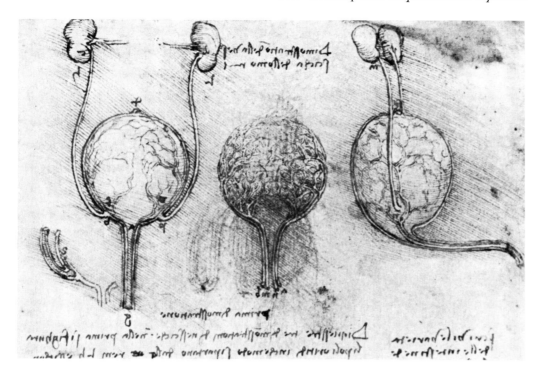

Figure 17.2.   Kidneys, ureter, urethra, and vascular supply of the bladder (W 19054 r; K/P 53 r).

The entrance of the ureter into the bladder was of particular interest to Leonardo. He writes, 'These ureters pour urine into the bladder from *pb* into *nf* in the way drawn at the side in channel *s* whence it is then poured through the pipe of the penis', shown on the left. And he draws special attention to 'the way the veins and artery go round the origin of the ureter, *mn*, at the position *n*', in the right-hand drawing.

Figure 17.3.   The kidney and its blood supply (W 19030 v; K/P 72 v).

This excellent drawing of the kidney underwent the fate of being obscured by script on an entirely different subject – a comparison between the acuity of the senses of man and animals. The only relevant comment is written within the outline of the kidney itself and suggests making a section of the kidney. None is extant.

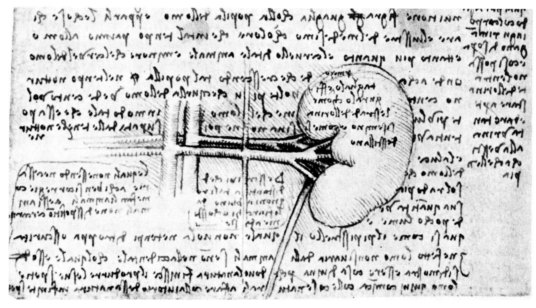

versely between one coat and the other. This oblique perforation was not made because Nature was doubtful whether the urine could return to the kidneys, for this is impossible from the 4th [Book] on conduits, where it says, "water which descends from a height through a narrow vessel and penetrates down to the bottom of a lake cannot be opposed by reflected movement unless the water of the lake is equal in size to that in the vessel which descends; nor if the height of the water is greater than the depth of the lake". And if you should say that the more the bladder fills the more it closes, to this I should reply that such perforations being closed by the urine which closes up the walls

would prevent the entrance of other urine which descends, which cannot happen according to the 4th Book mentioned above, which says that the narrow, high [column of] urine is more powerful than the low wide which lies in the bladder' (W 19054 r; K/P 53 r). No Books on conduits now exist.

A number of experiments on water in pipes (conduits) relevant to this question are to be found in *Madrid Codex* I. In *Madrid Codex* I 113 v Leonardo states, 'No quantity of water can be raised as high as was its descent. This applies to water conducted in closed channels where no air is enclosed.' This statement is based on a series of complex experiments described on Madrid I 115 r (Figure 17.4) 'On the nature of movement of water'. Here the proportion of the height of a column of water is related to its width and pressure. See also Figures 4.35 and 4.36, pp. 120, 121.

Leonardo amplifies his objection to the traditional valvular theory in W 19031 r (K/P 73 r, Figure 17.5). This time he asserts that with such closure of the valve the urine in the bladder 'would never exceed half the capacity of the bladder'. He then discusses at length the consequences to be derived from his 6th Book on waters, and the influences of various positions of the body in relation to urine entering the bladder. 'If a man lies down it can turn back through the ureter and even more so if he should be upside down'. 'If a man lies on his side one ureter remains above, the other below, and the entrance of that above opens and discharges urine into the bladder, and the other opening below is closed by the weight of the urine'. If a man lies in the prone position, 'since the ducts are attached to the posterior part of the bladder . . . the entrances of the urine remain open and give the bladder as much urine as will fill it'.

Ureteric reflux has of recent years assumed ever-increasing importance. Studies of the relation of posture to the occurrence of this phenomenon inaugurated here by Leonardo are perhaps deserving of more recognition than they have received.

The passage of urine along the urethra presented no problem to Leonardo. With a full bladder urine flows out according to hydrostatic pressure. This is illustrated in Figure 5.11 (p. 140). The problem that remained was 'the closure of the gateway'. This he 'resolved' in his discussion of sphincter action (see Figure 16.14, p. 341).

THE REPRODUCTIVE ORGANS

EARLY INVESTIGATIONS (1489–1494)

'Man does not have the power of creating any simple [elementary] thing except another like himself, that is his children' (W 19045 v; K/P 50 v). With Leonardo's intense interest in creation in all its forms, one is not surprised to find him giving a good deal of attention to the creative, reproductive parts of man's and woman's anatomy. His earliest exploration into this field is illustrated by the coitus figure, Figure 11.2 (p. 244). This constitutes an example of Leonardo's inexorable logic in carrying out the programme he set himself on, 'The order of the book'. 'This work', he writes, 'should begin with the conception of man, and should describe the form of the womb, and how the child lives in it, and to what stage it resides in it, and in what way it is given life and food, and its growth, And what interval there is between one degree of growth and another, and what it is that pushes it out of the body of the mother' (W 19037 v; K/P 81 v).

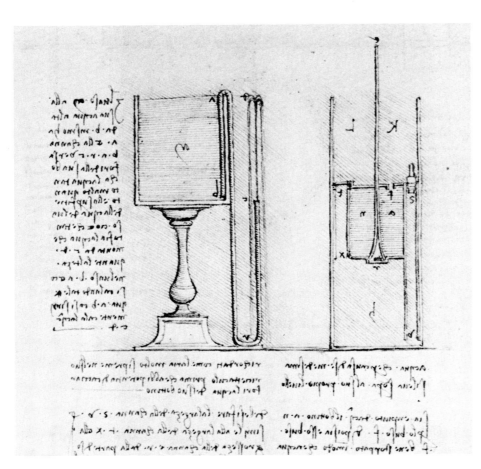

Figure 17.4. Experiments 'on the nature of movement of water' (Madrid I 115 r).

On the left Leonardo is concerned amongst other things to demonstrate that 'water rises as high at *cd* as is the height in the vessel *ba*', and not higher.

On the right he notes that 'air must be greatly compressed in its receptacle [*sv*] before it acquires the power of driving water out of its container', i.e., up the narrow spout *re*.

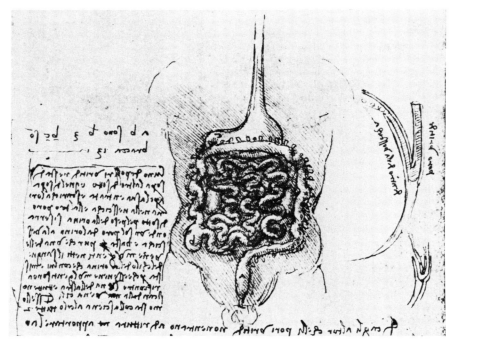

Figure 17.5. How urine enters the bladder (W 19031 r; K/P 73 r).

Left: above, Leonardo illustrates the traditional concept whereby urine enters the bladder high up. Below this he sketches a much shorter inner lip of the entrance placed much lower in the bladder wall, as shown in Figure 17.2 (p. 347).

Right: a series of four drawings illustrate the effects of different bodily postures on the flow of urine into the bladder. The top drawing is labelled 'upside down', the second, 'upright'; the third, 'on the side', the fourth, 'prone'.

The structure and function of the male genitals in relation to coitus is the logical starting point for this programme. In the coitus drawing (Figure 11.2) Leonardo shows little or no direct knowledge of the anatomy of the male or female reproductive organs. He is merely presenting a visual hypothesis of the traditional views expressed by his authorities, e.g., Plato, Hippocrates, Guy de Chauliac. It therefore contains a record more of anatomical errors than truths, thus constituting an interesting historical record and a good base-line from which to assess Leonardo's later progress.

Comments here will be confined to points concerning reproduction. The spinal cord, seen running down the whole of the spine, gives off a veritable plexus of nerves which cross the bladder to run the length of the penis. These nerves, according to Plato and Hippocrates, carry the semen from the spinal cord along a special urethral canal through the penis for ejection into the vagina in coitus. All this Leonardo illustrates. The urine, according to this view, passes from the bladder along a lower, separate canal. These two canals Leonardo illustrates by an imaginary cross section of the penis in two drawings below in Figure 11.2. The whole of the 'coitus figure' corresponds closely to the 'anatomy' described in Plato's *Timaeus*, 91 B, from which other authorities presumably derived it.

The Hippocratic idea of the derivation of the semen Leonardo himself records thus: 'Hippocrates says that the origin of our semen is derived from the brain, lungs and testicles of our forefathers; here the final decoction is made; and all the other parts transmit their substance to this semen by sudation, because one cannot demonstrate any channels which could come to this semen' (Forster III 75 v).

In this drawing a special artery from the aorta to the testes is shown carrying hence material for manufacture into semen according, this time, to Galen's ideas. The combination of both is to be found in Guy de Chauliac's *Anatomical Questions* (32 v), where he asserts, 'The sperm comes from all the body and especially from the principal members for the breeding of their vessels, as of the heart, liver and kidneys. And for the cause of delectation the brain has commutation therein for the nerves descend from the brain to the ballockes. Thus of all the body it taketh nature not by quantity but by vigour'. This seems to be where Leonardo left the subject of the anatomy and physiology of the male organs of reproduction about 1493.

His knowledge of the female generative organs was at this time equally rudimentary. Judging from this illustration the uterus is divided into several 'cells'; this accords with Michael Scott and Mondino. Mondino adds that the semen coagulates in these 'cells' from which arise the orifices of the veins 'which carry the material of blood which fails to be evacuated by the menses in a pregnant woman to the breasts for the formation of milk'. All this is so literally illustrated by Leonardo that one cannot but conclude that Leonardo used Mondino's *Anatomy* as his source. However, confusion is shown in the heading at the top of the page, where he writes, 'I display to men the origin of their first or perhaps second cause of existence'. This assertion refers to Avicenna, who took the view that 'vital spirit' is primary, generative spirit, secondary – a variation of Aristotle's concept of the soul.

Another subtitle to the drawing, Figure 11.2, is written beneath it: 'Here two creatures are cut through the middle and what remains is described'. Alongside the upper parts of the bodies of the two 'creatures' he notes, 'Through these figures are demonstrated the cause of many risks of sores and diseases'. From the next note below this he draws a line to the diaphragm, saying, 'Division of the spiritual parts from the material'. All these comments are to be found in Mondino.

Leonardo cites Avicenna's view of the soul in a note below: 'Here Avicenna claims that the soul begets the soul and body; the body consists of every used part'. Avicenna voices this view in his *Canon* (Volume 1, Chapter 6, sections 1–4).[1] Leonardo pursues the theme further, asking, 'What animal parts [of the soul] arise from any parts of the members of man – simple or compound?' Then he raises questions to which he returns years later: 'Note what the testicles have to do with coitus and the sperm . . . And how the infant breathes; and how it is nourished through the umbilicus. And why one soul governs two bodies as is seen in a mother desiring food and the infant remaining marked by it'. Finally he asks, 'How are the testicles the cause of ferocity?' One notices that almost all of Leonardo's comments on this drawing are couched in the form of questions or projects for further investigation.

LATER INVESTIGATIONS (1506–1513)

Male

On returning to anatomy Leonardo answers the question, 'How are the testicles the cause of ferocity?' thus, under the heading, 'Testicles witnesses of coitus. These contain in themselves ardour, that is, they are augmenters of the animosity and ferocity of animals. Experience shows this clearly in castrated animals; which is seen in the bull, boar, ram and cock, very ferocious animals, which after they have been deprived of their testicles are left very cowardly' (W 19030 r; K/P 72 r).

The mechanism of erection of the penis attracted Leonardo's attention repeatedly. He noticed that the penis is often erect in men who have died by hanging (W 19101 r; K/P 197 r). Maybe he observed this first at the hanging of Bandino Baroncelli, which he illustrated in Figure 10.7 (p. 236). Erection was attributed by his authorities (including Mondino) to inflation by wind. Leonardo disagrees. 'On the penis', he writes, 'This when it is hard is big, long, dense and heavy; and when it is flaccid it is small, short and soft, that is to say, soft and feeble. This makes one judge that here its flesh is not supplemented by wind but by arterial blood. I have seen this in dead men who have this member erected, for many die thus, especially those hanged. Of these I have seen the anatomy, all of them having great density and hardness, and being quite filled by a large quantity of blood which has made the flesh inside very red, and in others outside as well as inside. And if the adversary says that such a large quantity of flesh has grown through wind causing the enlargement and hardness, as in a ball with which one plays, such wind gives neither weight nor density but makes flesh light and rarefied. Besides one sees that the erect penis has a red glans which is a sign of the inflow of blood; and when it is not erect this glans has a whitish surface' (W 19019 v; K/P 39 v).

Similar views are expressed later, in W 19017 r (K/P 151 r), where he considers again the possibility of wind being responsible for the enlargement of muscles in action, as well as that of the penis. He makes particular reference to a 'mule which on seeing a mare, suddenly its penis and all its muscles became turgid'. Once more he denies the possibility of air or wind being the cause (see Chapter 13).

Leonardo sees a special relationship between the male sexual organ and the psyche. In a section entitled, 'On the penis', he writes, 'This disputes with the human intellect, and sometimes has an intellect of its own. And though the will of man may wish to stimulate it, it remains obstinate and goes its own way, sometimes moving on its own without the permission or intention of a man. Thus be he sleeping or waking it does what it desires. Often a man is asleep and it is awake, and many times a man is awake and it is asleep. Many times a man wants to use it, and it does not want to; many times it wants to and a man forbids it. Therefore it appears that this animal often has a soul and intellect separate from a man; and it appears that a man who is ashamed to name or show it is in the wrong, always being anxious to cover it up and hide what he ought to adorn and show with solemnity like a minister of the human species' (W 19030 r; K/P 72 r).

Though the scientific Leonardo sees such interesting and admirable features in the activities of the penis, one finds the aesthetic Leonardo saying, 'The act of coitus and the parts employed therein are so ugly that if it were not for the beauty of the faces, the adornments of the actors and the frenetic state of mind, Nature would lose the human race' (W 19009 r; K/P 143 r).

The anatomy of male and female genitalia are compared in Figure 17.6.

Female

Since Mondino in 1316 had described his own dissections of the uterus Leonardo obviously paid particular attention to his findings. Leonardo's interest in the female genitalia was focussed on (1) the reciprocal sphincter action of the vulva and anus, as part of his investigation of sphincter

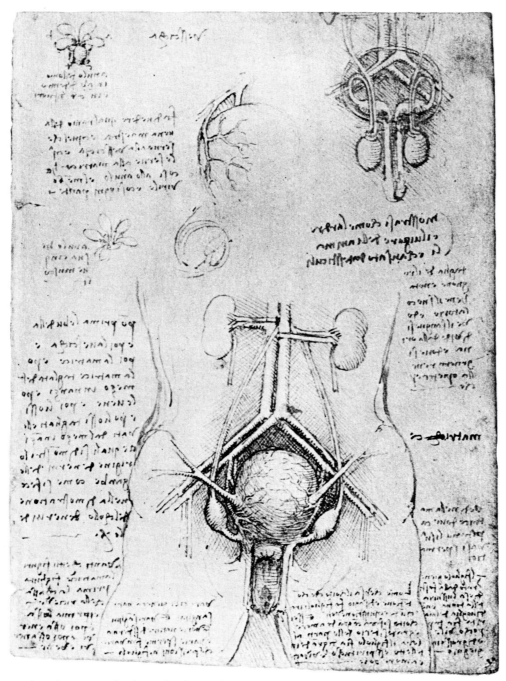

Figure 17.6. Male and female generative organs (Weimar Blatt; K/P between 54 and 55).

On the left, studies of the sphincters and vessels to the bladder are shown. On the right and below, female genitalia are shown. The ovaries in the female are seen as internal testicles connected to the uterus in much the same way as male testicles are connected to the seminal vesicles in the drawing above. Sphincter studies resemble those in Figure 16.14 (p. 341).

action in general, described in Chapter 16; (2) the relation between the size of the vagina and the penis in the mechanics of coitus; and (3) the part played by the uterus as a container of the embryo – this last aspect claimed by far the greatest amount of his attention.

Many drawings are devoted to the subject of the erect penis fitting into the vagina. Mondino gives relatively little attention to this, limiting himself to noting that both vagina and penis measure about 'a palm's breadth'. Leonardo tries to refine this into quantitative measurement. A long passage begins, 'A woman commonly has a desire directly opposite to that of a man. This is, that the woman likes the size of the genital organ of a man to be as large as possible, and the man desires the opposite in the genital organ of a woman, so that neither one nor the other ever attains his desire, because Nature, who cannot be blamed, has so fashioned it for the purpose of parturition. Woman has this genital organ larger than any other animal species in proportion to her body, which is usually, from the pit [illegible in the text; ? costal angle] to the anus, one *braccio* in length in a straight line. The bovine species has a body three times longer than that of a woman so that multiplying cubically the one body by the other you would have to say that 3 times [3 makes 9] and 3 times 9 makes 27. Therefore the body of the cow is 27 times greater than that of a woman.

'But such a multiplication has no place here because a cow would have such an organ 7 times larger. Experience in the dead shows that it is a quarter of a *braccio* in its greatest

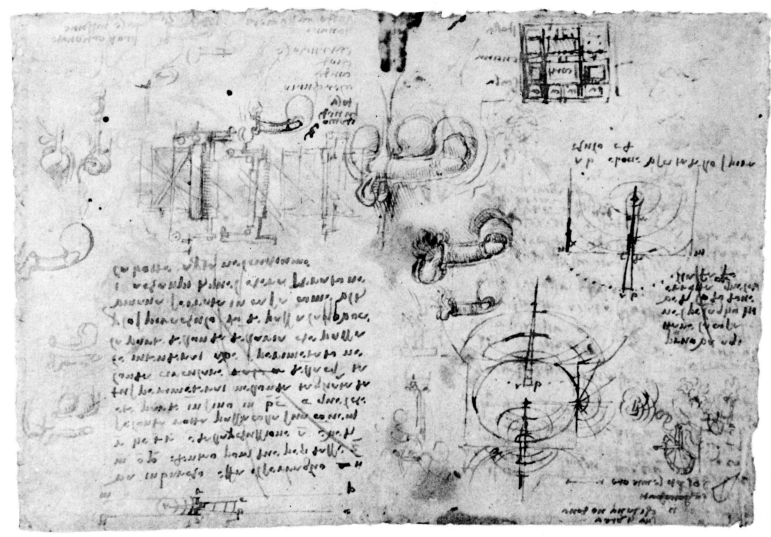

Figure 17.7.   External male and female genitalia (W 19106 v; K/P 126 r).

Studies of the relative sizes of the male and female genitalia are presented here without comment. The subject is discussed in some detail in W 10101 r (K/P 197 v) but is not illustrated there.

length in woman, just as in the bovine or horse species, these being the biggest animals in Europe. Still one can say by the rule of 3 that if one *braccio* of the body of a woman gives 1/4 of a *braccio* of the organ, how much will 3 *braccia* of the body of a cow give? If 4 quarters, that is one *braccio,* give one fourth of a *braccio* of the organ of a woman, how much will 12 quarters in the cow give me? They will give me three quarters of a *braccio.* Thus such an animal would have to have 3/4 of a *braccio* of member in proportion to a woman who has 1/4' (W 19101 r; K/P 197 v). This unfortunately mutilated passage reveals Leonardo's determination to apply measurement and his mathematical rules to biological problems. It particularly demonstrates his abiding faith in the rule of three, the arithmetical form of his pyramidal law (see Chapters 4 and 5).

The importance attached to the 'fit' of the male and female parts of the sexual organs is viewed by Leonardo as a problem akin to the fit of machine tools, about which we still use the term 'male and female' parts, as with the fittings of hose-pipes, etc. Leonardo's studies of the subject are represented more often in visual than in verbal language, as in Figure 17.7, where they are repeatedly drawn without comment. The illustrations in Figure 17.8 provide a good case in point. Of the two uppermost drawings on this page, one shows the hollow tube of the vagina leading to a spherical-shaped uterus beside which two globular 'ovaries' or female 'testicles' can be seen supplied by arteries from above, with ducts leading into the lower part of the cavity of the uterus. The similarity between this set of organs and the male set depicted below is striking. Here the penis is shown as a suitably designed male organ for fitting into the vagina. From each testicle, supplied by an artery, arises the ductus deferens, which carries sperm to be stored in two globular expansions (seminal vesicles); from these the semen passes by a duct into the penis. The duct is like that which passes from the ovaries to the uterus.

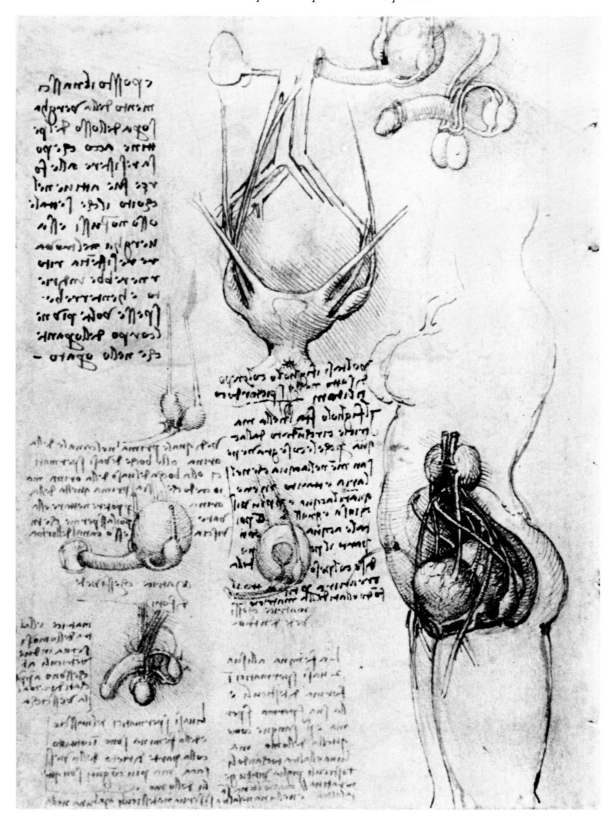

Figure 17.8. Male and female genital organs (W 19095 v; K/P 54 v).

Above is a comparison of the formation of both sets of sexual organs intended to emphasise their similarities in basic form. Below are drawings showing the uterus with its single cavity outlined. Below and to the right is a drawing of a pregnant woman; note the size of the uterus and breasts and the marked lordosis.

This search for homology between male and female organs, a common medieval tradition, appealed to Leonardo, who tried to fit his anatomical findings into the scheme, thus falsifying anatomical facts. Perhaps for this reason he failed to detect the Fallopian tubes, which in fact carry the ovum from the ovary to the top of the body of the uterus, not to its lower, cervical region, as constantly drawn by Leonardo.

There was keen conflict in Leonardo's day as to whether the embryo was formed from 'two seeds' (one from the male and the other from the female) or whether the male supplied 'seed' whilst the female supplied merely the nutritive bed in which the seed could grow. Galen supported the former, Aristotle the latter. Mondino held that the ovaries generate only a certain moisture like saliva which causes pleasure in woman. Leonardo supported the Galenic view that male and female both supplied 'seed'. He cites this in contradiction of Mondino's view in Figure 17.9, and in Figure 17.8 he states, 'The woman has her two spermatic receptacles in the form of testicles [ovaries] and her sperm is first blood like that of man. But both reaching the testicles take on generative power. But not one without the other'. He then goes on to make a mistaken homology. 'Neither the one nor the other keep [the sperm] in the testicles but one in the womb the other, that of the male, is kept in the ventricles *ab* [seminal vesicles] which are attached to the back of the bladder'. These vesicles can be seen labelled *ab* (not very distinctly) in the drawing in the lower left corner of Figure 17.8. Here Leonardo's false homology likens the seminal vesicles of the male to the uterus of the woman.

When one compares the early 'coitus figure' (Figure 11.2, p. 244) with the drawing of the uterus and its adnexa in the central upper part of Figure 17.8, one realises how much progress Leonardo has made. Here the ovaries, supplied by ovarian vessels, hang on the side of a uterus which is clearly shown to possess one cavity – not seven. Moreover, this cavity is shown passing downwards as the vagina to the vulval orifice. Leonardo's concern is still with the mechanics of the penetration of the penis into the vagina. In the passage to the left of this drawing he writes, 'The origin of the penis is placed on the pubic bone. It is so supported in order to resist the forces active in coitus. If this bone did not exist these forces would return the penis backwards on meeting resistance and often it would enter more into the body of the acting person than into that of the person acted upon'.

A side view of the uterus is shown in the middle drawing on the left margin. It is similar to that at the top of the page, but Leonardo labels it, 'Uterus seen from outside', in contrast with the drawing to its right, which is labelled, 'Uterus seen from inside'. Here the single human uterine cavity is very clearly shown.

The uterus in relation to its surrounding parts is shown in the right-hand drawing in Figure 17.8. It can be seen rising up higher than the distended bladder in front of it. The rectum is seen behind. Cords running to the upper surface of this enlarged, pregnant uterus are similar to those seen in the upper central drawing. Their nature is conjectural.

They would appear to represent Mondino's description of 'two strong and thick ligaments proceeding like horns from the head of an animal', a description which probably confused Leonardo about the anatomy of the broad and round ligaments attached to the uterus. This error he never overcame.

The last note on this page carries us to the next stage of Leonardo's investigation of the uterus: 'The child turns with its head downwards on the detachment of the cotyledons. The child lies in the womb surrounded by water because heavy objects weigh less in water than in air, and so much the less as the water is more viscous and greasy. Further, such water distributes its own weight with that of the baby over the whole base and sides of the womb'. (For Leonardo's observations on specific gravity, see Chapter 4.)

THE GROWTH OF THE EMBRYO IN THE UTERUS

There was no doubt in Leonardo's mind that fertilisation of the 'two seeds', male and female, takes place in the uterus and that as a result of this the uterus at once begins to enlarge. Leonardo was determined to study both processes. 'Your arrangement shall be with the beginning of the formation of the infant in the womb, stating which part of it is composed first and so on successively putting parts according to the duration of gestation until birth; and learning how it is nourished, partly from the eggs which hens lay'.

This note is written on Figure 17.9, the great double sheet which displays Leonardo's demonstration of the relationship between all the viscera in a pregnant woman. This is his own representation of a 'situs figure', a term used in traditional medieval anatomy for showing the 'sites' of the main organs. The anatomical errors and revelations contained in this figure have been described in previous chapters. Indeed, it serves as a good summary of Leonardo's anatomical knowledge of the viscera up to 1508–1510. The only organs represented on which he makes subsequent advances are the heart and great vessels and the foetus in utero.

The uterus is here represented with the 'two horns' mentioned by Mondino. Indeed Leonardo names Mondino himself on this page. 'You, Mondino, state that the spermatic vessels or testicles [ovaries] do not eject true semen but only a certain spittle which Nature has ordained for the delectation of women in coitus, in which case, if it were so, it would not be necessary for the origin of the spermatic vessels to arise in the same way in women as in men'. Mondino makes this assertion on the same page as he describes his 'two horned' uterus, which Leonardo depicts. Though Leonardo cleared up one of Mondino's errors, he failed to correct the other.

How early did Leonardo find a developing human or animal embryo? This is a difficult question to answer categorically, but in Figure 17.10 is a drawing of an early embryo enclosed in amniotic fluid with chorionic villi scattered round its circumference. The oval embryonic body

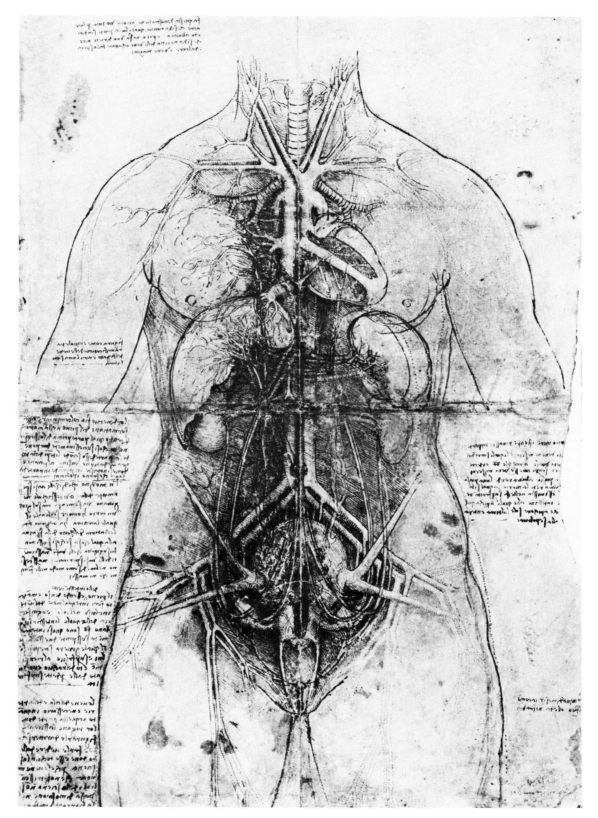

Figure 17.9. Leonardo's situs figure (W 12281 r; K/P 122 r).

This large drawing of the viscera summarises Leonardo's anatomy of the thoracic and abdominal organs. Pin-pricks round its edges show that tracings of it were made. The enlarged uterus once more denotes pregnancy. The liver, spleen and kidneys are very similar to those depicted in other places, particularly Figure 16.8, p. 335. This large picture is a composition of the organs into an integrated whole.

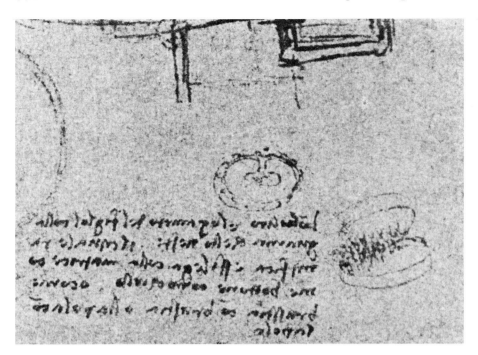

Figure 17.10.   Early embryo (CA 385 ra).

The central drawing shows an early mammalian embryo surrounded by amniotic fluid, attached to the uterine wall by a body-stalk. Below and to the right is a cow's cotyledon.

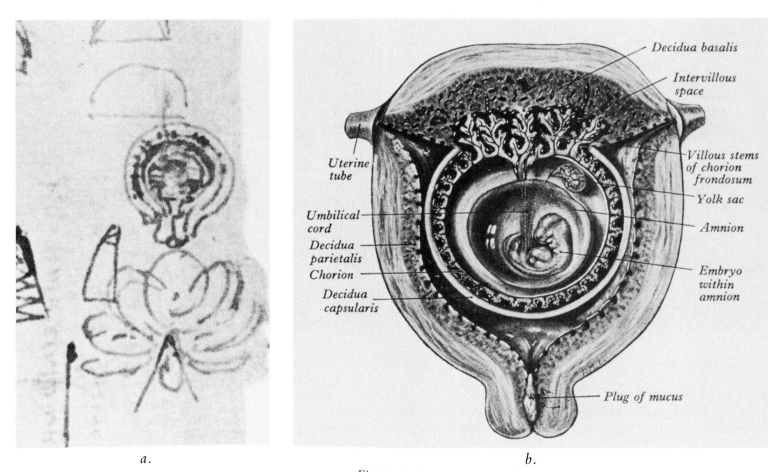

a.   b.

Figure 17.11

a. Study of an early embryo (CA 113 rb).

The embryo can be seen suspended in amniotic fluid. In spite of the blot obscuring detail, this drawing strongly resembles an embryonic development of some six to eight weeks, illustrated in Figure 17.11b.

b. Figure from *Gray's Anatomy,* 37th edition (Longman, London, 1980), figure 2.46.

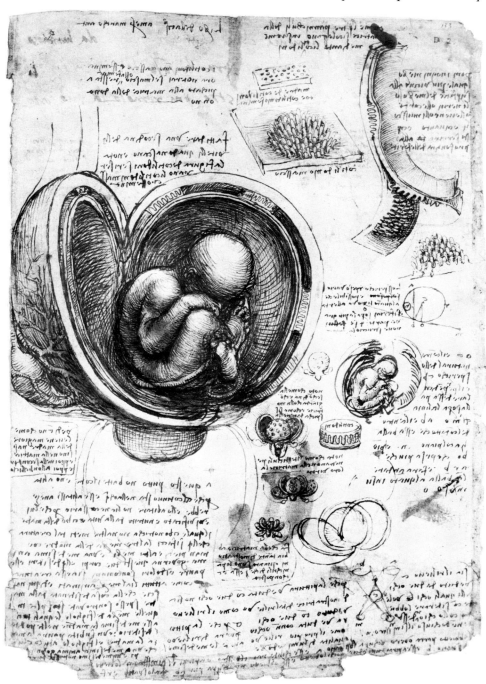

Figure 17.12. The foetus within the womb (W 19102 r; K/P 198 r).

The uterine wall has been cut open to show the foetus inside. The cow's form of cotyledons is shown in the wall and in other drawings on the page. To the right the foetus is shown within the amniotic fluid; and above this is a study of the mechanics of its rotation so that the head presents at birth.

is attached by a clearly defined body-stalk. Such an appearance, whether of animal or human embryo, can be dated to one to two months. The accompanying note describes 'How the umbilicus is a bond between the infant and the sheath that invests it. This branches and is bound to the uterus like a button in a buttonhole or like a brush with a brush or burdock with burdock'. The last sentence of this note might well apply to either of the drawings it accompanies, i.e., to the interdigitating chorionic villi shown on the right. The appearance of this early embryo may well have suggested to him that 'All seeds have an umbilicus which breaks when the seed is ripe. And in like manner they have a womb and membranes, as is seen in all seeds that grow in pods' (W 19103 v; K/P 196 v).

Another suggestive drawing is to be found in Figure 17.11a. Here the uterine cervix is shown plugged with mucus; the chorion is attached all round the foetal membranes; the embryo itself appears to float in amniotic fluid. Below, Leonardo shows the appearance after he has extracted the foetus and its membranes, slicing and opening them, like flower petals. He makes no relevant verbal note, but one cannot help comparing this appearance with Figure 17.11b, the drawing of a gravid human uterus in the second month.

In the famous drawing in Figure 17.12 Leonardo depicts a uterus containing a five months pregnancy as seen through the incised uterine wall. It has often been noticed that Leonardo here depicts the placental formation of the

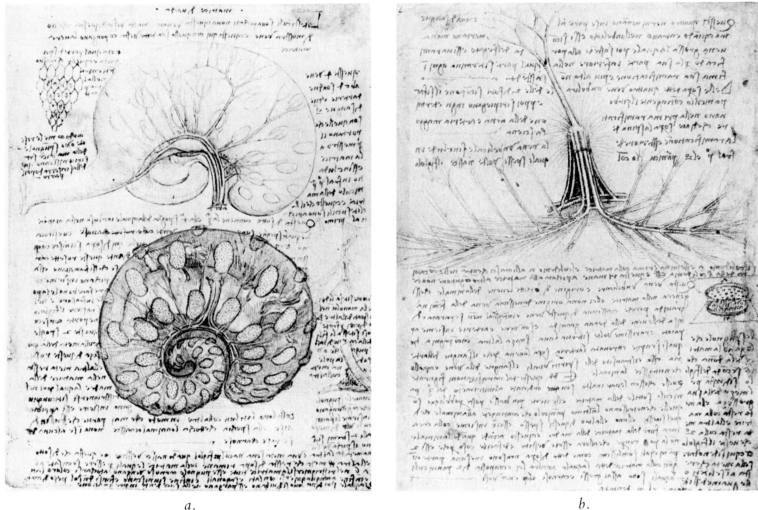

a.   b.

Figure 17.13.

a. The foetal calf within a cow's uterus.

The top drawing shows the uterine wall with its rich blood supply. Below, the wall has been removed so that the foetal calf can be seen lying within the transparent amniotic membrane and its contained fluid. The umbilical vessels are clearly shown (W 19055 r; K/P 52 r).

b. The umbilical vessels shown in Figure 17.13a are here enlarged so as to show their ramifications within the uterine wall as well as their mode of entry into the umbilicus of the foetus (W 19046 v; K/P 51 v).

cow in the discrete cotyledons scattered at intervals over the wall of the uterus. The large ovary to the left can be seen sending its imagined duct to the base of the body of the uterus, the cervix of which is beautifully drawn. The foetus within is in the breech position with the left heel tucked into the perineum. The umbilical cord winds round beneath it. There is much insight into the form of the foetus and the intra-uterine position of its head and limbs, features which predominate in this revelation of an hitherto unexplored 'secret of Nature'. Above the main drawing Leonardo studies placental formation, noting visually how the chorionic villi of the foetal membranes penetrate into the wall of the uterus; the upper pitted drawing shows what he calls 'female cotyledons'; the lower projecting villi he calls 'male cotyledons'. Their interdigitation is beautifully expressed in the drawing to the right.

Below and to the right of the main drawing in Figure 17.12 Leonardo shows the stages of his dissection as he slits open and peels off the membranes layer by layer, displaying in one drawing the foetus within the transparent amnion and its contained 'waters'. Lower down he shows 'how the membranes are joined to the uterus'. Then in a little drawing he shows how he reaches the flower petal arrangement of the peeled layers. Incidentally, a human discoid placenta is shown in Figure 17.16 (p. 361).

Halfway up the right margin of the page in Figure 17.12 is a diagram of a ball rolling up an incline by virtue of a lead weight (*n*) seen in a segment on its left. The note below explains in detail how 'since the lead *n* which weighs more than the rest of the ball *arb,* will make the ball go somewhat upwards towards *o*'. The drawing is related to the position of the foetus in the 'ball' of membranes in the uterus. It

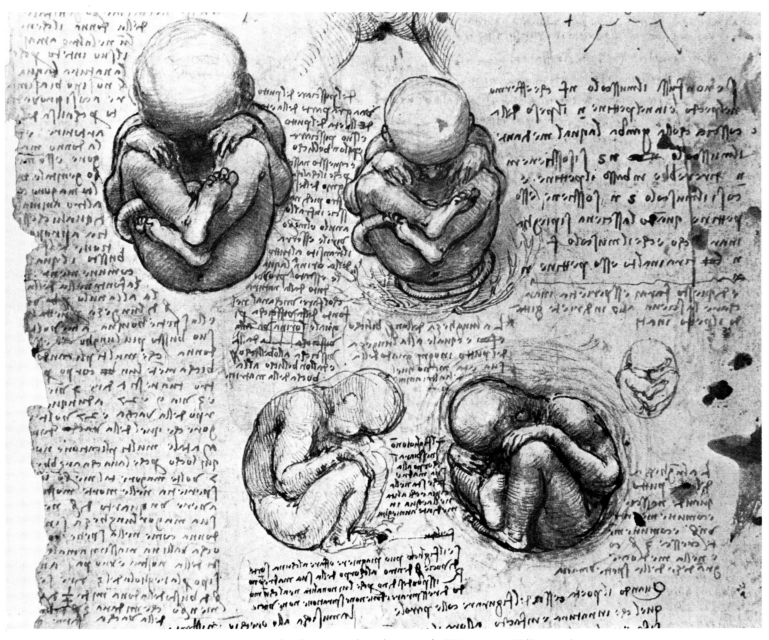

Figure 17.14.   The foetus within the womb (W 19101 r; K/P 197 v).

The position of the head, arms and legs of the foetus is here studied in detail. Drawings are made from the front and from both sides. Particular emphasis is made on the way the left heel presses into the perineum, thereby, according to Leonardo, preventing urine from passing along the urethra.

shows how the foetus may, and does, change its position within the 'ball of water' and how, when detached from the uterine wall, the foetus rotates to engage head first. Leonardo notes, too, that sometimes the newborn infant cannot rupture the foetal membrane and is born in the caul. 'Both one and the other of these membranes often emerge together with the creature out of the womb of the mother. This happens when the animal cannot rupture them, for then it emerges clothed in them. This easily happens because these two very thin membranes, as said above, are in no part united with the said uterus' (W 19046 v; K/P 51 v).

The long note in the left lower corner of Figure 17.12 raises questions of foetal physiology. 'In the case of this child the heart does not beat, nor does it breathe because it lies continually in water and if it breathed it would drown. And breathing is not necessary to it because it is vivified and nourished by the life and food of the mother. This food nourishes this creature not otherwise than it does other organs of the mother, namely her hands, feet and other parts. One and the same soul governs these two bodies; and desires, fears and pains are common to this creature as to all other animated parts. And from this it arises that a thing

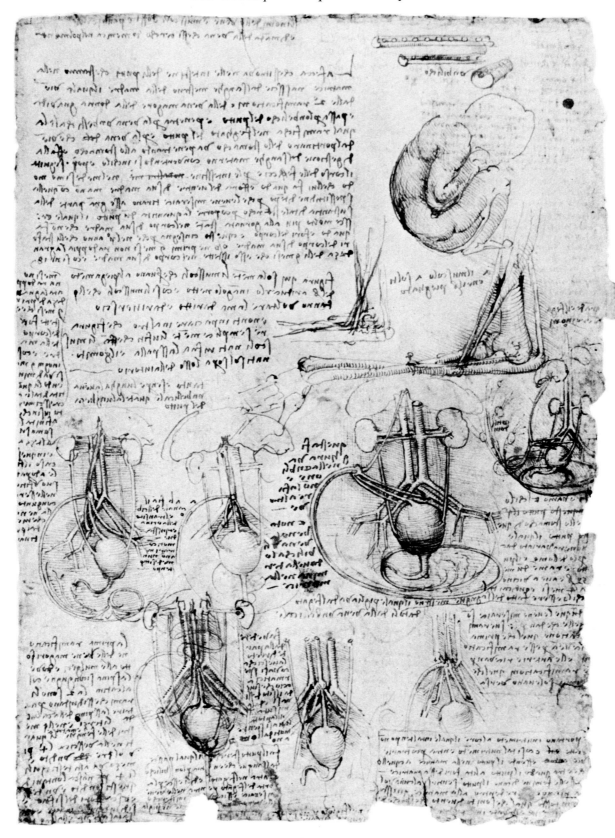

Figure 17.15.   Studies of the foetal viscera (W 19101 v; K/P 197 r).

Most of the drawings are concerned with the formation of the umbilical cord (right top and centre) and its entry into the foetus, which is always given umbilical arteries and veins (erroneously). The liver is symmetrical. The urachus takes urine from the top of the bladder into the umbilical cord. The umbilical cord is seen arising from a discoid placenta.

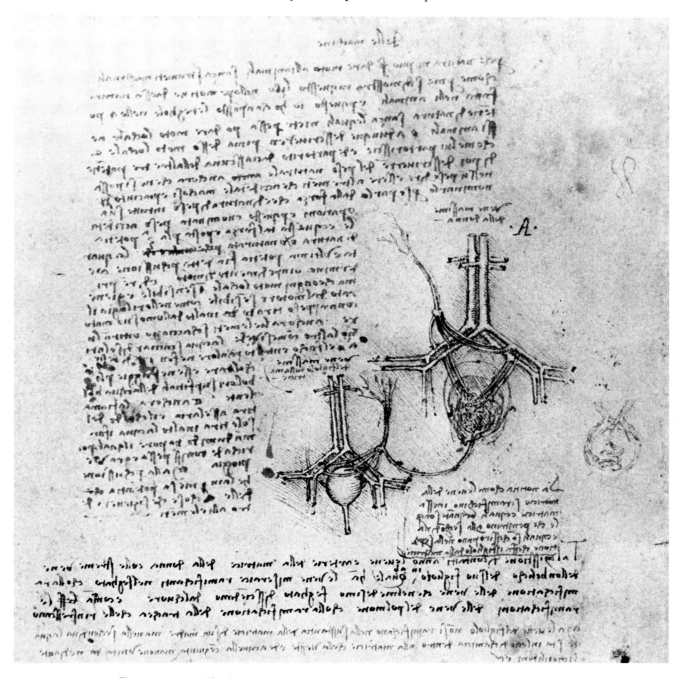

Figure 17.16. The bond between mother and child (W 19060 r; K/P 153 r).

The umbilical cord is shown arising from the discoid placenta in the mother, *A*, passing to the foetus on the left.

The whole upper portion of the page is headed 'On Machines' and describes at length 'Why Nature cannot give movement to animals without mechanical instruments' (see Chapter 7).

desired by the mother is often found imprinted upon those parts of the infant which have the same qualities in the mother at the time of her desire; and a sudden terror kills both mother and child. Therefore one concludes that one and the same soul governs the bodies; and one and the same soul nourishes the two bodies'.

Right at the top of the page is a note, 'Book "On water" to Messer Marco Antonio'. Whether this refers to Marc Antonio della Torre, the anatomist, or not, it reminds us

that Leonardo was keenly aware at this time of the infant or microcosm growing within the sphere of water like the macrocosm.

Leonardo did not confine his study of pregnancy to the human foetus. Even on this page he mentions and draws the membranes of the foetal calf which he was also studying. The preliminary small sketch of a foetal calf in the uterus is to be found in Figure 9.8, p. 222, amongst drawings of the lips.

In Figure 17.13a two superb illustrations of the gravid uterus of the cow are presented. The upper drawing shows the bicornuate cow's uterus, gravid, with its rich blood supply and the scattered cotyledons within faintly outlined. In the drawing below he has removed the uterine wall down to the depth of the decidual cotyledons, which lie outside the inner transparent foetal membranes. Thus the foetal calf can be seen lying within, its head to the left, its bent forelegs with cloven hoofs uppermost, and its body stretching across to the hind legs on the right of the drawing. This is a superb example of Leonardo's transparency technique, so subtle that often the foetal calf is not detected.

In Figure 17.13b the umbilical vessels of Figure 17.13a are enlarged and their distribution to the foetus and to the foetal membranes are detailed. The umbilical vein running up from the umbilicus to the foetal liver is clearly shown.

THE GROWTH AND FORMATION OF THE FOETUS

Leonardo's determination to describe the growth and development of the foetus has already been mentioned. His general view on this subject followed Avicenna's version of Aristotle's principle. All growth is subject to the guidance of the soul: 'Nature . . . needs no counterweights when she makes the limbs fitted for movement in the bodies of animals, but puts there the soul, the composer of the body, that is the soul of the mother, which first composes within the womb the shape of man and in due time awakens the soul which is to be its inhabitant. For this at first remains asleep under the guardianship of the soul of the mother who nourishes and vivifies it through the umbilical vein, with all its spiritual organs. And so it continues for such time as the umbilical cord is joined to it by the foetal membranes and cotyledons by which the foetus is attached to its mother' (W 19115 r; K/P 114 v; see Figure 17.14). 'The whole body originates from the heart insofar as its first creation is concerned. Therefore the blood, vessels and nerves do the same. Yet all the nerves are clearly seen to arise from the spinal cord, remote from the heart. And the spinal cord consists of the same substance as the brain from which it is derived' (W 19034 v; K/P 76 v).

He observes that during foetal life the left side of the liver atrophies: 'The centre of the liver of infants in the womb when they are small lies under the centre of the heart and above the umbilicus. And when it is born the liver is drawn to the right side', He relates this to the development of the spleen thus: 'And the spleen which was at first a viscous watery substance, pliable and supple, giving way to anything which pushed it out of its position, afterwards begins to contract and condense and form its essential shape, and it necessarily enters the place which was occupied by the left part of the liver, whence . . . it is drawn back to the right side, pressing and condensing this right part of the liver, and uniting with it. Thus the liver loses 7/8 of its left side and is withdrawn so that its centre is condensed on the right side' (W 19102 v; K/P 198 v).

With this shift of the liver goes the umbilical vein. 'When the umbilical vein is functioning in the way for which it was created it obtains the principle position in man, that is, the centre of the trunk of the body, of its length as well as its breadth. But when this vein was later deprived of its function, it was drawn to one side together with the liver which was created and then nourished by it' (W 19102 v; K/P 198 v).

The umbilical vein Leonardo looks upon as 'the gateway whence our body is composed' (W 19021 r; K/P 153 r). Its length is 'equal to the length of the foetus at all stages of its growth (W 19101 r; K/P 197 v; see Figure 17.15). The nature of this 'bond' between mother and child is beautifully expressed in a drawing in Figure 17.16. Here Leonardo draws the 'great vessels' of the mother (upper right part of the drawing, near the letter *A*). Following their bifurcation, their branches to the uterus are shown, ending in a discoid placenta. From this emerges the umbilical cord, passing in a loop to the drawing below and to the left, which represents 'the great vessels of the foetus in the womb'. As it enters the foetal umbilicus the umbilical vein can be seen, passing upwards to the foetal liver. The 'great vessels' of the foetus are also shown giving off arteries and veins in the pelvis. The arteries, erroneously paired with veins, together with the urachus (a channel from the top of the bladder) pass up to the umbilicus to enter the umbilical cord on their way to the placenta, where they receive 'vivification' (oxygen) and nourishment. The excretion of the metabolic residues, 'superfluities', of foetal metabolism pass from the bladder via the urachus to the mother's placenta. It will be noticed that Leonardo draws the mother's pelvic arteries again with 'veins' giving off branches to her umbilicus also. The arteries in the adult are closed fibrous cords. No doubt he is expressing here the fact that the mother, too, was once a foetus herself. In this diagram, therefore, he delineates his own visualisation of the mechanism of the continuity of the generations, and man's continued power 'to create another like himself'.

His verbal description of this situation runs thus: 'The veins and arteries of the uterus of the woman have a similar intermingling by contact with the terminal vessels of the umbilical cord of her infant at *ab* as the mesenteric [portal] veins ramifying in the liver have with the ramifications of the veins descending from the heart into that same liver [hepatic veins]; and as the ramification of the pulmonary vessels have with the ramifications of the trachea which refresh them. But the vessels of the infant do not ramify in the substance of the uterus of its mother, but in the placenta which takes the place of a shirt inside the uterus which encloses it and to which it is joined (but not united) by means of the cotyledons, etc.' (W 19060 r; K/P 153 r).

REFERENCES

1. Avicenna. *Canon of Medicine*, trans. by M. H. Shah, in *The General Principles of Avicenna's Canon of Medicine*. Karachi, Naveed Clinic, 1966, pp. 126–132.

Epilogue

THE PHYSICAL WORLD (MACROCOSM)

This attempt to construct a balanced picture of Leonardo's physical and physiological scientific research is necessarily an abbreviation. To do his work justice every chapter should be expanded into a book. Leonardo himself was intolerant of abbreviators. 'The abbreviators of works do injury to knowledge and love, for the love of anything is the offspring of knowledge', he writes (W 19084 r; K/P 173 r). However, the danger of this ideal position is revealed by his assertion on the very same page: 'Certainty is born from the integral knowledge of all those parts which being united together compose the whole of that thing which is to be loved'. And Leonardo's mind, continuously searching and discovering during the whole of his life, never presented his findings on any subject in the artistic form of 'integrated knowledge of all those parts which being united together compose the whole'. The reason for this is not far to seek. The movement of his mind throughout his life constituted a continuum, and 'Every continuous quantity is infinitely divisible' (BM 204). Recognising this, Leonardo makes his own apologia, describing his notes as 'a collection without order, made up of many sheets which I have copied here, hoping afterwards to arrange them in order in their proper places according to the subjects of which they treat' (BM 1 r). This composition of the whole he never achieved.

This present book is an attempt to analyse and integrate the basic ideas – the 'elements' of Leonardo's continually developing concepts regarding the physical world and the body of man. To achieve this with any hope of success one would ideally have to know and understand the whole of Leonardo's works. This is impossible – if only for the reason that in all probability the approximately 6,500 pages we possess represent only one-third of Leonardo's notes. And from the nature of Leonardo's own references some of the lost codices were indeed efforts at integration.

The position is not however as hopeless as the preceding comments might suggest. Recently (1974) two newly discovered codices, *Madrid Codex* I and II, were published. They have provided an increase of approximately ten percent in the aggregate knowledge available about the work of Leonardo. This new knowledge, remarkably enough, fits so well into principles enunciated elsewhere that no basically new Leonardian ideas have emerged. Rather do these volumes demonstrate the astonishingly detailed observations and experiments upon which Leonardo developed and applied the scientific principles described in other notebooks, such as MS A.

My own examination of the whole of Leonardo's extant notes reveals a remarkable integration of his science primarily based on the geometry of perspective. There is no doubt that Leonardo's ultimate goal was an integration of experience through a mathematical medium, chiefly that of geometrical transformation. This he attempted to achieve by an analysis of phenomena into perspectival, or what he called 'pyramidal', forms of the 'four powers' of movement, weight, force and percussion acting on the four elements of earth, water, air and fire, as summarised in Chapters 2, 3 and 4. These describe how he reached the generalisation, 'All the powers are to be called pyramidal' (CA 151 ra).

Whilst the powers of the sun, light and heat travel in straight lines, those in fluid 'elements', i.e., water, wind and fire, take the forms of curvilinear pyramids or cones, which he calls 'falcates'. For Leonardo the diminution of any power with distance is always in perspective or pyramidal proportion as the power spreads in circular waves away from its point of origin. Conversely, the power is similarly concentrated as it approaches the focal point of the pyramid. Thus, 'Proportion is not only found in numbers and measurements but also in sounds, weights, times, spaces and in whatsoever power there may be' (K 49).

Transformation of perspectival shapes led Leonardo to a thorough understanding of anamorphosis. By further progressive geometrical transformations such as those represented in Figure 5.34 one can see Leonardo entering that field of distorted improbable shapes more recently entitled topology.

As the years passed Leonardo accumulated a vast store of observations and experiments involving the powers of the macrocosm. Many were compatible with his geometrical transformations of perspective 'pyramids' and circles; some were not. In some cases his application of geometry was only partially successful. For example, he was correct in expressing the uniform 'pyramidal' acceleration of falling bodies, but he failed to detect that the distance covered by a falling body varies with the square of the time of its fall. Only relatively late in his life did he come to appreciate that his pyramidal law could include geometrical progression and a direct or inverse square relationship (see Chapter 5).

Leonardo's physical world-system was unique in that he postulated an invisible geometrical point of incorporeal energy or 'spirit' as its basic unit. This 'resided in space'; it did *not* occupy it. For Leonardo the atom was a 'natural point' occupying space and therefore infinitely divisible. The movement of a geometrical point created a line. The angle made by converging lines is a point. The lateral movement of a line makes an area, and the movements of areas in different planes constitute the volume of bodies. In his view the three other 'powers', weight, force and percussion, can be expressed geometrically as potential or actual movement; so he fits geometrical garments to physical reality.

In this basic view Leonardo was far from the materialistic atomism which came to hold the limelight on the scientific stage until the twentieth century. Leonardo describes force as 'A spiritual power . . . because in force there is an active incorporeal life; and I call it invisible because the body in which it is created does not increase either in weight or size' (B 63 r). His use of the word *spiritual* to translate 'incorporeal' force, power or energy is applied to both the physical world of the macrocosm and the physiological world of the microcosm of man. 'Force . . . is the grandchild of spiritual motion. Force with material motion, weight and percussion are the four accidental powers with which all the works of mortals have their existence and their death. Force has its origin in spiritual motion; this motion flowing through the limbs of sentient animals expands their muscles' (BM 151 r). Thus Leonardo's incorporeal 'geometrical point' comes to be identified with a centre of force. This definition which we may today apply to the nucleus of an atom was again reached after the early development of investigations of electricity by Boscovitch in the eighteenth century. Such a centre of force was expressed mathematically by André Ampère and Michael Faraday in the nineteenth century, and it has been only too destructively demonstrated in the twentieth century.

It is interesting to compare Leonardo's speculative concept of his moving geometrical point of force with the modern concept of a photon as a spinning mass-less particle. Such 'twisters' are abstract mathematical entities within the realm of geometrical concepts. Leonardo's description of the point as 'the first principle of geometry, and nothing in nature or the human mind can be the origin of the point' (TP I 1), is consistent with such a mathematical concept. Both raise the same critical question: Are our bodies and minds composed of abstract mathematical concepts?

MAN THE MICROCOSM

According to the context man may be looked upon physiologically as a machine or chemical factory, as a psychological entity, a social unit or theologically as an immortal soul. Leonardo was concerned with the physiological movements of man's body and his mind, i.e., his emotional or psychological aspects. 'I shall describe the function of the parts from each side, placing before your eyes knowledge of the whole shape and strength of man insofar as he has local movement by means of his parts'. His awareness of his neglect of the social aspect of man comes out when he adds, 'And would that it might please our Creator that I were able to demonstrate the nature of man and his customs in the way that I describe his shape' (W 19061 r; K/P 154 r).

From this statement, amongst others, it becomes clear that Leonardo's dominant motive in investigating human anatomy was to analyse and demonstrate the movements of this microcosm and explain how his 'four powers' brought these about. His physiology consists of demonstrating how these invisible powers produce transformations of shapes, how the body in exercising its infinite movements 'does not increase either in weight or size'.

Insofar as these transformations of movement occur in the body as a whole he draws and describes them, as exemplified in Chapter 6 of this book, 'The movements of man and animals'. Such drawings, he is well aware, depend for their element of truth not only on the actions of his 'pyramidal powers' but also on his capacity to record these according to perspectival experience. Thus his earliest anatomical drawings – descriptively some of his best – are studies devoted to an attempted analysis of the physiology of visual experience as related to perspective. For Leonardo perspective was the key to an objective reality which he verified by experiment and measurement, not a mere acquired visual convention. In the eye and brain perspective resulted from an organic processing of forms and colours. The main processing organ he called the *imprensiva,* and he finally hypothetically located this in the lateral ventricles which he had discovered by wax injection. One may still justifiably speculate that such an area of perspectival processing is present in the brain in the region of the cortex of the occipital lobes.

The centre of nervous force in man Leonardo located in the third ventricle, in what he called the *senso comune.* For him this geometrical point 'resided' in a space but, being a 'spiritual' power, was itself unextended, incorporeal, invisible and weightless. Such was his concept of the soul.

Movements of the organs of the body as well as those of the body as a whole were brought about by forces emanating from this source traversing 'hollow' nerves to muscles which acted on bony levers or visceral contents. In all cases such actions followed the laws of the 'four powers' of the physical world of the macrocosm. In all cases such actions brought about geometrical transformations of muscles and viscera, so that what was diminished in one dimension was added to another. No visceral or muscular part increases in weight or size as a result of its activity. Perhaps one of the neatest examples of this is to be found in Figure 14.4 (p. 293), where the space left by the descent of the diaphragm is shown to be equal to the space occupied by the corresponding forward movement of the abdominal wall.

For Leonardo the principle of transformation was a necessary law of movement – space acquired equals space lost (see Figure 4.10, p. 100). Therefore, if any muscle contracts in one dimension, e.g., its length, it necessarily acquires space by expanding in other dimensions, i.e., breadth and depth. Again, if a limb, e.g., the arm, flexes at the elbow, it shortens the flexed side; therefore it necessarily increases the length of the opposite, extensor side (Figure 12.11, p. 262). The same rule applies to flexion and extension of all parts of the body. In all their infinte variety, therefore, the movements of the human body bring about geometrical transformations whilst its surface-areas, volume and weight remain constant. Always the new space acquired exactly compensates for the space vacated.

From Leonardo's point of view there was no question of a muscle being inflated by air, pneuma or 'spirit'. This he specifically denies in discussing what it is that so quickly enlarges part of a muscle. 'They say it is wind. And where does it retreat to when the muscle diminishes so quickly? Into the nerves of stimulation which are hollow? Then there would have to be a very great movement of air . . . Indeed there would not be enough air in the nerves, not even if the whole body were full of air' (W 19017 r; K/P 151 r). Thus for Leonardo, who saw muscle contraction essentially as power produced by a change of shape, the experiments of Swammerdam and Glisson in the seventeenth century, showing that muscles did not increase their volume on contraction, were superfluous.

In his patient studies of the movements of the heart Leonardo brought to bear many of his unique techniques. He inflated the atria and ventricles with air, he cut transverse sections at different levels. He dissected out the tricuspid valves and reconstituted them. He made a glass model of the base of the aorta in order to see the movement of the valves. All these procedures he meticulously illustrated (see Chapter 15). For example, the closure of the aortic cusps in a vertical rather than in a horizontal plane is clearly illustrated (see Figure 15.22, p. 318). So accurate were these experiments with models that B. J. Bellhouse in 1970 confirmed Leonardo's findings and suggested that such a mode of closure diminishes wear and tear of the valve cusps, so contributing to the long duration of their functional competence.[1] These are but a few examples in which Leonardo was a great precursor in the science of man.

During Leonardo's lifetime the numerous channels of scientific enquiry into which he poured his passion for investigation aroused little interest. He found few friends capable of being colleagues, few supporters capable of appreciating his intensely individualistic approach to problems of the physical or physiological worlds. He truly exemplified his own assertion – 'The painter or designer must be solitary, and most of all when he is intent on those observations and considerations that continually appear before the eyes which give material to the memory to be well stored. If you are alone you will be all your own' (CU 31 v).

The intellectual loneliness of the artist–scientist Leonardo was not merely contemporary; it has lasted for centuries.

Recent movement towards interdisciplinary science would have met with Leonardo's support as a reaction against its fragmentation into separate disciplines so typical of the nineteenth and twentieth centuries. 'Every part', wrote Leonardo, 'is disposed to unite with the whole in order that it may thereby escape from its incompleteness. The soul desires to dwell within the body because without the parts of the body it can neither act nor feel' (CA 59 rb). Leonardo would have seen interdisciplinary science as a halfway house towards an integrated human science of which art was the creative expression. This conviction filled him with veneration for human life. Amongst his notes on the bodies he dissected he wrote, 'And you O Man who in this labour of mine considers the marvellous works of Nature . . . think how very wicked a thing it is to take away the life of a man. If this his composition appears to you a marvellous construction remember that it is nothing in comparison with the soul that dwells within this structure; for truly, whatever it may be, that soul is a thing divine. Leave it then to dwell in its work at its good will and let not thy rage or malice destroy such a life – for in truth he who values it not deserves it not' (W 19001 r; K/P 136 r).

Nowhere does Leonardo unite his science and art more harmoniously into creative expression than in his painting of *The Last Supper*. Using his science of perspective he elusively simulates visual reality. The words of Christ, 'One of you will betray me', are dropped into the pool of silence like a stone into still waters. Their sound waves and their meaning are simultaneously heard and expressed in the grouped gestures of the responding disciples, so recalling those waves around a stone which Leonardo had dropped into a pond some years before. Here surely we can see those separated incomplete parts of the human mind, art, science and religion, escape from their incompleteness towards harmonious perfection, however blurred the picture may now be. Here surely is an example of a deeper, more complete expression of the human soul than that of the accepted different sciences, even of interdisciplinary science. Here science, art and religion are boldly welded into a unique unity by a man who today we can hope to see as a unifying prototype of future human genius.

The only hint he gives us of his own evaluation of his life's work is found in *Madrid Codex* I 6 r, 'Read me O reader if you find delight in me because I am very rarely reborn in the world'.

REFERENCES

1. Bellhouse, B. J. 'Fluid Mechanics of Model Aortic and Mitral Valves', *Proceedings of the Royal Society of Medicine*, 1970, p. 996.

List of Illustrations

Chapter 7

Chapter 8

Chapter 9

Chapter 10

Chapter 11

Index